Karl A. Froeschl

Metadata Management
in Statistical Information Processing

A Unified Framework for Metadata-Based
Processing of Statistical Data Aggregates

SpringerWienNewYork

Mag. Dr. Karl A. Froeschl
Department of Statistics, Operations Research, and Computer Science,
University of Vienna, Austria

This work is sponsored by the Fonds zur Förderung der wissenschaftlichen Forschung, Wien

This work is subject to copyright.
All rights are reserved, whether the whole or part of the material is concerned, specifically those of translation, reprinting, re-use of illustrations, broadcasting, reproduction by photocopying machines or similar means, and storage in data banks.

© 1997 Springer-Verlag/Wien
Printed in Austria

Typesetting: Camera ready by the author
Printing: Druckerei Novographic, A-1238 Wien
Binding: Fa. Papyrus, A-1100 Wien

Graphic design: Ecke Bonk

Printed on acid-free and chlorine-free bleached paper

With 65 Figures

Die Deutsche Bibliothek – CIP-Einheitsaufnahme

Froeschl, Karl A.:
Metadata management in statistical information processing : a unified framework for metadata-based processing of statistical data aggregates / Karl A. Froeschl. – Wien ; New York : Springer, 1997
ISBN 3-211-82987-3 brosch.

CIP data applied for

ISBN 3-211-82987-3 Springer-Verlag Wien New York

geistiges eigentum,
an der ganzen welt fällt so einem knirps gerade jetzt dieses bisschen grüne auf, da verbreitet er sich, da macht er dann seine abhandlung da hat er dann seine basis, da steht er, da hat er geleistet, da wollen wir ihm doch diesen für ihn bestimmten tritt in den arsch geben, denselben der seit unserer geburt als möglichkeit in unserem bein mit uns die ganze zeit her grossgewachsen ist.

OSWALD WIENER [1969]

Preface

The concepts and ideas developed and presented in this book cannot be claimed honestly one humble man's accomplishment; rather I have got the feeling of having achieved little more than combining and making fit together the pieces and proposals being there already – perhaps adding an amount of "logical glue" and some eagerness to shuffle pieces into position. Many of my "original" contributions have, in fact, emerged in numerous discussions and previous project work I had the opportunity to participate in as well as other, less formal sources of reference I'm no longer aware of. Among those persons who inspired me and helped in one way or another to shape my conception of a formal data integration framework as expounded in Chapter 2 and illustrated by exemplary applications discussed in Chapters 3 and 4, I gratefully acknowledge the readiness to discuss, criticize, comment, make suggestions, motivate, and cooperate of the partners of the EC (EUROSTAT) DOSES research project *Modelling Metadata* which brought forth many of the functional aspects of metadata management and sparked a broader interest in metadata methodology – in particular, I would like to mention *Gerda van den Berg* and her "Leiden team" including *Paul de Greef* and *Ed. de Feber*, *Dennis Conniffe* (Dublin) as well as especially *Paul Darius* (Leuven) and his engaged assistant *Michel Boucneau* who gathered valuable experience in transforming the emanating metadata methodology into real software for the first time; of the varying participants of an informal working party centering around Statistisches Landesamt Berlin headed by *Günther Appel*, giving me the opportunity to exchange and develop ideas with *Norbert Kopp*, *Klaus Neumann* (Berlin), *Beate Glitza* (Wiesbaden), *Bo Sundgren* and *Erik Malmborg* (Stockholm); of the group of Italian researchers renowned for its pioneering work in statistical data modelling, consisting of *Fabrizio Ricci*, *Maurizio Rafanelli*, *Leonardo Meo-Evoli* and – as our human "long-distance communication link" – *Nico Pisanelli*; of *Giovanna D'Angiolini* (Roma) who especially drew my attention to the concept logic-based approach to statistical data modelling; of *Jana Meliskova* (Geneve), *Martin C. Fessey* (Undy/ Gwent), *Daniel Defays* (Luxembourg), *Joanne Lamb* (Edinburgh), *Heidrun Ortleb* (Oldenburg), as well as *Wolfgang Dorda*, *Marcus Hudec*, *Thomas Plank*, *eva*

Kühn, and *Ewald Bartunek* (Wien), all of whom have made valuable remarks and suggestions and gave me their best advice; of *David J. Hand* (London) who offered me an unexpected opportunity to explicate my metadata approach in a *Statistics & Computing* paper to which he, moreover, contributed a lot of improvements and helpful remarks; of *Hans-Joachim Lenz* (Berlin) who reviewed thoroughly and commented critically on an earlier manuscript version; of *Karin Lackner*, *Patricia Fahrngruber*, and *Sabine Pöschl* who helped me preparing the abundant source material available in Austrian and European labour force statistics, analyzing the structure of statistical tables published by statistical offices, and surveying statistical data models; of my Viennese colleagues *Hannes Werthner* and *Siegfried Mattl* who, in countless discussions, opened my mind to see and reflect the broader implications of my work, fuelled my arguments with rather contrasting (software) engineering and historical-philosophical viewpoints I would have hardly recognized or appreciated otherwise, and pointed out minor and major flaws of my manuscript; of *Werner DePauli-Schimanovich* (Wien and Gran Canaria), my "logic & soul"-compatriot, who always persuaded me compellingly to relativize bad criticism and do away with ill-minded third-party advice; and of *Anneliese and Herbert Raab* (Horn) who allowed me to spend my summer labour "vacations" at their lovely country domicile which turned out as truly contemplative work environment.

Writing a book, of course, depends on more technical matters as well. In this respect, I luckily benefited from support of *Claudia Schlager* who patiently and competently turned all of my scribbled drawings into the pleasing figures adorning the bulky text to follow; *Michael Schwab*, *Ewald Hellerschmid* and *Werner Hennrich* backed me in technical trouble-shooting and made all software and hardware systems I had to rely upon run smoothly whenever necessary; *Veronika Doblhammer* and *Jan Sramek*, Springer Verlag Wien, provided an obliging productive cooperation and did a fine job in preparing the publication of this book.

The deepest, after all, I am indebted to my thesis advisor *Wilfried Grossmann*; without his passionate encouragement, insightful support, and highly stimulating cooperation the project of compiling this thesis probably would neither have ever taken off nor got finished in finite time. His belief in the soundness of the proposed theoretical framework spurred me to continue with my research and made me feel strongly responsible for bringing it to a definite conclusion.

Finally, I would really like to express my deep thankfulness towards my family – instead of understandable complaints about my absorbing academic engagement during all these months of intensive writing I always received sympathetic motivation and an unreserved feedback of doing everything right.

Wien, August 1996 $K_\forall F$

Table of Contents

Index of Figures ix
Index of Tables xi
Prologue * Introduction **1**
1 Needs, Tools, and the State of Research **7**
   1.1 The Data Challenge of Information Society 8
   1.2 Contributions of Information Technology 11
   1.3 Statistical Data Modelling 17
      1.3.1 Statistical Data Structures and Data Semantics 21
      1.3.2 Data Modelling Approaches 27
   1.4 Systems, Prototypes, and Research Projects 58
      1.4.1 A Survey of Systems and Developments 61
      1.4.2 Current Trends in Research 69
   1.5 Statistical Metadata Management 74
      1.5.1 A General View of Statistical Information Processing 75
      1.5.2 Metadata: Roles, Typology, and Representation 81
      1.5.3 An Effective Approach to Data Harmonization 87
2 METASTASYS – A Formal Metadata Approach **91**
   2.1 Premises and Design Principles 92
      2.1.1 Notation Conventions 101
      2.1.2 Elementary Definitions 104
   2.2 The Data Model 112
      2.2.1 Populations, Scales, and Qualitative Characteristics 117
      2.2.2 Property Spaces, Frames and Frame Linkages 126
      2.2.3 Domain Concepts 133
      2.2.4 Data 142
   2.3 Data Source Mapping 144
      2.3.1 Source Views 150
      2.3.2 Mapping Definitions 155
      2.3.3 Source Images 161
   2.4 Data View Specification 164
      2.4.1 Target Concepts 168
      2.4.2 Summary Types 178
      2.4.3 METASPEL 187
      2.4.4 META Algebra 199
   2.5 META Derivation 203
      2.5.1 Basic METAGEN Operations 208
      2.5.2 METAGEN Operators 222
      2.5.3 METAGEN Derivation Schemata 245
   2.6 Distributed Request Processing 253
   2.7 The DÖS'chen Prototype 265

**3 Application Case I: Labour Force Statistics
– Harmonizing Official Statistics Data** **279**
  3.1 Social Change and Economic Welfare 280
  3.2 Measuring Employment 291
    3.2.1 International Development of Labour Force Statistics 294
    3.2.2 European Labour Force Statistics 300
    3.2.3 National Labour Force Statistics 317
  3.3 Labour Force Survey Integration 328
    3.3.1 An Employment Statistics Data Model 330
    3.3.2 Global LFF Edits 347
    3.3.3 Survey Mappings 352
  3.4 Harmonized Data 362
    3.4.1 Labour Force Concept Definitions 363
    3.4.2 Labour Force Aggregates 376
**4 Application Case II: Tourism Statistics
– Strategic Management Information** **391**
  4.1 Electronic Markets and Decision Making 392
  4.2 Traditional Tourism Data Structures 398
    4.2.1 Lodging Enterprise Statistics 400
    4.2.2 Bednight Statistics 401
    4.2.3 Statistical Tourism Aggregates 404
  4.3 Tourism Data Frame Definition 410
    4.3.1 Preparatory Considerations 410
    4.3.2 Frame Definitions: A Couple of Approaches 413
  4.4 Information Delivery at the Management Level 420
Conclusion * Epilogue **437**
Appendix A – METASTASYS Operation Definitions **445**
  A.1 Idendity, Equality, Structure Comparison 446
  A.2 Boolean Operators 449
  A.3 Ordering Relations 450
  A.4 Type Casting Operations 451
  A.5 Set Operations 452
Appendix B – Dependency Edit Processing **461**
Appendix C – Labour Force Metadata **471**
  C.1 Labour Force Quality Definitions 472
  C.2 Labour Force Frame Edits 480
  C.3 Labour Force Concept Definitions 484
  C.4 Labour Force Survey Mappings 496
  C.5 Sample Request Processing Minutes 508
Literature **515**
Definition Index **535**

Index of Figures

1.3.1	An Infological Model Instance [Sundgren, 1992]	23
1.3.2–1	A SUBJECT Graph [Lenz, 1993a] (adapted)	29
1.3.2–2	An Extended Object Graph [Lenz, 1993a] (adapted)	30
1.3.2–3	A GRASP-Type Object Graph [Lenz, 1993a] (adapted)	31
1.3.2–4	Conceptual vs. Stored Files [Sato, 1989]	33
1.3.2–5	Excerpts of SDM4S Concept Hierarchy [Sato, 1989]	35
1.3.2–6	Conceptual Object Hierarchy in SDM4S [Sato, 1989]	36
1.3.2–7	Semantic Net Excerpt of SDM4S [Sato, 1989]	37
1.3.2–8	Metaobject Graph [Sundgren, 1993] (modified)	38
1.3.2–9	Transparent Operation on Object Graph [Rafanelli and Ricci, 1990]	48
1.5.1	A General Statistical Data Flow Scenario	79
2.1	General Architecture of METASTASYS	96
2.2	Property Space Representation of a Population	114
2.2.1	SCALETYPE Hierarchy	119
2.2.2–1	FRAMEdit Structures	129
2.2.2–2	FRAME Linkage Types	131
2.2.3–1	(a) CONBOX Representation with CONEDits	135
	(b) CONBOX Representation without CONEDits	136
	(c) CONBOX Representation Including NULL	137
2.2.3–2	(a) Axis Rectifications – General Case	138
	(b) Axis Rectifications – Simplex Case	139
2.2.3–3	FRAMEdits in Axis Rectifications	140
2.3–1	Data Source Mappings	145
2.3–2	Data Source Mappings' Semantic Loci	147
2.3.1–1	Time Series SOTRAITs Representation	154
2.3.1–2	Relation Patterns between Data Sources and FRAMEs	156
2.3.2	A Sample SOVAR Mapping Using an IMAXIs	160
2.4.2	Sample FRAME Summation	184
2.5.1–1	Sample Unification Graph	218
2.5.1–2	(a) Sample TEXTures for Clustering	220
	(b) GRUSET Intersections by Dimension	221
	(c) TEXTure Clusters	221
2.5.2–1	IMAGE Selection	226
2.5.2–2	Patching up a Work Table	239
2.6–1	Virtual METANET Connectivity	255
2.6–2	METANET Topology	256
2.6–3	METANET Communication Flow	257
2.6–4	METANET Server Structure	263
2.7–1	*GME* Main Screen	269

2.7–2	*GME* QUalities Definition Screen	269
2.7–3	*GME* AGGLEvel Definition Screen	270
2.7–4	*GME* FRAMEdit Definition Screen	271
2.7–5	*GME* Domain Concept Definition Screen	272
2.7–6	*GME* Survey Definition Screen	273
2.7–7	*GME* Survey Variable Definition Screen	274
2.7–8	*AGen* Concept Selection Screen	275
2.7–9	*AGen* Concept Restriction Screen	276
2.7–10	*AGen* Cross-classification Screen	277
2.7–11	*AGen* Request Screen	278
3.1	Development and Forecast of Economic Sector Shares in Germany	288
3.2.2–1	Main Concepts Discerned in European Labour Force Statistics	307
3.2.2–2	Dependency Structure of CLFS Variables (Excerpt)	312
3.2.3	Four Different Austrian Unemployment Rates	325
3.3.2	Main *LFF* QUality Dependencies	348
3.4.1	Logial Hierarchy of Labour Force Domain Concepts	364
4.1–1	Main Tourism Communication Flows [Werthner, 1993] (adapted)	395
4.1–2	Tourism Management Information Capturing and Dissemination	399
4.2.1	Inventory Form Used in Austria's Tourism Statistics	402
4.3.2–1	Tourism Data Multi-FRAME Model	417
4.3.2–2	Sample Bednight FRAME	418
4.3.2–3	Ternary FRAME Linkage	419
B–1	A Sample Dependency Graph	462
B–2	Sample Dependency Graph after Elimination of Conditionally Nested Variable	463
B–3	A Sample RDG	464
B–4	A Sample Composite Dependency	465

Index of Tables

1.3.1–1	Case-by-Variate Structure of Statistical Observation Data	22
1.3.1–2	Statistical Macrodata Structure	24
1.3.1–3	Data Processing Stages vs. Data levels	27
1.3.2–1	(a) Sample Relation r of Relation Schema R [Ghosh, 1991]	41
	(b) Summation Output Relation R_1 [Ghosh, 1991]	42
	(c) Product-Sum Output Relation R_2 [Ghosh, 1991]	42
	(d) Frequency Distribution Output Relation R_3	43
1.3.2–2	Summary Relation for Category r	43
1.3.2–3	(a) Sample Table of Rates	46
	(b) Restricted Sample Table of Rates	46
1.3.2–4	Interpretation of $\mathcal{A}(P,R)$	54
1.4.1	Coarse Processing Stage Model of Statistical Information Processing	62
1.5.2	Functional Areas of Operative Metadata vs. Descriptive Dimensions	83
2.2.1	SCALETYPE Summary	120
2.3	Source Types and Data Levels	149
2.4.3	Native METASPEL Clauses	189
2.6	Ranking Criteria for Retrieval Set Formation	260
3.1	Synopsis of Symbolic-Analysis Job Titles	289
3.2.2–1	Reference Weeks and Sample Sizes of the CLFS	302
3.2.2–2	Discrimination Criteria Used in CLFS Concept Formation	306
3.2.2–3	Listing of the CLFS-92 Variables	308
3.2.2–4	Categories and Codes of Cardinal CLFS Variables	311
3.2.2–5	Relationship between CLFS Concepts and Variables	310
3.2.2–6	CLFS Variables and Main Population Groups	311
3.2.2–7	CLFS-83 and -92 Coding Schemes of Variable *"Main reason for leaving last job or business"*	315
3.2.2–8	The Employee Case of Category Association for Variable *"Main reason for leaving last job or business"*	316
3.2.3–1	Austrian Sources of Employment Data	319
3.2.3–2	Categories of Question 13 of the Portuguese Labour Force Survey	320
3.2.3–3	Mapping of Portuguese Labour Force Survey Questions to CLFS-92	321
3.2.3–4	Reference Dates Summary of National Employment Statistics	323
3.2.3–5	AMZ Questions and their Relation to the CLFS-92	324
3.2.3–6	Comparison of AMZ and CLFS-92 Categories for Variable *"Relationship to reference person in the household"*	326
3.3.1–1	Qualities of *Labour Force Frame*	333

3.3.1–2	MO Dalities of *LFF* QUality *work status*	340
3.3.1–3	AGGLEvel 'CLFS-92' for *LFF* QUality *professional status*	342
3.3.3–1	SOVAR Mappings to *LFF* QUality Q_{12}	356
3.3.3–2	AMZ Variables Needed for *work status* Mapping	357
3.3.3–3	AMZ *work status* Mapping	358
3.3.3–4	P-LFS Mappings to *LFF*-QUality *professional status*	359
3.3.3–5	Q_{50} MODalities and CLFS/P-LFS Code Mappings	360
3.3.3–6	Mapping CLFS-92 Variable No. 50 to the *LFF*	361
3.4.2–1	Sample CLFS Table [EUROSTAT,1993a]	379
3.4.2–2	Sample Three-Way Breakdown Table of Rates	381
3.4.2–3	Sample CLFS Table with Numerical Agggregate [EUROSTAT, 1989c]	385
3.4.2–4	Sample Compound Table	390
4.2.1–1	Tourism Enterprises and Beds in Austria 1992/93 [ÖSTAT, 1994]	403
4.2.1–2	Room Standards in Austrian Tourism at May 31, 1993 [ÖSTAT, 1994]	405
4.2.2	Austrian Bednights by Communities, 1986–1993 [ÖSTAT, 1994]	407
4.2.3	Foreign Share of Tourist Bednights in Austria 1993 [ÖSTAT, 1994]	409
4.3.2	Sample FRAME Definition	413
4.4	Sample AGGLEvels for QUality *accommodation type*	422
B	Sample Edits	421

Prologue * Introduction

Post-industrial society is of strikingly complex nature; progressive division of labour and changes in production and goods structures inducing a deepening interdependency of actors and economic market participators and giving rise – both enabled and enforced by powerful transportation and tele-communication media – to the globalization of markets and, hence, of society as well is accompanied inevitably by increasing information demands and more complex control and decision tasks. Whether we concern governmental planning and policy, top level management decision making of large and particularly multi-national companies, or scientific research: more than ever before, rational action crucially depends on our *data image* of the world, on the inferences we (can) draw from bodies of data – numbers, in fact – representing reality. In its 17th century beginnings conceived of as *disciplina status* – a description of state affairs as exemplified by polyhistorian Gottfried Wilhelm Leibniz's „Staatstafeln" or political arithmetician John Graunt's mortality tables – in order to have a means *to reason about* reality, the role of data representing (views of) the world has undergone transformations since – by proposals such as Lambert Adolphe Jacob Quetelet's *Physique sociale* (which stipulated the *homme moyen*-concept) in the 19th century – of so profound a nature that reality in our age *is* data in a very definite sense: there simply is no other way to get hold of our complex world mentally. Thus, preparing and maintaining data images of the world has become a crucial factor of economic prosperity and social self-awareness, and those who take care of these data images are considered to be better off in world-wide competition and political leverage.

Recognizing the relevance of data for society, political decision making and economic wealth, *information technologies* have gained growing attention. In particular, digital communication – data exchange – and symbol processing facilities – computer systems and networks – provide the technical infrastructure of present and forecasted societies, and considerable political efforts and investments are made all over the world to improve these infrastructures at national as well as supra-national levels. Next time, broadband tele-communication media will make possible the transmission of huge amounts of data to almost any point on the globe at an instant of time, thus providing a nearly unbounded capacity of data dissemination and distribution.

Technically, we take already note of a global flow of data (of enormous and yet breathtakingly increasing volumes); in fact, there seems to be an abundance of data – yet *information* still is a scarce resource. This malicious discrepancy results from a lack of data *context* and, particularly, context-preserving data *integration*. Apparently, data is produced in local contexts – but once going to be used outside its originating context as secondary data or in transborder mileux, measures to provide proper data *interpretation* have to be taken consciously. For example, the emergence of global markets tremendously decreases the efficacy of national economic policies whence trade, payment balance as well as social and economic planning data must be consolidated at supra-national levels such as the European Union (note that here data integration is an immediate consequence of political integration) despite the obvious differences between national statistical systems which, essentially, have been established independently in each member country. Now, national differences can either be flattened out (nominally, at least) by mandatory legal regulations or resolved by making intelligent use of information technology. Practically, the former approach incurs – political and monetary – costs too high to merit further consideration in most application domains, so, essentially, we are left with a data *harmonization* problem: the development of an EU umbrella statistics system, on top of still autonomous (subsidiary) national statistical systems, actively accounting for semantic heterogeneity and data production peculiarities of local (national) data contexts. As a matter of fact, a closer look unveils the very same situation with enterprise and scientific data though, clearly, the pragmatic point of view is changing considerably; formally, however, it turns out as rather immaterial whether data to be consolidated originate from different national offices, different production sites or branches of a large multi-national company, or scientific measurements or observations from different places and periods, possibly motivated by various concerns and conducted under quite different conditions. As data production is very costly most of the time, under-utilization of data due to lacking means of effective data integration is a little more than a technical nuisance: forgoing the decision making, competitive, and scientific potential of data integration simply is a waste of resources and also misses to partake in a qualitative "quantum leap" in rational policy making; for instance, without data integration most of historical data such as socio-economic time series – the feedstocks of economic model building ! – tend to be devalued seriously because of semantic contexts fading away so quickly.

Before proceeding further, a concise definition of what the term *data* should mean and, hence, what kind of *data integration* will be talked about subsequently is in place. First of all, only *statistical* data will be considered, that is, empirical data not collected for its own sake but for its capability to represent features of interest as typically carried by individuals of some well-defined population. This is meant to rule out *transaction* data kept for purposes other than providing descriptions or permitting inferences of feature distributions (although, of course, some data may serve either purpose, operational and statistical); quite in contrast, disregarding the peculiar individual entity, emphasis is laid on *statistical aggregates* obtained by combining the individual entities' characteristics. In this respect, this definition fully conforms to common statistical usage. Additionally, however, the scope of statistical data is further constrained to *unique* data which, basically, refers to the spatio-temporal dependency of empirical data or the observational/experimental set-up: by definition, data is unique if the repeatability of the data generating process (regardless of being pure observation or controlled experiment) is in doubt; in particular, data is unique if it cannot be reproduced (stochastically) either for principal reasons or simply because of a lack of approved, convincing structural models. Applying this definition of uniqueness, typical cross-sectional experimental data is *not* unique because – although each particular repetition of the experiment,

INTRODUCTION 3

of course, may give a unique result – it is claimed (according to Newtonian physics, at least) that *in the long run* results are governed by some natural (time-invariant, stochastic, etc.) law, the persevering *harmonia mundi*.

Obviously, the definition of data uniqueness is a rather weak one, making this property dependent on context and circumstances and giving ample room to subjective argument. Its sole purpose, however, is to help drawing a pragmatic distinction between data relevant in the present context of investigation and data which is not. To give an example, despite the sound belief that a particular economic indicator is indeed governed by some stationary or non-stationary process the uniqueness assumption may be sustained, that is: empirical data is kept, since continued observation and analytic scrutiny may very well reveal different interpretation patterns (models) and further theoretical insight. The uniqueness concept applies particularly in cases where observation, or measuring, covers rather lengthy time intervals – such as longitudinal studies with repeated measurements taken on screened individuals – because in those situations it is usually not possible to repeat empirical data collection for practical reasons (that is, the notion of repeatability is empirically meaningless).

The uniqueness property of data only, of course, justifies permanent storage; apparently it would be rather devoid of interest to keep data which can be reproduced – albeit stochastically – on demand. Despite its uniqueness, however, *pooling* of recorded data either in spatial or temporal (or both) respects is a frequently required task. Without concerning the statistical (more precisely: the stochastic) underpinnings of such an operation for the time being, any kind of pooling of unique data will be called *data integration* provided that collated datasets still permit consistent semantic interpretation (as contrasted with mere syntactic concatenation). Obviously, semantic data integrity is not a property inherent to data but to the models used to interpret data. Although the processing of data has been mechanized thoroughly by using electromechanical and, later on, electronic computing devices, creating and handling data models still has remained a deeply human domain. Now, as we have started to recognize, it has also become a fundamental obstacle to effective data processing on a global scale: the attained mobility of data calls for a mechanized processing of data contexts unless we are willing to take the risks of dealing with meaningless data and, hence, meaningless conclusions drawn thereof. In other words, ways and means have to be found to couple data and data context in formal representations amenable to joint dissemination as well as mechanized processing.

Since the envisioned type of data integration appears to be a fairly general problem occurring in a multitude of domains of practical relevance the endeavour of devising a general framework – a theory as well as actual tools – of data integration proves worthwhile. In fact, current information technology offers quite sophisticated media and theoretical proposals, in particular *semantic data modelling* and the concept of *metadata* to start with: the former aims at symbolic representations of real-world entities and real-world relationships between entities at machine (software) level, whereas the latter serves to capture formally the structural properties of semantic data models by mapping this information to ordinary data representations (hence the name *meta*data). Thus, metadata materializes – denotes extensionally – the intrasymbolic content of semantic data models; viewed procedurally, metadata represents the physical output of semantic data model design. However, in contrast to conventional data dictionaries used to describe database structures, metadata is destined to provide much richer domain representations addressing model features beyond data retrieval and extensional consistency.

A little scrutiny reveals that semantic data models hitherto developed belong to either of two hemispheres none of which happens to address data integration requirements as sketched above. On the one hand, data models are an outcome of (relational) database research oriented towards commercial and administrative use of databases with its predominant interest in record-level data (tuple-management, record identification, normalization, expression of relationships, etc.); on the other hand, in statistical data processing – particularly in official statistics – the data modelling problem as such has been hardly recognized yet and is, at best, dealt with very pragmatically (reminiscent of the generally low level of formalization in statistical data handling). Quite recently, statistical data modelling is receiving increasing attention, reflecting apparently the changing role of data management as a whole: had organizations made use of database technology internally to streamline clerical work for the sake of cost reduction and improved efficiency, this is no longer of significant competitive advantage (since, meanwhile, everyone does it that way); in fact, improved *decision making* – affecting the external relation between an organization and its environment – is the logical next step to regain leadership in competition. Since decision making depends largely on aggregate (that is: statistical) data, traditional data modelling approaches are augmented correspondingly. From a public point of view, data integration is a prerequisite for achieving the intrasocietal information infrastructure, or coherent communal digital self-image, upon which social forecasting, planning measures as well as assessments of future effects of current policies and socio-economic development can be based. Similarly, scientific world description – model building at large – in terms of complex and, hence, intricate dynamic systems calls for integration of partial systems and, as immediate consequence thereof, for utilization of rather disparate, heterogeneous data bodies.

By now, it has been realized and accepted quite generally that aggregate data management requires rather specific data modelling approaches. In particular, the conventional relational database technique fails to capture aggregate data semantics adequately and, therefore, supports data manipulation poorly either. This holds even more in view of the data integration task to be solved.

Conversely, to date no comprehensive attempt has been undertaken to design semantic aggregate data models addressing the data integration aspects as well. Referring to the concept of metadata, this book sets out to expound a theoretical framework, METASTASYS, for semantic integration of statistical data aggregates taking into account both, grown data infrastructures and the expected usage of future information systems operating on a global scale. More specifically, information systems of the conceived type will have to support the tasks of *locating* data aggregates and of *retrieving* data aggregates – with data integration taking place

INTRODUCTION

implicitly if necessary – on the data consumption side; on the data production side semantic interfaces to existing data pools and data production systems have to be established (in addition to the technical coupling of systems). In particular, the suggested framework leads to a *statistical data environment* devised to integrate large existing data volumes with different semantic, statistical and storage structures in a simple, usage-oriented way by applying advanced concepts of computer science (especially semantic data modelling and metadata techniques) to achieve a *uniform conceptual data view* in front of an inherently heterogeneous, distributed, and non-stationary data repository. Among the distinguishing features of this framework are to be mentioned:

- the increase of efficiency (both temporally and in terms of cost) by effectively integrating data from (physically) different sources which, in practice, will mean that in most cases those data sources become really joined up for the very first time at all;
- simplified and cost-reduced dissemination of available data throughout the whole environment without the technically compulsory necessity of any kind of pre-aggregation of record-level data (except for voluntary reasons of data privacy, suppressing more detailed cross-classifications, etc.);
- the provision of user-defined data aggregates instead of all-or-nothing data delivery, that is data access and aggregate definition are collapsed into a single output definition step – in fact, data integration is achieved possibly unnoticed by the end-user since the data environment itself spots those cases where data integration becomes necessary and, if so, how it can be brought about effectively;
- a data-independent data consumer interface, that is navigation-free non-procedural specification and composition of data aggregates with respect to a universal semantic (conceptual) data model entirely independent of physical storage organization;
- a single, all-encompassing aggregate data structure reducing all of the semantically different aggregate structures to a single data type to be processed automatically by formal-symbolic metadata represented as part of aggregate structures (that is, internal processing of aggregates is controlled by metadata basically without requiring further human intervention);
- a taxonomic framework based on simple mathematics to link local data sources to the overall data environment which, at the same time, provides a concise documentation of data and data production processes in terms of relative, or "first derivative", semantics (that is, expressing the relative semantic difference between concepts used in different data sources).

In order to illustrate structure and functionality as well as to convey evidence in favour of the viability of the suggested data integration framework, its theory is worked out to some detail for several selected application domains. Moreover, a brief report on a software prototype implementation, *DÖS'chen*, is given to round off the presentation with respect to indicating how to turn theory into practice.

In several respects, the proposed data integration approach draws on previous research in statistical data modelling as well as general database research; irrespective of its novel overall architecture, the framework has proven to operate as a platform for amalgamating a couple of hitherto non-integrated partial proposals and lines of pertinent research emanating from such differing scientific communities as the SSDBM (Statistical and Scientific Data Base Management) meetings, special EUROSTAT (the statistical office of the European Union) meetings (particularly, those of EUROSTAT's DOSES programme), VLDB (Very Large Data Bases) conferences as well as the UN/ECE (the United Nations' European Commission for Economics) working groups. In summing up, the METASTASYS framework is not intended as yet another proposal for statistical database management; quite on the contrary, it has been conceived consciously as an approach to statistical data integration exploiting potential synergies with existing technology and research to the best of it.

* * *

The material presented in this book divides into two parts. The first, *theoretically* oriented part is devoted to a detailed description of the METASTASYS framework; in order to set the stage, this is preceded by a summary of the state of research in statistical data modelling as well as a summary of developments in relevant information technologies (Sections 1.1–1.4), followed by a discussion of the general model of structure representation and processing of statistical data aggregates (Section 1.5). Chapter 2, as the core chapter of the entire book, presents the theoretical set-up of METASTASYS in rather formal terms and summarizes data structures used to represent and manipulate metadata technically, too. The chapter closes with a brief description of the DÖS'chen software prototype, following roughly METASTASYS design principles. The second, *empirically* oriented part presents two specific data integration application cases, viz. Labour Force statistics (Chapter 3) and Tourism statistics (Chapter 4) discussing preconditions and consequences of establishing METASTASYS in these different areas. The final chapter, Conclusion, tries a provisional evaluation of the merits of the proposed data integration framework, sketches some of its still unexplored capabilities and makes an assessment, with reference to the discussed examples, of the advantages of this approach as opposed to conceivable alternatives; it also investigates very shortly implications and chances of this powerful new technology if introduced on a larger scale.

The five chapters of the book are followed by three appendices. The first one presents an excerpt of METASTASYS's symbolic function definitions encoding metadata operations. The second appendix summarizes some preparatory steps required for integrating a data source into the METASTASYS data model. Finally, the third appendix comprises a polished transcript of selected Labour Force statistics metadata (discussed in Chapter 3) as it has been used in DÖS'chen demonstrations.

Chapter 1

Needs, Tools, and the State of Research

The basic tenet of the following considerations is the widely-adopted conviction of a generally increasing demand for information in all planning and decision tasks affecting individual businesses and society as a whole alike, as an improved and timely access to information as well as advanced methods of information processing have already become tangible factors of competitiveness, economic prosperity, and social stability and welfare. However, despite the abundance of data and the presence of quite powerful information technologies for data processing, the effective conversion of data into information is in fact the really critical, and limiting, issue. In particular, if data of different origins has to be merged, implying that different data contexts have to be merged either, current information technologies, by and large, still fail to meet the pressing demands for interlacing and fusing data *semantically* such that combined data are amenable to a coherent interpretation.

Before a viable concept of semantic data integration can be formulated the state of development in formal symbol processing and digital information technology supplying the fundamental principles, methods, and technical devices must be reviewed. Likewise, another point of departure is the state of the art in statistical data modelling, especially semantic data models aiming at a faithful formal representation of semantic relationships both within and between empirical datasets. Apparently, semantic statistical data models are the primary vehicles of any prospective approach to data integration.

A further prerequisite of the data integration methodology expounded in Chapter 2 is an overview of current research in statistical information processing with respect to the proposals' and systems' contributions to semantic data modelling and statistical information management techniques some of which have already been implemented in practice; this survey is conducted mainly from the institutional data providers' point of view and emphasizes the concept of *metadata* and its formal management, building the basis of the research programme worked out technically in Chapter 2 and illustrated by selected application examples in Chapters 3 and 4.

1.1 The Data Challenge of Information Society

"TO SAY that the advanced industrial world is rapidly becoming an Information Society may already be a cliché. In the United States, Canada, Western Europe, and Japan, the bulk of labor force now works primarily at informational tasks such as systems analysis and computer programming, while wealth comes increasingly from informational goods such as microprocessors and from informational services such as data processing. For the economies of at least a half-dozen countries, the processing of information has begun to overshadow the processing of matter and energy."

This is the introductory paragraph of James Beniger's [1986] seminal analysis of what he calls the "Control Revolution"; essentially, his thesis claims that the tremendously increasing utilization of machines and mechanized processes in industrial goods production during the pushes of the Industrial Revolution of especially the 19th century incurred the critical problem of *control* which became accentuated and aggravated further with the growing particularization of industrial society (specifically, the increasing division of labour) and the systemic interdependency of its functional and decision units such that *information* – in both its meanings of external control and self-regulation – assumed the role of a cardinal explanatory paradigm, named *cybernetics*, from the 1930ies onwards, resting on communication technology, especially the electrical principles of remote effect, and mathematically stated models of cycle-relational (self-referential) causation systems. This new discipline, destined to encompass biological and physical phenomena alike, contained already the "genes" of what later was to become the so-called science of computing which, however, could not take really off and establish its own identity before the advent of improved electronics making possible self-controlled – that is, programmed – digital computing devices, both direct offspring mainly of war-time research predominantly carried on in the USA and the United Kingdom. Apparently, computer science has outranked cybernetics in the meantime as the digital paradigm turned out superior in practical respects compared to the basically continuous, or *analogue*, modelling of cybernetics and control theory. This correlates with the decline in importance of controlling energy-converting processes – which basically receive now attention with respect to optimizing technical efficiency for the sake of sustentation and saving ecological balance of the natural environment – relative to a tremendous increase of information-converting processes determining the macro-development of globalizing society. Therefore, digital communication and information technologies have attained a key position in our world, and the primary focus has shifted from mechanical engineering and electronics towards the science and art of symbolic problem solving or, in less mundane words, from the intentional design of material structures towards the intentional design of informational structures.

The crux of digital information processing consists in the materialization of information – namely: *data*, that is formal-symbolic objects with linguistic structure – replacing the process-oriented information concept used in biology and communication theory (where, basically, information is a dynamic property of the receiver of a message, stimulus, or input, depending on its internal state). In a sense, data is an "abbreviated" mode of textual representation amenable to rigorous logical manipulation, thus extending more narrative discursive methods of arguing by strictly formal-deductive (yet still discursive) symbol transformation methods which can be carried out by machine, in this respect fitting perfectly the logic of development of literate societies [Logan, 1986] using alphabetic languages – that is, languages of an algebraic nature – as the main mediator between active subject and the experienced object world. In particular, symbol-coded data is extricated from the course of time, it can be stored, organized (arranged, or rearranged), and moved from place to place; in a word, it becomes *manageable* which constitutes its real value (in this respect, data is quite similar to money [Foucault, 1966] which is appreciated just because of its multiple use). In view of its terseness, of course, the semantics of data must be defined precisely such that, in particular, discursive processing is reflected faithfully by syntactic rules of symbol rearrangement which is to say that there must not be semantic relationships not representable by symbol placement lest this extra-symbolic meaning is in danger to get lost. Apparently, from a pragmatic point of view, some care in defining symbol structures encoding the meaning to be conveyed is in order to safeguard the information content of data. Practically, this functional role of data is emphasized implicitly noting that it is remarkably difficult to find succinct definitions or explanations of the term "data". In the computer science literature, at least, *data* (the plural form of *datum*, now used as both singular and plural) usually is defined by structural properties or operations that can be carried out on it rather than in qualitative terms delineating it from other entities. After all, Knuth [1968] defines data as "*[R]epresentation in a precise, formalized language of some facts or concepts, often numeric or alphabetic values, in a manner which can be manipulated by a computational method*" which, obviously, barely resembles its original Latin meaning of "gift" or "donation" being derived from the verb *do* (*dare, dedi, datus*), to give, to represent, but also: to bring forth; in a wider sense, *data* simply refers to "something given".

Digital control involves two cardinal preconditions, viz. a discrete symbolic representation of processes and systems to be controlled, and functional models of these processes and systems; while the former deal with static spatial symbol structures, the latter focus on the dynamic (stepwise) transformation of static symbol representations in terms of operations modifying symbol placements. As a typical feedback of technology, the symbolic re-engineering of (excerpts of) reality evokes directed adaptations in order to improve the possibilities of digital control; "digitizing" is yet another instance of redefining reality in accordance with the dominating technological paradigm – this time, digital information processing – leading up, quite consequently, to both virtual digital worlds ("cyberspaces") [Rheingold, 1991] and a strong bias towards "digital" perception of and interaction with reality [Schachtner, 1993]. As society depends on and is determined heavily by the technical communication means it has at its disposal and, moreover, as technical communication media exert a subtle though undeniable influence on what and how is communicated – McLuhan's [1964] thesis – a profound change of Western society, towards the so-called Information Society, must be envisioned indeed. Plastic money, flexible manufacturing, CAD-ed products, tele-work, home-shopping, "paperless" offices, electronic leisure amusements, computer-aided medical care, etc. may be only the first faint shadows of the things to come. Recent estimates indicate that by now, for instance, around about 60% of all employment in the European Union is conditioned or influenced by information technologies, and for the year 2000 about 10% of the world gross domestic product is forecast to be contributed by information technological products and services [Daser, 1994]. So much is life already depending on digital information and communication infrastructures that even a partial breakdown would immediately cause terrific disaster, at least for all developed societies, as virtually all value chains would be interrupted, logistics would collapse, and probably even the most elementary kinds of demand like food, energy, basic medical treatment, and so on, could no longer be served effectively. Even in a less horrible scenario, commerce, for instance, cannot be sustained any longer without tele-communication (in fact, commerce has early been a driving force in establishing tele-communication systems [Flichy, 1991]) just as the global capital flows – in magnitudes of daily turnover equivalent to the gross domestic product of countries like Germany or France – would be completely unthinkable; dramatically decreasing transaction costs have spurred a vigorous re-engineering of business value chains; the shift from product to information competition calls irreversibly for different management structures and business processes.

To cope with these challenges, the new branch of *information economics* has emerged [Parker and Benson, 1988] investigating the interrelation between business performance and information technology mainly from a micro-economic point of view which, as a matter of fact, applies to macro-economics as well in dealing with the utilization of informational tools and procedures to meet increasing demands of competitiveness, responsiveness, flexibility, and, generally speaking, managing planning and decision complexity. To some extent, perhaps, information technologies are also believed to improve the world (in terms of *progress*), helping to shape a "better" future. In fact, originating from a reasonable interest in sustaining individual or societal welfare, information economics enables a *rational* dealing with the future by having a computational means to project the future courses of the world, making tomorrow's business today's concern and leading, in turn, to a replacement of the future with a "stretched" presence [Nowotny, 1993]. This, after all, might be the quintessence of the control revolution: not the management of present processes is economically and socially decisive anymore but the optimal design of the future or, more specifically, of future decision possibilities. It is generally expected that the management of these scenarios and forecasts itself is about to become a main economic subject generating added value, thus assigning information economics a pivotal role in this development. As this creates a new information market, *information resources* become a critical factor, and data could be thought of as a new type of "raw" material feeding a new industry branch. The data challenge of information society, hence, consists in meeting its present and future information demands, calling, in turn, for directed measures to keep and extend its collective data repositories as well as the information infrastructures and symbol processing methods necessary for turning "inanimate" data into productive information *on demand* and *just in time*. Basically, this is the theme the following considerations will be devoted to.

1.2 *Contributions of Information Technology*

To meet the data challenge of the information society, a widening application of advanced symbol processing and information technologies seems inevitable. In particular, three areas of digital information technology will continue to contribute significantly to this societal evolution, viz.

- *data management* addressing the practical organization of symbol structures on account of improved semantic data models such that powerful computation processes can take place on top of them;
- *symbol processing techniques* providing the tools to devise and express these computational processes in concise formal-linguistic, though looking "natural", terms; and
- *digital tele-communication* supplying the means for moving symbols over distances and accomplishing remote effects of symbol processing especially with respect to organizing distributed systems and the synchronization of remote computation processes.

Jointly, these three areas create a global digital symbol manipulation and information infrastructure becoming irresistibly and irreversibly the pivotal backbone of the emerging world economy and, in its wake, world society. This on-going development, of course, owes its tributes to historical roots shaping Western information technology from its very beginnings [Flichy, 1991].

Turning to *data management* first, the approach still favoured bears witness of its origins in the early days of digital electronic computing, and especially of the memory devices then available. A lasting influence on digital computing has been exerted by the machine design ascribed mainly to John von Neumann [Goldstine, 1972] and first explored in the famous EDSAC/EDVAC projects and especially in the computing machine built at Princeton's Institute of Advanced Study under his own supervision shortly after World War II [Williams, 1985]. A physicist first and foremost, John von Neumann had in mind the potential of electronic computing in solving intricate numerical (differential) equation systems which led to a machine design focusing on computation structures, and giving comparatively little attention to memory structure; in fact, at that time, memory was a limiting factor in machine design forcing engineers to resort to costly and unreliable experimental solutions.

From a cognitive point of view, digital storage – an instance of *artificial memory* [Yates, 1966] – is a temporary representation of state information, such as numbers being interim results of a computation; simply speaking, parts of (mental) memory are externalized and denoted symbolically (for instance, by putting calculi on abacus columns or by leaving marks on checked paper) so that the human mind can concentrate on *state transitions* – the next operation to be carried out – without being burdened by remembering state information details. Thus, in traditional computing technology, mind remains the driving force of reckoning operations. With the advent of mechanical inventions – the progenitor of them all being, of course, the mechanical clock – state transitions became increasingly mechanized as well giving rise to the first ingenious designs of calculating machines (like those of Schickard, Pascal, and Leibniz; cf. [Bischoff, 1990; Williams, 1985]), paralleled by inventions to code and store state transitions (punched card machines like the Jacquard loom, automatic music instruments, etc., [Froeschl *et al.*, 1993]). Despite this progress in mechanizing processes, the formal coincidence of state and state transition data escaped recognition until Alan Turing's design of the Universal Turing Machine [Hodges, 1983] 1936 which promoted a radically different view on computation processes and effectiveness to delineate, in a rigorous fashion, the notion of computability [Rogers, 1987]. Essentially, in this set-up some part of data is interpreted as control information ("parameter data") governing state transitions while another part is interpreted as basic state descriptive information including "input data" (if any). The procedural core of the Universal Turing Machine reduces to an elementary command interpretation device providing the machine's internal clock pulses triggering state transitions – being, in a literal sense, the machine's "balance" supplying its physical propulsion. John von Neumann – who was aware of Turing's epochal work – turned this conception into a real computing machine design for the first time. In so doing, he managed to create a truly ingenious blueprint of virtually all generations of computing machines built ever since, from the biggest supercomputer down to the meanwhile ubiquitous microprocessors – the so-called "John von Neumann-architecture".

In von Neumann's memory model, (numerical) state information, mathematically speaking, represents a *state vector* changed at discrete ("clocked") time ticks by fully predefined state transition data ("programs"). Apparently, this view complies perfectly with numerical computing tasks like the simulation of continuous physical processes expressed in terms of systems of differential equations (with analytic solutions difficult or impossible to obtain): memory represents the states, one by one, of the simulated system whereas system dynamics is mimicked by state transitions coded into programs. Basically, von Neumann considered first-order programs (programs not modifying themselves during execution) only, and storage is an unstructured sequence of uniform memory words, that is a vector of scalar numbers (such as reals in floating point representation with fixed word length). In formal terms, letting x_0 denote a program's starting, or initial, memory state and F the program operator encoding the state transition function such that, for each state x_i, its successor state x_{i+1} is obtained as $x_i = F(x_{i-1})$, for $i > 0$, then the comput-

.2 Contributions of Information Technology

ing machine realizes a universal operator – "the" von Neumann machine – $G(x_0, \varepsilon) = F^{n_0}(x_0)$ where $F^n(x) = F(F^{n-1}(x))$, $n > 1$, and $F^1 = F$ such that $|F^n(x_0) - F^m(x_0)| < \varepsilon$, $n, m > n_0$ (fixed point semantics); apparently, for approximating numerical computations, $0 < \varepsilon < \infty$, with ε chosen close to zero (convergent sequences or series) while, typically, $\varepsilon = 0$ for algebraic computations (path searching systems or reduction systems based on Noether relations), and – as a specific case – $\varepsilon = \infty$ for discrete event simulation (where computation termination is controlled exogenously in terms of n_0). It should be noted that this definition applies to both address and associative memory systems [Baron, 1987] on condition that computations take place in a strictly stepwise, synchronized fashion.

Completely in line with this predominant view, the emergent computer science devoted its interest to the formal-linguistic structures for determining "naturally" and efficiently state transition operators, or programs, *F* which induced a wealth of programming languages [Wexelblat, 1981] and gave birth to the *art* of programming [Knuth, 1968] as the discipline of squeezing out the formal flesh of mental problem solving to mould it into deterministic *algorithms* stated in digital typographic calculi executable by machine. Software engineering and the improvement of automatic translators (so-called "compiler" programs) converting "problem-level" program statements into internal microcommand structures of digital computing machines became, accordingly, the main issue of computer science, accompanied by a theoretical branch investigating the general principles of these endeavours. For quite some time, conversely, little attention has been paid to formal structures for modelling the *x*-component of the von Neumann-machine except as far as it was considered helpful in reducing the programming effort by, for instance, implicit memory management replacing loops over vector components with dynamic recursive program and (stack) data structures.

Meanwhile, software crises and different requirements, especially those emanating from system simulation and business data processing, have caused a tangible shift of focus. Particularly the application of computers in simulation studies has stressed the necessity of expanding the expressiveness, or data semantics, of symbol memory structures; it simply turned out that no longer only *F* was complex but *x* was as well. Accordingly, flat vector structure was to give way to more flexible, user-definable data structures, followed by even more powerful concepts of semantic data modelling such as *data abstraction* and *object orientation* (OO, in short) which, in a sense, gradually turned original memory management upside down by assigning state transition code to local data objects now, that is the global control of state transition in terms of *F* is replaced by local threads of process control interconnected by a general communication protocol (message passing), and the once monolithic state transition function becomes segmented into almost independent parts interacting in standard fashion (see below). Excepting purely numerical applications, computers – more precisely: software systems – could now be legitimately seen as *data machines*, that is logical representations of partial worlds denoting faithfully real-world entities and relationships between these entities upon which computation processes are defined. A particularly important aspect of these data machines is the *additive* memory organization allowing a distributed storage of objects among physically dislocated memory units which, evidently, is a fundamental prerequisite for computation processes spreading over objects stored in distributed memory without additional data exchange interfaces.

Memory may in fact not only be used for temporary symbol storage in state representation between consecutive state transitions; in many cases, a *persistent* storage of symbols is called for. In computer science, this is the domain of databases emphasized particularly in electronic data processing (transaction processing) where, in general, operations on data are of simple nature though being repeated often at stunning rates. Moreover, since persistent data typically represents persistent real-world entities, escorting in a sense these entities' life-cycles, data "records" must be preserved as the lifetime of storage objects exceeds the duration of computing processes. This places yet higher demands on adequate data modelling (symbol arrangement patterns) met best by specific *data languages* as counterparts to programming languages. A data language, in turn, comprises two interrelated components, viz. a *data definition language* (to establish and maintain semantic data models at the intensional level of a database) and a *data manipulation language* (to establish and maintain a database's extensional level) where the latter is responsible for expressing basic operations such as symbol insertion, modification (update), and retrieval. Like programming languages, data languages have undergone an evolution from hardware-oriented concepts to higher-level, abstract semantic data modelling concepts concealing the underlying physical memory structures. In particular, as mathematics provides a well-proven language of state description mathematical objects such as relations have been proposed successfully for use in data modelling although, despite research into the representation of information structures for some 30 years now [Taylor, 1992], the ultimate data language has not been discovered yet (should it exist at all).

A special case, and a quite important one in what follows, of data machines are *data recording machines*, that is databases destined to record empirical states (observation "snapshots" of reality) or longitudinal traces for later reference. As these observation datasets come with a non-standard semantics and are processed in rather specific ways, it comes perhaps as no surprise that general data modelling approaches and languages will not suffice and more custom-tailored variants are required (cf. Section 1.3). The semantic complexity and diversity of non-standard data modelling applications – like scientific and statistical databases – have stressed the necessity to extend semantic data modelling approaches to include semantic *data model representations* in addition to data representations, realizing that – suitably formalized – data models, like programs, are yet another kind of data. Consequently, this has led to the introduction of *metadata management* in order to mechanically support ordinary data management in scientific and statistical data storage systems. Metadata management, in turn, assumes a key position in establishing *active* symbolic data storage systems facilitating conceptual – more specifically: *deductive* – interfaces and adaptive internal storage regimes for relieving both system maintenance and usage.

This borders on the second contributing area of information technology, viz. *symbol processing*. Formally, any kind of computing is in fact symbol processing and vice versa [Krämer, 1988]. In practice, however, numerical computing and data processing are discerned branches mainly because of historical reasons, but also because of significant structural differences with respect to data management as discussed above. Mechanical symbol processing apparently rests on two preconditions: a machine ("mechanical" device) capable of performing symbol processing procedures and a formal means of expressing explicitly symbol processing (term rewriting) procedures. Since any kind of computing amounts to term rewriting, it suffices to use general-purpose computing devices (Turing machines); leaving the delicate problem of how to ordain mechanized symbol processing activity effectively and unambiguously. For principal reasons, a problem is amenable to translation into an effective procedure if its mathematical structure is sufficiently simple; practically, this is equivalent to the following cumulative conditions:

- unless the problem amounts to an arithmetical computation executable directly by wired operations, it must be reducible to a representation by *discrete* symbol structures (this analysis is the main concern of Applied Computer Science);
- the problem structure so obtained must be *finite*, at least as far as it is expanded extensionally to derive actual solutions;
- the problem must not be intractable, that is its run time and space complexities ought to be polynomially bounded or, otherwise, there must be shortcuts (encoded expert knowledge, *heuristics*), reducing genuinely intractable problems to *tractable* ones (this analysis is the main concern of so-called Artificial Intelligence).

To date, two broad avenues have been followed in algorithmic programming: the first one presupposes a succinct anticipation of state transitions encoded – perhaps implicitly – in a sequence of elementary commands to be executed one after another by the machine (*imperative* programming style) while the other one states only single state transitions delivering the (possibly indeterministic) concatenation of state transitions up to a general hull operator (*declarative* programming style). In the latter approach, the global state transition operator F is segmented spatially analogous to the memory segmentation accomplished by abstract data types and in OO programming; probably this coincidence explains the relative success of OO methodology: segmented procedures – such as local functions, term rewriting rules, etc. – can be attached directly to segmented memory chunks, thus combining powerful data semantics with more "compact" procedure coding (that is to say, the OO software paradigm strikes the hitherto most satisfactory balance between both programming and data language features). In particular, the hull operator – interpreting $G(.,.)$ as discrete cybernetic feedback loop to remove the distance between x_0 and $F^{n_0}(x_0)$, or to "reduce" x_0 to $F^{n_0}(x_0)$ – can be viewed as a "general" problem solving machine [Newell and Simon, 1963] where, in all but the most trivial cases, general means: guided by problem specific control information cutting down computational complexity to evade combinatorial explosion of search spaces. A particular strength of the declarative programming style is the parsimony – and formal elegance – achievable in defining intensional computation spaces compared to the explicit determination of computation sequences in imperative programming which tends to be both tedious and error-prone, a handicap partly neutralized by considerable gains in runtime efficiency though.

An important class of symbol processing problems apparently meeting the threefold condition of discreteness, finiteness, and tractability are database systems or, more generally, *information systems*. Basically, an information system might be defined as a collective digital *electronic memory* [Bolter, 1984] used for keeping persistently information in some given object domain, including the utility functions to import (store) and export (retrieve) information as well as the physical devices necessary to accomplish this. From a cultural point of view, memorization has always been the predominant aide in establishing societal continuity, whether by oral or literate traditions, in that it helps to preserve contexts by connecting the past to the future (by the way, literate traditions are not always superior to oral ones as the deciphering of the celebrated Rosetta Stone has made apparent – in fact, an early example of ceased context). Biologically, memory makes up an internal representation layer of the "outer" world received by sensory inputs with the eminent function to decrease behavioural determinism (stimulus-response patterns) and, thus, gain better control over the surrounding environment (improve fitness to survive). Digital memory systems, in turn, re-externalize this representation layer in terms of material symbol storage which, to be really useful, need to be equipped with suitable communication, or interaction, *interfaces*. While the *cognitive* interface design aims at improving the mental grasp of an information system's content by introducing specific ordering systems (table of contents, etc.), symbol structures (semantic data models; see above), and indices (semantic networks, thesauri, hyperlinks, etc.), *functional* interface design addresses the mediating computing logic in storage and particularly retrieval operations. Especially with respect to the latter, advanced symbol processing techniques – such as deductive query answering systems – permit and support the logical interconnection of stored symbol patterns, thus increasing significantly the efficient usage of information systems although at the price of incurring tractability (and, sometimes, finiteness) problems inherent to all kinds of intensional programming.

At present, the ultimate convergence of all memory media is nearly accomplished as virtually all kinds of information become "digitized" and, hence, prepared for uniform storage. This propensity must be seen, of course, in close connection with the development of tele-communication systems also going digital. Historically, tele-communication technologies always supplied the driving paradigms of computing technology, at least since the advent of electricity and electromagnetism giving birth to electrical engineering applications in signal transmission (vacuum tubes, relay switches, semiconductors, RADAR sensing, etc.) and remote control of various devices. In fact, it has been the switching logic necessary to operate telephone networks which not only determined connection routes but also stored dialled numbers in temporary memory until connection routes could be switched successfully and, in doing so, might justly be called computers in today's very meaning of the term [Flichy, 1991]. Quite logically, the first electronic computing machines have been designed by electrical and radio engineers using the building blocks and circuitry of their familiar domains [Randell, 1982]. If communication technology was used to build computing devices, it was even more natural to use it for tele-processing (which practically started in the 1950ies with the SAGE – Semi-Automatic Ground Environment – system of the U.S. Dept. of Defense used for air reconnaissance; [Flichy, 1991]) and inter-computer communication in connecting devices by trunk lines. Today, this basic infrastructure of telephone networks is augmented with even more powerful and higher capacity signal transmission infrastructures using fiber-optics and satellite channels permitting a virtually instantaneous intercontinental data exchange of even breathtaking volumes of data including all types of symbol patterns (text, sound, image data, etc.). As computing machines can perform both symbol processing and transmission switching, computing and tele-communication technologies are about to fuse entirely, turning digital computers into mutually interconnected *multi-media communication machines* creating an encompassing uniform digital data infrastructure, a first glimpse of which is the nowadays oft-cited "data highway".

For obvious reasons, digital networks are a natural precondition of distributed computing and the distributed organization of shared symbol storage. In particular, by virtue of digital communication links the data recording machines being of predominant interest in statistical information processing become interconnected electronic memories which, however, must not be confused with *integrated network information systems*; quite on the contrary, at present the digital data infrastructure admits little more than moving symbols – mostly out of context – between dislocated memory units as an effective metadata management and the principles for designing and implementing efficient cognitive and functional information system interfaces still are in a truly infant stage of development. Hence, improving semantic data modelling, devising metadata management schemes, and elaborating advanced symbol processing techniques will be the key issues of future research at least in statistical information processing, lest the collective utilization of empirical data collected world-wide will be markedly less than optimal.

1.3   Statistical Data Modelling

Trivially, any formal reasoning based on empirical observation data presupposes a symbolic encoding of observations; in practice, the very notion of *data* is equalled with the outcome of such an encoding. Simply speaking, the resolution of observations into their logical constituent parts denoted graphically by unique symbols – from some given finite alphabet – amounts to a (naive) logical atomism. Insofar the symbolic representations of observations encompass the structural relationships between (micro-)observations entailed by the data generating observation method, it is reasonable to call them *data structures*; however, this interpretation differs somewhat from the term's common use in computer science where data structures are thought of as "... *collections of variables, possibly of several different data types, connected in various ways*" [Aho et al., 1983] understanding variables as names for discrete locations of addressable physical symbol memory (cf. Section 1.2). While statistical data structures emphasize the more static aspect of faithfully representing observed reality, computer science stresses the more dynamic operative aspect of data structures (that is, the set of applicable operations). Apparently, in statistical information processing both views must be combined sensibly in order to facilitate an effective use of faithful empirical data images of (parts of) the world.

Statistical data structures naturally subdivide into two layers of structural relationships, viz. relationships between observations belonging to one and the same conceptual *dataset* and relationships between datasets. For simplicity's sake, these layers could be referred to as *within*-dataset and *between*-dataset relationships. First and foremost, within-dataset structures need to be formalized because it is the conceptual unit of dataset entering statistical computations; less important, at first sight at least, might appear the formal capturing of between-dataset relations. From a system point of view, however, this – certainly useful – distinction is more a matter of discerning a microscopic from a macroscopic level than a principal one: in both cases, the formal representation of structural relationships amounts to establish explicit *statistical data models*.

In general terms, a data model can be identified with a set of concepts to describe, in a formal-symbolic way, the structure of a coherent body of data, including a rigorous description of operations applicable to incorporated data [Aho and Ullman, 1992]. In computer science, the prototypical instance of such a data body is a *database*. According to contemporary views (for instance, cf. [Ullman, 1982; Teorey, 1990]), the basic pillar of data modelling consists of a *logical* specification of data models, that is a formal symbol structure equipped with well-defined algebraic properties. In particular, good data modelling abstracts from the contingencies of physical memory structures and algorithmic access functions. Since the early 1970ies, when the relational database model had been introduced [Codd, 1970], the separation of physical and logical database models has become accepted practice; probably, this is not only due to the compelling hardware independence reducing adaptation costs and raising database availability but to the plain fact that relations are suited ideally to represent the typical within-dataset structure of most commercial and administrative data – in short, a *tuple* as an instance of a relation can be identified simply with a data record comprising the descriptive elements of a conceptual domain entity (individual) like a person, an order, a bill, a transaction, etc. Thus, relations represent intensionally collections, or populations, of structurally equivalent individuals and denote extensionally particular finite sets of individuals belonging to such (possibly infinite) collections. Operationally, relations are dealt with rigorously by relational algebra or calculus furnished with clear-cut semantics based on sets and Cartesian products.

By separating internal data models (dealing with physical aspects of data organization) and external models, usually called "database views", the overall logical database model is amenable to a demand-driven adaptation to specific needs without compromising its structure or contents; this enables not only physical data independence but helps significantly to accomplish a usage-oriented information supply. However, the mere existence of external data models does not contribute anything at all to context preservation: without further provision, a database is not "aware" of its own external model's logical structure. In Date's [1986] words, "... *most database systems – relational or otherwise – really have only a very limited understanding of what the data in the database means. They typically 'understand' certain simple atomic data values, and certain many-to-one relationships among those values, but very little else (any more sophisticated interpretation is left to the human user)* ...". Although context preservation might not be critical in many commercial or administrative settings where data life-cycles may be short and semantic content self-evident within a closed community of data users, statistical data holdings typically face a serious problem in self-documentation as soon as data producers (such as statistical offices) and data consumers fall apart [Froeschl, 1993]: statistical data usually outlives persons and even institutions, and increasing *data mobility* renders the interpretations so clear within the data-originating context rather elusive to all data consumers outside this context. Thus, in order to retain at least the indispensable minimum of meaning attached to statistical data, symbol storage structures must be extended to comprise context denotations linked to plain data either (cf. [Dolby *et al.*, 1986] with respect to a stimulating approach – the "Language of Data" project – to context preservation).

In a sense, traditional data modelling has responded to this deficiency by extending the relational modelling domain with *semantic* models: this explicit *entity-relationship modelling* (ER modelling, for short) conceptualizes the domain to be data-modelled in terms of (i) data objects, called entities, and (ii) relationships between entities. ER models serve several purposes at once; first, they are, of course, data structures of themselves – usually called *schema definition* – which can be stored and processed like any other type of data structure; secondly, they can be transformed into actual database designs (which is economically important in computer-aided software engineering); thirdly, they enable quite suggestive visualizations of external model structures at the user interface level, thus supporting database usage and retrieval. As to context preservation, the schema definition aspect certainly is the most important one; this is reflected by using the term *information modelling* instead of the more mundane data modelling. Although being mapped to ordinary (first-order) data structures, schema definitions must not be confused semantically with first-order data stored in a database; since schema definitions are, in a sense, data *about* data [McCarthy, 1982], they constitute a kind of second-order data for which the term *metadata* has become the widely preferred synonym (cf. Subsection 1.5.2).

A closer look, however, makes it clear that general information modelling – driven by the practical demands of commercial and administrative databases – copes rather insufficiently with statistical needs of self-documentation and context preservation. Even worse, since the semantics of statistical data structures deviate considerably from standard semantics aimed at by ER modelling techniques, traditional data models, including the relational model, are not particularly suited to represent either statistical within-dataset or between-dataset relationships. Basically, ER modelling focuses on two things, viz. the dependency structure between an entity's attributes (functional, multi-valued, and join dependencies and the normal forms resulting thereof; cf. [Ullman, 1982]) and the arity of associations between two or more entities of the same or of different entity types (relations). In statistical information processing, both within-dataset and between-dataset relations are of a somewhat different kind though, requiring a shift of focus accordingly. First of all, entities are instances of observation units (statistical populations of carriers of observable characteristics) which may occur repeatedly with or without identical feature patterns; conversely, even different entities may share the very same feature pattern. All of this, of course, requires a corresponding mechanism of unique keys identifying entities rather different from standard methods for creating primary and secondary access keys in databases. Furthermore, in addition to the within-dataset structure of observation dimensions (variables, as the term 'attribute' is not popular in statistics) there is another within-dataset structure comprising a collection of entities belonging together in a stochastic sense such as originating from the same experiment, or random sample, etc. Apparently, there could be several datasets of entities of the same population, with identical or different instances of entities in each dataset. In addition to these specific between-dataset relationships there may, of course, still exist between-dataset relations linking different entity types (albeit these are of minor relevance, in general). Moreover, statistical information is by no means restricted to observation data; in particular, as observation data is transformed into data *aggregates*, yet another specific type of entity must be dealt with efficiently. Altogether, it must be concluded that in the statistical information storage domain quite different and more complex entities are encountered which call for adequate representation means. Another peculiarity of statistical information processing is the importance to distinguish between what *is* observed and what *could have been* observed. Quite in contrast to standard data modelling resorting to a "closed world" approach where extensional non-representation can be identified faithfully with the non-existence of entities (for instance, if there is some white-collar worker really employed in a company than there are corresponding records in the staff and payroll relations for this person) this by no means holds for statistical entities: Sato [1989] remarked correctly that in statistics and statistical data modelling it is absolutely necessary to differentiate conceptually between a terminological frame – delimiting the discourse as to what can be expressed or named – and the practical realizations which might cover only more or less extensive portions of this frame. In plain words, what is may not have been observed (yet) but it nevertheless may be reasonable to talk about it by, for instance, taking recourse to some model assumption.

In view of the "non-standard" requirements of statistical data processing [Shoshani, 1982; Shoshani and Wong, 1985; Chen *et al.*, 1995] it comes as no surprise that research in statistical databases and statistical data modelling has departed from the mainstream of database research [Michalewicz, 1991]. Apparently, the differences in data semantics lead to more specific conceptual data models

.3 STATISTICAL DATA MODELLING

[Lenz, 1993a; 1994b] which, in turn, call for custom-tailored operations [Lenz, 1994a] and access methods. There is a variety of models suggested and approaches pursued (some of which will be outlined briefly in Subsection 1.3.2) showing a general tendency towards an increasing functional importance of metadata either in supporting data access (browsing facilities, sometimes nicknamed "data mining" facing the huge data repositories mankind has compiled over generations with increasing eagerness) or deductive retrieval expecting a qualified (data) response to a given *statistical* database query. This is particularly evident concerning the *problem solving* view of statistical database querying which – contrary to traditional, non-deductive database retrieval – implies, in general, an intermediary transformation of stored symbol sets into statistical aggregates involving some kind of statistical-methodological information processing in addition to straightforward symbol retrieval. In this respect, statistical data modelling touches on the research domain of statistical expert systems [Froeschl, 1989] which, for apparent reasons, also puts strong emphasis on the role of metadata [Hand, 1993; 1994]: as symbol transformation casts a heavy burden – especially of context assurance – on a data consumer's shoulders, metadata attains a pivotal role in advanced statistical information processing in providing deductive query processing support mainly based on the second-order data context attached to first-order observation or aggregate data.

1.3.1 Statistical Data Structures and Data Semantics

The most elementary statistical data structure in statistical data processing is the rectangular number schema called *case-by-variate* matrix listing observed cases – instances of some defined population – in a column vector such that the observed characteristics – the statistical variables – are entered by row in a predetermined ordering. This data structure, depicted schematically in **Tab. 1.3.1–1**, can be easily identified with array structures provided by virtually any higher-level computer programming language from FORTRAN [Backus, 1981] onwards. Hence, observation matrices are chosen predominantly as basic data structure in statistical programming systems (for instance, the GENSTAT statistical software package introduced quite a general matrix-based "table calculus" in the early 1970ies, cf. [Nelder, 1974]; [Malmborg, 1992] proposed a matrix-based data interchange format; etc.). Moreover, cases can be identified with relational tuples either; thus, observation matrices are also serving as a blueprint for relational statistical data modelling. Despite these superficial similarities, however, the semantics of observation matrices cannot, in general, be captured adequately by arrays or relations alone. In particular, with respect to both rows (cases or tuples) and columns (variables or attributes) there usually are several within-dataset relationships external to the two-dimensional array or relation arrangement (analogous arguments apply to higher-dimensional m-way arrays). For instance, cases may be related by specific experimental or observation designs or partake of a spatial, temporal, or spatio-temporal super-structure [Plank, 1994]; variables may be partitioned into controlled factors, responses, and covariates; some variables may depend conditionally on observed values of other – conditioning – variables; some value combinations may not be admissible or feasible at all (structural zeroes); between cases and/or variables there may exist additivity constraints inhibiting particular types of statistical aggregations, etc. An illustrative case in point is longitudinal data featuring both semantics of temporal observation schemes and a high degree of structural variation (unbalanced designs) of observation records [Froeschl and Grossmann, 1988] hardly accounted for by plain matrices. As a major consequence of these additional semantic relations, uncritical application of general array or relational operations may turn out hazardous (for instance, by creating "non-observable" relations [Wittkowski, 1989]); instead, operation applicability is highly context-sensitive. In particular, relational *join* operations are rendered meaningless in statistical data management except in a few specific cases such as equi-joins on case identifiers (observation numbers, etc.). Furthermore, traditional operations are not prepared to preserve semantic data contexts (not to speak of their inability to keep track of changes to semantic contexts effected by operator application) and, therefore, place the burden of maintaining coherence and consistency of data transformations entirely on the analyst driving the processing sequence. Conversely, many tuple-based operations of relational algebra (for instance, single tuple retrieval) are of little interest in statistical data processing as its focus is on summarizing and filtering average characteristics (suggesting a variable-oriented instead of a tuple-oriented data access; cf. [Shoshani, 1982]).

Table 1.3.1–1: Case-by-Variate Structure of Statistical Observation Data

Case	*Observable*		
	O_1	...	O_m
1	:	...	:
2			
:	:	...	:
n			

The relational model and particularly ER modelling, however, are useful for interrelating different types of statistical populations by establishing explicit links between (conceptual) case-by-variate matrices. For instance, assuming an official statistics context comprising continued surveying with respect to four kinds of populations, viz. *households*, *persons*, *regions*, and *employers* (note: should be read as *enterprises*), **Fig. 1.3.1** reproduces an instance of Sundgren's [1992] *infological* model. Within each of the four populations, cases (sampling or observation units) are identified by unique IDs (attributes typeset boldface); populations are linked by binary relations of *1:n* or *m:n* types ("crow's foot" [Everest, 1986] is used to symbolize the 'many'-side of binary relationships). In particular, one household may consist of one or more persons, a person may work for zero or more employers

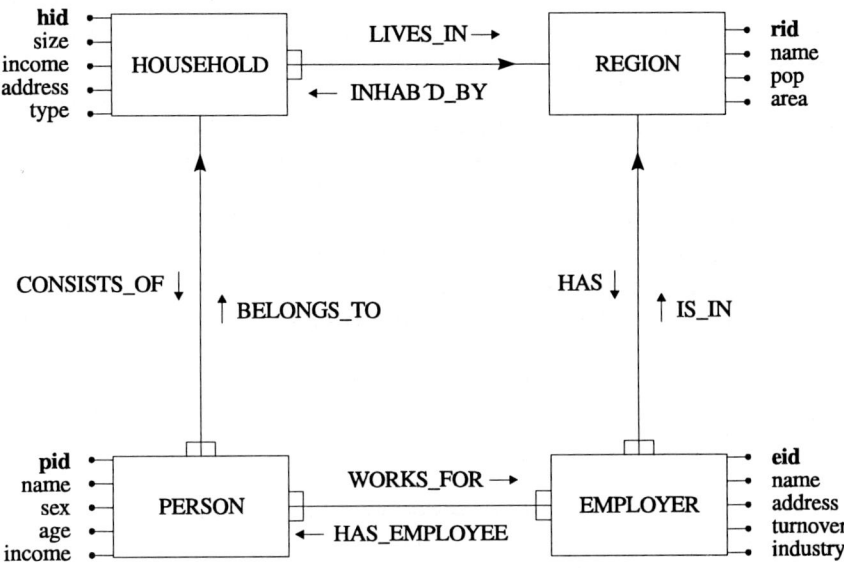

Figure 1.3.1: An Infological Model Instance [Sundgren, 1992] (adapted)

while an employer may have zero or more employees (persons), etc. In addition to structural relations (constraints), such as

PERSON.address ←
    BELONGS_TO.HOUSEHOLD.address

also quantitative relationships can now be stated explicitly; for instance,

HOUSEHOLD.size ←
    CONSISTS_OF.PERSON.**count**

or

HOUSEHOLD.income ←
    CONSISTS_OF.PERSON.**sum**(income)

where **count** and **sum** are, of course, statistical aggregation functions. A particular virtue of this infological approach is the conceptual clarity of query specification which can refer directly to the ER model. For instance [Sundgren, 1992], the number of persons with income less than 100,000 currency units and their average income earned, broken down by region and sex, is specified simply by

PERSON (**with** income < 100,000) (**by** REGION.rid * sex).
    (**count**, average_income ← **sum**(income)/**count**);

Since observation records are gathered directly in experimental settings or from survey responses (questionnaires), this type of data has been dubbed *microdata* as opposed to the statistical aggregates derived from it which, naturally, are called *macrodata*. Contrary to microdata, macrodata do not represent a case-by-variate structure fitting raw or pre-processed observation data; quite on the contrary, macrodata refer to structures resulting from applying statistical aggregation functions to microdata. Hence, macrodata consist of statistical summary values (such as counts, sums, averages, percentages, indices, etc.), possibly broken down with respect to one or more taxonomic dimensions. Shoshani [1982] refers to this distinction by discerning *summary* and *category* attributes of a macrodatum generating a data structure as depicted schematically in **Tab. 1.3.1–2**.

Table 1.3.1–2: Statistical Macrodata Structure

Category attributes		*Summary attribute*
C_1 ... C_k		f
⋮ ... ⋮		⋮
⋮ ... ⋮		⋮
⋮ ... ⋮		⋮

Practically, this is an information-equivalent [Özsoyoglu et al., 1989] relational representation of a k-way (contingency) table where the category attributes $C_1,...,C_k$ are derived from observables $O_1,...,O_m$ and f denotes the aggregation function ranging over the cases of a (possibly pre-processed) microdatum. In particular, if $k=1$ and C_1 is a temporal attribute with (equidistant) time values in its domain, the macrodatum is called *time series*. In the example given above, two macrodata structures would result, viz. one for f=**count** and another one for f=**sum**(income); in both structures, the category attributes would comprise 'region' and 'sex'. In relational (retrieval) terms, the category attributes $C_1,...,C_k$ function as the minimal composite key for the summary attribute f implying that the k-tuples of $C_1,...,C_k$ appear uniquely within a macrodatum. Moreover, as typically all k-tuples that can be generated from the value domains of $C_1,...,C_k$ occur, the extension of the relation induced by category attributes amounts to the full Cartesian product of attribute domains (which, by the way, could then be represented more space-efficiently for evident reasons; cf. [Shoshani, 1982]).

In general, statistical information processing proceeds from microdata to macrodata by applying summary functions to microdata. However, sometimes macrodata may be the real point of departure for further processing, particularly if original microdata are not available anymore or because microdata have been prepared specifically by information "condensing", with or without applying statistical aggregation functions. Apparently, in either case further statistical aggregation functions can be applied formally to macrodata on condition that they are adapted appropriately as discussed below.

Although still representable in a "flat" two-dimensional relational arrangement, macrodata bear even more semantics than microdata do. For one thing, summary attributes have a semantic quality entirely different from category attributes which, in turn, do not represent individuals (observations) anymore but *collections* of observations sharing the same patterns of selected observable features; taxonomically, category attribute tuples denote sub-populations of the underlying microdatum's sampling or survey population. Adopting a more conceptually oriented approach, Sundgren [1992; 1993] coined the term "*e*-message" to describe the semantics of (sample survey) macrodata; according to his view, an *e*-message consists of three major components, viz. (cited from [Sundgren, 1993]; boldface words originally in italics): "

- an **object component**, indicating
  - a **population of objects of interest**; which is sometimes
  - **restricted to a subset** by means of a **selective property**; and which is usually subdivided into
  - a set of **(sub)domains of objects of interest**; often by means of
  - a combination of **variables**; the value sets of which **cross-classify** the objects in the population;

- a **property component**, indicating
  - a **parameter**, or **statistical characteristic**, which is estimated for the population as a whole, as well as for the domains of interest within the population; the parameter is usually defined in terms of
  - an **aggregation operator** (count, sum, average, correlation, etc.) operating on one or more **aggregation arguments**, defined in terms of microlevel **variables** of the statistical units (objects, entities) in the population;

- a **time component**, indicating the (point or interval of) **time** at (during) which the population and its (sub)domains of interest existed and had the estimated parameter value."

On top of this basic distinction, Sundgren proposes an *alfa-beta-gamma-tau* structure of macrodata, or *e*-messages, such that the population part, including the selective property (if any), is referred to as *alfa*-component; the property description part is referred to as the *beta*-component; the cross-classification of the population into (sub-)domains is referred to as *gamma*-component; and the temporal reference is making up the *tau*-component. For reasons of symmetry, this structure could be amended by adding a *sigma*-component holding the spatial reference of an *e*-message. With respect to the macrodatum example cited above,

- 'PERSON (**with** income < 100,000)' denotes its *alfa*-component,
- '(**by** REGION.rid * sex)' denotes the *gamma*-component, and
- '(**count**, average_income ← **sum**(income)/**count**)' are two different *beta*-components – referring to the aggregation operators **count** and **sum**, respectively – gathered in a single expression

(noting that, in this particular example, the *tau*-component of the *e*-message has been omitted). Apparently, the *gamma*-component coincides with a macrodatum's set of category attributes while the summary attribute *f* determines its *beta*-component.

A similar, though more descriptive than functional approach has been proposed by Olenksi [1992] who conceptualizes macrodata in terms of *statistical indicators*. In his generic model of statistical indicators, a macrodatum is a pair consisting of a *name* and a *value* component; the name component, in turn, comprises a metadata description of the indicator composed out of a *unit* description, a *semantic attribute*, and a *measurement attribute*. Essentially, unit description and semantic attribute can be identified with Sundgren's *alfa*-component whereas the measurement attribute combines *beta*- and *tau*-components. In Olenski's terminology, an elementary statistical indicator refers to a single population or domain concept; however, elementary statistical indicators can be lined up to composite indicators by extending the unit description component; this way, cross-classifications – the *gamma*-component – are also dealt with representationally. The *value* component of elementary or composite indicators contains the summary attribute values in terms of scalars, vectors, or (multi-way) matrices depending on the structure defined in the respective indicator's name component.

In terms of statistical modelling, summary attributes, or *beta*-components of macrodata can be identified with (empirical) distribution parameters or estimates thereof. Hence, macrodata – when viewed as database relations – obviously must not be processed by standard relational operators unless these are redefined suitably; this redefinition or, rather, extension of operations regards particularly the semantically correct handling of the summary attribute. To this end, operations generally depend on further second-order data external to the plain relational macrodata representation (such as sampling and distribution characteristics, measurement units, summary types, model assumptions, etc.) of *beta*- and *gamma*-components.

On its passage from microdata to macrodata, statistical information processing runs through a sequence of stages achieving the transformation of raw observation data into final aggregates in a stepwise fashion. Particularly in official statistics, but also in many branches of natural and physical sciences producing data (such as meteorology, remote sensing, medicine, etc.) there is usually a long way from data inception to eventual data consumption. Therefore, it comes as no surprise that, at different processing stages, datasets are encountered at varying processing *levels* (for instance, there may be legal reasons of data privacy etc. calling for particular data levels prior to dissemination). In order to enable a rough separation of these data levels, the following terminology compiled in **Tab. 1.3.1–3** is introduced. Initially, of course, empirical data structurally is at microdata level irrespective of data cleaning, editing, or plausibility checking measures. Next, microdata may be converted into macrodata structure albeit without losing information what, essentially, implies that summary functions are not yet evaluated; formally, this amounts to replace the rightmost column of **Tab. 1.3.1–2** with a *set-valued* attribute gathering all cases with identical feature pattern $C_1,\ldots,C_k$ such that $C_j = O_j$ for $1 \leq j \leq k = m$.

Like microdata, pre-macrodata may be aggregated statistically by any applicable (summary) function. Pre-macrodata are transformed into lower-level macrodata by grossing-up observation data if pre-macrodata (and microdata) have been obtained by some kind of sampling procedure; otherwise, of course, pre-macrodata and lower-level macrodata simply coincide. For economic reasons (to get rid of sparse datasets, for instance), lower-level macrodata are stored in compressed format by collapsing breakdown dimensions, selecting category attribute combinations, etc., yielding so-called *summary sets* [Shoshani, 1982]. Finally, lower-level macrodata are condensed further to higher-level macrodata corresponding to familiar statistical table and time series layouts as disseminated in printed or electronic formats.

Table 1.3.1–3: Data Processing Stages vs. Data Levels

Processing stage	*Data level*	*Dimensionality*
raw dataset	micro	–
edited, imputed, cleaned dataset	micro	–
(optional) anonymized dataset	micro	–
structure conversion	pre-macro	high
elementary counting	pre-macro	high
(optional) anonymizing	pre-macro	high
weighting/estimation	lower-level macro	high
(optional) anonymizing	lower-level macro	high
(optional) pre-aggregation	lower-level macro (summary set)	medium
final aggregation (stockpile)	higher-level macro	low
post-processing	higher-level macro	low

According to present practice in statistical database management, datasets are available either at microdata level (possibly anonymized), as summary sets (database segments), or – the most frequent case – as higher-level macrodata (cf. Section 2.3). The rightmost column of **Tab. 1.3.1–3** indicates roughly the dimensionality of macrodata in terms of available or remaining breakdown axes.

1.3.2 Data Modelling Approaches

Taking into account the power, versatility, and wide reception of relational databases and relational data models, the practical management of statistical microdata will resort to more or less specifically adapted relational structures in most cases. Of course, insufficiencies of bare relational data models must be bridged by added metadata which, in turn, could be represented quite straightforwardly in terms of relational structures either. With respect to macrodata, however, demands are tougher, as the discussion of Subsection 1.3.1 has made markedly apparent. Correspondingly, statistical data modelling stresses the importance of adequate macrodata representations, pursuing a variety of data modelling approaches and goals. At present, two broad categories of approaches might be discerned, viz.

- "factual" data models aiming at improved data *access*, with an apparent emphasis on database indexing and search facilities supporting data retrieval, and
- "conceptual" data models aiming at improved data *manipulation*, focusing on algorithmic structures to support the transformation of stored (micro- or macro-) data into specified (macrodata) output structures.

Although both approaches have in mind eventually to facilitate data retrieval, the latter one is considerably more ambitious as the derivation of macrodata structures might require more or less intricate *symbol transformations* of both algebraic and statistical-methodological types which is tantamount to – human or mechanized – problem solving using special symbolic inference schemes based on formal data contexts (metadata). To date, there is no statistical data model of either category which could claim honestly to cover all facets of relevance in statistical information processing. The following overview tries to summarize very briefly the major lines of research undertaken in the last couple of years to give an impression of current activities in this field (for more comprehensive surveys, cf. [Rafanelli, 1991] and [Grifoni et al., 1993]).

Graph-based Data Access Models

A rather evident approach to simplify data access in statistical data holdings comprising particularly summary sets and higher-level macrodata uses semantic network structures representing subject-matter and structural relationships both within and between stored data objects. An exceptionally appealing feature of graph-based modelling of database contents is its simple conversion into visual representations enabling non-verbal ("point-and-click") interaction modes between user front-ends and data access software. Exploiting the capabilities of up-to-date graphical user interfaces (GUIs), semantic relationships between represented objects can be made visible easily by using different contrasts, hues, lines, shapes, symbols, and texts in scrollable and resizable windows outfitted with interaction modes corresponding to what is exhibited in respective panes. Furthermore, semantic relationships can be used as backbone link structure in hypertext-based interaction models.

One of the first proposals to use visualized semantic graphs was the SUBJECT model [Chan and Shoshani, 1981]. In this model, each macrodatum is represented by an "object graph" denoting the within-dataset structure by discerning two types of nodes – labelled either C, for Cluster, or X, for Cartesian product – linked to a tree as exemplified in **Fig. 1.3.2–1**. Apparently, this graph represents a statistical table concerning employment statistics and providing absolute counts of employees broken down by sex, year, and professional category, as can be inferred immediately from the labels attached to the graph's nodes. While the terminal nodes are labelled with elementary categories used in the table breakdown, C-nodes represent

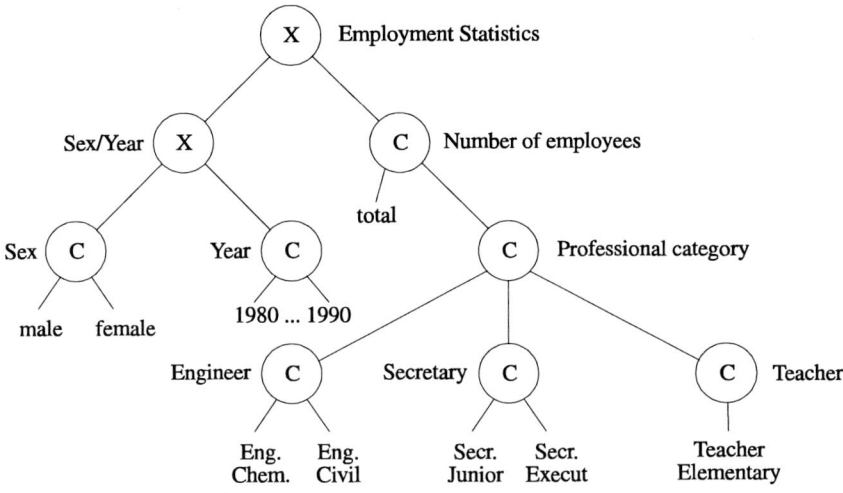

Figure 1.3.2–1: A SUBJECT Graph [Lenz, 1993a] (adapted)

either statistical variables, "aligned" collections of variables, or structural information of table design (the number of employees is given as both marginal distribution and breakdown by a selection of professions comprised in the three main profession groups of engineers, teachers, and secretaries; likewise, the presence of the X-node labelled 'Sex/Year' indicates that employee counts are available in this macrodatum for each combination of 'Sex' and 'Year' terminal categories), and X-nodes represent Cartesian multiplication of sub-structures. Operationally, nodes are furnished with type-dependent functionality. This way, by simple pointing at nodes information about individual nodes, the data extension they represent, or the whole macrodatum they stand for (node at root) can be retrieved and inspected.

Apparently, the SUBJECT graphs are but a first step in arranging a coherent semantic description of database contents. Emphasizing an object-oriented (OO) point of view, an object graph's nodes can be regarded as software objects (abstractions) in the OO-sense to which further individual descriptions might be attached. More specifically, node types could then be identified with object classes the instances of which are equipped generically with certain description elements. For instance, Oldford [1988] proposed to model statistical observations, variates, and datasets in terms of hierarchically composed OO classes and instances strongly matching the "natural" structure of statistical objects such as datasets, tables, etc. In this model, a table object could consist, inter alia, of category and summary attributes (that is, instances of object classes, say, 'CategoryAttribute' and 'SummaryAttribute', respectively) comprising description elements such as 'ValueList', 'UnitOfMeasurement', etc., a table header, a reference to a statistical population of sampling units, and so on.

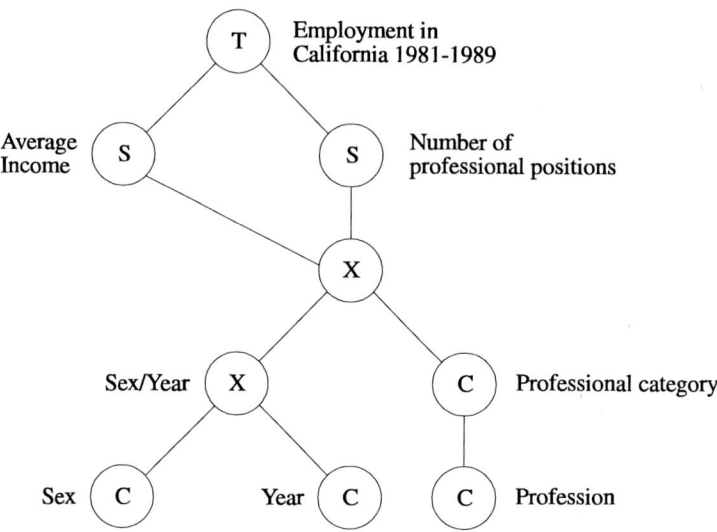

Figure 1.3.2–2: An Extended Object Graph [Lenz, 1993a] (adapted)

Another straightforward extension [Rafanelli and Ricci, 1984] proposes an embedding of object graphs into a network of thematic relations. To this end, the structural descriptions of macrodata (object graphs) are linked to "topic" nodes (*T*-nodes) as shown in **Fig. 1.3.2–2**; in this example, there are two structurally identical tables with breakdown dimensions sex, year, and professional category, one providing absolute counts of persons, and the other one providing average incomes of employed persons for topic 'Employment in California 1981–1989'. Note that, in this graph, *S*-nodes denote summary attributes, thus adding further semantic expressiveness. *T*-nodes, in turn, may now participate in a (hierarchical or heterarchical) *topic network* relating macrodata in terms of subject-matter criteria; masking out lower-level object graph details, this network facilitates a convenient tool to browse the contents of a statistical data holding by tracing paths between *T*-nodes and, if pleasant, zooming into object graphs for further description details. An even further detailed association structure, based on seven different link types discerned explicitly, has been suggested by Su [1983].

The object graph approach lends itself to several obvious extensions. First, object graphs can be re-expressed algebraically in a quadruple format <N,C,S,*f*> [Rafanelli and Shoshani, 1990; Bezenchek *et al.*, 1994] where N is the name of a statistical object (macrodatum), C a finite set of category attributes, S a single summary attribute, and *f* the aggregation function mapping from the Cartesian product of C attributes to the domain of S; apparently, these *statistical objects* paraphrase Olenski's composite statistical indicators with formal rigour, adding a

.3 STATISTICAL DATA MODELLING

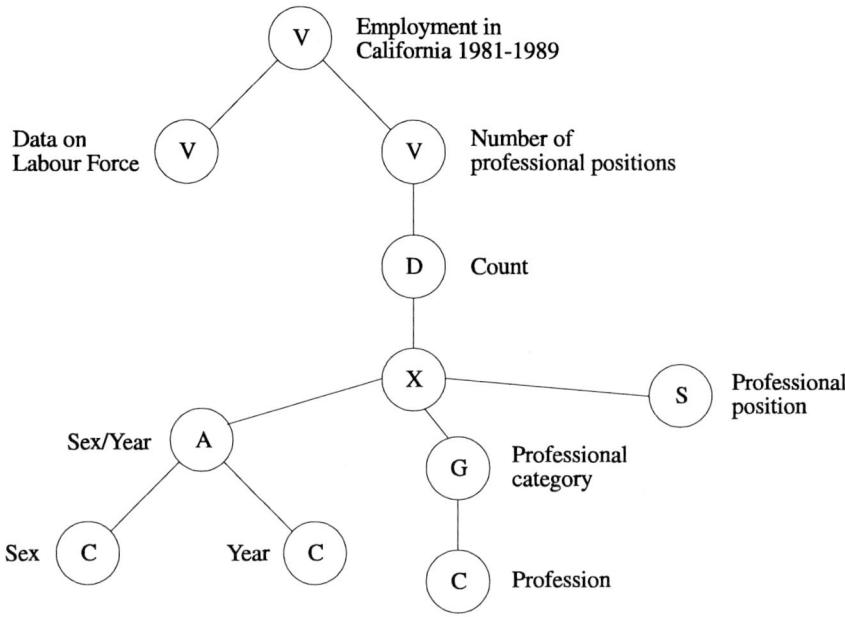

Figure 1.3.2–3: A GRASP-Type Object Graph [Lenz, 1993a] (adapted)

means to visualize indicator structures and relationships. Stated this way, statistical indicators become first-class objects of functional manipulation in specific statistical operator algebras. Secondly, the nodes of object graphs could be linked functionally with definite database views, or external models, established in terms of ER modelling at microdata level. By means of linking relational microdata models with data aggregates derived from it, much data semantics is conveyed consistently in a "bottom-up" fashion ensuring context preservation quite elegantly. For instance, in the GRASP system [Catarci and Santucci, 1990], picking up the threads of an earlier proposal of interactive query specification in relational statistical databases [Wong and Kuo, 1982; Wong and Yeh, 1983], microdata queries can be specified interacting with a graphical ER model representation of the schema definition of a statistical database [Di Battista and Batini, 1988] by means of a GUI, which – at the macrodata level – is translated into an object graph such as the one shown in **Fig. 1.3.2–3**. In this graph, V-nodes denote "data views" linked to the underlying ER model of the database, D-nodes the summary types of macrodata, X-nodes statistical classifications (breakdowns), and G-nodes groupings of more elementary categories. S-nodes denote classes of statistical objects.

In statistical data holdings, particularly those maintained in official statistics, data objects are not only related by subject-matter or structural similarities. Quite frequently, data objects are linked together *implicitly* by common terminology, more or less extensive parts of which reappear, whether processed or not, in different subsets of, for instance, a data holding's macrodata collection. From a practical point of view, it is much more reasonable to expect that database users will be familiar (or willing to familiarize themselves) with agreed-upon and wide-spread terminology rather than with specific and sometimes peculiar data contents of a particular data holding; hence, locating and accessing data is facilitated primarily by exploiting terminological relationships interconnecting stored data objects. However, since object graphs are destined to reflect a data object's internal structure only, they do not take automatically account of any implicit terminological linkages. Recognizing this severe deficiency, Sato [1989; 1991] proposed a frame-based [Minsky, 1975] multi-level taxonomic network existing independently of actual data objects which, of course, are linked to this semantic web of terms in multifarious ways. For example, there may be several real datasets of somewhat different structure contributing empirical information to the very same *conceptual* dataset as illustrated in **Fig. 1.3.2–4**. Although not fully equivalent, the existing datasets are tied together taxonomically by using either the same terminology or terminologies that are related meaningfully. If, for instance, population totals for some region and some time period are of interest, a *conceptual* structure like the one shown in **Fig. 1.3.2–4** will be referred to primarily in scanning a data holding for real datasets providing the figures looked for. On condition that the respective semantic links have been established, datasets related to the conceptual structure can be sifted out irrespective of minor divergencies such as different value domains in breakdown dimensions, summary types, etc. In the example shown, the dataset 'PERSONS-FOR-CENSUS' is conducted every fifth year only while the 'PERSONS-FOR-SURVEY' dataset is delivered annually; nevertheless, both datasets may still be eligible for answering the population data request. In order to arrive at a clearly defined semantic network structure, Sato [1991] proposes a "Statistical Data Model based on a 4 Schema Concept" (SDM4S) comprising four levels of abstraction, viz.

- a *data model level*, which discerns three generic types of domains: statistical objects, category values/attribute, and summary values/attribute;
- a *conceptual level*, which describes statistical objects, category and summary attributes available in principle (cf. **Fig. 1.3.2–5** overleaf);
- a *DB schema level*, indicating which of the conceptual statistical objects, category and summary attributes actually occur in real datasets stored in the database; and
- an *instance level*, which contains the actual data values of statistical aggregates.

1.3 STATISTICAL DATA MODELLING

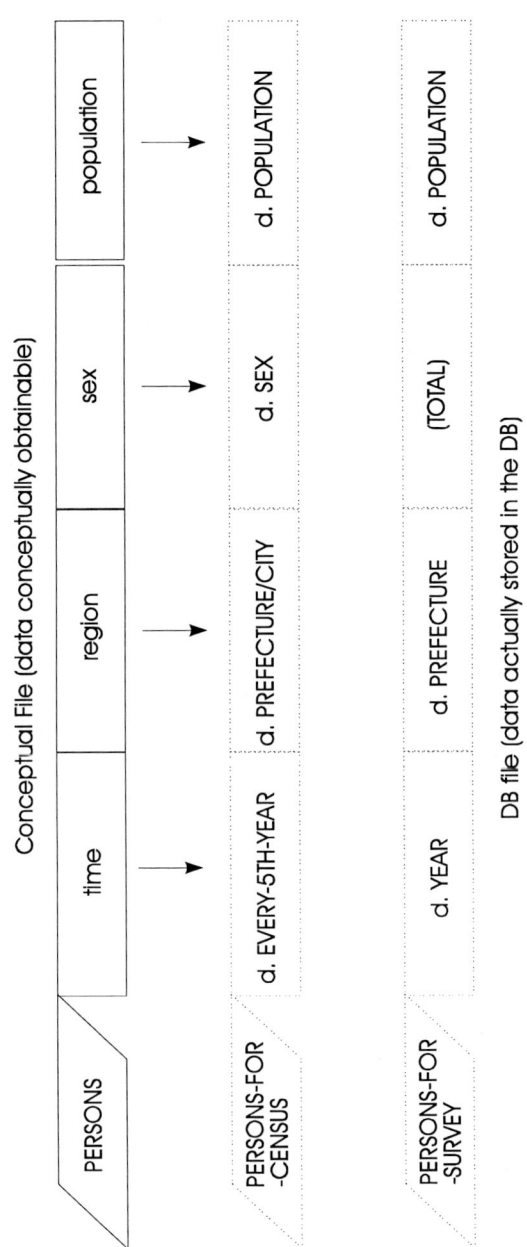

Figure 1.3.2–4: Conceptual vs. Stored Files [Sato, 1989]

Omitting the instance level, **Fig. 1.3.2–6** (overleaf) illustrates how the sample datasets appearing in **Fig. 1.3.2–4** might be integrated into a generalization hierarchy of statistical objects related to the statistical unit 'person'; the *'ako'*-links represent subtype-supertype relationships between statistical object frames. At the instance level, an instance frame like

> **TOKYO-1980-MALES**
>   *is-a*: PERSONS-FOR-CENSUS
>   *time*: 1980
>   *region*: TOKYO-prefecture
>   *sex*: MALE
>   *population*: 5856

(adapted from [Sato, 1989]) would be linked, by the *'is-a'* slot, to the DB-level frame 'PERSONS-FOR-CENSUS'. Likewise, *'time'*-, *'sex'*-, and *'region'*-slots are defined as category attributes associated to statistical objects of 'PERSON' type (**Fig. 1.3.2–5**) from where, for instance, it could be inferred that *time* is measured ("classified") in years.

Using an analogous network approach, semantic relationships between terms are arranged as exemplified in **Fig. 1.3.2–7** (overleaf). Starting from CATEGORY and CLASSIFICATION prototypes at the data model level, various categories and classifications are arranged as instances at the "metadata" level which, essentially is a name space of value collections the members of which are defined at the "data" level. Both categories and classifications are interrelated in quite complex a fashion denoted explicitly by semantic links of five different types, viz. *'ako'*, *'is-a'*, *'part-of'*, *'finer-than'*, and *'member-of'*. While *'ako'*-links denote instance links connecting extension elements with a term's intension (for example, cf. **Fig. 1.3.2–7** ①), *'is-a'*-links associate constituent elements with type terms or, in plain words, associating values with value set names (for example, cf. **Fig. 1.3.2–7** ②). *'part-of'*-links denote subset-superset relationships in terms of semantic enclosure (for example, cf. **Fig. 1.3.2–7** ③) reflected by *'finer-than'*-links between corresponding value sets (for example, cf. **Fig. 1.3.2–7** ④). Categories are associated formally with classifications by *'member-of'*-links (for example, cf. **Fig. 1.3.2–7** ⑤) which, in turn, might give rise to further links induced by *'part-of'* relationships defined between classification instances; these induced links – such as the *'member-of'*-link connecting the category 'MANUFACTURING' to classification 'STANDARD-INDUSTRIAL-CLASSIFICATION' in **Fig. 1.3.2–7** ⑥ – could, of course, be inferred automatically.

.3 STATISTICAL DATA MODELLING

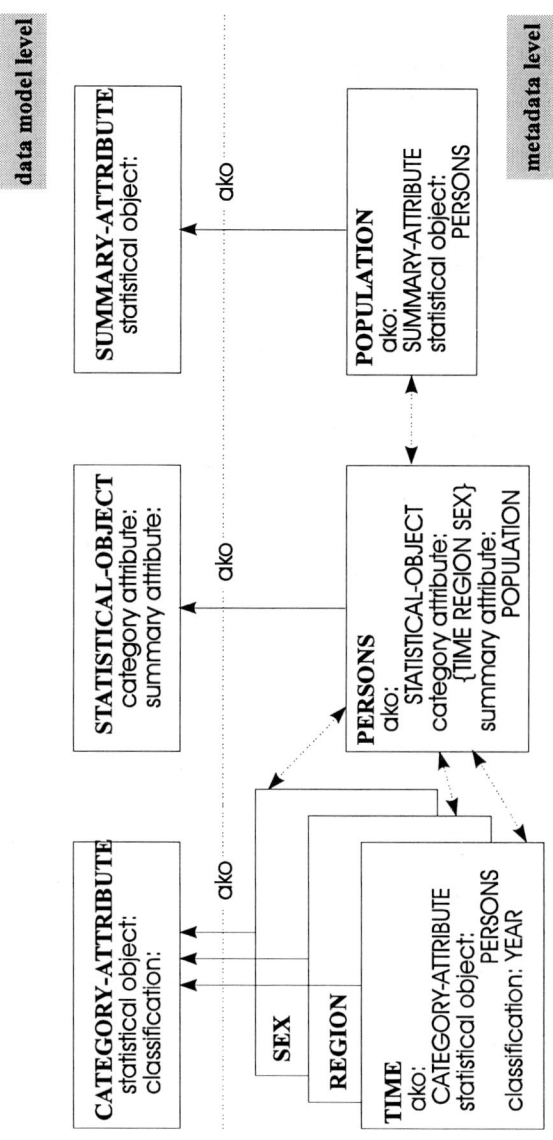

Figure 1.3.2–5: Excerpt of SDM4S Concept Hierarchy [Sato, 1989]

In SDM4S, terminological and statistical object sub-networks are tied together by attaching classifications of the term network to instances of the generic domain type 'CATEGORY-ATTRIBUTE' as shown in **Fig. 1.3.2–5**. If, for instance, a data request refers to datasets related to the term 'MACHINERY', all datasets comprising category attributes with classifications like 'STANDARD-INDUSTRIAL-CLASSIFICATION'

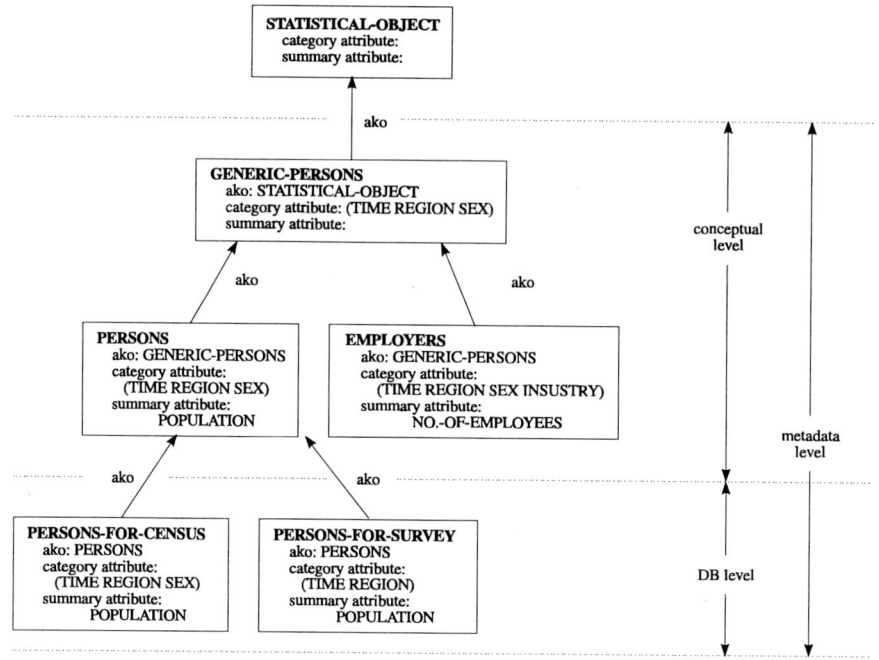

Figure 1.3.2–6: Conceptual Object Hierarchy in SDM4S [Sato, 1989]

or 'INDUSTRIAL-CLASSIFICATION-IN-LABOUR-FORCE-SURVEY' (noting that 'MACHINERY' is *'part-of'* 'MANUFACTURING') are sifted out simply by checking whether their instance frames (such as the one shown above) comprise a pertinent category attribute slot. Apparently, search/retrieval conditions could be combined logically in order to select datasets only fulfilling several semantic criteria simultaneously. Both terminological and statistical object networks, of course, could be converted into graphical representations displayed on GUIs enabling direct manipulation interaction modes.

Sato's SDM4S approach replaces the more simple-minded natural language-oriented database indexing approaches – like "flat" thesauri or linked word lists – with a vigorous semantic-terminological characterization of a statistical data holding's content. However, there still are further semantic relationships between (real) datasets not yet captured in this framework. In particular, the SDM4S model fails to take account of the *production structure* of datasets as it has been pointed out by Sundgren [1992]. In many domains, especially in official statistics, data production proceeds usually according to a periodical, fairly stable schedule repeating semantic dataset structures with but minor modifications. Adopting this production view of statistical datasets, the major semantic relationships between relevant data objects amount to a three-layer structure, called *metaobject graph*, as depicted in **Fig.**

.3 STATISTICAL DATA MODELLING

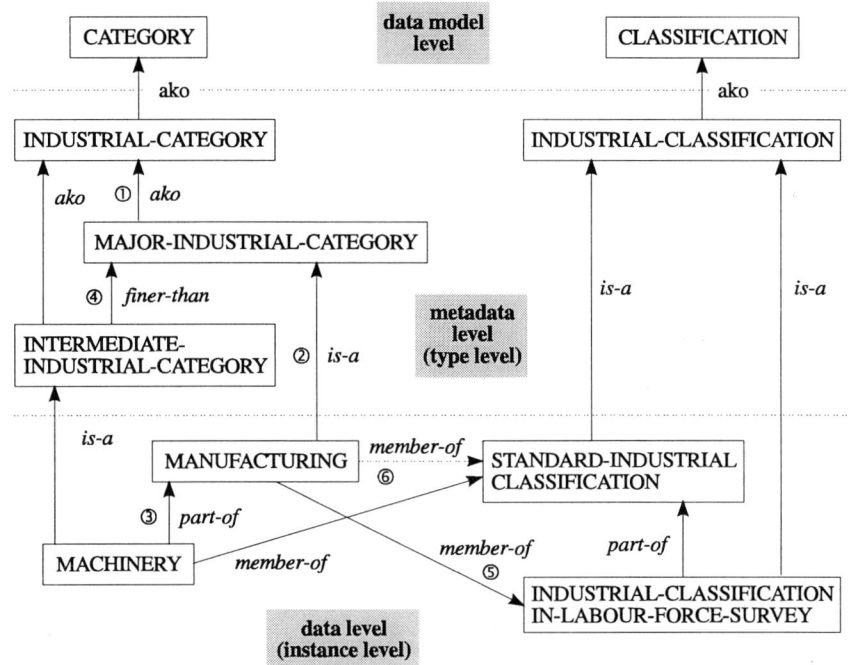

Figure 1.3.2–7: Semantic Net Excerpt of SDM4S [Sato, 1989] (adapted)

1.3.2–8 [Sundgren, 1993] (overleaf). Although biased towards survey processing from a microdata perspective, the common structure elicited in this metaobject graph discerns a "type layer" of conceptual objects, such as statistical POPulations (of sampling units), distribution or model PARameters, observation VARiables, SURveys, etc. which, in general, come in *series* of instances; hence, with each conceptual object type there is associated a set of identically repeated *occurrences*. Quite naturally, occurrences are described at the "occurrence layer" while the structural and semantic features of identically repeated occurrences are represented at the "series layer" of the metaobject graph.

It is easy to see that the metaobject graph structure complies well with the SDM4S model by introducing, at the data model level, new generic types of domains (such as POP and SUR or, perhaps, some generalized version of it), and associating formally the three layers of the metaobject graph with the conceptual, DB schema, and instance levels of the SDM4S model, respectively. In so doing, the VAS (VAlue Sets) objects become identified with CLASSIFICATIONs, VALue objects with CATEGORYs, and BOXes (that is, *alfa-beta-gamma-tau* structures) with STATISTICAL-OBJECTs; by and large, VAR objects conform to CATEGORY-ATTRIBUTEs, and PARs correspond to SUMMARY-ATTRIBUTEs. In addition to POPulation and SURvey features, the metaobject graph comprises a SAMple object

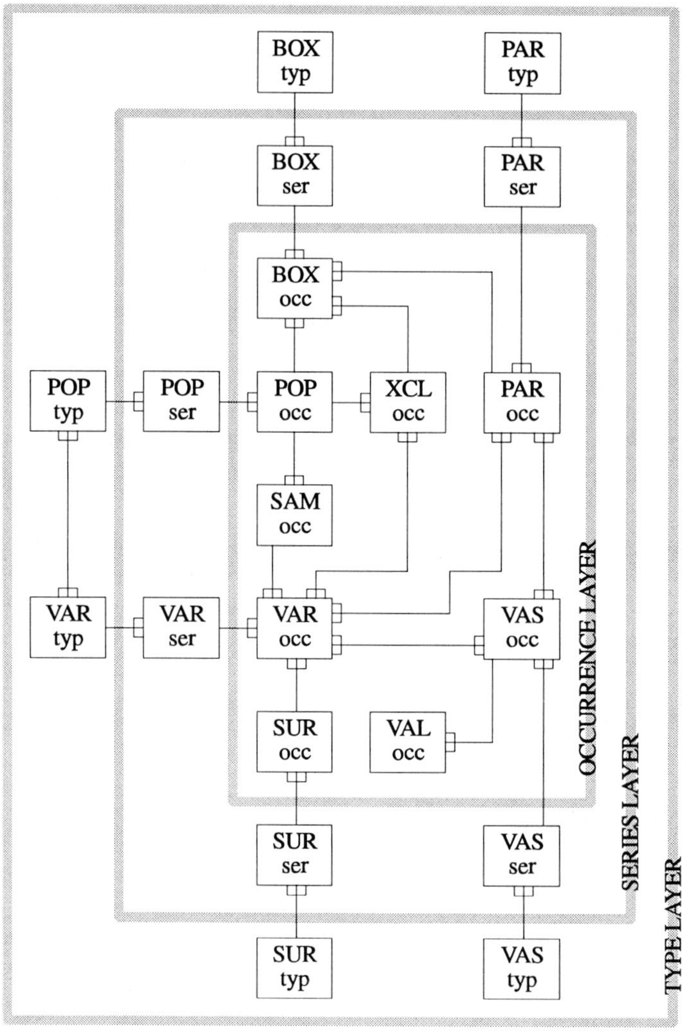

Figure 1.3.2–8: Metaobject Graph [Sundgren, 1993] (adapted)

to which information about the sampling structure (POP individuals included in the sample, sampling scheme, etc.) becomes attached; as samples may possess series-specific features (for instance, in rotation sampling schemes), SAM objects should certainly be provided at the series level either; like POP and SUR, SAM would constitute yet another generic domain type at the data model level.

A similar remark applies to the XCL (cross-classification) object of the metaobject graph which describes the *gamma*-component of BOX objects in closer detail and, in a sense, reflects the core structure of SUBJECT object graphs as discussed above. Again, as certain cross-classifications may be typical for SURveys or other data sources, XCL objects may occur at type and series layers of a metaobject graph – together with corresponding relationships between SUR, VAR, POP, and BOX objects – as well.

To sum up, graph-based data models provide very useful tools to arrange semantic representations of statistical database contents simply convertible into direct manipulation or hypertext-oriented user interaction models. The models proposed to date, however, by and large do not attempt an integrated view of statistical data modelling although, as brief analysis has made apparent, they are more or less complementary and could be combined quite easily. In particular, Sundgren's data production-targeted metaobject graph complies well with Sato's SDM4S model focusing on statistical taxonomies and coherent semantic database description by using a frame-based network approach. Moreover, if adapted suitably, object graphs and linkages between ER microdata models and macrodata representation as proposed in the GRASP model or the infological model could be incorporated in order to yield a fairly powerful semantic modelling tool to characterize statistical databases meeting a multitude of practical data retrieval demands of statistical information processing albeit being still restricted to search for and plain access to stored data.

Extensions of the Relational Model

In order to close the gap between a conceptual description of macrodata and the *effective* derivation of conceptually described macrodata – unless accomplished by a simple retrieval of stored data (symbol sets) – a recourse to the extensional (first order) data layer is indispensable. Hence, basically, the discussion returns to the relational data model and relational database operations once again. However, in regard of the semantic peculiarities of statistical datasets (cf. Subsection 1.3.1), especially macrodata, the ordinary – tuple-oriented – relational data model needs to be augmented and furnished with specific extensions to reflect the operational constraints of statistical retrieval and aggregation operations in a relational setting. These extensions concern two different aspects, viz. the transformation – often termed "summarization" – of (relationally represented) microdata into (again relationally represented) macrodata, and the further processing of macrodata within the class of (relational) macrodata structures.

With respect to summarization, the conversion of microdata to macrodata format amounts to introducing some kind of "grouping" operation pooling the set(s) of microdata records (or *tuples* in database terminology) to which some statistical aggregation function is then applied to get the (numerical) aggregate, or *summary value*, of interest. The basic principle of aggregate formation in a relational context has been stated explicitly for the first time by Klug [1982]: letting **E** denote a set of extended relational algebra expressions comprising statistical aggregation functions; $Attr(r)$ the set of attributes (variables) of relation r of a relation schema R with $Deg(R)$ ordered attributes; and Agg a set of aggregation functions; then, for $e \in \mathbf{E}$, $X \subseteq Attr(e)$, and $f_i \in Agg$ where f_i returns a simple summary value obtained from applying f to the i-th attribute, $1 \le i \le Deg(e)$, of e, $e\langle X, f_i \rangle \in \mathbf{E}$ such that

$$e\langle X, f_i \rangle(I) = \{t[X] \circ y \mid t \in e(I) \text{ and } y = f_i(\{b \mid b \in e(I) \text{ and } (b[X] = t[X])\})\}$$

where I denotes a particular instance of the database schema, b and t are tuple variables ranging over relation $e(I)$, and $[.]$ denotes the projection operator as usual; '∘' denotes the concatenation operation. Apparently, due to the summary attribute y added to the resulting relation, $Deg(e\langle X, f_i \rangle) = |X| + 1$.

Aggregate formation can be extended easily to *aggregation-by-template* [Özsoyoglu et al., 1987] to account for breakdowns: letting $e_1, e_2 \in \mathbf{E}$, $Y \subseteq Attr(e_1)$, $Z = Attr(e_2)$, where $|Y| = |Z| \ge 1$ and each attribute of Z is *set-valued*; letting further Y_a be the set of simple-valued (atomic) attributes in Y, $Y_n = Y - Y_a$, and Z_a and Z_n denote those attributes of Z corresponding to Y_a and Y_n, respectively; then $e_1 \langle X, Y, f_i \rangle e_2 \in \mathbf{E}$ is defined as

$$e_1 \langle X, Y, f_i \rangle e_2(I) =$$
$$\{t \circ f_i(G_t) \mid (\exists t_1)(\exists t_2).(t_1 \in e_1(I) \text{ and } t_2 \in e_2(I) \text{ and } t[X] = t_1[X] \text{ and } t[Z] = t_2[Z])\}$$

where

$$G_t = \{t' \mid t' \in e_1(I) \text{ and } t'[X] = t[X] \text{ and } t'[Y_a] \in t[Z_a] \text{ and } t'[Y_n] \subseteq t[Z_n]\}$$

such that $Deg(e_1 \langle X, Y, f_i \rangle e_2) = |X| + |Z| + 1$. This aggregate-by-template operation groups tuples of $v \in e_1(I)$ by combining the attributes of X and Z such that for each combination of $t[X \cup Z]$ the summary function f is applied to the set G_t of tuples of $e_1(I)$ coinciding with $t[X]$ and either $v[Y_a] \in t[Z_a]$ or $v[Y_n] \subseteq t[Z_n]$. The pooling of attribute values is facilitated by a *pack* operator [Özsoyoglu et al., 1987] defined as follows: letting $e \in \mathbf{E}$ with $Deg(e) = n$, $X \in Attr(e)$, and $CX = Attr(e) - \{X\}$; then $P_X(e) \in \mathbf{E}$ with $Deg(P_X(e)) = n$ such that, for each tuple $g \in e[CX](I)$, where $Deg(e[CX]) = n - 1$, tuple w_g is computed as

$$w_g[CX] = g$$

.3 STATISTICAL DATA MODELLING

$$w_g[X] = \begin{cases} \{t[X] | t \in e(I) \text{ and } t[CX] = g\} & \text{if } X \text{ is atomic} \\ \{y | (\exists t).(t \in e(I) \text{ and } t[CX] = g \text{ and } y \in t[X])\} & \text{otherwise} \end{cases}$$

and, thus,

$$P_X(e)(I) = \{w_g | g \in e[CX](I)\}$$

The explicit use of aggregating operations and set-valued relations leads to rather clumsy operation specifications. This can be circumvented by extending relational database query languages, such as SQL (see, for instance, [ISO, 1989]), by specific aggregation syntax structures as suggested, for instance by Johnson [1981] or Ghosh [1986; 1991] ([Özsoyoglu and Özsoyoglu, 1985] and [Tansel, 1991] give succinct overviews of statistical database query language proposals). Provided that a relation consists of both categorical and numerical attributes (variables), Ghosh introduces a POWER(p)-AGGREGATE clause applicable to a numerical attribute V such that the attribute values $t[V]$ are raised to the p-th power prior to summation. For instance, if a relation r with relation schema

$$R(\text{COUNTY}, \text{CITY}, \text{POPULATION}, \text{BUDGET})$$

is given as shown in **Tab. 1.3.2–1 (a)** such that COUNTY and CITY are categorical and POPULATION and BUDGET are numerical attributes, the query

    SELECT COUNTY, POPULATION, BUDGET
    FROM R
    POWER(1)-AGGREGATE(BUDGET)
    INTO R_1
    WHERE POPULATION < 400;

would result in the output relation ("table") shown in **Tab. 1.3.2–1 (b)**.

Table 1.3.2–1 (a): Sample Relation r of Relation Schema R [Ghosh, 1991]

COUNTY	CITY	POPULATION	BUDGET
Dixon	Dodge	250	35.5
Dixon	Orange	250	45.0
Dixon	Clara	500	37.2
Marin	Uba	350	45.2
Marin	Sun	350	32.0
Alameda	David	750	86.0

Table 1.3.2–1 (b): Summation Output Relation R_1 [Ghosh, 1991]

COUNTY	POPULATION	BUDGET
Dixon	250	80.5
Marin	350	77.2

The POWER(1)-AGGREGATE operation is easily generalized to products of attributes by introducing a further PRODUCT-AGGREGATE(V_1, V_2) operation where V_1 and V_2 name the numerical attributes (factors) to be multiplied prior to summing up the products. Thus, modifying the sample query given above, a covariance calculation between attributes POPULATION and BUDGET of relation r would require to specify

> SELECT COUNTY, POPULATION, BUDGET
> FROM R
> PRODUCT-AGGREGATE(POPULATION, BUDGET)
> INTO R_2(COUNTY, PR-AGGR)
> WHERE POPULATION < 400;

returning an output relation as shown in **Tab. 1.3.2–1 (c)** where PR-AGGR is introduced as a new attribute name holding the obtained aggregate values.

Table 1.3.2–1 (c): Product-Sum Output Relation R_2 [Ghosh, 1991]

COUNTY	PR-AGGR
Dixon	22375.0
Marin	28620.0

In a similar way, it is also possible to create new summary attributes (such as, for instance, frequency distributions) by replacing the standard SELECT clause of SQL with a more specific statistical aggregation function. As an example, Ghosh proposes a clause termed

$$\text{AGGREGATE-DISTRIBUTION ON } V_1, \ldots, V_k$$

to realize a counting of tuples $t \in r$ for identical projections $t[V_1, \ldots, V_k]$. Hence, a query like

> AGGREGATE-DISTRIBUTION ON COUNTY
> FROM R
> INTO R_3(COUNTY, FREQUENCY)
> WHERE POPULATION < 600;

would result in an output relation as shown in **Tab. 1.3.2–1 (d)**.

Table 1.3.2–1 (d): Frequency Distribution Output Relation R_3

COUNTY	FREQUENCY
Dixon	3
Marin	2

In addition to elementary summing and counting operations, Ghosh proposes several extensions of SQL grammar accounting for more specific statistical computations like recoding numerical attributes' values into ranks, rescaling numerical domains, ordering tuples by attribute values or ranks, and determining run lengths as well as *cusums* on ordered observations, etc.

Once summarization has transformed microdata into macrodata, no further reference to original microdata is necessary unless aggregation levels lower than those provided by pre-computed macrodata are asked for. For evident reasons, the computation of statistical parameters speeds up considerably if the derivation of elementary aggregates such as counts, sums, or sums of products is replaced by direct retrievals of cut-and-dried macrodata. To this end, Chen et al. [1989] proposed a relational summary data model of macrodata attaching several summary statistics to so-called "categories" which, basically, are definite extensions of relation schemas comprising (at least) the attributes over which summation is ranging. If, for instance, R is a relation schema, $X_1=Age$ and $X_2=Income$ are two numerical attributes such that $X_1 \in R$ and $X_2 \in R$, respectively, and r is an instance of R, or category, consisting of the tuples $\{g_1, g_2, g_3, ...\}$, then a possible summary relation could have a shape as exhibited in **Tab. 1.3.2–2** [Chen et al., 1989].

Table 1.3.2–2: Summary Relation for Category r

Category	Cardinality	ΣAge	$\Sigma Income$	$\Sigma Age * Income$
g_1	2	82	135	5550
g_2	1	55	35	1925
g_3	2	51	70	1835
⋮	⋮	⋮	⋮	⋮

As most statistical parameters (of linear statistics) are composed of summary values like those shown in **Tab. 1.3.2–2**, the derivation of summary values reduces to retrieving categories corresponding to the domain concepts in mind; unfortunately, to determine the derivability of a particular category from a set of categories stored in a statistical summary relation database is an *NP*-hard decision problem [Chen et al., 1989], thus calling for a careful preparation of "generating" category sets partitioning the space of categories (with respect to a given relation schema R) such that the union of generating categories is reflected homomorphically by adding the attached summary values on account of the linearity property of aggregation functions.

A major weakness of all extensions to relational data models discussed so far is their failure to represent explicitly the distinction between category and summary attributes – reflecting the distinction between *gamma-* and *beta*-components of statistical macrodata [Sundgren, 1993] – within the relation format. As a rather inconvenient consequence, the consistency of relation semantics must be maintained manually whenever aggregation functions are applied successively. Apparently, this is an error-prone method putting high demands on the care of database users. One possible approach to remedy this deficiency is to hide the internal relation structure by providing user-level macro-operators restricting the "visible" part of relational operators to category attributes. For instance, Sadreddini *et al.* [1992], in devising their MIMAD (micro-/macro-) data model, introduced a canonical summary relation schema structure, dubbed MAOB (for MAcro OBject),

$$R(C_1,\ldots,C_k\,;\,P_1,P_2\,;\,S_1,\ldots,S_m)$$

where $C_1,\ldots,C_k$ is a set of $k \geq 0$ category attributes, P_1 and P_2 are numerical attributes serving as arguments to statistical aggregation functions, and $S_1,\ldots,S_m$ are $m \geq 1$ summary attributes associated with (additive) summary functions such as cardinality, sum, sum of squares, sum of products, etc., ranging over the domains of P_1 and P_2 understanding that univariate and bivariate statistics are of predominant practical interest (in fact, MAOBs could be viewed as a further refinement of Su's [1983] concept of G-relation). To be valid, MAOBs must fulfil several internal consistency requirements (for instance, if both P_1 and P_2 are instantiated then there must be at least one bivariate summary attribute among $S_1,\ldots,S_m$), and macro-operations – such as MPR (macro-projection), MSL (macro-selection), MAG (macro-aggregation), MUN (macro-union), etc. – now correspond to specifically restricted SQL queries obeying the semantics of category, numerical, and summary attributes. For instance, MPR is forced to refer to summary attributes such that all category and numerical attributes of the argument MAOB of a macro-projection are retained in the output MAOB; dropping a category attribute is achievable only by applying a MAG operator implying a summation with respect to all summary attributes over all argument MAOB tuples.

A more radical view has been adopted in the MEFISTO data model [Falcitelli *et al.*, 1989; Rafanelli and Ricci, 1991; 1993] framing statistical summary relations in so-called *statistical entities* which, essentially, are statistical tables encapsulated formally in data objects, that is abstract data types manipulated and accessed by a set of custom-tailored table operations. In this model, each statistical entity s representing an individual macrodatum is an instance of some statistical entity scheme $S = \langle \mathbf{C}, t \rangle$ where $\mathbf{C} = \langle C_1,\ldots,C_k \rangle$ is a set of category attributes and t denotes the summary type of S, that is the statistical data type of summary values (like count, sum, average, percentage, etc.) represented by S. Letting $D(C_i)$ denote the domain of C_i, $1 \leq i \leq k$, and defining the statistical entity space of S as Cartesian product

.3 STATISTICAL DATA MODELLING

$r_C = \times_{i=1}^{k} D(C_i)$, a statistical entity s is now described in terms of a pair $s = \langle \mathbf{D}, g \rangle$ where $\mathbf{D} \subseteq r_C$ is the statistical entity's relational extension of category tuples and g is a function mapping from $\mathbf{D}$ to the domain of t. Thus, given some tuple $d \in r(s) \equiv \mathbf{D}$, $g(d)$ returns the summary value associated with d in s. In a sense, statistical entities resemble the quadruple structure <N,C,S,f> already mentioned, as C can be identified with $\mathbf{C}$ and f denotes a summary function the range of which is summary type t of summary attribute S. In contrast to summary category relations or MAOBs, however, summary types t are not restricted to additive aggregation functions f. Moreover, due to the explicit object structure, statistical entities can be furnished simply with additional semantics such as, for instance, references to parent entities of statistical entities with non-additive summary types.

In MEFISTO, operations on statistical entities can be stated in a rigorous algebraic format. For instance, at entity scheme level, unary operations obey the template

$$\varphi: \langle \mathbf{C}_0, t_0 \rangle \mapsto \langle \mathbf{C}_1, t_1 \rangle$$

where φ, in general, is a parameterized operation. In practice, $t_0 = t_1 = t$ for all elementary operations and, thus, $\varphi(s_0)$ is effected actually by transferring the symbolic operations on $\mathbf{C}_0$ to the statistical entity's extension $r(s_0)$ such that g_0 is transformed into g_1 automatically. Taking the operation of deriving marginals (marginal distributions) as an example, $\varphi \equiv \Sigma_{C'}$ implying that the parameter $C' \in \mathbf{C}_0$ denotes the category attribute being dropped from $\mathbf{C}_0$ and

$$\Sigma_{C'}: \langle \mathbf{C}_0, t \rangle \mapsto \langle \mathbf{C}_1, t \rangle$$

such that $s_1 = \langle \mathbf{D}_1, g_1 \rangle = \langle \pi_{C_1}(\mathbf{D}_0), g_1 \rangle = \Sigma_{C'}(s_0)$ where $\mathbf{C}_1 = \mathbf{C}_0 - \{C'\}$, $\pi_{C_1}(\mathbf{D}_0)$ is the relational projection of $\mathbf{D}_0$ over $\mathbf{C}_1$, and

$$g_1(w) = f_t(\{g_0(v) | v \in r(s_0) \text{ and } w = v[\mathbf{C}_0 - \{C'\}]\})$$

for $w \in r_C(s_1) \equiv r_{C_1}$. Apparently, this is a special case of Klug's *aggregate formation* operation (see above); f_t, in general, must be chosen depending on t and, hence, denotes in fact a *family* of aggregating functions. As additivity is a fixed quality of a summary type t, summary value management in MEFISTO is facilitated automatically [Falcitelli et al., 1989]. For instance [Rafanelli and Ricci, 1991], assume that a table s_0 of rates is given as shown in **Tab. 1.3.2–3 (a)**; restricting the modality set of category attribute 'state' to '{Florida,Texas}' and noting that, in this particular case,

$$g_1(w) = f_\%(g_0(w))$$

for $w \in r(s_1)$, and $f_\%(x) = xp^{-1}$ where $p = \sum_{w \in r(s_1)} g_0(w)$, a "rescaled" table of rates as shown in **Tab. 1.3.2–3 (b)** would result.

Table 1.3.2–3: (a) Sample Table of Rates

%	Industry			
State	agriculture	metal	other	total
California	1.2	5.4	9.1	15.7
Florida	2.1	10.1	12.1	24.3
Oregon	2.4	11.5	7.5	21.4
Texas	6.4	12.7	19.5	38.6
total	12.1	39.7	48.2	100.0

(b) Restricted Sample Table of Rates

%	Industry			
State	agriculture	metal	other	total
Florida	3.34	16.06	19.24	38.64
Texas	10.17	20.19	31.00	61.36
total	13.51	36.25	51.24	100.0

Analogously to the computation of marginals of statistical entities, other elementary operations are defined in terms of relational operations on statistical entity spaces. In fact, [Meo-Evoli et al., 1992] give a proof of relational completeness of a set of macrodata operators (comprising S-projection, S-union, S-selection, and S-aggregation), generating a so-called S-algebra, in that the corresponding operations on statistical entity spaces are relationally complete with respect to the usual definition of relational completeness of database query languages.

The relational data model, in particular its extensions to non-first normal form relations including set-valued attributes, provides a fundamental prerequisite in establishing a continuous statistical data processing chain from raw microdata capture through the dissemination and graphical display of macrodata structures. By exploiting the virtues of relational database technology, statistical data models benefit from principles like physical data independence and can fully reap the efficiency of relational database management systems (for instance, the possibility to extend data models to distributed databases; cf. [Sadreddini et al., 1992]) as well as the elegance, power, and simplicity of formal retrieval languages. However, choosing the relational framework for statistical data modelling does by no means coincide automatically with a semantically adequate data representation, especially with respect to macrodata. To this end, specific semantic "super-structures" must be defined upon the bare relational model or, probably even more appealing in the longer run, logical data views must be separated from the internal relational representation of statistical data models; moreover, such an object-oriented conceptualization of statistical entities opens remarkable opportunities to furnish logical operative units with the additional statistical context denotations so badly needed in advanced statistical information processing.

Transparent Management of Statistical Aggregates

Taking account of the vast amounts of data and datasets typically comprised in statistical and scientific databases, graph-based data models and semantic network approaches are very useful for browsing database contents and locating data of interest. However, this kind of interaction is good only for bulk data access; if conceptual data views diverge from the structure of stored datasets, some kind of transformation must be applied additionally. In other words, finding relevant data in a statistical database is but a first step of conceptual data retrieval which, in general, is immediately followed by a data processing step. Hence, what in fact is desired is the extension of data access models with data *manipulation* models within uniform end-user interfaces.

In principle, there are several approaches to achieve direct manipulation man-machine interaction modes in visual data manipulation and database query processing (for instance, the proposals of [Viehstaedt and Ambler, 1992] or [Angelaccio *et al.*, 1990]). At present, two paradigms seem to be of major interest in statistical database querying, viz. the visual composition of (extended) relational queries by specifying relation templates which are transformed directly into standard query languages like SQL, and the direct manipulation of statistical object graphs such that visual operation primitives coincide semantically with elementary data manipulation operations drawing particularly on the concept of automatic summary value management. In either case, the internal data model must be equipped with a suitable "operative" surface such as relational algebra; with respect to non-standard data models like object graphs or statistical entities, this surface must be devised specifically to meet the demands of visual computing.

The direct graphical realization of relational querying is based on the original Query-by-example (QBE) model of Zloof [1977] in which database queries are represented and composed – by entering constraints, variables, and command operators into respective attribute columns – on a terminal screen depicting, essentially, the involved *microdata* relation schemata ("table skeletons") visually. For instance, to obtain the relation shown in **Tab. 1.3.2–1 (b)** from data as comprised in **Tab. 1.3.2–1 (a)**, it would suffice to enter the condition '< 400' into the 'POPULATION' column of relation schema R and to state (i) the *summation* operator in the 'BUDGET' column as well as (ii) the *print* operator in column 'COUNTY'. Özsoyoglu *et al.* [1989] have adapted and enlarged the QBE model to their Statistical-table-by-example (STBE) model based on extended relational algebra operations (like aggregate formation, aggregation-by-template, etc.) encompassing set-valued attributes, nested relations, and a well-contrived sub-querying mechanism.

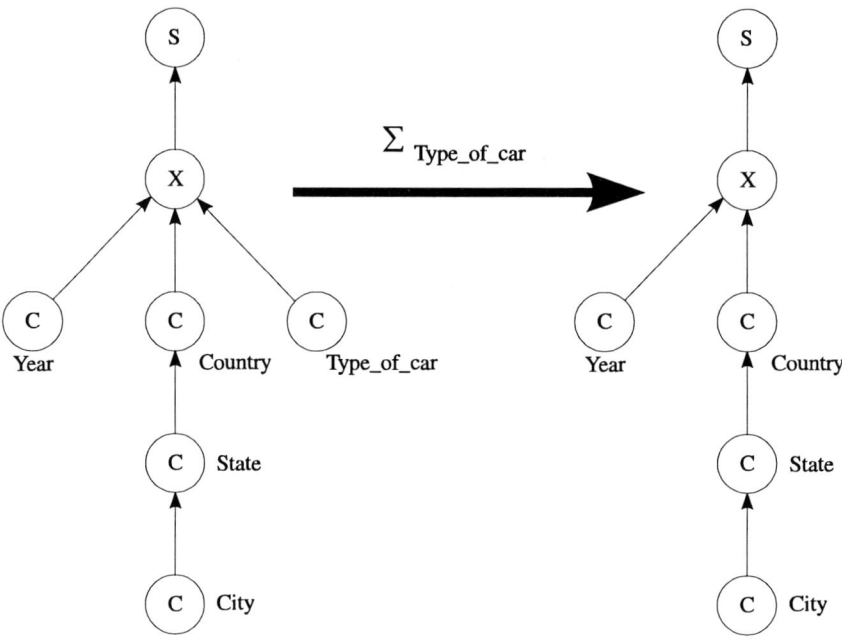

Figure 1.3.2–9: Transparent Operation on Object Graph [Rafanelli and Ricci, 1990] (adapted)

Despite the visually appealing appearance, its flexibility and its power, however, the interactive specification of STBE queries – simple cases excepted – quickly becomes very complex and annoying; this becomes apparent particularly considering that quite often the tuple sets to be summarized (the *alfa*-component of macrodata in Sundgren's [1993] terms; cf. Subsection 1.3.1) must be composed disjunctively from several sub-queries. Moreover, as relational join operations are rather untypical for statistical database query processing, many of STBE's virtues in fact miss the point.

A different approach to direct manipulation of statistical data relies on graph-based models used to represent database contents visually. Particularly the concept of statistical entities, by separating logically statistical entity spaces from summary values, lends itself favourably to direct manipulation of macrodata: using ME-FISTO-like operation primitives, the specification of data manipulating operations – being restricted to category attributes (the *gamma*-component in Sundgren's [1993] terms) thanks to automatic, or *transparent*, summary value management – reduces to changing the structural description of data objects (statistical entities); in other words, operation primitives can be re-expressed in terms of changes in the structure of object graphs. Hence, a faithful conversion of operation primitives into visual praxemes for modifying the object graph representation of statistical entities suffices to arrange an effective statistical direct manipulation database interface like the VISTA language proposed by Rafanelli and Ricci [Rafanelli and Ricci, 1990; Meo-Evoli *et al.*, 1994]. In this model, all visual operation primitives are resolved into two kinds of praxemes only, viz. *adding* or *deleting* edges to or from object graphs, respectively, where the meaning of each praxeme is determined by the its local operative context, that is the meaning of the edge. For instance, in order to obtain a marginal of a statistical entity with a three-way breakdown as shown in the left-hand object graph of **Fig. 1.3.2–9** [Rafanelli and Ricci, 1990], simply an edge connecting a *C*-node to the object graph's *X*-node must be removed. In the example given, deleting the edge incident with the 'Type_of_car' node implies a summation over this breakdown dimension – corresponding to a MEFISTO $\Sigma_{\text{Type\_of\_car}}$ operation – returning the object graph of the resulting statistical entity as shown at the right-hand side of **Fig. 1.3.2–9**.

Neither transparent management of statistical data nor direct manipulation modes are data models by themselves although both presuppose specific data models since – while operative semantics refers to extensionally represented data – conceptual operations apply to structural data object representations (intensions) instead of actual data objects (extensions). By means of this structural data object representation formal consistency of data manipulations and transformations can be maintained mechanically to at least some degree; for instance, it is easy to inhibit summations over rates or averages as both summery types are declared non-additive. Conversely, automatic consistency preservation implies a regression to internal sub-operations (in the case of summing over, say, averages structurally compatible tables of absolute frequencies are required to provide the summation weight factors) which apparently pushes the focus towards the intensional layer of statistical database processing: inadvertently, a requested data transformation may call for auxiliary *intensional* data objects with given structural descriptions for which an extension has to be derived from the data stored in the database. Thus, aiming at transparent aggregate management eventually demands some kind of formal derivability of data object (macrodata) extensions. At this point, in fact, it is reasonable to change the overall point of view and redefine the problem of statistical data retrieval in terms of specifying intensional data objects for which extensions are sought against a given stock of data. Such an intensional space of data objects (that is, macrodata described within a pre-determined terminological frame of discourse) could be delimited by, say, a taxonomic grid as embedded in Sato's SDM4S model providing empirical terminology independent of the actual database extension. From a user's perspective, this approach is remarkably advantageous in that it replaces the rather tedious and inefficient database browsing mode with a succinct structural description of the data extension looked for whenever such a clear description can be submitted. In plain words, the intensional approach enables a sound and efficient logical *query-answering* mode of database retrieval. This, by the way, might be an elegant method to circumvent the arbitrariness of graph-based data models demanding quite a lot of patience of prospective users to become acquainted with both, data model structure and interaction modes. Furthermore, by specifying *what* is asked for instead of *how* it should be obtained, transparent data management can be extended to comprise both algebraic and statistical conditions and criteria determining the eventual derivability of a particular data extension. Clearly, given an effective notion of formal data derivability, it is no longer reasonable to store reproducible macrodata at any price; quite on the contrary, it is a matter of economy which "generative" elements are stored extensionally such that a concisely delineated set of extensions – to be determined by practical considerations, of course – is derivable on demand.

In concluding, graph-based data models and direct interaction techniques are suited particularly for searching data (extensions) in case of weak or imprecise preconditions ("informative search") or for simple processing tasks got by with applying a few primitive operations to extensionally available datasets. Otherwise, recourse must better be taken to more sophisticated approaches incorporating the concept of formal derivation of data extensions from intensional request specifications.

Models Based on Concept Logic and Reducibility

In most general terms, data retrieval amounts to referencing a specified memory location of a segmented ("addressable") symbol store. To each memory location is assigned a unique name (that is, unique in some limited reference frame) representing the symbol configuration stored contiguously at this location. Thus, individual symbol configurations, or data objects, can be accessed *directly* by name.

Apparently, direct access is efficient only if the sole data object to be retrieved is known precisely. If a retrieval operation refers to a larger object set, an intensional description of the members of this set offers a more efficient and conceptually convenient way of processing. By means of a structural description of target objects, the repetitive comparison of stored objects with target objects can be mechanized (by executing a *loop* over stored objects) such that only those stored objects are sifted out matching the specified structure description. Essentially, this is the approach chosen in relational database retrieval processing which, evidently, can be extended easily to include minor transformations of the objects selected; in statistical databases, a typical transformation applied to selected objects is some kind of aggregation (sum, mean, etc.) or set function (count, min, max, etc.).

The intensional retrieval method can be generalized to incorporate a class of transformations of stored objects in terms of logical implication; in doing so, the notion of database retrieval includes both extensionally stored objects and objects effectively derivable from stored objects (hence, stored objects are a subset of retrievable objects). In such a deductive database [Gallaire and Minker, 1978; Gallaire *et al.*, 1984] the name space of objects comprises a (possibly infinite) set of intensional data objects deducible by formal term rewriting rules from the database extension. Basically, there are two approaches to object derivation, viz.

- interpreting the object name as a *procedural description* (computable expression) prescribing an algorithm to evaluate the extension of the object, and

.3 STATISTICAL DATA MODELLING

- interpreting the object name as a purely *structural description* from which an algorithmic derivation procedure for the object extension is, in principle, inferable effectively.

While the former approach – used particularly in relational retrieval languages – starts from a structural description of object classes (relation schemata) upon which a computable (recursive) function is defined, the latter one stipulates an interpretation (model) associating implicitly each valid (=well-formed) object name with a computable function returning the object's extension – provided that it exists at all – such that, in a sense, the derivation function expression is "encoded" in the object's name structure.

In statistical information processing, both micro- and macrodata must be dealt with efficiently. In case of microdata retrieval, it is the object sets, or *alfa-*components of conceptual statistical entities (cf. Subsection 1.3.1), which have to be stated; as these sets of observation units (subsets of some well-defined population of carriers of observable characteristics) tend to be quite large, direct database access methods are actually out of question. In case of macrodata, in turn, most of the time a single object is chosen; however, this usually is a choice from a huge repository of (extensional) objects rather tedious or even impossible practically to screen by hand. Moreover, extensional macrodata by themselves give rise to an intensional space of derivable macrodata objects, given a set of macrodata transformation (rewriting) rules. Hence, there is every reason to argue in favour of a conceptual data access in statistical information processing and promote inference-oriented retrieval approaches based on intensional, or structural, target macrodata descriptions. Incidentally, this line of argument meets perfectly with the requirements and opportunities emanating from direct manipulation approaches as discussed above.

To date, essentially two more or less tentative proposals pursuing deductive database retrieval approaches in statistical data processing have been put forward. One approach, the "StEM" deductive query processor [Basili and Meo-Evoli, 1992], based on the MEFISTO data model, seeks to derive a statistical entity's extension – against a database comprising both (relational) microdata and macrodata (the latter represented in statistical entity format) – for a given statistical entity scheme $S_* = \langle C_*, t_* \rangle$ from either a macrodatum (statistical entity) $M_0 = \langle C_0, t_0 \rangle$ or a microdatum m_0 such that $S_* = \kappa(M_0)$ or $S_* = \kappa(m_0)$. In fact, the κ-operator amounts to a *resolution* plan composed of elementary MEFISTO operators reducing, in a stepwise fashion, M_0, or m_0, to S_*. Which elementary operators have to be applied eventually to obtain S_* is determined during a query *feasibility verification* stage. Roughly, for given S_*, κ is synthesized as follows [Basili and Meo-Evoli, 1992]: suppose that there is some macro-operator Ω transforming (given) C_0 into $C_* = \Omega(C_0)$; now,

if t_0 is additive,
then simply compute $\Omega(M_0)$
elseif there is some $M' = \langle C_0, t' \rangle$ such that t' is additive
    then compute $\Omega(M')$ as final result
    elseif there is some $M' = \langle C', t' \rangle$ such that t' is additive and $C_0 \subseteq C'$
        then apply the MEFISTO *restriction* operator to M', returning
$$M_r = \langle C_0, t' \rangle, \text{ and compute } \Omega(M_r) \text{ as final result}$$
        elseif C' is reducible to C_0 through *taxonomic links*
            then apply the MEFISTO *reclassification* operator
to M', returning $M_c = \langle C_0, t' \rangle$, and compute
$\Omega(M_c)$ as final result
            elseif there is some suitable m_0
                then generate $M'' = \langle C_0, t'' \rangle$ from m_0 such that t''
is additive to compute $\Omega(M'')$ as final result
            else *fail*

In this procedure, operator Ω – determined jointly by C_0 and C_* – still must be supplied as query parameter although, from a purely deductive point of view, it could also be determined dynamically depending on the structural difference between S_* and suitable candidate objects selected from the underlying extensional (macro-) database as well. Admittedly, as Ω may comprise several instances of elementary MEFISTO operators, query processing is simplified considerably, if Ω is supplied as a pre-determined parameter (note that, by virtue of MEFISTO's automatic summary value management, Ω adapts itself to its argument's summary type, provided this summary type is an additive one). Furthermore, the proposed method assumes that all candidate entities M' are generated from one and the same microdatum (which is assured simply by selecting only entities connected accordingly by semantic links). In StEM, *taxonomic links* are used to state *1:n* associations between modality sets (co-domains) used as alternate and hierarchically nested domains $D_i(C_j)$, for $C_j \in C$, of category attributes participating in statistical entity schemes.

The deductive power of StEM is quite limited as S_* is derivable from a single operand entity only; starting from a definite data object M_0 stored extensionally in the database, StEM supports the derivation of S_*, in case of t_0 not being an additive summary type, by substituting M_0 with a suitable additive "predecessor" object M' (in terms of the derivability relation); if no such statistical entity can be found, recourse to the microdatum generating M_0 is taken, if possible. An M' is suitable, if its entity space $\mathbf{C}'$ is reducible to $\mathbf{C}_0$ either by restriction (that is, by selecting a subset of entity space tuples of $\mathbf{C}'$) – called "simple" reducibility – or by reclassification using taxonomic links – called "semantic" reducibility (cf. [Ahn et al., 1990] with respect to a similar suggestion), both transformations of a fairly simple type. From the description given in [Basili and Meo-Evoli, 1992] it is not entirely clear why reclassification cannot be extended to include also projections (that is, dropping category attributes from $\mathbf{C}'$) although this would enlarge the class of derivable macrodata considerably. Unfortunately, the published accounts of StEM do neither indicate, how M'' is obtained from m_0 as this operation, in general, requires yet another parameter determining the *alfa*-component of M'' which, of course, might depend on further attributes not among those in $\mathbf{C}_0$.

Thus, despite being an interesting proposal to statistical database retrieval processing, StEM can hardly be regarded as a full-fledged deductive query method. Its major deficiency certainly is the implicit (or, rather, non-) treatment of the *alfa*-component of statistical entities not managed by the MEFISTO macrodata algebra. Apparently, once S_* is stated regardless of the extensional existence of some M_0, there is no way to avoid an explicit reference to S_*'s *alfa*-component in addition to the references to *beta*- and *gamma*-components. This, in turn, implies a formalized language to express statistical concepts in terms of statistical domain terminology defining a taxonomic space subdividing the set of observation units belonging to a well-defined statistical population, as it has been proposed by Catarci et al. [1990] based on concept logics, particularly KL-ONE [Brachman and Schmolze, 1985]. The fundamental tenet of concept logic approaches to the representation of information content of statistical databases is a faithful reproduction of structural relationships both within and between datasets by symbolic structural relationships between corresponding data element *denotations*. More specifically, concept logic aims at a coherent algebraic denotation of a whole database's extensional contents upon which a concise and rigorous notion of formal derivability (the database intension) can be established. In other words, while semantic network approaches facilitate a (human-driven) navigational exploitation of database contents, concept logics facilitate an inferential exploitation. Due to the inherent semantic conciseness, concept logic-based approaches enable comparison and integration of statistical databases in a well-defined way.

The statistical concept logic proposed by [Catarci et al., 1990] rests on a couple of familiar statistical concepts such as population, sub-population, variable, and aggregate. Formally, the model discerns *concepts* (abstractions of sets of real world entities) and binary *relationships* (associations between concept pairs), insofar resembling the infological model of Sundgren [1992]. However, quite in contrast to standard practice in ER-modelling, statistical populations (well-defined sets of observation or reference units), specified subsets thereof (sub-populations), and value sets (variable domains) are modelled as concepts whereas variables and se-

Table 1.3.2–4: Interpretation of $\mathcal{L}(P,R)$

$I(Empty) = \emptyset$

$I(T) = D$

$I(not\ P) = D - I(P)$

$I(C_1\ and\ C_2) = I(C_1) \cap I(C_2)$

$I(C_1\ or\ C_2) = I(C_1) \cup I(C_2)$

$I(some\ S.C) = \{x \in U \mid (\exists \langle x,y \rangle \in I(S)).y \in I(C)\}$

$I(all\ S.C) = \{x \in U \mid (\forall \langle x,y \rangle \in I(S)).y \in I(C)\}$

$I(atleast\ n.S) = \{x \in U \mid \|\{\langle x,y \rangle \mid \langle x,y \rangle \in I(S)\}\| \geq n\}$

$I(atmost\ n.S) = \{x \in U \mid \|\{\langle x,y \rangle \mid \langle x,y \rangle \in I(S)\}\| \leq n\}$

$I(F.C) = \{x \in U \mid \langle x,y \rangle \in I(F) \wedge y \in I(C)\}$

$I([F_1:C_1,\ldots,F_m:C_m]) = \{\langle F_1:v_1,\ldots,F_m:v_m \rangle \in U \mid v_i \in I(C_i), 1 \leq i \leq m\}$

$I(setof\ C) = \{v_1,\ldots,v_m \in U \mid v_i \in I(C_i), 1 \leq i \leq m\}$

$I((v_1,\ldots,v_m)) = \{v_1,\ldots,v_m\}$

$I(S_1 \circ S_2) = \{\langle x,y \rangle \in U \times U \mid (\exists u \in U).\langle x,u \rangle \in I(S_1) \wedge \langle u,y \rangle \in I(S_2)\}$

$I(S^{-1}) = \{\langle x,y \rangle \in U \times U \mid \langle y,x \rangle \in I(S)\}$

$I(S:C) = \{\langle x,y \rangle \in U \times U \mid \langle x,y \rangle \in I(S) \wedge y \in I(C)\}$

mantic associations between (sub-) populations are modelled as relationships. Moreover, relationships are subdivided into *functions* and *roles* where the latter are used to denote binary relations; the former are used to link variable (co-) domains to populations. Starting from two – countably infinite – sets of primitive concepts P and relationships R, respectively, each comprising a collection of unique atomic symbols naming real-world entities, a formal concept language $\mathcal{L}(P,R)$ is generated bottom-up by two grammar rules, viz.

```
<C>   ::=   P | not P | <C> and <C> | <C> or <C> |
            some <S>.<C> | all <S>.<C> |
            atleast n.<S> | atmost n.<S> | setof <C> |
            F.<C> | [F₁:<C₁>,...,Fₘ:<Cₘ>] | (v₁,...,vₙ)
<S>   ::=   R | <S> ∘ <S> | <S>⁻¹ | <S>:<C>
```

.3 STATISTICAL DATA MODELLING

where $P \in \mathbf{P}$, $R \in \mathbf{R}$, $n > 0$, F and F_j are function symbols, and v_i a value. The meanings of concept and relationship constructors are intuitively clear, and can be given a precise semantics as listed in **Tab. 1.3.2–4**. The interpretation (model) for $\mathcal{L}(P,R)$ is built by establishing recursively a universe U of structured terms from a basic domain D comprising all elementary (undefined) objects, or names, of interest such that U is the smallest set containing D and, if $u_1,\ldots,u_m \in U$ and $F_1,\ldots,F_m$ are function symbols of $\mathcal{L}(P,R)$, the labelled tuple $\langle F_1:u_1,\ldots,F_m:u_m \rangle \in U$, too. I, the interpretation of $\mathcal{L}(P,R)$, is a function mapping each concept to a subset of U, and each relationship to a subset of $U \times U$. 'Empty' and 'T' are two distinguished concepts contained in set P.

On top of the concepts definable in terms of $\mathcal{L}(P,R)$, aggregates and "goals" can now be specified. Basically, aggregates are either of simple or complex type. Simple aggregates are denoted by a *set* of value sets s_i, $\|s_1,\ldots,s_n\|$ (the s_i might be intervals in case of numerical value sets), whereas complex aggregates are denoted by a term

$$agg\ C * S_1.B_1 * \cdots * S_k.B_k$$

where C is a concept, $S_1,\ldots, S_k$ are relationships, and $B_1,\ldots, B_k$ are simple aggregates expressed as variable co-domains. Semantically, aggregates are interpreted as follows:

$$I(\|s_1,\ldots,s_n\|) = \{I(s_1),\ldots,I(s_n)\}$$
$$I(agg\ C * S_1.B_1 * \cdots * S_k.B_k) =$$
$$\{\langle S_1:b_1,\ldots,S_k:b_k,\Xi:s\rangle \mid b_j \in I(B_j), 1 \leq j \leq k\}$$

where $s = \left\{ x \in I(C) \mid \bigwedge_{j=1}^{k} \left(\left(b_j \in I(B_j)\right) \wedge \left(\langle x,y \rangle \in I(S_j)\right) \wedge \left(y \in b_j\right) \right) \right\}$; Ξ, denoting the "target" of the aggregate, is a distinguished function symbol. Finally, goals are terms of the form

$$goal\ OP\ on\ V\ of\ AGG$$

such that OP denotes a statistical aggregation operator, V a summary variable, and AGG an aggregate term. As 'summary variable' is used in this proposal in a somewhat restricted sense referring to numerical magnitudes only, goal terms appear also in simplified format, viz.

$$goal\ OP\ of\ AGG$$

if the statistical aggregation operator does not have a numerical argument variable. The semantics of both terms can be defined as

$$I(goal\ \mathrm{OP}\ on\ \mathrm{V}\ of\ \mathrm{AGG}) =$$
$$\{\langle\Xi{:}s,\Phi{:}r\rangle\,|\,(\exists\langle S_1.b_1,\ldots,S_k.b_k,\Xi{:}s\rangle \in I(\mathrm{AGG})).r = \mathrm{OP}(I(\mathrm{V}.s))\}$$
$$I(goal\ \mathrm{OP}\ of\ \mathrm{AGG}) =$$
$$\{\langle\Xi{:}s,\Phi{:}r\rangle\,|\,(\exists\langle S_1.b_1,\ldots,S_k.b_k,\Xi{:}s\rangle \in I(\mathrm{AGG})).r = \mathrm{OP}(I(s))\}$$

where, again, Φ is a distinguished function symbol flagging the numerical output component.

In order to illustrate this concept logic-based approach, think of a goal statement like

$$goal\ \mathrm{COUNT}\ of\ \bigl(agg\ \mathrm{Person} * \mathrm{Sex}.\|\,(male),(female)\,\| * \mathrm{Age}.\|\,1..50,51..100\,\|\bigr)$$

assuming that COUNT is the standard statistical counting operator, 'Sex' and 'Age' are two variables observed for sampling units of population 'Person', and 'a .. b' is a special range notation abbreviating the usual set enumeration; letting, furthermore, hold that

$$I(\mathrm{Person}) = \{a,b,c,d\}$$
$$I(\mathrm{Sex}) = \{\langle a,male\rangle,\langle b,male\rangle,\langle c,female\rangle,\langle d,female\rangle\}$$
$$I(\mathrm{Age}) = \{\langle a,20\rangle,\langle b,30\rangle,\langle c,20\rangle,\langle d,60\rangle\}$$

it follows that

$$I\bigl(agg\ \mathrm{Person} * \mathrm{Sex}.\|\,(male),(female)\,\| * \mathrm{Age}.\|\,1..50,51..100\,\|\bigr) =$$
$$\left\{\begin{array}{l}\langle\mathrm{Sex}{:}\{male\},\mathrm{Age}{:}\{1..50\},\Xi{:}\{a,b\}\rangle,\\ \langle\mathrm{Sex}{:}\{male\},\mathrm{Age}{:}\{51..100\},\Xi{:}\varnothing\rangle,\\ \langle\mathrm{Sex}{:}\{female\},\mathrm{Age}{:}\{1..50\},\Xi{:}\{c\}\rangle,\\ \langle\mathrm{Sex}{:}\{female\},\mathrm{Age}{:}\{51..100\},\Xi{:}\{d\}\rangle\end{array}\right\}$$

and the *goal* term, eventually, yields the interpretation

$$\{\langle\Xi{:}\{a,b\},\Phi{:}2\rangle,\langle\Xi{:}\{a,b\},\Phi{:}0\rangle,\langle\Xi{:}\{a,b\},\Phi{:}1\rangle,\langle\Xi{:}\{a,b\},\Phi{:}1\rangle\}$$

.3 STATISTICAL DATA MODELLING

It is fairly simple to embed the concept formation language in a convenient database definition language such that populations, sub-populations, variables, functions, and roles can be specified suitably (cf. [D'Angiolini, 1992]). For instance, using a *def* construct, a sub-population like 'PersonsOver14' could be arranged easily by forming the concept expression

$$\text{PersonOver14} \quad def \quad \text{Person } and \text{ Age.}(14 .. 100)$$

and so on. In that concept formation can resort to all variables and value domains (that is, variable co-domains) as well as functions and roles defined between populations, the language provides full flexibility in setting up domain concepts (in fact, expressiveness equals that of the extended relational algebra proposed by [Özsoyoglu et al., 1987]). Similarly to concepts, even aggregates and goals could be *de*fined; assuming appropriate definitions of variables, roles, and concepts, for instance,

$$\text{A3} \quad def \quad agg \text{ Household } and \text{ }(some \text{ Member-of}^{-1}.\text{UnemployedPerson}) *$$
$$\text{Possess: (Habitation } and \text{ Habitation-kind.}(primary) \circ$$
$$\text{City} \circ \text{Province} \circ \text{Region}$$

$$\text{G3} \quad def \quad goal \text{ COUNT } of \text{ A3}$$

such that, if evaluated, 'G3' amounts to a contingency table of absolute frequencies for households with at least one unemployed person such that only households in primary habitations are included broken down by regions where the habitations are located in. Likewise,

$$\text{G4} \quad def \quad goal \text{ SUM } on \text{ Income } of \text{ A3}$$

would denote the total 'Income' of selected households (assuming, of course, that 'Income' is a variable observed on households).

The rationale of using concept logic to represent the information content of a statistical database is not, of course, delimited to giving a succinct content description; the real virtue of this approach is the formal reasoning that can be based on it. First, concept definitions may suffer from inconsistencies which are easily detected by determining a concept C's model $M(C)$ which must not be empty (that is, $M(C) \neq M(Empty) = \emptyset$) for a concept definition to be consistent. Similarly, subsumption relations between concept definitions can be investigated formally. More important, obviously, is the formal derivability relation for checking if a concept A is derivable from another concept B. On condition that there is a total *1:n* correspondence between instances (tuples) x of concept A and $y_1,...,y_n$ of B, $x \mapsto \{y_1,...,y_n\}$, such that $x = y_1 \oplus \cdots \oplus y_n$ ("additivity") where

$$\langle F_1:v_1,\ldots,F_n:v_n,\Xi:r\rangle \oplus \langle F_1:w_1,\ldots,F_n:w_n,\Xi:s\rangle =$$
$$\langle F_1:v_1\cup w_1,\ldots,F_n:v_n\cup w_n,\Xi:r\cup s\rangle$$

then A is said to be derivable from B. Apparently, in view of this derivability relation it is possible to *deduce* mechanically if a (macro-) data request can be satisfied without taking recourse to the statistical database's extension. Moreover, this derivability relation entails a fully transparent *alfa*-component management of statistical macrodata (aggregates) which is of utmost importance as to context preservation in statistical information processing. As it stands, however, the proposed calculus does not consider the derived aggregates' (macrodata) summary types.

Since formal reasoning remains restricted to the symbolic denotation layer of $\mathcal{L}(P,R)$, references to microdata are expressed as descriptions as well; more specifically, the case-by-variate matrix (**Fig. 1.3.1–1**) is now interpreted as a "case-by-function" matrix: for each observed case i, multivariate observations are lined up as row vectors holding the function values $F_j(i)$. Thus, cases must be furnished with unique labels (identifiers) explicitly to be operable in the calculus. For obvious reasons, the symbolic representation of the calculus's deeply nested term structures makes rather high demands on the technical data modelling environment which presumably could be met best by resorting to some kind of object-oriented approach.

Altogether, to date the concept logic-based statistical database model of Catarci et al. [1990] must be considered the most advanced and promising one. In particular, with respect to a formalized management of database queries and in view of the urgent needs of data integration and harmonization this approach (if any at all) seems prospective. Exactly for this reason, the theme will be taken up anew in Chapter 2.

1.4 Systems, Prototypes, and Research Projects

The essence of more than one decade of work in statistical data modelling lies in emphasizing the role of a *functional* approach to data documentation. Going beyond semantic data models – which by now have become a kind of standard in information processing and database management – the efficient organization of huge, centrally administrated statistical data holdings depends crucially on methods to process data context explicitly *by machine*. Conversely, non-formatted, human-readable on-line documentation comprising purely descriptive metadata (sometimes dubbed "metatexts") loses importance as it falls short of supporting effectively automated data and information processing. Correspondingly, the focus of research has shifted away from modelling primary (observation) data towards the modelling of secondary data structures – metadata (cf. Subsection 1.5.2) – used to denote and preserve (primary) data context in automated statistical information processing sequences. Within metadata, in turn, the formalized terminological and pragmatic context descriptions characterizing taxonomic population structures, measurement scales, and observation settings, occupy a pivotal position, at least as long as advanced methods of (inference) statistics are not taken into account (which, additionally, require more or less comprehensive formalized model contexts, too).

This overall development trend is reflected, of course, in the everyday practice of statistical information processing and even more so in recent scientific research endeavours briefly reviewed in the following paragraphs. Apparently, the new and unprecedented opportunities offered by advanced information technologies have to be exploited beneficially to meet the increased and still increasing societal demands for statistical data supply:

- For one thing, data production and dissemination is forced more and more to operate economically efficient, focusing at a consumer-oriented data supply, flexible production according to customer prescriptions (this way paraphrasing flexible manufacturing in commodity production replacing high-volume production of standardized products), and production of timely information ("just in time") on demand. Particularly in official statistics there is a growing pressure of justifying its privileged position by tremendously improved public data services while, on the other hand, this position could be threatened markedly by competitive private information providers more responsive to information markets.
- The "product" statistics itself is changing as the traditional static (= status) descriptions are replaced more and more by dynamic (state-change) descriptions. Against the background of competitively speeding up market and societal processes decisions have to based increasingly on anticipated scenarios and forecast models consisting of both, a cross-sectional and a longitudinal model component (where, apparently, the latter feeds dynamic data into statistical forecast models). In the foreseeable future, traditional official statistics will turn into a basic supplier of cross-sectional data (such as population censuses, administration registers, etc.) which either may not be producible profitably at all or remain a privilege of official authorities for juridical reasons (data privacy, etc.) and, hence, represent a public measure to maintain society's statistical information backbone. Additionally, a sharply raised product quality is expected implying that statistics data must be available promptly, reliably, at high granularity, and well-documented.

Responding to these emerging requirements, the research and development of next-generation statistical information systems centers around the following interrelated major themes:

- The improved, simplified, and more efficient access to statistical data and information presupposes a *logical view of data* upon which conceptual retrieval can be based. Conceptual retrieval refers to abstract data models equipped with either non-verbal (direct-manipulation) or formal data language (concept logic) interfaces as discussed in Section 1.3. Furthermore, depending on application domains, even more intentional modes of data access are conceivable as illustrated by the EXPLORA example [Klösgen, 1986; 1990] filtering data by specifying criteria applying to analytical-inferential parameters derived from stored observation data. In any case, improved data access encompasses an unrestricted possibility to process stored data (within defined legal boundaries) instead of a simple retrieval of pre-aggregated datasets not amenable to further modification except truncation or yet higher aggregation. The logical data view enables an intensional database access sheltering retrieval specifications from explicit navigation and physical data organization, and delimiting query specifications to semantic elements denoting *what* is asked for irrespective of how a query target might be derived from actual database extensions. In particular, intensional data retrieval implies an unrestricted transparent combination of stored data, thus touching on the issue of effective integration – or "harmonization" – of data originating from different sources and/or production contexts. This, in turn, stresses the role of logical data independence [D'Atri and Ricci, 1989] as the cardinal criterion of an information-economic usage of data. However, data harmonization, as opposed to plain data collating, poses the problems of statistical consistency of data holdings (cf. [Malvestuto, 1989; 1991]) and model-based extrapolation of database extensions to fill in those areas of the database intension lacking empirical data [Rowe, 1991]. Conceptual retrieval in statistical databases presupposes an extensive metadata layer denoting database content; hence, data production and metadata production ought to go hand in hand. For this reason, data producers will turn into metadata producers as well, thus contributing much to their own benefit as, in general, data producers functionally are also data consumers reaping the advantages of established metadata.

- Conceptually oriented data retrieval is facilitated by an effective translation of manipulations on the formal representation of data objects onto the extensional layer of object storage by arranging *transparent statistical database operations*. Ideally, operational transparency is realized in a context-sensitive way such that pragmatic operation details can be inferred by the information processing system autonomously. Evidently, transparent operators rely on data context provided by formalized context preservation means; this entails two cardinal consequences, viz. each operator must convert context *pre*-conditions into context *post*-conditions (in order to update local contexts in a coherent way), and a basic data context must be supplied by establishing a continuous processing chain passing on the data context from stage to stage parallel to primary or processed extensional data. Given such a data context and a set of re-formulated database operators such that data and metadata are processed in strict correspondence [Darius et al., 1993] all the time, an algebraically closed information processing system is obtained furnishing each derived data object with its associated data context. As a matter of fact, this dual data/metadata methodology of statistical information processing is an outgrowth of statistical expert system research [Hand, 1994] where data context turned out indispensable in providing methodological guidance to data consumers.
- Both, conceptual retrieval and transparent database management, place specific requirements on the underlying extensional database structure. In particular,

with respect to the non-standard semantics of statistical data, custom-tailored *extensional data models for primary as well as secondary data* (metadata) of maximal generality need to be devised such that the whole statistical data processing chain is supported effectively. As far as conventional database modelling approaches are not really adaptable to these requirements, fresh approaches must be pursued which, for instance, could be stimulated by proposals in related domains such as geographic information systems (cf. [Maguire *et al.*, 1991; Plank, 1994]. Since, in general, statistical data of relevance reside on different database systems hosted at decentral computing sites, data models must be prepared to handle distributed data [Sadreddini *et al.*, 1990; 1992; 1993] – and, of course, metadata – both flexibly and efficiently as well.

- In the wake of the changing demands and technological preconditions of statistical data production, *statistical systems* themselves undergo a thorough assessment of function and purpose, leading as a rule to profound *business re-engineering* of both organization structure (labour division schemes) and statistical information processing chains in order to improve efficiency, product quality, and service levels, thus assuring the systems' own functionality. In particular, an information-economically oriented re-design of data production processes in official statistics will challenge grown horizontal as well as vertical organization structures in favour of more integrated, network-like overall production systems lowering internal transaction costs, facilitating trans-sectoral synergies, and enabling a tremendously increased capability to interrelate and combine sectoral data. Noting that, by and large, a statistical office's information-technological infrastructure determines its overall organization structure, re-engineering the former just means re-engineering the latter, too (this is particularly highlighted by the – pioneering – introduction of an integrated information infrastructure at Statistics Canada [Graves, 1992; Hutton and Graves, 1993; Graves *et al.*, 1993]).

Currently, quite a lot of proposals and recommendations addressing either of the broad topics mentioned is put forward; as yet, to a lesser degree attempts have been undertaken to turn theory into practice. In view of the many and intricate questions awaiting viable answers, it is not reasonable, for the time being, to expect that there will be comprehensive solutions within the reach of this decade. Thus, the following subsections will give only a scanty and incomplete account of recent developments and research efforts which, hopefully, will lead up some day to a convergent statistical information processing system design.

1.4.1 A Survey of Systems and Developments

Traditionally, in statistical information processing three major functional areas, corresponding to the main data flow stages, can be discerned as summarized in **Tab. 1.4.1**'s left column; these stages may be juxtaposed to Sundgren's [1994] functional three-phase model reflecting a typical data producer's view of a statistical information system.

Table 1.4.1: Coarse Processing Stage Model of Statistical Information Processing

data capture	1	input acquisition
data storage	2	aggregation
data dissemination	3	output delivery

In terms of a data communication model, this three-phase model is condensed further to a two-phase model comprising a data *production* (sender) and a data *consumption* (receiver) stage, separated functionally by an intervening data store (transmission channel with delay) buffering captured data on its way to dissemination. While data capture and dissemination are activities loading data into the store or retrieving it from the store, respectively, the store itself is a physical memory structure operated by symbol import and export routines. Contrarily, Sundgren [1994] views the information system very much as an industrial production enterprise where input acquisition can be identified with procurement of raw materials, aggregation with the value-adding stage moulding raw materials into finished products, and output delivery with a sales department responsible for shipping products. While both models have their undeniable merits, the production enterprise metaphor already undergoes a definite revision as data producers see themselves more and more in a position to supply moderately pre-processed data or semi-finished products (aggregates) being transformed into finished products by consumers, or customers, on their own. This is but a matter of historical continuity considering that, initially, data storage owed its importance to the bare fact that automated data processing simply presupposes some kind of intermediary storage (such as punched cards in Hollerith's days, but note also the concept of "system files" in statistical software packages) whereas, later on, database systems and tabulating or statistical analysis software became separated functionally for good reasons. Due to this separation, the roles of data provider and data consumer/analyzer depart logically and, as far as data analyzer and data producer do not coincide anymore, explicit and documented end-user interfaces to data holdings arose quite naturally. The advent of remote data access by tele-communication links and time sharing operating systems stressed this functional separation even further, giving rise to specific database query languages (particularly in the relational database domain; cf. [Froeschl, 1989; Lackner, 1990]). Essentially, this circumscribes the present practical state-of-the-art in statistical information systems (for instance, [EUROSTAT, 1994] lists some 100 statistical databases throughout Europe providing on-line access at least to a restricted class of end-users; [Staud, 1988] counted some 600 economic databases supplying statistical data, mainly aggregates). Typically, end-user databases are organized hierarchically by partitioning successively subject domains down to a "segment" level comprising more or less high aggregated macrodata, sometimes accompanied by metadata such as category lists, classifications, footnotes, and comments (for instance, the ISIS database of Austrian's Central Statistical Office [ÖSTAT, 1980]). The hitherto available statistical databases, however, enforce rather terse interaction styles and, in general, do not admit data manipulation except plain retrieval, selection of data subsets, or projections (*marginals*) of stored tables. This supply of predefined statistical aggregates is felt increasingly insufficient and counterproductive given that many data consumers have at their command quite powerful computing and statistical analysis software systems equalling or even surpassing, in many cases, those of the data producers. Thus, functionally there is no longer a marked difference between a data producer's internal data usage and the external data consumer's data usage. Hence, with respect to Sundgren's three-phase model, the aggregation phase by and large tends towards the data consumption phase, favouring the two-phase production/consumption view of statistical information processing (although the term 'data consumption' may look odd, it should be kept in mind that, in fact, from an inference-statistical point of view, empirical data get worn out by repeated usage; cf. Bonferroni's inequality).

These emerging changes in requirements are, at least partially, accounted for by recent statistical information system designs, leaving behind the "closed shop" philosophy of previous system designs and abandoning the concept of canned statistical products ready for delivery; quite on the contrary, virtually all of them have in mind an end-user oriented support of data access and data presentation. To this group of statistical information systems belong, for instance, the systems PC-AXIS [Nordbäck, 1992; World Systems, 1995], DUVA [Appel, 1993ab], and, particularly, GENESIS [Glitza, 1994].

Among the systems mentioned, PC-AXIS probably is the least ambitious in that it serves as an interactive (GUI) front-end for the generation of statistical tables, time series, and graphics, using summary sets – that is, medium- to high-level macrodata – down-loaded either from an on-line back-end mainframe system (specifically, the host computing system of Statistics Sweden) or some intermediary storage (like a CD-ROM). PC-AXIS receives back-end data such that the arrays of plain cell data values and associated metadata are packaged in message-like information units enabling a metadata-based handling of datasets. PC-AXIS's set of operators comprises several table management functions such as table editing (deleting, interchanging, nesting, etc. of table rows and columns), table arithmetics (creation of new variables, adding, subtracting, multiplying, dividing, etc. of rows or columns, and so on), and table aggregation (marginals, collapsing of value sets) all of which make use, though rather modestly, of available metadata encoding, essentially, structural table descriptions. In particular, all operators transform cell data values and the created tables' metadata synchronously furnishing consistently all output structures with attached descriptive metadata elements like table headers and stubs, and footnotes. Although PC-AXIS recognizes a few attributes of a table's summary variable, the system lacks a summary variable management as proposed in the ME-FISTO system (cf. Subsection 1.3.2); in particular, PC-AXIS is not aware of additivity restrictions in summation operations. Likewise, PC-AXIS is not intended to provide support in locating datasets of interest, that is data access still is fully navigational.

Contrary to PC-AXIS, the DUVA system strives to encompass the whole statistical information processing chain from data and metadata capture up to the final data dissemination. DUVA's basic tenet is the superiority of a natural language interface for data access established on top of a pointer network linking (key-) words, or semantic items, to statistical data objects; furthermore, these semantic items are classified in terms of several item categories (catalogues) such as measurement units, value ranges, keys, variables, definitions, surveys, etc. The DUVA data model comprises two elementary data structures, viz. so-called *basic files* (microdata) and *macro-files* (macrodata) where macro-files by and large correspond to Özsoyoglu *et al.*'s [1989] primitive summary table concept, that is counted or summed up pre-macrodata (cf. **Tab. 1.3.1–3**) obtained by elementary aggregate formation from underlying basic file-microdata. By definition, each basic file refers to a single statistical population and is indexed by links to (global) temporal and geographical keys as well as a link to the type of data source (for example, a survey) spawning it. Prior to the final preparation of output tables and graphs, basic files are processed into macro-file format mainly by applying microdata filters (case selection) and "reference tables" (re-classifications, or change keys, induced by arranging taxonomic links in Basili and Meo-Evoli's [1992] sense; cf. Subsection 1.3.2); temporal and spatial links enable the aligning of "disjoint" basic files on condition that, of course, all component basic files share the same record structure. In deriving macro-files, the pointer network of semantic items is automatically and consistently updated such that new macro-files can be referenced in a fully transparent way via the keyword access interface. As DUVA is prepared to store any kind of textual documentation of statistical data (metatexts), macro-files and derived statistical products are accompanied coherently by (unformatted) data context throughout. The functional use of metadata, however, remains restricted to (i) locate data objects by evaluating keyword-triggered pointer references and (ii) apply (user-) predefined change keys automatically during macro-file derivation (this amounts to an instance of the *aggregate-by-template* operator discussed in Subsection 1.3.2).

The GENESIS system, a joint development of the German Bund and Länder coordinated by the Statistisches Bundesamt (Wiesbaden), aims to merge the virtues of semantic (macro-) data modelling with the data access approach of DUVA. Like DUVA, to which GENESIS owes much of its conception, the system offers a keyword-oriented data access in addition to a more conventional navigational directory-oriented browsing mode. More specifically, a distinction between *information enquiry* and *data enquiry* is introduced such that the former is based on the system's metadata layer while data enquiries apparently involve the data layer as well. GENESIS's core data model consists of a "block" model where each block represents a multi-way table the cells of which may comprise a vector of summary attributes, including different variants thereof carrying data quality tags (such as preliminary values, forecasts, estimates, final results, etc.); basically, a block corresponds to a set of macrodata, or tables, sharing the same breakdown structure, and conceptually restates the DUVA macro-file structure. Semantically, blocks may represent both cross-classified tables and time series. For efficiency reasons, the pool of basic blocks is accompanied by another pool of ready-made "standard tables" (print tables with a two-dimensional graphical layout) derived from blocks to respond quickly to frequently occurring data requests. Somewhat in contrast to DUVA, GENESIS discerns explicitly, for the sake of semantic conciseness, three different variable (attribute) types, viz. summary variables, cross-classifying (that is, category) variables, and identifying variables. The latter type is used to indicate temporal references (key date, reporting period) of a block while, as usual, the former denote the structural table description including, if need be, temporal and regional breakdowns. Unlike DUVA, GENESIS, in general, does not store marginals, mainly because of the exponential order of storage requirements (moreover, since many aggregates are no longer additive, marginal sums often would not be derivable from stored sum blocks at all). If derived sum blocks are stored at all, they can be joined to "block families" attached to their respective parent basic blocks to speed up data enquiries. In any case, all data objects derived from basic blocks are interconnected fully with the internal metadata network structure. In addition to its dual data access interface, GENESIS provides a PC-AXIS-like tabulation and table management interface enabling end-users to define and customize requested statistical products in a variety of presentation formats. As an alternate retrieval mode, GENESIS admits the structural specification of tables by iteratively refining a dummy table layout until a source block or standard table is identified unambiguously which then is used to feed the output table's cells automatically. According to its end-user orientation, GENESIS strongly emphasizes data access validation, safeguarding confidentiality at all levels down to individual cell values within blocks. Being targeted primarily at data delivery, GENESIS relies on data production systems like DUVA feeding both data and metadata into its database. Although GENESIS is prepared to compile unified metadata representations from different input sources by adapting its metadata network to the added basic blocks as the database grows, it does not specifically support data harmonization which is a task still left to the system's end-users.

To date, of the three systems presented only one (PC-AXIS) can be deemed a finished product; DUVA, although in use for a couple of years, is still under development, and GENESIS is now put into action for the very first time.

As far as statistical information systems are confined functionally to data storage and retrieval, they depend on preceding data capture and pre-processing systems feeding their outputs into the databases. Adding metadata to the primary observation data or aggregates derived thereof naturally calls for a re-engineering of the preparatory input stage such that data and accompanying metadata is interlinked and provided coherently. Typically, this revision coincides with a general information-technological reshaping of data capture and data validation procedures aiming at a continuous digital chain of data processing right from the preparation of survey programmes and actual observation or measurement. As most measuring instruments operate on a digital basis by now, data can be passed on from one processing stage to the next in purely electronic formats tapping and transmitting synchronously major parts of data context. An illustrating case in point is the BLAISE system [Bethlehem and Hundepool, 1992; Schuerhoff, 1993] of The Netherlands's CBS (Central Bureau of Statistics) replacing paper questionnaires by electronic ones installed on portable computing equipment used in survey interviewing. Not only does BLAISE avoid most of subsequent data validation measures as it assures proper responses to questions from the outset, its output files are fully documented observation records (microdata) providing a concise definition of all survey design and questionnaire elements such as questions, code lists, value ranges, data dependencies, non-response, etc., without any need for keying in this metadata in a separate pass (quite on the contrary, survey definition – and, hence, metadata capture – *precedes* data capture). In view of the dynamically increasing density of digital tele-communication networks many traditional modes of data capture will become substituted by more efficient, time and effort saving telematics applications, particularly where original statistics data itself is a by-product of information processing (for instance, in business statistics many of the reported datasets can be readily gained from commercial transaction processing systems, office automation systems, etc.; cf. [Mesenbourg and Ambler, 1992] with respect to an early telematics application in official statistics data collection). Thus, in various domains fully automated data capture is within reach using advanced auxiliary data processing methods like automatic coding procedures [Miller, 1992; Creecy, 1992], systems for classifying the output of remote sensing devices, voice recognition, image processing, etc.

The organization of data capture and data tapping routines more or less amounts to a data producer's internal affair. In contrast to this, output delivery and data dissemination receives much public attention whence, in order to meet increasing demands, most data suppliers have adapted and "streamlined" their services. Basically, there are two opposing views on statistical data supply, viz.

- a producer, or "depositor", view concerned with the arrangement of an information infrastructure and distribution system permitting a wide scope of end-users easy and efficient access to data, and
- a consumer, or "enquirer", view seeking easy and efficient means to spot and retrieve data.

Contrary to earlier days when statistical data served narrow-defined purposes and information technology delimited processing possibilities severely, today statistical data holdings are regarded as major *public* resources that ought to be accessible by the general public except perhaps for a small portion of sensitive data demanding specific protection measures. Indeed, given the huge repositories of accumulated observation and administration data, the practical value of this asset can be hardly overestimated with respect to, for instance, policy making and evaluation or hypotheses generation in scientific research.

For an individual data provider, information technology offers a variety of new media useful for both, making data accessible and distributing data actively. Many statistical offices have decided, for instance, to package statistical products together with suitable access and presentation software on CD-ROMs (cf. [Marske and Zeisset, 1992; Pricking, 1992; World Systems, 1995], just to mention a few ones) noticing that many potential data consumers already possess PCs equipped with CD-ROM drives; after all, in a few years this will be included in the standard configuration of personal multi-media work-places. In the longer run, CD-ROMs will replace most of printed statistics publications because of their enormous storage capacity and, even more relevant, the opportunity to process statistical data further. Off-line services like CD-ROMs (which, in a sense continue the tradition of magnetic tapes and data diskettes) will, of course, not abandon established on-line services. Quite on the contrary, as the bandwidth of tele-communication links is stepped up successively, large amounts of data can be transferred over long distances almost instantly, opening an unprecedented avenue to retrieve remotely even the latest mass data. Thus, telematics services are extended from plain teletype on-line access to mainframe systems (data hosts) to client server-links providing *telnet* and *file transfer protocol* services (such as, for instance, the U.S. Bureau of the Census's file server 'ftp.census.gov' [148.129.129.15]) and to network services like *Internet* or *WorldWideWeb* [Krause, 1995]. In a broader perspective, of course, the question arises to which degree statistical information services should be offered cost-free, especially by official statistical agencies, and how pricing and marketing policies could contribute beneficially to improve both, service levels and appropriateness of data supply, without sacrificing the public responsibility of producers of general-domain source data [Thygesen, 1993].

For an individual data consumer, advanced information technology offers promising ways to locate and access statistical data, too. As more and more datasets are compiled world-wide, many of which may be relevant, for instance, in a given research context, it becomes increasingly challenging to hunt them up. What in fact is needed amounts to an electronic *agora* where data supply and data demand can meet. A very stimulating attempt in this direction has been undertaken by the GENIE project [Newman *et al.*, 1992; 1993]; this meta-information system proposal is based on an open approach to capture both verbal descriptions of supplied datasets and data requests of enquirers such that a dynamic terminological linkage structure is created and maintained adaptively by attaching "concepts" (word lists) to datasets corresponding to the researchers' actual usage of terminology. Although, at a superficial level, resembling traditional bibliographic thesauri and keyword reference systems, a self-organizing distributed architecture is envisaged allowing and encouraging, for the sake of network load balancing, the establishment of embedded interest clusters and requiring only rather modest maintenance efforts. Particular emphasis is devoted to a three-way indexing of available and offered datasets (topic reference, spatial coverage, temporal coverage) in order to achieve well-targeted responses to enquiries without swamping enquirers with bulks of irrelevant dataset references. As it stands, the proposal is deliberately restricted to meta-information management and does not handle primary observation data by itself. Evidently, data auditing systems like GENIE will gain importance particularly in interdisciplinary research (the project's background is research in Global Environmental Change) as there is little impetus and economic justification to set up costly data integration systems.

In technical terms, the dissemination of statistical data calls for agreed upon standards of electronic data exchange such that data transmission protocols ensure proper documentation of conveyed data, that is ensuring context preservation. A major stride in this respect certainly is the implementation of the GESMES (GEneric Statistical MESsage) data exchange standard [Lebaube, 1992; GESMES, 1993] and its sectoral refinements (such as the ECOSER message format used for transmitting time series data; [GESMES, 1995]). Essentially, GESMES comprises an array data part surrounded by several metadata items describing the array arrangement and pointing to codebooks and nomenclatures shared by both message sender and receiver. The GESMES standard certainly will replace more system-specific message formats as, for instance, the one used in PC-AXIS (see above) or the U-file format proposed by [Saijets, 1993] (all of these formats are easily converted into each other). A thorough attempt to investigate the structure of metadata needed to provide a sufficient documentation escorting primary statistical data is the long-term endeavour of METIS [UNDP/ECE/SCP, 1984ab; Sundgren, 1991] under the auspices of the Economic Commission for Europe of the United Nations (UN/ECE) proposing a unified metadata terminology as well as a more general guideline for devising statistical meta-information systems, both of which have exerted a considerable influence on the development of GESMES and the re-engineering of official statistics information systems in general.

The primary aim of data exchange standards like GESMES is a smooth technical transmission of datasets between computing systems using established telecommunication networks. From a subject-matter point of view, however, the message contents of data transfer protocols like GESMES are rather terse, admitting, at most, the attachment of plain textual comments (like footnotes to array elements). Conversely, due to the clumsiness of the GESMES format conventions there is little room to extend the scope of transmitted meta-information packaged with the data shipped in a message. A more appealing approach has been proposed by Bisdorff [1992] who devised a formal grammar describing quite exhaustively the contents of contiguous sets of statistical data (such as a survey). Basically, Bisdorff suggests that each description is encoded into one sentence of the formal language induced – recursively – by the defined grammar's rewriting rules where a sentence (regular expression) comprises a variety of meta-information items ranging from the informal background of a survey (author, publication references, etc.) through a concise denotation of survey design, survey variables, taxonomies, questionnaire design, the data itself, data transformations applied, post-stratifications, statistical analyses envisaged or already carried out, presentation views, etc. Parsing such a sentence by the grammar's equivalent finite-state device recovers all its information elements; since formal sentences are plain symbol strings (just like GESMES) they are ideally suited for encapsulation in standard message protocols (for instance, electronic mail). Moreover, it is fairly easy to adapt the language's syntax to other types of statistical data (say, time-series) as this amounts to add simply further rewriting rules to the grammar. Such a flexible message concept may turn out crucial in the longer run, considering the tremendously increasing necessity to exchange statistical data, at varying processing levels, between data *providers* in creating harmonized distributed statistical information systems as well as the naturally growing meta-information demands of data consumers and front-end data analysis systems.

1.4.2 Current Trends in Research

The roots of statistical information systems are twofold, viz. in computing on the one hand, supporting the automated sequential processing of tedious arithmetical number crunching, and in the clerical data processing tasks such as sorting, counting, and summing of huge amounts of filed business or administration records on the other hand. Initially, in a typically traditional attitude, statistical data processing has been viewed predominantly as a technical tool used to do more of the same though in less time and cheaper than before; strategically, these new data processing tools were fitted into existing organization structures leaving, by and large, everything else unchanged. In contrast to this, presently there is a thorough change in progress, induced and enforced by advanced information technology and telecommunication media, affecting the whole organization structure of statistical data production (and, of course, data consumption and analysis) processes and, as a further consequence, calling for a fundamental re-design of involved institutions. In particular, conservative statistical offices and agencies understanding themselves as governmental authorities will soon vanish since it is easily foreseen that they will fail to meet society's information demands adequately; mainly, the sectoral internal business organization founded on strict hierarchies and domain responsibilities will obstruct the effective cross-sectoral data integration and, thus, impair the organizations' flexibility and timeliness to deliver custom-tailored statistical products. Additionally, due to the erosion of national concepts – underlying the self-understanding of many national statistical institutes – these primary data providers will turn, to a high degree, into data-talebearers of supra-national statistical systems taking account of the increasing international business concentration and political and societal globalization tendencies; apparently, these statistical systems will no longer be defined in territorial but in *functional* terms.

The organization and function changes in statistical data production and dissemination systems are escorted by the emergence of new subject domains – for instance, environmental statistics – which need to be provided with high-quality and timely data for both policy making and monitoring as well as scientific research. Compared to more traditional subject areas of statistical information processing, data in these domains tend to be considerably more voluminous (cf., for instance, [Chen *et al.*, 1995]) and complex in structure (particularly demanding the establishment of *spatio-temporal* information structures [Plank, 1994]), this way challenging grown practices in data management and storage organization. The problems are actually aggravated as in these domains the unrestricted integration and interlacing of data is a topic of utmost urgency (just think of regional development policies relying on, say, labour force statistics, tourism statistics, mobility statistics, family budget statistics, environmental statistics and, possibly, balance of payment statistics all at once).

A specific challenge of official statistics is the on-going process of EUropean integration which must be accompanied by the development of a consolidated EUropean statistical system on top of the established national statistical systems of the Member Countries of the European Union in a time of rather strained governmental budgets. Apparently, this ambitious aim is elusive without resorting to a bundle of efficiency gaining measures and increased utilization of information technologies, that is: massive office and business process automation. Moreover, for reasons of tradition, national statistical agencies, as a rule, are equipped insufficiently with manpower and financial resources to promote genuine research programmes targeted at business re-engineering and internal streamlining of operations. Consequently, the development of a EUropean supra-national statistical infrastructure and research in advancing statistical information processing altogether have been formulated as high-level cohesion and RTD targets of Community interest by the European Commission.

A first pertinent initiative has been started within the Community's second research framework programme which launched a multi-annual sub-programme called "Development of Statistical Expert Systems" (DOSES) in June 1989. The DOSES effort was focused on the research topics (cited from [EUROSTAT, 1993c])

- *identification of information from the data producer or the user requiring formalisation to preclude incorrect use of statistical data;*
- *design of intelligent dissemination systems;*
- *design of instruments, such as interfaces between statistical software packages, expert systems, and computer-assisted learning concepts;*
- *practical experiments such as studying the application of existing expert systems in fields which have already undergone significant development, e.g. compilation and management of questionnaires, encoding of replies, data validation and estimation of missing data.*

Evidently, the programme was biased heavily towards the then much attention arousing expert system technology, as the typical artificial intelligence hype fuelled optimistic expectations in knowledge-based techniques in virtually any application domain. Specifically, four development themes were addressed eventually, viz.

- Theme 1: the preparation of complete systems for automated information processing;
- Theme 2: the documentation of data and of statistical metadata;
- Theme 3: access to statistical information;
- Theme 4: forecasting.

From all research proposals submitted, seven projects (two for each theme except theme 3) have been elected for funding, with actual work commencing during 1990.

From an operational point of view, the most rewarding projects were those of theme 1. The projects CASIP [Saris *et al.*, 1992; Aluja-Banet *et al.*, 1993], dealing with family budget survey processing, and ESIA [Volle, 1992; Augendre and Hatabian, 1992], aimed at the expert system-aided generation of customized business reports, brought forth practically usable software systems. Both domains feature quite clear-cut operative targets, and – as there is much software engineering experience in devising and implementing self-contained software projects applying established methodologies – these projects' success is not really a surprise. Especially the CASIP system is a closed one, comprising all data processing components from data capture (by providing an electronic housekeeping-book to record expenditures interactively) through an internal data management system down to a tabulation and statistical analysis and presentation module without depending in its functionality on any interfaces to external software components. Despite the apparent success of CASIP and ESIA, both systems demonstrate compellingly how far current methodology is away from a general-purpose solution for devising encompassing (instead of sectorally isolated) statistical information processing systems.

Although the terms "metadata" and "meta-information" have been around for quite some time (cf. Subsection 1.4.1), the theme 2 projects of the DOSES programme probably established this topic as mandatory for any ensuing serious discussion about statistical information systems. Contrary to DOSES theme 1, this theme's projects did not bear practical software packages; however, the propotype systems devised and implemented emphasized the necessity of metadata management and, moreover, highlighted the vast potential of metadata in automating statistical information processing chains, particularly with respect to improved data access and data management.

One of the theme 2 projects, termed "Expert Interface to Statistical Information" (EISI) [de Vaney *et al.*, 1992; 1993; Lamb, 1993] considered specifically capture and provision of metadata necessary to describe the contents of an official statistics database holding typical survey data, following mainly the suggestions and terminology of METIS (see Subsection 1.4.1). In addition to its overall GUI-based hypertext interaction mode, the EISI approach supports a problem-oriented utilization of metadata by so-called usage scenarios encoding procedural templates to access and combine metadata chunks to effectively scrutinize the statistical information of a given database. However, EISI does not provide any access to primary data and, hence, cannot trigger actual data access or data manipulating operations.

The second theme 2 project, "Modelling Metadata" [Darius *et al.*, 1993], focused more on formal metadata structures [de Feber and de Greef, 1992], seeking to devise a general metadata language amenable to incorporation into statistical information processing systems such that – by augmenting elementary database and statistical analysis operators – especially data transformations in the widest sense can be controlled actively by metadata in a transparent way. With respect to macrodata management, the envisaged methodology comprised both summary value *and* summary type management on account of operational metadata elements attached inherently to conceptual data objects containing either of micro- and macrodata (so-called "*md/ds*", or metadata/dataset, structure). In doing so, metadata contribute not only to a more responsive, context-sensitive system behaviour but especially to safeguard semantic consistency of data manipulations and data transformations by preventing unsound operations. As the project stressed the operative usage context of metadata [Froeschl, 1992b; 1993; 1996], special emphasis was laid on the analysis of a generalized statistical information processing model [van den Berg et al., 1992; van den Berg and de Feber, 1992] (see also Subsection 1.5.1) to infer, where and which metadata is generated by and fed back into the process. As a by-product, project work made it clear that traditional views of metadata (cf. Subsection 1.5.2) are much too narrow as these usually have been confined to more or less extended data dictionaries providing additional textual connotations at best. Apparently, proper handling and interpretation of data does not only rely on elementary data descriptions like variable names or codebooks but depends on background information as well (cf. [Richter, 1994] arguing, with respect to the compelling example of production statistics, that productivity is always measured relative to theoretical models seldom set out in detail although this model knowledge is crucial in interpreting data aggregates).

Within the "Modelling Metadata" project the pragmatic aspect of metadata in supporting the harmonization of datasets originating from different sources but relating to the same subject domain – specifically, labour force statistics – was pursued further leading to yet another software prototype [Grossmann and Froeschl, 1992; 1994], dubbed "DÖS'chen", which also explored the possibilities of using metadata to arrange a distributed multi-server approach to data integration respecting the grown data organization structures of participating statistics data servers [Froeschl et al., 1996]. This prototype is described to some detail in Section 2.7 with examples of internal metadata structures reproduced in Appx. C. Furthermore, the DÖS'chen development provides the point of departure for the considerations expounded in Chapter 2.

The ATIIS project [D'Aubigny et al., 1992; 1993], being the only one under the heading of DOSES theme 3, access to statistical information, investigated a way to assist data consumers in finding data, getting at both data and metadata, and supporting data analysis in a self-contained statistical working environment providing automatic analysis management (such as protocolling [Cowley and Whiting, 1985; Froeschl, 1990; 1992a]) and process reproduction features; system design was based on available tools for object-oriented data modelling, user interface building, and statistical analysis (the *New S* system [Becker et al., 1988]). The goals of the ATIIS project have been very ambitious; thus, in spite of the fact that the system could be worked out only partially, its overall design is no doubt a very promising one and comprises a lot of innovative features (not the least being its object-orientation) pointing the way to a truly new generation of integrated computing environments for use both inside and outside statistical offices.

DOSES theme 4, finally, stressed the growing importance of forecasting models especially in governmental policy making. One of the systems, LIKELY [Talbot et al., 1992ab], a prototype system based on influence diagrams, was aimed at the automated monitoring of forecasts providing interactive graphical tools using judgemental expert forecast knowledge encoded in a knowledge-base. Mainly theoretical in orientation (bringing forth a series of very stimulating research papers), the project suffered from the difficulties encountered in merging theoretical and applied research threads. The second forecast theme project of DOSES was AMIA [Ollivier et al., 1992] targeted at the computer-aided management of large simulation models helping forecasters to construct and adapt simulation models (in intricate technological development domains like energy or water supply). While the projects of themes 1 through 3 center around a coherent set of deeply interrelated ideas, both projects of theme 4 are somewhat outside the main line of DOSES research in that they are capitalizing on developments and research in econometrics, forecast model building, belief propagation, etc.; although the devised systems necessarily have to be fed with (econometric) data down-loaded from statistical information systems, their design and development give rise to yet another kind of self-contained working and modelling environments, the interfacing of which to more general statistical information systems seems not to have been addressed at all.

In evaluating the DOSES programme [EUROSTAT, 1993c] consensus has been achieved that, despite the limited financial resources available, research has been successful. On condition that no ultimate solutions could have been expected given the state of the art when research projects commenced, the programme helped to establish a research infrastructure for the area of statistical information systems in official statistics for the very first time; moreover, many new ideas have found their way into the general discussion permeating further research and creating a stimulating background of yet more promising developments. A major effect of DOSES certainly must be deemed the broad recognition of metadata and the increasing attention dedicated to the role of metadata in all stages of statistical information processing. Conversely, the rather narrow perspective of DOSES on expert system methodology might be assessed critically and must be considered as mistake a posteriori. Likewise, the programme failed to recognise the true importance of telematics and distributed systems whence only a single project (ESIA) gave proper attention to the opportunities in data collection via electronic networks.

A major consequence of the DOSES effort has been the 1994 decision by the European Commission to continue this line of research within the Fourth Framework Programme. Under the slightly modified heading of "Development of Statistical Information Systems" (DOSIS), RTD in statistical information systems has been subsumed under the Software Technology branch of the Framework's RTD in Information Technologies, Tasks 1.11 and 1.12 [EC-DG III, 1994]:

1.11 Development of tools for statistical workstations, to provide statisticians with a homogeneous information and software environment to process and analyse statistical data. The research will aim at integrating statistical packages, implementing libraries of reusable statistical tools, improving statistical algorithms, and prototyping tools for automated data/survey processing.

1.12 Development of techniques for the representation, visualization and formal description of statistical data of varying complexity, ranging from numerical data to unstructured text, spatial data, metadata, time series and tables. Further development of numerical and symbolic methods. Development of knowledge intensive data analysis methods, knowledge extraction from statistical databases, and cognitive skills in data analysis. Development of appropriate statistical knowledge-based systems addressing the complexity of statistical data.

Apparently, the research goals of DOSIS are stated in a more neutral fashion as to methodology compared to the preceding DOSES programme. Practically, three major thematic areas appear in outlines, viz.

- Area 1: data analysis,
- Area 2: integrated statistical processing systems, and
- Area 3: statistical confidentiality.

Except area 3 which was not on the DOSES agenda at all, areas 1 and 2 roughly comprise the former DOSES themes 3/4 and 1/2, respectively, and clearly indicate the persisting interest to take up the DOSES line of research again. At the time of writing, nine project proposals – including two telematics programme projects – are considered for funding, most of which are tightly geared to metadata methodologies although the scope of topics covered by these proposals comprises no less than forecasting, data mining, analysis support for data with complex structure, database searching facilities using intelligent retrieval filtering methods, formal management of data harmonization in distributed environments, integration of statistical meta-information and the statistical (micro-)data production process providing data documentation necessary for data aggregation and dissemination, electronic data reporting and capturing based on electronic data interchange formats, development of statistical client-function server networks for sharing software components among statistical offices, and safe data provision by advanced confidentiality measures in data dissemination. Actual project work commenced, by and large, in late 1995/beginning of 1996.

In addition to this research in advanced statistical information processing methods and systems funded by the European Union there will, of course, continue to be both academic research in and applied development of statistical information processing methodology at universities, research centres, and institutions concerned with actual data production, dissemination, and analysis, all of which keeping on to exert major and vigorous influences on the discipline. However, as statistical data modelling and the design and implementation of statistical information systems is evolving from its infant stage into a (information systems) branch of computer science in its own right, the constant in-flow of ideas and innovations will face an increasing back-flow of domain-specific results and solutions, especially with regard to data integration and distributed aggregate data processing in large-scale heterogeneous commercial and administrative information networks.

1.5 Statistical Metadata Management

At the end of this century, mankind certainly does not suffer from a lack of data; quite on the contrary, it is the amazingly low rate of collected data ever getting analyzed that causes concern. A possible explanation for this malady could be, in part at least, a serious lack of information *about* data which impedes both finding relevant data in the huge piles of numbers allocated and investigating it sensibly. In

1.5 STATISTICAL METADATA MANAGEMENT

different words, a tangibly improved management of metadata is in place lest our valuable data resources turn into what is called despicably "data cemeteries". In fact, the survey of statistical data modelling approaches and research programmes aimed at advancing statistical information systems technology pointed out that statistical data processing more and more is augmented with statistical *metadata processing* to support data access, retrieval, transformation, or even analysis by inference-statistical methods. In the narrower domain of statistical database systems, metadata is seen as a means to reconcile data and data documentation in electronic formats, and to preserve data contexts by tightly packaging data and metadata. However, a well-considered approach to metadata processing presupposes a profound analysis of statistical information processing, only parts of which have been undertaken to date, and with different and, apparently, not yet converging goals in mind. Entirely missing, unfortunately, is a coherent *theory of metadata* on the grounds of which an orderly information system development could proceed. Without pretending to give a conclusive account of metadata methodology, the following paragraphs make a try to sketch a fairly general framework of such an emerging metadata theory to set the stage for the specific approach to statistical information processing promoted in Chapter 2.

1.5.1 A General View of Statistical Information Processing

From a subject-matter oriented point of view, statistical information processing rests on two complementary pillars: more or less general *structure models* for the representation of statistical data – as they are scrutinized and developed in statistical data modelling – and more or less general *processing models* for the organization of typical statistical data processing sequences, into which structural models are embedded and from which the requirements of structure models are derived after all, respectively. Particularly in the procedural respect, traditional data models have been called into question mainly because of their insufficient capability to preserve data contexts or, in other words, because they lack any tangible metadata management. As a matter of fact, structure models must be prepared to account for a variety of data processing aspects such as production, storage, aggregation, retrieval, and – last but not the least – statistical data analysis of both micro- and macrodata. Apparently, such a comprehensive approach to statistical data modelling presupposes a continuous and coherent data description which can be gleaned from a thorough system analysis of statistical information processing only.

Perhaps a major stride in this direction has been achieved with the preparatory work on metadata by van den Berg *et al.* [1992] which brought forth a phase model of statistical *survey* processing largely independent from a particular participator's view of process and data organization (thus contrasting, for instance, with Sundgren's [1973; 1991; 1992; 1993; 1994] analysis which restricted its focus to a statistical data producer's reception of data processing):

- Essentially, this phase model comprises four cardinal processing stages of survey processing, viz. *conceptualize, obtain data, analyze, interpret*, such that in between the stages there are placed "information objects" conveying both data and data description, in terms of metadata, to the subsequent stage(s). In addition to more conventional data descriptive elements, the metadata concept has been extended particularly to comprise a survey's *goals* as pragmatic metadata component of information objects.
- The phase model emphasizes the functional role of metadata in survey processing, that is: where is metadata used in the processing sequence, which metadata must be available, where is it obtained from ? Investigation of these questions led up to a (data) production-oriented compilation of metadata elements accumulated and passed on from stage to stage in information objects.
- As this research in metadata was a genuine part of the *Modelling Metadata* project of EUROSTAT's DOSES initiative (cf. Subsection 1.4.2), the phase model has been tuned to metadata relevant for a single, self-contained statistical survey and its descriptive and inference-statistical analysis. Apparently, this is but a rather specific case of general statistical information processing.

Thus, broadening the view on metadata in statistical information processing, three aspects specifically must be taken into account:

- Usually, data is not collected for a single purpose (although there might be a single driving motive to set out for data collection); rather, it is deliberately open to *data sharing* [French, 1991]. Especially in scientific research, comparably few data sources feed a multitude of sometimes not even related research projects. As a consequence, in many cases data collector/producer and data consumer might be different persons or institutions. This, in turn, calls for a *generic* processing model defining the baseline of metadata requirements of statistical information processing.
- Furthermore, and especially in official statistics, data is procured by established production schemata with, in general, few and minor changes between repetitions. The repetitive nature of data production evokes a hierarchy of datasets necessarily reflected by the logical organization of metadata – as depicted exemplarily by Sundgren's metaobject graph (cf. **Fig. 1.3.2–8**) – describing coherently and consistently the stock of a data producer's, or data provider's, data. Typically, the dataset hierarchies created by production schemata are interrelated additionally by shared definitions, nomenclatures, sampling schemes, etc., at least within individual data producers; in a range of domains nomenclatures are standardized at national and, increasingly, at international levels. For apparent reasons, these semantic interrelations between datasets ought to be represented faithfully by metadata.
- Frequently, data originating from different sources, be it data producers or data production schemata, need to be integrated, or – more neutrally – interlaced *statistically*. To this end, of course, metadata must be interlaced either, presupposing, in turn, a kind of standardized data documentation framework all dataset, or data source, descriptions are obliged to refer to.

With respect to data interlacing, two different kinds can and must, in fact, be discerned, viz. *data integration* and *data harmonization*. While data integration, used with this more specific meaning now, addresses the aspect of interlacing, in a meaningful way, data from different subject domains (for instance, combining production statistics with social statistics), data harmonization is concerned with the aspect of interlacing, in a meaningful way either, data from the *same* subject domain though from different data sources (for instance, labour force statistics from different countries); naturally, both aspects may also overlap occasionally. Despite the somewhat deceptive similarity of both aspects, each of them really implies rather divergent consequences as to metadata requirements, viz.

- in case of data integration, the main issue is the composition of a joint stochastic model out of given overlapping partial models, where "overlapping" basically refers to statistical model structures (such as common marginals or factorization assumptions) on account of which datasets might be fused analytically;
- in case of data harmonization, the main issue is the assurance, or achievement, of semantic coincidence of the datasets to be combined which, essentially, refers to the coherence of meanings as determined by the terminological frameworks of the respective datasets' production schemata (including sample design, measurement methods, stochastic models, etc.), concept definitions (units, populations, etc.), and nomenclatures.

Thus, in short, while data integration mainly depends on statistical (stochastic) criteria, data harmonization mainly depends on taxonomic criteria. In particular, data harmonization presupposes a succinct documentation of a dataset's semantics in order to locate the "semantic distance" between different datasets; a (formal) scrutiny of this distance determines if and how data harmonization can be accomplished for a given set of datasets.

In a generalized view, hence, the dynamic-pragmatic aspect of metadata with respect to a closed processing model restricted to a single, isolated survey is of less importance; on the contrary, the statistic-semantic aspects of metadata in a wide sense have to be emphasized to elucidate the *meaning* of statistical quantities encoded in observation datasets independent of the local contexts and dynamics of individual processing threads. Put differently, the central question is, how can a data source or a dataset be documented formally such that *any* kind of subsequent statistical data processing – even one totally unforeseen – gets optimum support as to data context ? And, what specifically has to be provided to facilitate data integration and data harmonization ?

Having in mind data integration and especially data harmonization, one might be tempted to plead for standardization to get rid of semantic mismatches but, conversely, this probably would be a grossly misleading advice because (i) future requirements cannot be anticipated reasonably and (ii) this would tremendously depreciate the practical value of existing data failing to meet the proposed standards. "*Heterogeneity in data ... is a fact of life.*" as French [1991] has pointed out; moreover, semantic coincidence by no means is a direct consequence of nominal standards – from a statistical point of view, harmonization aims at the true meaning of statistical figures irrespective of superficial identities or literal conformation to standardized definitions which may or may not conceal real semantic discordances. Therefore, a more appropriate response to data heterogeneity consists in exploiting information technology, particularly metadata management, to facilitate the interlacing of empirical data and statistical information processing in general.

Taking up again the distinction of structural and processing models in statistical information processing, there obviously are two sorts of metadata, viz. *static* and *dynamic* metadata (cf. Subsection 1.5.2). The former are coupled quite statically with primary, or first-order (observation or aggregate) data and reside – physically packaged with primary data – in statistical symbol storage; naturally, static metadata belongs to the structural model of statistical information systems. Procedurally, of course, static metadata must be created somehow, and will be retrieved and used during the data consumption phase (*analyze* and *interpret* phases in van den Berg et al.'s [1992] model). Thus, from a process-oriented view, dynamic metadata are converted ("frozen") into static metadata during data production, and re-converted to dynamic metadata during data consumption; by and large, van den Berg et al.'s [1992] information objects could be viewed as containers carrying dynamic metadata. Moreover, if a statistical analysis or data aggregation yields an output fed back to the permanent storage of the information system, dynamic analysis metadata is again converted to static metadata attached to the macrodatum generated. Thus, in a sense, dynamic metadata provide the backbone of process management in a metadata-driven statistical information processing system.

Emphasizing the static representation of statistical information (that is, both data and metadata) and swapping the focus from processing phases to information objects, **Fig. 1.5.1** depicts a general *data flow scenario* of statistical information processing. The double horizontal line separates logically the functional spheres of data production and data consumption implying an individual data consumer's point of view who may face, in general, a multitude of data producers. The rectangles denote states coinciding, essentially, with information objects; arrows denote processing phases. The shaded area indicates the portion of the data flow scenario constituting the actual statistical information system. Schematically, starting with given objectives, data producers – for instance, national statistical offices – develop their own conceptual views of investigated domains whereupon particular data production schemes are devised to collect, or obtain, data. Since, in general, the dissemination of raw data is economically or technically inefficient or even restrained by law, a pre-processing stage is appended converting observation data (microdata) into an intermediary general storage format called *mesodata* (see below). Using a business metaphor, mesodata is a kind of semi-finished product open to a multitude of finishing procedures. Conversely, a data consumer typically investigates a specific statistical problem and will therefore develop a specific conceptual view, too. Since high-quality data production is costly, a data consumer will look for useful data available at accessible sources provided that conceptual views converge. To this end, a comparison of conceptual views must be carried out which, apparently, capitalizes on static metadata supplied by data producers. Depending on the outcome of this comparison, the data consumer will decide to use

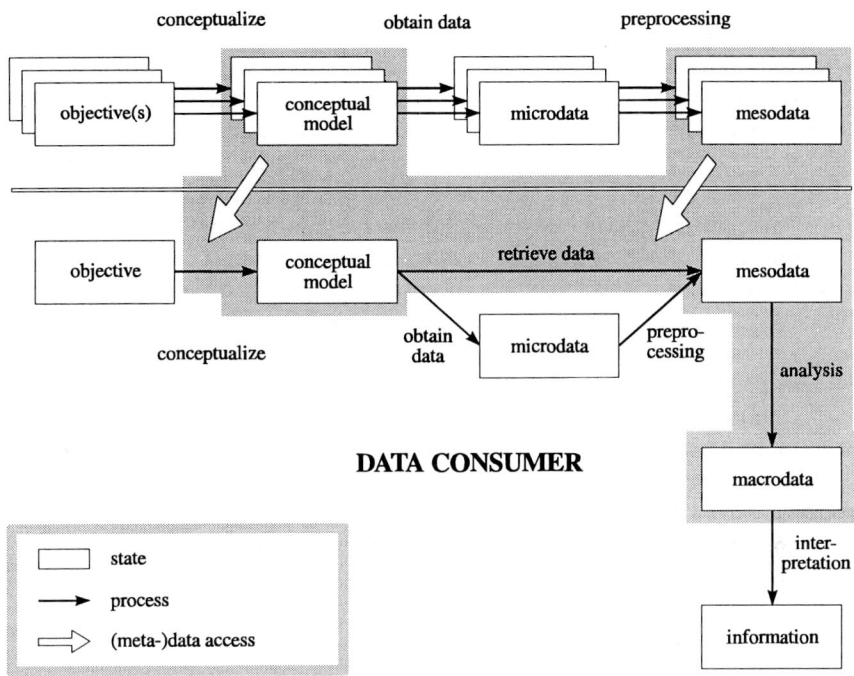

Fig. 1.5.1: A General Statistical Data Flow Scenario

the data producers' off-the-shelf data or, instead, collect data on his own; also a mixed strategy is conceivable, of course. If data readily available is going to be used, actual data retrieval is performed such that attached metadata is supplied concomitantly. Hence, from a metadata perspective, any data access is in fact preceded and accompanied by metadata accesses. In doing so, the statistical information system acts as a symbol storage platform to which all processes – viz., data producing as well as data consuming ones – are coupled. By the way, this data flow scenario highlights that the very same metadata may serve rather different functions: for instance, for a data producer the metadata describing the objectives of data collecting will provide a fundamental backbone of data production (say, survey organization, data tapping, editing, etc.); for a data consumer, this metadata will indicate the usefulness, resolution, accuracy, etc. of stored data.

Although there is common agreement upon the general process structure of statistical information processing as shown in **Fig. 1.5.1**, the opinions about structural models seem less conforming; all recent practical information system developments, however, favour – mainly for formal reasons – some uniform statistical data model (for instance, *statistical entities* of the MEFISTO approach, cf. Subsection 1.3.2; the *block* model of GENESIS, cf. Subsection 1.4.1; or the "tandem" *md/ds* structure devised in the *Modelling Metadata* project, cf. Subsection 1.4.2) around which system functionality is organized: apparently, as the tight connection between primary data and associated metadata must not be injured at any data manipulation or dissemination stage, a structural model as parsimonious as possible is advantageous. This, in turn, is largely determined by the assignment of the structure conversion (from micro- to macrodata) stage – cf. **Tab. 1.3.1–3** – to the data consumer or the data producer side of the processing model. If the data producer is committed exclusively to structure conversion – as it is in fact suggested in **Fig. 1.5.1** – a single generalized data structure is fully sufficient (this is the strategy adopted in the GENESIS system, for instance); if both producers and consumers are permitted to carry out structure conversion (alias aggregate formation, in this case) at least two structures must be provided, viz. one for microdata and another one for macrodata. However, this unpleasant distinction can be easily done away with – in the dominant share of standard cases at least – resorting to a kind of "intermediary" *pre-macrodata* representation (cf. Subsection 1.3.1) aimed at keeping all information contained in microdata, yet coinciding formally with macrodata. Quite naturally, this hybrid structure might be termed *mesodata* as it truly realizes a data representation half-way between information-equivalent microdata representation and not yet macrodata representation. Formally, letting a microdatum consist of n records or cases, $\{c_1,...,c_n\}$ such that, for $1 \le i \le n$, $c_i = \langle id_i, \mathbf{q} \rangle$ is a pair composed of a record (case) identifier id_i (observation argument) and a data vector $\mathbf{q}$ (comprising observation, factor level, etc. values) where $id_i \in \mathbf{ID}$ and $\mathbf{q} \in \mathbf{Q} = \{\mathbf{q}_1,...,\mathbf{q}_k\}$; **ID** denotes a (not necessarily finite) set of unique names and **Q** is the universe of *m*-tuples of possible outcomes of measurements gathered from empirical observation of observables $O_1,...,O_m$ on case (sampling or population unit) i. Note that, contrary to standard function definitions, $id_i = id_j$, for $i \ne j$, is perfectly admissible, although observation (function) values may differ provided that additivity with respect to all observables involved is assured. Now, such a microdatum is simply converted to a mesodatum by grouping cases by shared observation values, that is, by generating the set $\{y_1,...,y_k\}$ – which is equivalent to a generalized, *m*-dimensional "primitive summary table" [Özsoyoglu et al., 1989] – such that, for each $\mathbf{q}_j \in \mathbf{Q}$, $y_j = \langle \mathbf{q}_j, \{id_i \mid c_i = \langle id_i, \mathbf{q}_j \rangle\} \rangle$. Obviously, given a statistical aggregation function $f \in \Phi$ (with Φ denoting a repository of aggregation functions) with co-domain $\mathcal{N}_f$, macrodatum formation is facilitated by the mapping

$$f : 2^{\mathbf{ID}} \to \mathcal{N}_f$$

(for instance, f could be the cardinality function with co-domain $\mathcal{N}_f$ of all nonnegative integer numbers). By the way, this mesodata structure fits neatly the concept logic approach proposed by Catarci *et al.* [1990] in that it conforms to "basic" complex aggregates (cf. Subsection 1.3.2).

As mesodata is a very versatile data structure reconciling the requirements of both microdata and macrodata, it ought to be considered as a first-class candidate for metadata-oriented structure models in statistical information systems (which is not to say that in special cases more semantics might have to be represented than provided by the plain mesodata structure). Quite favourably then, in fact, structure conversion can be assigned to data producers throughout, as this no longer implies a loss of microdata information while still maintaining the clear functional separation between data production and data consumption in a statistical information (database) system.

1.5.2 Metadata: Roles, Typology, and Representation

As highlighted by the discussion in preceding subsections, present and, even more so, future statistical information processing is going to be based on *metadata* concepts. Although there is little controversy about the increasing importance of metadata in statistical data management, at the same time it cannot be overlooked that the very term "metadata" is associated with quite a lot of differing interpretations and apparently means different things to different people except that "... *there is general agreement that most* [statistical] *databases suffer from incomplete metadata*" [French, 1991]. Therefore, a short review of metadata definitions seems in place before exploring the concept in closer detail.

Many of the metadata definitions are, in fact, rather superficial and, for this reason, not very useful. For instance, one definition assumes that "... *metadata is data describing the content of a database*" [Lübbe, 1990] while another states that metadata "... *may be loosely defined as data about data rather than data itself*" [Lundy, 1984], neither of which is very illuminating. Somewhat more precisely, Costa *et al.* [1986] define:

> "A dictionary is an ordered collection of references to terms relating to a given field or a given activity, defining their meaning and describing their uses. A data dictionary, naturally, contains «data about data» or, as is commonly said, «metadata»."

Similarly, Stephenson and Clowes [1989] note:

> "A problem facing users of statistical databanks is obtaining secondary information relating to the statistics. Such information includes methods of collection, problems with individual observations, precise definitions and the conceptual framework underlying the data."

A more normative aspect is stressed implicitly in the definition attempt by Drewett *et al.* [1989] delineating meta-information by what currently is missing in statistical information systems in stating that

> "... the empirical researchers suffered from ignorance about the context of the data being utilized. When it was available, information about information, or meta-information, reflected the information provider's priorities. These may not have coincided with the information user's needs."

With respect to the interpretative interface function of metadata these authors conjecture furthermore that "[An] *accessible, comprehensible and standardized source of meta-information, incorporating the experience and knowledge of both producers and users of data series, would form the basis of such an interface.*"

Cubitt [1990] points out that metadata in fact is indispensable for both retrieval and analysis of statistical data: "*Retrieval: how to find data, is it available, quality, appropriateness for certain types of analysis, combination with data from other sources, ...*" and "*Analysis: knowing what processing can be validly undertaken on the data and what data can be compared and combined.*" This definition emphasizes the functional role of metadata; similarly, also Ghosh [1988] distinguishes functional and operational properties of metadata where the former refer to data usage while the latter describe structural elements of primary data (somewhat oddly, Ghosh's metadata definition actually has in mind what more accurately should be called summary set, or summary statistics, as used, for instance, in the MIMAD model or in Chen *et al.*'s [1989] category format; cf. Subsection 1.3.2).

From a database perspective, the definition recommended by McCarthy [1982] is among the more useful ones in that it states:

> "*Metadata is systematic descriptive information about data content and organization that can be retrieved, manipulated, and displayed in various ways. Metadata may be simple and unstructured, such as a typewritten narrative describing a datatape, or structured and complex, such as an active machine-readable DBMS dictionary used to control multiple databases.*"

In summing up, neither of these definitions is really satisfactory since each of it addresses some specific facet of metadata only. Moreover, previous considerations (cf. Subsections 1.5.1 and 1.4.2, respectively) suggested a more comprehensive extension of the notion of metadata. As a matter of fact, a richer picture of metadata is obtained by simply listing what might be subsumed practically under this term; for instance, French [1991] enumerates

- general identifying information such as: who collected the data, when the data was collected, and where the data was collected;
- characteristics of the device(s) that collected the data;
- transformation operators (e.g. calibrations) applied to the data;
- programs used to manipulate or modify the data;

- models used in processing or interpreting the data;
- documentation relating to or derived from the data along with technical manuals related to the data source and relevant publications, reports, and bibliography.

A yet more informative picture is gained, as shown in **Tab. 1.5.2**, by introducing a coarse four-fold subdivision of metadata domains according to major functional areas opposed to a couple of descriptive dimensions relevant for both, produced primary data (static view) and the analysis of primary data in the data consumption stage (dynamic view).

Table 1.5.2: Functional Areas of Operative Metadata vs. Descriptive Dimensions

Description dimension	FUNCTIONAL AREA			
	Production structure	*Taxonomic structure*	*Temporal structure*	*Spatial structure*
organizational foundations	*	*	*	*
coverage		*	*	*
periodicity			*	
reference period			*	
concept definitions		*		
classifications		*		*
observation variables		*		
production mode	*			
sample design	*	*	*	*
response characteristics	*			
statistical methodology	*		*	*

This descriptive view of metadata can be contrasted with a more usage-oriented view [Hand, 1993] stressing, inter alia, data interpretation, data validation, imputation, guidance in statistical analysis (reminiscent of developments in so-called statistical expert systems), but also data production (note that in systems like BLAISE or CASIP metadata precede actual observation data) as well as data consumption in general – for one thing, locating a specific portion of data is not so easy anymore in a time when data is accumulated at gigantic rates ([Waldrop, 1990] reports some 10 to the power of 12 bytes a day, for 15 years, which are scheduled alone for NASA's Earth Observing System).

This functional attitude towards metadata is also emphasized by Bretherton's [1994] "reference model" proposal replacing the rather sterile attempts at defining metadata by relating the metadata concept to symbol interpretation contexts:

> *"Metadata becomes the additional data that must be invoked to implement the change in context. The prefix meta does not attach to the data itself, but derives from the circumstance of change."*

As a matter of fact, inside (local) contexts, there tends to be little ambiguity or it is easily removable; if, however, spatial or temporal "boundaries" intervene, proper interpretation becomes a function of "boundary crossing". More philosophically, the vigorous emergence of metadata could indicate a practical consequence of revising, in a sense, the purely empiristic programme claiming that measured quantities – primary data – by themselves give a faithful image of the "true" reality without any bias of interest. Metadata can be seen as a means to handle *semantic relativism*, or *pluralism*, accepting that reality is a little more than just what the case is (Wittgenstein) by re-embedding empirical evidence into a narrative context, attributing the "bare facts" (= numbers) of observation the function of relative co-ordinates marking mental locations on the cognitive road maps of positivist discourse which – compared to plain prose – permit, of course, definitely more precise referencing than fuzzy linguistic codes ever could.

The development of a practical and effective metadata methodology in statistical information processing presupposes some elementary distinctions helping to sort out basic requirements and types of metadata useful in meeting them. As there is no generally accepted metadata taxonomy, the following three-way distinction should be considered as a first rough approach which certainly needs further elaboration. Basically, it is suggested to discern the following broad metadata categories:

- *Off-line metadata*: this category encompasses all traditional, printed matter data documentation usually separated from stored primary data; in general, this type of metadata is termed "meta-information" because usually it is available as text in clear describing, more or less exhaustively, a vast variety of data production aspects such as definitions of sampling units and populations, data reporting procedures, sampling methods and sample designs, legal bases of data collection, subject domain concept definitions, statistical grossing-up methodologies to estimate population totals from sample figures, nomenclatures, indications of data validity and reliability, measurement accuracy, known or supposed sources of bias, background comparability, etc.
- *On-line unformatted metadata (metatexts)*: although being equivalent in content to off-line metadata, texts stored on electronic storage media are amenable to tremendously improved symbol processing methods; in particular, metatexts offer better indexing capabilities facilitating information retrieval and cross-referencing (hypertext-structures). Moreover, on condition of suitable symbol organization regimes, descriptive metatext segments (documents, reference items, explanatory entries, footnotes [Silver, 1993], etc.) can be associated formally to stored primary data such that on-line data documentation systems operate as *data browsing* systems ("datiographies") as well. Conversely, once primary data is retrieved, the links established between data and descriptive

metadata can be used to export primary data together with all formally associated meta-information.
- *On-line formatted metadata*: contrary to unformatted metadata, formatted metadata refers to formal data description elements closely associated to primary data such that any operation applied to primary data involves the attached metadata either and vice versa. More specifically, formatted metadata is used to control information processing threads *by machine*. This principle might be expressed by the pithy "formula"

statistical information processing = statistical (data + metadata) processing

From an electronic data processing point of view, metatexts represent the category of unstructured *descriptive* metadata elements (sometimes called "guide metadata") whereas formatted metadata represent *operative* metadata (sometimes called "control metadata") in that this metadata enters computations and inter-machine communication just as primary data does, that is operative metadata is algebraically *transformed* by computations carried out by the statistical information system; in this sense, data processing is strictly paralleled by metadata processing. A distinctive criterion to tell metatexts from operative metadata is provided by different symbol structures: while metatexts are organized in terms of separate information structures connected to primary data structures by various links, operative metadata necessarily need to be *integrated* deeply into statistical data models.

Both descriptive metatexts and operative metadata split naturally into three subcategories [Froeschl, 1992b], viz. *taxonomic* or terminological metadata (semantic metadata in the narrower sense), inferential-*statistical* metadata, and *technical* metadata. Taxonomic metadata concerns the meaning of terms, concepts, and nomenclatures used in data production schemata, sampling schemes, and population definitions; statistical metadata refers to the stochastic features of primary data (statistical model assumptions, estimation methodology, non-response, etc.); technical metadata covers the more mundane aspects of storage organization, access paths, graphical layouts, and so on. Among these sub-categories, technical metadata are the least controversial and mostly belong to the category of operative metadata (data dictionaries, variable and value labels, missing value codes, etc.). Statistical metadata has been concerned particularly in connection with statistical expert systems [Haux, 1986] where it supplies the main input to formalized analysis strategies [Grossmann and Froeschl, 1988; Froeschl, 1989; 1992c] aimed at both expert and non-expert user guidance [Froeschl and Grossmann, 1988; Dorda *et al.*, 1990; Froeschl *et al.*, 1992; Froeschl, 1995]. In the realm of statistical information systems, statistical metadata plays a major role especially in data integration and in estimation methods used to gross up sample data to population figures; apparently, the development of statistical information systems will benefit from the experience particularly with operative inferential-statistical metadata gathered in statistical expert systems research. Taxonomic metadata, finally, have received little attention yet although they occupy a key role in data harmonization as will be shown in Chapter 2.

For practical purposes, the top-level distinction of metatexts and operative metadata certainly needs considerable refinement. For instance, Hand [1993] proposed to distinguish *context-free* from *context-sensitive* metadata in order to isolate operative metadata elements holding independently of specific situations (such as admissible scale transformations) in contrast to metadata elements admitting less generalization. Another distinction of functional relevance is the one between *static* and *dynamic* metadata already introduced in Subsection 1.5.1. A further useful distinction could be achieved by separating *producer-supplied* and *consumer-supplied* metadata; basically, statistical database systems are expected to provide, first of all, producer-supplied metadata but, particularly if the results of statistical analysis can be fed back into storage, it might be interesting to add metadata different from those obtained from producer-supplied metadata by ordinary metadata processing as these enrichments – though possibly restricted to metatexts (remarks, connotations, etc.) – may benefit later data usage.

In present research and development of statistical information systems, heavy emphasis is devoted to metatexts (cf. the survey given in Section 1.4). This is understandable since the introduction of so-called *meta-information systems* capitalizes on well-established and proven information technologies. Conversely, despite the undeniable advantages reaped by on-line data documentation in data access, retrieval, and processing the overall potential of this type of statistical information systems should be assessed reservedly as they will remain limited substantially in functional scope, mainly because of the loose coupling between data and metadata. In particular, metatexts can do no more than, at best, reflect the *between*-dataset relationships (cf. Section 1.3) by establishing the respective interconnections between metatext elements linked to primary data (essentially, this is aimed at in systems like DUVA, cf. Subsection 1.4.1, or EISI, cf. Subsection 1.4.2). The vast potential of operative metadata, however, has not yet been explored systematically although some of the proposed and investigated statistical data models have achieved considerable progress in adapting and extending traditional data modelling approaches to better fit statistical information processing requirements; to date, the most outstanding examples are systems like MEFISTO, StEM, GENESIS, and the *Modelling Metadata* project described in Sections 1.3 and 1.4, respectively. For obvious reasons, the *deep integration* of metadata into the structure models of statistical information processing – as opposed to an *additive* combination of traditional structure models for representing primary data on the one hand and metatext data models on the other hand – calls for a complete re-engineering of both data structures and the operations defined on top of them all of which poses a multitude of technical problems by no means answered finally or satisfactorily by current software technology. This must be deemed the price to be paid for more powerful data modelling concepts prepared to deal adequately with both between-dataset and within-dataset relationships and aimed at giving stronger support to automated and, hence, more efficient and semantically reliable information processing.

Given the premature state of development in operative metadata modelling it is hardly surprising that there is not yet any compulsory or proven methodology to follow. A conceivably prospective point of departure could be Sundgren's *alfa-*

beta-gamma-tau analysis (cf. Subsection 1.3.1) which is quite easily extensible to a general macrodata structure model – like the *statistical entity* or *md/ds* structures – taking care of both the internal data object semantics and the external between-dataset semantics as indicated in the metaobject graph of **Fig. 1.3.2–8**. Apparently, this approach suggests an OO design as this facilitates, by virtue of the object concept, a natural representation of within-dataset relationships in terms of internal object structures, and of between-dataset relationships by references to static metadata defined globally, that is outside local objects [Grossmann and Froeschl, 1992].

1.5.3 An Effective Approach to Data Harmonization

Once the importance and potential of operative metadata are recognized, it is just reasonable to explore what can be gained practically by developing a definite operative metadata methodology. In continuing the line of research having commenced with the *Modelling Metadata* project described in Subsection 1.4.2, the aspect of data harmonization – as defined in Subsection 1.5.2 – is scrutinized further (in what follows, the terms 'harmonization' and 'semantic data integration' will be used synonymously). This choice is motivated particularly in awareness of the pressing need to support data harmonization tasks as they occur frequently, for instance, in a EUropean context where, as a rule, national statistics of the Member States must be consolidated, within rather tight due dates, to meaningful EU-wide aggregates involving huge amounts of figures. More specifically, the role of operative taxonomic metadata in the automated derivation of generalized statistical tables (macrodata) will be investigated such that aggregates can be specified intensionally over a given space of objects nameable in terms of a formal grammar composed of defined statistical nomenclatures, and actual data retrieval and data transformations from source data into the specified macrodata structures are accomplished mechanically to the farthest possible degree on account of established formal metadata relations denoting syntactically the subject-matter relationships holding both within and between extensionally stored datasets. Moreover, special attention is dedicated to the algebraic closedness of the devised system of operators effecting data retrieval and data transformations, thus enabling multi-stage processing sequences.

Coarsely speaking, the proposed metadata methodology proposed in Chapter 2 is based upon a logical (that is, physically fictitious, or *virtual*) super data model incorporating a system of *federated heterogeneous* data holdings interconnected by digital tele-communication links; in contrast to past practice, heterogeneity is not assumed to be restricted to hard- and software but, properly reflecting the actual situation, is extended to include the inherent heterogeneity of local statistical data logics as well (in fact, technical data heterogeneity appears to be the lighter problem). It is particularly the specific approach to handling semantic heterogeneity which distinguishes the devised system from other attempts to achieve remote access to data (for example, via 'ftp' or WWW-services), or effective database system interoperability [Sheth and Larson, 1990; Daruwala *et al.*, 1995]. A basic premise of the proposed structure model of data representation is the existence of something like "relative structural stability" of data generating sources in spite of the real dynamics of empirical data collection and the underlying theoretical frameworks. In particular, the envisioned model assumes a taxonomically "semi-closed world" in that it is possible effectively to establish a structural description of empirical observation, or measurement, systems of more or less global validity and pragmatic agreement which can be maintained – perhaps by slight amendments and adaptations – for an elongated period of time though being subject, of course, to interpretative shifts and an inevitable continuous historical reassessment.

Formally, the model draws deliberately on the concept of *sample space* worked out in statistics (particularly, probability theory) as this is a well-proven methodology to capture the semantics of random experiments specifically suited for statistical modelling and inference. Although sample spaces are designed usually with a specific and more or less short-lived research interest in mind, nothing opposes the arrangement of unified, or *harmonized* sample spaces to which given sample spaces can be mapped formally, without any loss of empirical information. Basically, harmonized sample spaces are introduced as abstract "context-mediating" devices fulfilling the sole purpose of providing common mapping platforms to make explicit the structural differences of contributing source sample spaces. With respect to data harmonization, it is the *taxonomic* structure of source sample spaces which receives primary interest, and effective semantic data integration reduces to the creation of harmonized taxonomies spanning a super sample space such that all source sample spaces of interest can be "embedded", by formal mappings, into the emanating super sample space. Apparently, this emphasizes the specification of *taxonomic* metadata (nomenclatures, concept definitions, observables, temporal and spatial observation structures, sample designs, etc.) giving formal accounts – metadata functions, in fact – of the relationships between source structures and joint super sample space structures. In other words, source structures (observation sample spaces) are not merely interrelated by a loose reference network of semantic indices but are related formally by taxonomic metadata expressions denoting the source sample spaces' relative semantic positions – *semantic loci* – within the joint super sample space they are contributing empirical data to. This serves as a particularly efficient method of data documentation either, as this semantic mapping needs to be done just once by the data producers, thus preventing the determination of semantic loci over and over again each time data is consumed (however, this argument presupposes a pragmatic invariance of semantic placements or, rather, an additional denotation of a data source's observation pragmatics by either descriptive or operative metadata). In turn, the data producers' responsibility changes in that now data production terminates functionally by providing the semantic placements of produced data relative to supported super sample spaces; any kind of data usage then belongs to the data consumption sphere even if data producers are consuming their own data.

In order to enable an effective statistical information processing, both source and super sample spaces need to be represented at symbol level in a structure model convertible to real software implementation and computation structures. To this end, a specific OO based data model is introduced which offers high flexibility to account for a variety of sample space structures of practical interest. As a matter of fact, a whole *data modelling tool* is developed to define both super sample spaces and the metadata functions defining the taxonomic mappings from source sample spaces to super sample spaces. Unfortunately, this incurs quite an extensive set of basic definitions in order to build up incrementally all data model components required; although stated verbally, these definitions – mostly contained in Section 2.2 – should be read as (abstract) data type declarations omitting, however, most of the operation definitions (an excerpt of elementary operation definitions is contained in Appx. A). A particular virtue of the devised structure model certainly is the coincidence achieved between stochastic sample space structure, taxonomic (nomenclature) structure, and database model structure in a unifying framework.

This framework, in turn, builds the platform for both conceptual and structural data access and aggregate formation, as described in close detail in Chapter 2, by exploiting the functionality of taxonomic operative metadata in a formal database query reduction system embodying a metadata calculus operating on the super data models defined in terms of the provided data model structures. In stressing statistical data modelling and data management, the proposed software architecture is quite different from usual expert systems although data harmonization – and even more so data integration – cannot, in general, be accomplished without taking recourse to stochastic models and inference statistics (for instance, estimation procedures) based on them. Nevertheless, the devised statistical information processing methodology does not aim at giving support in statistical data analysis; it merely seeks to provide the meta-information needed for any sound subsequent statistical analysis of retrieved data. This implies, by the way, that any statistical information system should also be capable to supply microdata if necessary – and available, of course – as most analytical methods of statistical inference presume microdata or produce possibly biased results otherwise (however, even microdata access may resort to statistical models in order to prevent the risk of disclosure of confidential data).

To justify the expense in setting up a statistical information processing architecture as suggested in Chapter 2, two specific examples will be used to both

- argue that the nature of actual real-sized data harmonization problems indeed calls for an approach like the one promoted and
- give evidence that the adopted approach, in turn, meets these real-life problems' requirements.

The first, and larger, example is taken from the official statistics domain and addresses the task of harmonizing European *labour force* statistics while the second, and smaller, example of *tourism* statistics is used to demonstrate a couple of further aspects of data modelling not fitting nicely into the framework of the first example. Both domains are first explored from a subject-matter perspective to reveal the essential prerequisites of effective data harmonization before these are turned into corresponding data models used for demonstrating how statistical aggregates are derivable from formal aggregate specifications.

While both examples mainly seek to highlight different merits of the devised information system architecture in two subject domains kept separate, it should be clear that, without compromise, even a combined European tourism data model could have been arranged merging, for instance, the specific characteristics of national tourism statistics at a European level (cf., for instance, [EUROSTAT, 1993d]); likewise, it would also have been possible to interrelate labour force and tourism applications in a single data model, for instance by including a suitable economic sector statistics linkage structure. However, such an example would have been difficult to present in a coherent way given the limited space available, and would have enforced the suppression of many crucial facets more lucidly expounded in two independent and, hence, smaller application examples. Additionally, setting up this type of integrated statistical data models entails a considerable effort in itself which, to the best of the author's knowledge, has not been tackled anywhere before, despite the apparent utility of such data models. Perhaps the envisaged kind of large-scale data integration is yet to come for two reasons, viz. the emergence of statistical information processing systems providing the required technological support, and the overall re-engineering of data production and dissemination schemes without which these new facilities cannot really be taken advantage of. Anyway, the chosen application examples are intended to illustrate the flexibility of the proposed metadata methodology which certainly transgresses the borders of official statistics, as many problems of data integration and data harmonization reappear, in different guises though, in various domains such as scientific databases and business statistics calling for composing data aggregates from "lower-level" primary data. Evidently, the devised methodology applies to all of these problem areas on condition that the data processing tasks can be fitted into the basic structure model presented. This might turn out increasingly important also in coping with the staggering rate of data produced especially in scientific domains where statistical information systems could be coupled with automated data screening systems used to filter observation data for "interesting" features that probably would escape attention despite their severe implications; for instance, French [1991] cites the perilous hole in the ozone layer that might have been detected a decade earlier if available satellite data would only have been looked on at all.

Chapter 2

METASTASYS
A Formal Metadata Approach

The systems analysis of statistical information processing (comprising data production, data storage, as well as data delivery and consumption) of Section 1.5 leads up to a *grand design* of a generic statistical information system. This design aims, first of all, at a rather general description format – an encompassing language to talk about empirical data, in fact – in order to explicate formally the linguistic structure of the datasets piled up in fairly huge data stocks which, thus, provide a digitized world image, or mirror, open to analytic scrutiny. However, unless condensed appropriately to *supercodes*, or "reduced" statistically [Ehrenberg, 1975], the bulks of data remain rather incomprehensible cognitively. Therefore, data need to be viewed conceptually or, metaphorically, through *statistical windows* expressing particular investigative interests or research questions. In pragmatic terms, hence, a statistical information system behaves as a question-answering device returning answers, that is: statistical aggregates, to specific questions (data views).

Adapted suitably, this statistical question answering complies exquisitely with axiomatic proof systems since the operations of statistical inference (on account of which data are reduced, or aggregated) exclusively draw on *positive* evidence, that is, data really gathered. Obviously, no conclusion is to be drawn from phenomena not (yet) observed. On account of this logical positivism, a special-purpose *calculus* is devised the language of which is providing a (finite) universe of names for statistical (domain) concepts algebraically composed of empirically validated taxonomies; the calculus's inference rules, in turn, induce a (decidable) *data theory* from a set of specific names – the axioms denoting real datasets in the conceptual (global) statistical database – such that each derivable name (that is, concept, or statistical data view) is accompanied by its corresponding statistical aggregate, viz. the numerically transformed datasets represented by the axioms involved in this name's – or theorem's – derivation.

As expounded here, admittedly, the proposed METASTASYS approach is an outgrowth of purely theoretical research; it remains to be inferred from real applications (such as those of Chapters 3 and 4), how far this paper-work proves useful as it stands and where, if need be, amendments are in place.

2.1 Premises and Design Principles

METASTASYS, an acronym for MEtadata-based TAble-oriented STAtistics SYStem, is to be understood as a metadata approach to statistical query processing specifically addressing the problem of large-scale data integration and, more specifically, *data harmonization*. Having in mind the rather urgent statistical data processing demands of contemporary scientific research (such as the Global Environmental Change project) as well as corporate decision making (of, for instance, multinational companies, or governments), the potential benefit of applying rigorously formal meta-information methodology to the design of next-generation statistical information systems is explored. Acknowledging the presence of already huge and yet – at tremendous growth rates – increasing data volumes mostly kept in local, autonomous, and disconnected data holdings, all of which maintaining their own, "traditional" standards as to data models, terminology, statistical and data processing methodologies, software, hardware, and organization structures, the cardinal obstacle to an effective exploitation of these information assets is a serious lack (of supply) of meaningful, consistently interpretable relationships within stored data. Likewise, technology readily enables, by means of global digital tele-communication links already operative to a large extent, the transmission and dissemination of (high volumes of) data at the purely *syntactical* level. As discussed in Chapter 1, the de-contextualization of information incurred from unrestrained shipping of data through space and time calls for conscious counteractive efforts to preserve data context, viz. surrounding data with appropriate interpretation frames, or meta-information, captured in terms of metadata. Moreover, since tracing – or simply keeping in mind – the origin of data becomes, from a global perspective, an almost insurmountable task, these interpretation frames will, in particular, be responsible for effectively achieving *semantic* data integration, that is deciding if data of different origin in fact comply with each other and, thus, can be combined meaningfully in statistical data processing. In different terms, the envisaged semantic data integration approach attempts to establish a *virtual statistical database* composed of hitherto disintegrated (local) data holdings. To this end, METASTASYS is devised as a rather general meta-information framework for statistical data processing which, mainly using symbolic reasoning methods, specifically aims at:

- a *conceptual* data access abstracting from the physical distribution of data (principle of physical data independence admitting navigation-free query specifications);
- a *structural* data access shielding also from the logical data organization as far as it is irrelevant for statistical analysis and model building (logical data independence);

- a convenient tool for *aggregate formation* resolving aggregate requests, if need be, into algebraic and statistical operations deriving specified aggregates from available source data in a mechanical as possible way.

In addition to this predominantly usage-oriented view, of course, the issues of creating and maintaining such an integrated statistical data processing system must be addressed and solved. Just as the system's front end user interface (for data consumption) *appears* as a cohesive, uniform global statistical database, its back end interface has to provide gateway structures flexible enough to really accommodate for a variety of existent data sources and formats lest composing the virtual global database fails because of mere technical incompatibility and/or economic intractability. In general terms, although not fitting particularly well into Sheth and Larson's [1990] taxonomy for federated database system architectures, METASTASYS comes fairly close to what they classify as *single-federation, tightly coupled heterogeneous multi-database system*. Since the specific type of heterogeneity dealt with regards primarily the different *semantics* of statistical data conceptualizations ("ontologies") engendered in different data sources or, technically speaking, component databases participating in such a "federation", with respect to these particular preconditions and requirements a few assumptions and guiding principles are set forth as follows.

The pivotal component of METASTASYS is a generalized statistical data model providing, in fact, a common language ("meta-terminology") for expressing empirical data languages which, basically, consist of the notions and terminologies used in local data sources for the description of sample spaces, experiments, or data generating – usually: observation – processes. In other words, the common language provides a linguistic frame for establishing *shared ontologies*, expectedly powerful enough to (re-) express, possibly after some intermediary but information-lossless transformation steps, virtually any conceivable statistical domain data model. To this end, a couple of structural assumptions is introduced to allow for the functionality METASTASYS aims at; these assumptions support METASTASYS's design pragmatically and could, of course, be replaced with different, or weaker, ones (though some structural assumptions will be indispensable anyway). The resulting generalized statistical data model, nicknamed METAMOD, is a rather straightforward implementation of the statistical systems analysis's findings expounded in Section 1.5.1. In principle, METAMOD is destined to provide a general framework for representing specific *domain data models* (METAMOD instances) each of which, in turn, hosts a variety of data sources to become integrated semantically. Stressing the role of taxonomies in establishing effective semantic data integration, METAMOD centers around the notion of *mesodata* which, in fact, is the sole data-level data structure used throughout METASTASYS. Thus, once local data sources are transformed into mesodata format, they can be linked to global domain data models by formally stating the semantic relationship between local terminology (viz., local empirical domain languages) and global domain taxonomies (viz., "theoretical" domain language). The formal *image* of a local data source obtained from *mapping* its extent expressed in local empirical domain language to the corresponding global domain language indicates the *semantic locus* of the data source and, thus, enables an explicit formal treatment of semantic relationships between different data sources with respect to a (domain) universe of discourse defined in terms of the global domain language. Compared to traditional statistical data models, formal spatial and temporal semantics are in no way treated differently from other (viz., domain specific) criteria of discernment; though being, perhaps, a somewhat unusual design feature, it enormously facilitates spatial and temporal data integration without requiring additional functionality. Explicit representation of data sources' space and time references as integral components of data source images clearly relates MeTaMod to specific spatial and temporal data models as proposed for and used in, particularly, geographic information systems (GIS; cf. [Burroughs, 1989; Samet, 1989]), temporal databases [Snodgrass, 1990], and combinations thereof (see, for instance, [Langran, 1992; Worboys, 1992] and especially [Plank, 1944]). Organizationally, irrespective of semantic mappings to a central domain data model, data sources – the component databases, technically speaking – remain fully autonomous and independent from the global federated database system except that they are, of course, committed to really supply statistical data within mapped local source data images.

In MeTaStaSys, taxonomies are not only used to reconcile local domain languages at a global level; in addition to this vital role in semantic data integration, global domain languages serve, simultaneously, as unified statistical model language *and* database query language as well. In connection with a set of general syntactical structures, the global domain languages define a formal specification language, called MeTaSpeL, for statistical aggregates. Thus, in MeTaStaSys, taxonomies assume a functional role in composing algebraic designators of well-defined objects within a logical universe of discourse which, in turn, is a crucial prerequisite of the mechanical processing of statistical aggregates. MeTaSpeL is conceived as a general aggregate specification language encompassing a wide variety of practically relevant aggregate structures, especially including conventional statistical tables and time series (which, in fact, are interpreted formally as just a variant of a generalized table format; see below). Adopting this aggregate specification approach also blurs the traditional distinction between (database) retrieval, table management, and data analysis; although the latter is not considered as a genuine part of MeTaStaSys, aggregate derivation, in general, implies a statistical processing of source data in addition to pure retrieval and (relation-) algebraic database operations and, hence, necessarily involves some (multivariate) statistical methodology as well.

According to its conscious design parsimony, MeTaStaSys internally knows merely of a single, though generally applicable, aggregate data structure: any aggregate nameable at all is represented in a generalized table format called MeTa. Apparently, this rules out a representation of microdata in standard observation record format – that is to say, MeTaStaSys is *not* intended to function as a substitute for traditional record-oriented, raw data providing statistical databases (but remember that, as a matter of fact, the mesodata representation of microdata does by no means imply a loss of information; cf. Subsection 1.5.1). Throughout,

METAs are (conceptual as well as implementation) *objects* consisting of a data *and* a metadata component. While the metadata component, essentially, is responsible for describing the semantic scope of a statistical aggregate (with respect to a global domain data model) the data component comprises the actual data in either "raw" mesodata or already statistically/stochastically processed mesodata (that is, macrodata) format.

Based on METAMOD, a *table calculus* for the metadata-driven derivation of target METAs expressed in terms of METASPEL expressions (sentences) is established. This formal calculus, METAGEN, hosts an inference engine effectively generating derivations of target METAs from data source images. In logical terms, METAGEN

- computes the transitive closure of METASPEL expressions for a given set of axioms (METASPEL expressions denoting source images) and
- tests the set membership of the target META's METASPEL denotation with respect to this transitive closure.

To this end, METAGEN employs specific inference rules the metadata part of which is built of term rewriting rules transforming METASPEL sentences; the data part of these inference rules holds the algebraic and/or numerical data-level operations corresponding to the data transformation implied semantically by the metadata-level rewriting rule. Thus, given a *target* META and a set of mapped data sources, formally denoted by *source* METAs, METAGEN acts as *semantic distance reduction operator* seeking to find a derivation composed of inference rules eventually rewriting a target META to a (sub-)set of source METAs. Apparently, the effective power of METASTASYS is determined by the nature of its inference rules. The derivability of a META (within the set of METAs nameable in terms of a given METAMOD domain data model) may depend on algebraic *and* stochastic transformations of source METAs; in the present set-up, however, table management operations are considered mainly and, hence, little attention is paid to non-algebraic inference rules. In spite of its unfinished stage of development it should be remarked that, in principle, METAGEN's inference rules represent a rigorous denotation of operations occurring in almost every statistical analysis though typically being brought about in an *ad hoc* fashion most of the time.

In order to give an idea of the overall component structure of METASTASYS, **Fig. 2.1** (overleaf) sketches the main flow of statistical data. Statistical observation data, possibly being pre-processing appropriately, reside at database systems external to METASTASYS but are integrated logically in METAMODel terms and linked to METASTASYS by a rather special communication protocol. End-user access to METASTASYS data is facilitated by METASPEL and, more specifically, by META-SPEL's syntactical structures enabling the definition of customized views on METAMOD contents. The analysis and formal reduction of METASPEL expressions denoting statistical aggregates in terms of a generalized table format is carried out by METAGEN, the "dynamic" part of METASTASYS responsible for data retrieval and effective data integration using a couple of hard-coded elementary, though high-level, problem solving operators. In principle, all data structures occurring

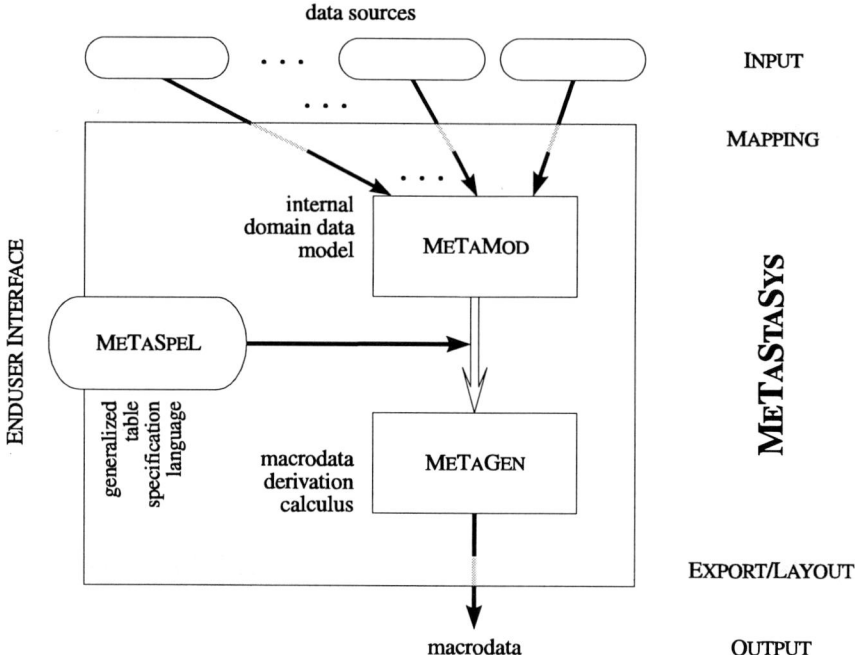

Figure 2.1: General METASTASYS Architecture

between **Fig. 2.1**'s grey-shaded bars conform to METAs or to parts (sub-structures) of METAs; these bars depict the "sluices" to be passed by all datasets entering or leaving the closed area of META-ness. **Fig. 2.1** omits deliberately the feed-back loop of returning derived macrodata to (internal or external) data repositories as this feature is a matter of economy rather than functionality.

A Survey of METASTASYS Design Resources

The METASTASYS approach to statistical data processing relates to several proposals and projects in the statistical database domain as reviewed in Chapter 1. Although METASTASYS is focused on data integration – an aspect hardly acknowledged in previous research – and, thus, does not directly draw on earlier proposals in statistical data modelling, it certainly addresses many – and most of the crucial and still unsolved – problems in research on statistical and scientific databases. The following summary lists briefly the most important influences on METASTASYS design and indicates each reference's particular contribution:

- D'Atri and Ricci [1989] pointed out the advantages of the "logical independence" approach to statistical query processing achievable by *universal relation interfaces* to databases;

- Sato [1989] emphasized the importance of distinguishing between data conceptually available and data actually stored in a statistical database, and proposed a semantic network approach to specify terminological relationships in statistical domains independent of particular data samples;
- Sundgren – in a long-time effort producing papers such as [Sundgren, 1973; 1992; 1993] – provided, inspired mainly by entity-relationship modelling methodology, a thorough systems analysis of data processing and database management in the official statistics domain and, in particular, contributed a four-way distinction (object reference, summary attribute, cross-classification, time reference) of *statistical table descriptions*;
- Fortunato *et al.* [1987], Falcitelli *et al.* [1989], Meo-Evoli *et al.* [1992], Rafanelli and Ricci [1993] have developed, in a series of papers, the "S-algebra" of *transparent macrodata operators* (which manage the statistical tables' summary types automatically) to transform both table descriptions (category attributes) and table contents (summary attribute) simultaneously in terms of extended relational operators;
- Klug [1982] and Özsoyoglu *et al.* [1987] introduced *set-valued attributes* to the relational database model to denote aggregations over attribute domains, thus laying a formal foundation of mesodata representation;
- Catarci *et al.* [1990] and D'Angiolini [1992] suggested a statistical *concept description language* for data modelling supporting, in particular, symbolic reasoning about concept specifications such as concept consistency, concept subsumption, and *aggregate derivability*;
- Basili and Meo-Evoli [1992] proposed a *deductive query processor*, "StEM", for statistical databases converting a high-level, declarative, end-user oriented macrodata query language into elementary database and table management operations automatically; in a sense, StEM's κ-operator can be viewed as a rudimentary core of METAGEN;
- Chen *et al.* [1989] devised a summary data model emphasizing the derivability of statistical aggregates from lower-level aggregates for additive summary types;
- Malvestuto [1989] employed a universal table model (limited to categorical data) in order to *combine information* contained in partial tables by means of statistical disaggregation of interconnected marginal distributions (facilitated by acyclical decomposition of category attribute subsets) into joint distributions (although, in the present setting, Malvestuto's assumption of nominal consistency of marginal distributions cannot be maintained);
- Rowe [1991] contributed the quite interesting concept of *statistical inheritance* as a means to migrate – even crossing the borders of classical model-based statistical reasoning, if need be – stochastic models into "dataless" areas of sample spaces on account of *a priori* structural model assumptions;
- Sadreddini *et al.* [1990; 1992] devised a framework for distributed processing of statistical aggregates in a distributed statistical (relational) multi-database management system developed of their own which particularly emphasizes macrodata structures and problem decomposition reflecting the physical topology of

the – heterogeneous – database systems hosting both statistical micro- and macrodata.
- Daruwala *et al.* [1995] proposed and developed a *context interchange network* aiming at interrelating different data holdings by using operative metadata for what they termed "context mediation" which, essentially, facilitates automatic nomenclature and scale conversion by establishing formally so-called "shared ontologies" on top of pre-existing local data source ontologies, or data models.

Apart from these specific references, the METASTASYS design rests on a couple of more fundamental results of theoretical and applied computer science, viz. theories of *effective computation* [Rogers, 1987], *databases* [Ullman, 1982; Date, 1986], *deductive databases* [Gallaire and Minker, 1978], and *data modelling* [Chen, 1976; Teorey, 1990]. In particular, METASPEL draws on the *universal relation* concept developed in connection with relational databases, and the splitting of statistical aggregate derivations in a symbol-level part and a data-level part essentially amounts to a – considerable – extension of the *deferred query evaluation* concept (introduced, for instance, for the sake of query optimization in relational databases), that is the processing of set *in*tensions (symbolic expressions) instead of set *ex*tensions (evaluated expressions) as far as possible. From the data modelling field, METAMOD primarily borrows the entity-relationship approach to represent semantic relations between (observable) populations, viz. collectives of sampling units, and instances (that is, individual sampling units), respectively.

The Cardinal METASTASYS Design Principles

Paying its tributes to *computability* and computational *tractability*, METASTASYS is conceived with a few simple principles in mind. Quite radically and, at first sight, in contrast to statistical requirements, METAMOD incorporates only a single type of scale used for modelling the ranges of observable values of sample space dimensions (variables). Although this design decision blatantly contradicts the smoothness assumption of real-valued random variates in mathematical statistics it must, however, be kept in mind that *real* numbers, in their totality, are neither computable entities nor observable in reality (except, possibly, for the Omnipotent Observer); what is made use of, in fact, remains restricted to a finite portion of rational-scaled values. Furthermore, METAMOD does not distinguish between categorical and numerical scales for the very same reason, since – whether ordered or not – there always can be only a finite collection of possible values, isomorphic to a subset of natural numbers, to be concerned. As an immediate consequence of this, a generalized scale model prepared to represent a *finite ordered* set of discerned scale values (that is the categories at most one of which a sampling unit must assume with respect to a considered observation variable) suffices fully. Additional semantics governing the various relationships among scale values is modelled on top of this basic scale by, for instance, introducing several *scale types* as genera specifica. Occasionally, it is quite difficult to indicate the numerical range of scale values ahead of time; making use of the concept of *lazy induction* (cf., for instance, [Henson, 1987]), METAMOD accounts for the definition of nominally infinite scales by adapting lower and upper bounds of the empirical range of scale values internally.

Reducing the representation of statistical scales to a simple single basic structure gains considerable formal design advantages since it saves a lot of case distinctions unavoidable otherwise. Recognizing further that both, sample spaces as well as statistical aggregates (notably, cross-classified tables) consist of Cartesian products of factor ranges, the representation of METAs favourably employs a factored denotation implicitly generating the Cartesian product of factor combinations so denoted. This approach complies ideally with a measure-theoretical view associating probability distributions with an event algebra generating the Boolean lattice of n-dimensional intervals on a sample space of dimensionality n (note also that, due to the sample spaces' finiteness property, measurability is of no concern). Apparently, each of these intervals – which might be termed *hypervals* – is representable in factored form as well and, hence, the event algebra is induced directly by multiplying out the Boolean algebras of a sample space's individual factors. Exploiting this data structure given naturally, hypervals are used internally as METASTASYS's central formal device; as a consequence, all algebraic computations refer to boxes and box transformations, that is computations concerning hypervals are reduced to computations on hyperval dimensions. It certainly does not need to be pointed out further that, letting k denote the average number of scale values in a factor's range, shifting to box representation of hypervals also reduces the space complexity from $O(k^n)$ to $O(kn)$, that is from exponential to polynomial.

Conceptually, the internal symbol structures of METASTASYS are cast into *objects* according to the object-oriented (OO) paradigm of software engineering (cf. Section 1.2). Since this paradigm defines computing as passing control between object entities (in terms of messages) and thus completely separates internal object structure from externally visible computational behaviour, OO is the methodology of choice to facilitate the integration of data from a distributed and highly heterogeneous environment [Pitoura et al., 1995]. It is only a matter of consequence to model all of METASTASYS consistently in an object-oriented way as well. Although there is no real implementation of METASTASYS as yet, OO definition provides both, a succinct account of intended object functionality and a clear indication of how to translate specifications into actual object code.

Trivially, data integration is faced with spatially distributed data (although, of course, from a logical point of view the physical location of data is not essential). As a practical consequence, METASTASYS necessarily operates in a *network environment* comprising, presumably, a variety of heterogeneous physical data sources. Following an established route in distributed processing, this environment is conceived as a (non-standard) *client/server* network such that the physical data sources participating in the network are assigned the role of data servers and the possible sites of data consumption where computations are actually initiated take the role of clients. In this arrangement, clients and servers communicate in terms of a specific network protocol, METACOM, such that clients submit computation requests, basically as METASPEL messages, to server objects which, in turn, respond with METAs supplying the requested data (if any).

A Road-map to the Structure of METASTASYS Description

The subsequent sections of this chapter are devoted to a detailed description of METASTASYS's functional components, viz. the METAMODel (Section 2.2), the formal integration of data sources (Section 2.3 and Appx. B), the aggregate specification language METASPEL (Section 2.4), the META derivation calculus (Section 2.5), and the envisaged approach to distributed META processing (Section 2.6). Section 2.7, finally, will give a brief account of a "proof of concept"-prototype implementation, DÖS'chen, of a subset of METASTASYS functionality. As the presentation of Chapter 2 is kept somewhat terse, one might like to switch, every now and then, to the examples worked out in Chapters 3 and 4, respectively, to better grasp the intended meaning of all the notions and concepts introduced.

The following subsections summarize the notational conventions used throughout this chapter as well as in Chapters 3 and 4 and provide the basic object classes used to define the salient symbol structures of METASTASYS.

The description of METASTASYS given in this chapter is by no means complete. For one thing, the following paragraphs expound little more than a skeleton of core functions of a full-blown data integration system. Moreover, some important aspects aren't even addressed; to mention just a few of them:

- The presentation of the aggregate derivation calculus, METAGEN, focuses on the algebraic derivability in its full scope at the cost of a profound discussion of (somewhat intricate) statistical and mixed inference rules. As a direct consequence of this, the formal representation of *inferential-statistical* metadata (cf. Subsection 1.5.2) and its algebraic processing in METAGEN's inference rules is given little attention.

- Likewise, although an in-depth elaboration of the formal structure of data source mappings is presented (as far as variables and their associated value ranges are involved) the existence of appropriate *import interface manager* modules is simply assumed without actually specifying the technical data organization of import data; in particular, the presentation omits any indication of tapping statistical metadata (such as, for instance, sampling information, probabilistic edits, etc.) relevant for non-algebraic aggregate derivations. Practically, establishing digital interfaces to existing database and data capturing systems will be of utmost relevance in really connecting METASTASYS to existing data sources.

- Even in view of distributed, decentralized data processing, METASTASYS needs some *internal* database to keep intermediary META structures which is not yet specified.

- As it stands, the DÖS'chen prototype implementation of METASTASYS lacks an up-to-date *user interface* (favourably a kind of GUI concealing the rather terse, clumsy, and hence error-prone specification of METASPEL sentences). Furthermore, in many cases the derived aggregates are not the final outcome of data requests but serve as operands of further arithmetic operations or layout improvement (for instance, table editing). Excepting elementary capabilities of METASPEL this fertile field of supplementary functionality is completely left open to further consideration and development, including links to existing tabulating and table processing software.

In addition to ground-breaking research already reviewed in Chapter 1, some of the salient features of METASTASYS emanated from previous project work, in particular the ESPRIT (II)–DOSES project B41 *Modelling Metadata* [Darius et al., 1993; Grossman and Froeschl, 1992; 1994] which investigated the role of metadata in statistical survey analysis from a task-oriented point of view (cf. [van den Berg and de Feber, 1992]). A definite output of this project was the above-mentioned software prototype DÖS'chen marking a major point of departure for METASTASYS design.

2.1.1 Notation Conventions

The semi-formal description of METASTASYS in the following sections of this chapter is given in terms of an *applicative*, that is: functional, calculus used as definite computation metaphor. This implies the arrangement of conceptual object domains, or types, and functions defined on top of object domains. In contrast to mathematical conventions, sets are introduced as computational entities, viz. ordered configurations of individual elements (names) [Manna and Waldinger, 1985]; this way of set representation will be called *sequence* notation, embracing a *list* of elements in angle brackets. Concurrently with sequence definition, of course, the ordering of the sequence elements is arranged implicitly by enumerating elements or stating an effective procedure generating the elements since, formally, a sequence is a *pair* composed of an element taken from the sequence domain and another sequence (of same sequence domain); sequences always terminate with the *empty* sequence. Despite these deviations from standard notation, mathematical symbol usage is maintained by and large, thus giving rise, occasionally, to somewhat odd-looking formulae (without, however, compromising well-defined semantics). The conscious introduction of all objects amenable to computation is captured in a positive *valueness* property: basically, starting from elementary objects defined at the outset, (composite) objects are well-defined if and only if they recursively consist of *values* (of proper type). As usual in logical calculi, this bottom-up definition reduces the scope of functions and thus delineates the functions' semantics to narrow, well-founded domains of syntactic objects.

In view of METASTASYS's addiction to the OO-paradigm, virtually any entity is introduced as an OBJECT, and the caps/small caps-font will be used to signal that a METASTASYS object is designated by the name used; moreover, the first occurrence of an OBJECT designation will be underlined. However, in most cases objects are defined only partially to the extent of making clear their structural role and computational behaviour. In particular, object compositions – sequences and sequence constructs excepted – are left unspecified technically. According to OO conventions, object *classes* are defined first such that *instances* receive all properties of its class's generic object representative. Additionally, classes may be stacked in hierarchies refining object structure and behaviour passed on downwards the class hierarchy. Implicitly, all effective function code is assigned to object classes (in terms of "methods") although no such formal assignment is introduced in subsequent descriptions.

The FOOL Pseudo-language of METASTASYS Method Definitions

Abstracting from definite functional programming languages, function definitions will, if at all (cf. Appx. A), be given in a pseudo-language called FOOL (for Functional OO Language), a deliberate merge of applicative and relational specification elements resembling the MIRANDA system [Turner, 1986] and specifically inspired by the FOOPlog language [Goguen and Meseguer, 1987]. Note that all domains used implicitly contain an *undefined* value, denoted as '$\perp$', which is returned by functions if applied outside their domain. In particular, accessing object components not (yet) initialized properly at access time will result in '$\perp$'.

Computable expressions are stated in a *message* format, that is, computations are conceived as messages sent to objects returning answers to messages. Conforming to OO notation standards, the object receiving a message is always stated first, followed by a (simple or composite) message actually defining the operation; a message consists of a message selector followed possibly by an argument (simple message) or a predefined sequence of message selector/ argument pairs. Simple messages are either unary (postfix; without an argument following) or binary (infix; positioned between receiver object and argument). As a general message selector naming convention, simple message selectors end in a question mark if used as a predicate (that is, as *Boolean* messages returning '$\rceil$' – TRUE – or '$\lfloor$' – FALSE) or in inquiry/access mode, and unary imperative messages in an exclamation mark; other keyword message selectors are always designated by a trailing semicolon. To highlight messages, they are typed in italicized boldface ***font***.

Operation definitions make use of so-called *type polymorphism*, that is, the same operation names are introduced for like operations irrespective of the objects they are applied to. Following common usage, the different stances of an operation are termed *methods*. This convention favourably reduces the number of operation names by building classes of operations with comparable effect though for different (yet related) object types.

The general format of a method definition is as follows: first, the message selector (without arguments) is stated followed by the method *preamble* defining the message type (either of 'infix', 'postfix', or 'arg'), the argument domains including the domain of the receiving object – this information resolves the methods' type polymorphism – as well as the domain (range) of the returned method output (function value); a method applies only in case of all input argument domains matching its preamble definition. The preamble *text*, delimited by brackets, does not belong to the functional method definition and uses standard mathematical notation not to be confused with clause definitions possibly looking rather similar. The preamble terminates with a semicolon.

Most of the time, methods (functions) distinguish several *cases* governing the definite output generation; thus, method definitions consist of one *clause* for each of its distinguished cases. A clause consists of three parts: an *instantiation pattern*, a *body*, and a *guard* (which may be omitted formally, if it represents a Boolean expression reducing to '⌐', that is being constantly TRUE). Instantiation pattern and clause body are separated by a '⇒'; guards always begin with a vertical bar, '|'. Clauses are ordered by definition; in order to enhance readability of method definitions, clauses, in general, are ordered with instantiation patterns ranging from most to least specific.

The formal, or *execution part* of a method definition consists of a sequence of clauses, each of which terminates with a full stop. Clause bodies may consist of any well-formed expression composed of defined messages including explicit and implicit object constructors. In case of nested message expressions, the ordering of argument evaluation is indicated by parentheses wherever suitable to avoid ambiguity and enhance clarity of expression meanings. Although not distinguished typographically, parentheses around message arguments are strictly discerned from tuple parentheses (for instance, assuming a fictitious binary message *msg*, in the expression '((a *msg* b),c *msg* d)', the parentheses around sub-expression 'a *msg* b' are argument parentheses and thus dropped whereas the outer parentheses are *tuple* parentheses and thus determining, in fact, a resulting tuple structure as output argument. Guards may be composed just like clause bodies albeit being restricted to expressions returning a Boolean value eventually.

FOOL Method Execution

During method execution, as a first step the *active* clause of the method is determined; the active clause is the first one in the defined clause sequence for which

- the instantiation pattern complies with the arguments of the received message in terms of *term unification*, and
- the attached guard (with arguments instantiated to values obtained from the instantiation pattern unification step) evaluates to '⌐', that is TRUE.

Formal arguments in instantiation patterns not referred to any further, may be designated by '_', meaning that such a formal argument matches any actual value in the calling message. Next, the body of the active clause is executed, that is, after instantiation of the formal body arguments the current object submits the active clause's body as a message to be processed. Upon receipt of the value returned in response to this message, its value is passed on to the calling object. Frequently, the output value of a method is augmented with one or more object attributes which, in general, are passed on from the set of the input arguments' attributes. Which attributes a method assigns to its output objects is indicated in the method preamble as a subscript to the mapping arrow separating the domains of input and output arguments. In a few cases, the attribute value passed on to the output argument must be computed from the input arguments' attributes; these, however, will not be stated explicitly because the necessary transformation steps are easily inferred from the method context throughout.

It should be noted that messages lacking a suitably defined method always return '⊥'. Moreover, methods are *strict*, that is, if during clause execution any message returns '⊥', '⊥' is passed on immediately to the calling object.

Conceptually, METASTASYS itself is considered as an object receiving messages directly from the user interface and returning computed outputs back to this user interface. Submitting a message to METASTASYS for evaluation is denoted by an "enter" arrow, '↵'.

Frequently, METASTASYS objects are *labelled*. These labels are strictly distinct from unique "internal" object identifiers. In order to ease notation and enhance (pseudo-)code readability, object labels are used deliberately in place of objects, letting the – mental – interpreter infer at run time if a label needs to be replaced by the referenced object during expression evaluation. Thus, at "interface level", an object's label represents the named object's *definition* whereas evaluating the referenced object yields its *value*.

2.1.2 Elementary Definitions

The information structures of METASTASYS are established in a bottom-up fashion starting from lowest-level elements, viz. primitive values and primitive value sets, upon which everything else is based as introduced in the following paragraphs. Though informal in style, these definitions could and should be viewed as skeletal (abstract) data type definitions; without aiming at a complete exposition, most of these definitions are accompanied by a set of typical operations applying to the object types defined which, however, are stated rather allusively in general (cf. Appx. A with respect to a couple of fully stated operation definitions).

To begin with, the most fundamental object type of METASTASYS is that of *base value*:

Definition 2.1.2–1:
 BASVAL is an OBJECT representing a *scalar*, *atomic* value introduced as "undefined" (in the logical sense) primitive subject–matter term. Formally, the class of BASVALues provides a source of (unstructured) names, or literal designations.

Practically everything in METASTASYS is based on BASVALs. With respect to the subsequent definition of more complex objects, it must be noted, however, that BASVALs, first of all, are used as *names* and may, in general, designate extensionally different objects in spite of homonymous naming; in those cases, context will be responsible for disambiguating homonymous names properly. Wherever critical, of course, homonymity of names is prevented by syntactical rules.

Definition 2.1.2–2:
 A COMVALue is an OBJECT composed of an ordered set (sequence) of component BASVALues.

Formally, letting $b_1, \ldots, b_n$ ($n > 0$) denote a sequence of BASVALues, the COMVALue composed of these BASVALues is denoted as $b_1.b_2.\cdots.b_n$ (*dot* notation).
 For *unique* primitive OBJECTs reserved names are used which are called *special values*:

Definition 2.1.2–3:
 A SPECVALue is a distinguished, *unique* value (that is, syntactically different from any BASVALue) furnished with a predefined *constant* meaning.

The following definitions introduce a couple of SPECVALues:

Definition 2.1.2–4:
 UNDEF is the SPECVALue returned by a function (such as a message sent to an OBJECT) if applied inadmissibly (because, for instance, it is not understood by the addressed OBJECT). Symbolically, UNDEF is denoted as '$\bot$'.

Definition 2.1.2–5:
 HINFinity is a SPECVALue denoting the empirical upper bound of a numerical domain (such as measuring scale); likewise, LINFinity is a SPECVALue denoting the empirical lower bound of such a numerical domain (see below, Subsection 2.2.1). HINF is denoted as '∞', LINF as '$-\infty$'.

For the sake of simplified reference, BASVALues, COMVALues, and SPECVALues can be combined to a general VALue type:

Definition 2.1.2–6:
 A VALue is either a BASVALue, a COMVALue, or a SPECVALue.

Obviously, isolated VALues are of rather limited use; however, out of VALues set-like structures can be built.

Set-like Structures

The most fundamental of these set-like structures in METASTASYS are *base sets* defined as follows:

Definition 2.1.2–7:
 BASET is an OBJECT representing a well-distinguished *ordered* set of structurally homogeneous VALues (that is, VALues either are BASVALues or COMVALues of identical component structure throughout) which, at any time, is extensionally

finite; however, a BASET *must not* contain any SPECVALue. Prinicipally, BASETs are mutually disjoint irrespective of occurrences of homonymous VALues.

Formally, letting $b_1,\ldots,b_n$ ($n \geq 0$) denote a set of BASVALues, a BASET B could be defined as $B = \langle b_1,\ldots,b_n \rangle$ (*sequence* notation) implying the ordering

$$b_1 \triangleleft b_2 \triangleleft \cdots \triangleleft b_n$$

of elements (note that, at calculus level, $\langle b_1,\ldots,b_n \rangle =_{def} \langle b_1, \langle b_2,\ldots, \langle b_n, \langle \rangle \rangle \ldots \rangle \rangle$; sequences always terminate with the empty sequence, $\langle \rangle$). With respect to this ordering, b_1 is called *minimal*, b_n *maximal* element, respectively. If a BASET consists of COMVALues (composed, in turn, of components taken from "lower-level" BASETs), its element ordering is induced lexicographically at component level.

As the intentional name suggests, BASETs are – almost – static definitions of VALue collections to be left untouched after their initial arrangement; however, since in real life barely anything except formal definitions remains constant, BASETs may be changed with care every now and then, but in a monotonically increasing way only by either extending the domain (that is, adding new VALues) or by replacing individual VALues *in situ* such that a replaced VALue's descendants cover the previous VALue's *semantic scope* without disturbing or compromising the semantics of the remaining VALues of the respective BASET.

BASETs provide the terminological baseline of METASTASYS in that other sets and objects are derived inductively from BASETs. In this restricted view, all other classes of OBJECTs essentially are but more complex structures composed of elementary BASETs.

Frequently, instances of OBJECT types are labelled; these *labels*, naturally, are elementary names necessarily unique within their local domain. For this reason, a sub-type of BASETs is prepared:

Definition 2.1.2–8:
A LABSET is a BASET of BASVALues or COMVALues used to label instances of some given OBJECT type.

Some collections of values share a rather special status in that they are used to define the *functional* structure of METASTASYS. In contrast to BASETs, which are defined according to domain-dependent considerations, these special value collections cannot be redefined unless the core functionality of METASTASYS is preserved by accompanying adaptations. These special, or functional, value sets of METASTASYS are defined as follows:

Definition 2.1.2–9:
DEFSET is an OBJECT comprising a *definite*, static set of SPECVALues; any incidental ordering of elements is immaterial.

.1 Premises and Design Principles

The prototypical example of a DEFSET is the collection of OBJECT types META-STASYS can discern and handle:

Definition 2.1.2–10:
OBTYPE is a DEFSET comprising the SPECVALues: BASVAL, COMVAL, SPECVAL, BASET, LABSET, DEFSET, PLASET, GRUSET, BASPART, XVAL, BAXSET, PLAXSET, GRUXSET, XGROUP, BAXDOM, PLAXDOM, GRUXDOM, BLOCK, BLOCKCOLL, TAG, and BOX, superclass types comprising elementary types (such as VAL, SET, XSET, XDOM) as well as the more specific OBTYPEs introduced in subsequent sections.

This nominally circular definition assigns, as a matter of fact, a definite operative semantics to each of the thus defined OBJECT types. Consequently, each OBJECT *instance* carries an attribute describing its (most specific) OBTYPE to be interrogated by an *obtype?* message. For instance, the OBJECT type of OBTYPE is 'DEFSET' (because OBTYPE is an instance of type DEFSET) whereas the OBJECT type of (the instance) 'DEFSET' is SPECVAL since (the type) DEFSET is a collection of SPECVALues. By virtue of this definition, SPECVALue is of its own type self-referentially which, however, does not cause any harm because the name 'SPECVAL' (as instance of type SPECVAL) is well-distinguished semantically (and pragmatically) from the OBJECT type SPECVAL. A new OBJECT, conversely, is typed by sending ⬜, the generic OBJECT, the *obtype:* message with an appropriate OBTYPE 'VAL' as argument value.

The following definition introduces the DEFSET of logical truth values:

Definition 2.1.2–11:
BOOLE is a DEFSET with two SPECVALues only, TRUE and FALSE, denoting the logical values of truth, $\top$, and falsity, $\bot$.

In the formal arrangement of METASTASYS two strands of set-OBJECTs are distinguished: *plain* and *grouped* set-OBJECTs; in actual data processing operations, in fact, grouped set-OBJECTs are predominant. In preparation of the following definition, the power set of a BASET is to be understood as the usual set-theoretical notion of power set applied to the elements contained in BASET, yet with the subsets of elements "inheriting" the VALue ordering established in the BASET.

Definition 2.1.2–12:
PLASET is an OBJECT representing an element of the power set of some BASET; ⟨⟩ denotes the *empty* PLASET, and by virtue of this definition, BASET itself is a PLASET as to semantic content (formally, PLASETs and BASETs are of different OBTYPE for principal reasons).

Symbolically, PLASETs, like BASETs, are stated in sequence notation. A set of mutually disjoint PLASETs such that the union of all elements of these PLASETs equals the BASET the considered PLASETs are derived from, is called a (set) partition. Those partitions are fundamental in defining grouped set-like OBJECTs.

Definition 2.1.2–13:
A <u>BASPART</u>ition is a (proper or improper) partition of a BASET into *disjoint non-empty* PLASETs. The ordering of PLASETs in a BASPARTition is derived from the BASET's VALue ordering such that, by taking each PLASET's *minimum* VALue, these minimal VALues preserve the original BASET ordering.

Symbolically, if a BASET B is partitioned into $n \geq 1$ PLASETs $B_1,\ldots,B_n$ with $B_i = \langle b_{i1},\ldots,b_{in_i} \rangle$, $1 \leq i \leq n$, B_i precedes B_j ($B_i \triangleleft B_j$) iff $b_{i1} \triangleleft b_{j1}$. In order to simplify notation, BASPARTitions are denoted by enumerating the PLASETs VALues within brackets such that PLASETs are separated by semicola: [...;...;...] (*group notation*).

Definition 2.1.2–14:
A <u>GRUSET</u> is an OBJECT containing a (non-empty or empty) subset of a BASPARTition.

Formally, a GRUSET is sequence of PLASETs; a BASPARTition is a special case of a GRUSET covering all elements of the underlying BASET, that is, a BASPARTition is a sub-type of OBTYPE 'GRUSET'. In contrast to BASPARTition, a GRUSET may be empty; for the sake of distinction, the empty GRUSET is denoted as []. Apparently, a particular GRUSET could be contained in several BASPARTitions of a BASET. By definition, a GRUSET must not contain an empty PLASET. In sequence notation, a GRUSET in group notation changes to

$$[b_{11},\ldots,b_{1n_1};\ldots;b_{m1},\ldots,b_{mn_m}] =_{def} \langle\langle b_{11},\ldots,b_{1n_1}\rangle,\langle\ldots,\langle\langle b_{m1},\ldots,b_{mn_m}\rangle,[]\rangle\ldots\rangle\rangle.$$

For many practical purposes, a PLASET $\langle b_1,\ldots,b_n \rangle$ and a GRUSET consisting of singletons $[b_1;\ldots;b_n]$ being also elements of the PLASET are "equivalent" as to semantic content.

Definition 2.1.2–15:
A <u>SET</u> is either a PLASET or a GRUSET. {} denotes the *empty* SET, that is, either ⟨⟩ or [].

Each SET instance carries a specific attribute, BASE, holding the associated BASET; this attribute can be inquired by a ***base?*** message. Furthermore, a MODE attribute tells – via message ***mode?*** – whether a set is plain or grouped by returning either of the SPECVALues 'PLAIN' or 'GROUPED' of DEFSET MODESET (alternatively, the *boolean* messages ***plain?*** and ***grouped?*** may be used instead of the latter). As

.1 Premises and Design Principles

usual, both messages can be used also for set definition in their assertive variant. A GRUSET may be converted to a PLASET (of same BASET) by dropping the grouping structure sending it the unary *plain!* message. SETs respond to a couple of specific messages such as *empty?* and *disj?:* (with evident meanings), *full?* (returning TRUE, if *all* VALues of the BASET occur in the SET), =, *eq?*, ⊂, ≺, ≅, ≈ (for SET comparisons), ∈, *enc?* (for testing VALue and SET inclusion), + (for adding elements to SETs), − (for subtraction for elements), *co!* (for SET complement), ∩, ∪, #, *, \ (for binary SET operations); cf. Appx. A for concise definitions. In order to facilitate algorithmic processing of SETs, two access functions, *top* and *tail*, are provided, returning a SET's first (minimal) VALue and all VALues except the first, respectively, such that the axiom

$$s = s\ top + s\ tail$$

holds for any non-empty s (if the addressed SET is an empty one, both messages return UNDEF).

Product Set Structures

The formal development of the METASTASYS data model draws heavily on Cartesian products of BASETs; these products, however, can be described efficiently in terms of the *factor* BASETs or PLASETs derived thereof.

Definition 2.1.2–16:
Given some ordered collection of $n \geq 1$ (not necessarily different) BASETs (non-empty PLASETs) $B_1, \ldots, B_n$, a <u>BAXSET</u> (a <u>PLAXSET</u>) is an OBJECT $(B_1, \ldots, B_n)$ *implicitly* denoting the Cartesian product $\times_{i=1}^{n} B_i$. If at least one of the generating PLASETs is empty, the PLAXSET is empty as well.

Note that $(B_1, \ldots, B_n)$, called *tuple* notation, is resolved into a pair structure at calculus level, that is, $(B_1, \ldots, B_n) =_{def} (B_1, (B_2, \ldots, (B_n, ())\ldots))$ where () denotes the *empty* tuple.

Similarly, BAXPARTitions and GRUXSETs are introduced in

Definition 2.1.2–17:
Given some ordered collection of (not necessarily different) BASPARTitions (non-empty GRUSETs) $B_1, \ldots, B_n$ where, for $1 \leq i \leq n$, each B_i comprises a set of PLASETs $B_{i1}, \ldots, B_{in_i}$, a <u>BAXPARTition</u> (a <u>GRUXSET</u>) is an OBJECT $(B_1, \ldots, B_n)$ denoting the set of Cartesian products $\left\{ \times_{i=1}^{n} B_{ij_i} : 1 \leq j_i \leq n_i \right\}$ im-

plicitly. If at least one of the generating GRUSETs is empty, the GRUXSET is empty as well.

Evidently, a BAXPARtition is also a GRUXSET. For $n = 1$, of course, PLAXSET, BAXPARtition, and GRUXSET coincide *semantically* (but not formally) with PLA-SET, BASPARtition, and GRUSET, respectively. To make definitions fully analogous, the notion of XSET is added in

Definition 2.1.2–18:
A <u>XSET</u> is either a PLAXSET or a GRUXSET. ∅ denotes the *empty* XSET.

Analogous to SETs, each XSET instance comprises two attributes BASE and MODE which, again, can be inquired by the messages *base?* and *mode?*, respectively. Furthermore, all of the SET messages (excepting the access functions *top* and *tail*) apply to XSETs as well though internally adapted accordingly; in view of the somewhat more complex structure of XSETs, several additional messages are available such as the binary relations $=, \doteq, \hat{=}$ reporting on the *structural* relation between XSETs; cf. Appx. A for specification details.

Quite frequently, SETs and XSETs are extended to GRIDs and XGRIDs, respectively, by adjoining SPECVALues to corresponding BASETs. However, to keep terminology simple, no distinction between SETs and GRIDs (XSETs and XGRIDs) will be made, letting context clarify which sort of object type is meant in fact. Moreover, adjoined SPECVALues often receive a rather specific treatment in operations which is explained as appropriate; in particular, in GRUSETs SPECVALues *always* go into singletons.

Related to XSETs is also the following

Definition 2.1.2–19:
An <u>XGROUPing</u> is a sequence of all PLAXSETs induced by a GRUXSET such that the PLAXSETs components are selected from the respective component-GRUSETs of the GRUXSET in all possible element combinations. Within an XGROUPing the resulting PLAXSETs are ordered lexicographically by components on the ordering of PLASETs within components.

Anticipating later usage, another special instance of XSET-related data structures is introduced in

Definition 2.1.2–20:
<u>BLOCK</u> is an OBJECT consisting of a PLAXSET as well as TAGS. A <u>BLOCKCOLLection</u> is a sequence of *disjoint, non-empty* BLOCKs referring to the same BAXSET as indicated by a BLOCK's and a BLOCKCOLLection's, respectively, <u>BASE</u> attribute. BLOCK ordering within a BLOCKCOLL is induced lexicographically from the BASE's BASET orderings. <u>WRAPPING</u> is the *smallest* possible PLAXSET (of same BASE) completely containing *all* BLOCKs of a given BLOCK-COLLection.

.1 Premises and Design Principles

In sequence notation, a BLOCKCOLLection is denoted as $\langle B_1,...,B_n \rangle$, $n>0$, where, for all BLOCKs B_i, $1 \le i < n$, $B_i \triangleleft B_{i+1}$. The *empty* BLOCKCOLLection is denoted as $\langle \rangle$. BLOCKCOLLections understand *base?*, *top* and *tail* messages. Individual BLOCKs can process *base?* and *=?* messages; the latter returns the BLOCK's PLAXSET. A BLOCK is initialized by the *=:* message the argument of which must be a PLAXSET of compatible BASE.

Definition 2.1.2–21:
TAGS is a set (sequence) of TAGs where TAG is an OBJECT labelled by a TABLAB (taken from a LABSET called TAGLABSET) basically conveying a varying number of attributes depending on a TAG's TAGTYPE; TAGTYPEs are comprised in a DEFSET named TAGTYPESET.

The TAGS of a BLOCK can be inquired by sending the BLOCK a *tags?* message. By a *tags:* message, a BLOCK's TAGS is replaced with the argument value; *tag:* just adds another TAG to the TAGS of a BLOCK (if not included in TAGS). Likewise, TAGTYPE and TAGLAB as well the attribute values conveyed by a TAG are accessed by *type?*, *lab?*, etc. messages, and these messages, of course, also have an assertive mode.

Box Structures

The fundamental METASTASYS data structure is the so-called *box*. Formally, a box is a pair of an XSET and a BLOCKCOLLection as defined in

Definition 2.1.2–22:
A BOX is an OBJECT consisting of a TEXTure and EDITS such that TEXTure is an XSET defining the extension of a set of XVALues decomposed into factors of a Cartesian product and with reference to an underlying BAXSET. EDITS is a BLOCKCOLLection, with BASE referring to the same BAXSET as TEXTure, denoting some *proper* subset of XVALues implicitly included in the set extension of TEXTure. Depending on whether TEXTure is plain or grouped, a BOX is either a PLABOX or a GRUBOX.

Corresponding to general usage, BOX components can be accessed by *text?* and *edits?* messages; *base?* returns the BAXSET being the BASE of both, TEXTure and EDITS. Again, *text:*, *edits:*, and *base:* denote the assertive message variants.

A BOX is in *standard representation* if and only if its TEXTure is chosen as small as possible; formally, standard representation implies that augmenting either of a TEXTure's defining BASETs by some VALue enforces a corresponding augmentation of the EDITS's WRAPPING with exactly *all* XVALues added to the BOX by augmenting its TEXTure (that is, by adding another BLOCK to EDITS such that this new BLOCK ranges over all VALues of all BASETs except the augmented BASET where the BLOCK includes the augmentation VALue only). Unless stated otherwise, all BOXes will be assumed to be in standard representation.

Tuple Domains

Sometimes an explicit reference to the set extensions as defined implicitly by XSETs becomes indispensable. To this end, yet another OBJECT type is envisaged in

Definition 2.1.2–23:
   A BAXDOMain (PLAXDOMain, GRUXDOMain) is an OBJECT containing the set extension of a Cartesian product as defined by a BAXSET (PLAXSET, GRUX-SET); the XVALues of a BAXDOMain etc. are written in *tuple* notation. Conforming to former usage, XDOMain is either a PLAXDOMain or a GRUXDOMain. '{ }' denotes the *empty* XDOM; more specifically, '⟨⟩' denotes the empty PLAXDO-MAIN, and '[ ]' denotes the empty GRUXDOMain, respectively.

Each of these tuple sets bears a definite element ordering derived from its constituent SETs: sets of XVALues are always ordered lexicographically by XVALue component VALues whereas sets of sets of XVALues are ordered with respect to the minimal elements of the former within-SETs ordering. Since PLAXDOMains (GRU-XDOMains) are sequences of XVALues (PLAXDOMains), PLAXDOMains (GRUX-DOMains) are stated in sequence (group) notation. Again, all of the messages defined for SETs, such as *mode?* and *base?*, apply adapted suitably. A BAXSET (PLAXSET, GRUXSET) is converted into the respective (BA)XDOM by sending it the *dom!* message.

2.2 The Data Model

In the inductive development of METASTASYS, its data model occupies a fundamental position. Basically, this data model serves a three-fold purpose: first it has to take account of the terminology used in statistical analysis domains; secondly, it has to organize the ontological entities made discernible by domain terminology in a way suited to formal statistical analyses (that is, arranging entities in terms of probability models); and finally, it has to provide the operational backbone of actual data processing. Apparently, development of a data model achieving this dauntingly complex purpose and meeting all involved preconditions simultaneously can be brought about in a top-down approach only, lest not to get trapped in a simple-minded, one-dimensional view of data model design optimizing, rather inadvertently, only single aspects in isolation.

Any formal data modelling attempt starts, whether consciously or not, with an orderly linguistic frame providing a definite point of reference for a cognitive discourse in some given subject-matter field. In philosophical terms, the *ontology* of such a linguistic reference system – Strawson [1959], for example, defines ontology

.2 THE DATA MODEL

as "the conceptual definition and cataloguing of entities and relationships composing the mind representation of the real world" – essentially results from *a priori* discernments based on the tableau of prevalent cultural ordering codes of (some) society which conceptualizes objects, or entities, in terms of sensible – that is: perceptible, observable – signatures to classes of similarities and dissimilarities (however, note that, following Polanyi [1966], "sensible" by no means implies "verbally explicable" or "cognitively transparent"). Interestingly, as has been pointed out by comparative linguistics (for instance, [Whorf, 1956]), collective experience not only shapes linguistic frames of discernment allowing for expressing and communicating what needs to be expressed and communicated but, by reverting the direction of argument, once established linguistic frames in fact define cognitive universes *generating* distinctive societal world views and, thus, predetermine or, rather, pre-*structure*, the space of further collective experience (as an example, tracing back the tremendous upswing of technology in modern times based on the "new" language of natural sciences highlights the cognitive-cultural importance of language for so profoundly changing the world by the so-called technological *progress*). To restate in more mundane words, statistical analyses remain restricted necessarily to notions expressible in a given observation, or data, language and the expressiveness of available (empirical) data languages determines what is amenable to statistical analyses at all. However, it should be kept in mind that empirical, or statistical, data languages are set up not to their own ends but as a means to reason about the world and, particularly, to change (part of) this world, that is: making decisions, or to re-structure perception by gaining structural insight (modelling, pattern recognition) which, in linguistic terms, amounts to the adaptation of (research) language and, hence, of cognitive reference frames, or world views, as well.

Despite the intricate and obviously far-reaching consequences of definite ontologies for symbolic data representation models as well as perception, conceptualization, and cognition processes, the development of the METASTASYS data model – METAMOD henceforth – is based on a rather naïve positivism assuming the empirical presence of a conceptually well-ordered statistical world of objects and stipulating the availability of readily applicable *precise operations* (as termed by Foucault [1966]) actually constructing this world ordering in a straightforward manner. In view of the practical problems METASTASYS sets out to solve, however, the design is carried out in awareness of two ever-present complicating factors, namely

- the concurrent existence of different and, sometimes, even rather dissimilar ontologies which need to be harmonized, and
- the dynamic evolution (or, less euphemistically, temporal instability) of ontologies and associated terminological reference frames.

Though the latter factor is taken into account in principle, METASTASYS is destined to address the former one's implications fully by providing a well-tailored formal data model which is explicated, in a stepwise fashion, in this section. Before this rather formal exposition commences, however, both main assumptions and general

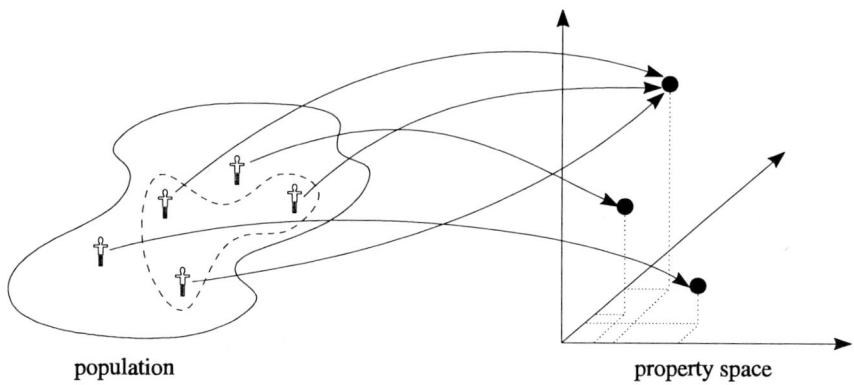

Figure 2.2: Property Space Representation of a Population

data model structure are summarized to aid the understanding of subsequent considerations.

Conceptual starting point of the METAMODel is the notion of *population*: it is simply assumed that the empirical world can be partitioned effectively for statistical purposes into a set of structurally similar "objects", or *individuals*, without requiring a formal statement how similarity classes are established or how "objectship" can be attributed. Of course, different kinds of similarity may apply and the one actually chosen after all will depend on research interest and possibly other pragmatic circumstances remaining outside METASTASYS's formal frame-up. Ideally, however, in view of attempted data integration, a data model comprising as few populations as possible seems favourable since particularly overlapping populations may disturb succinct data semantics as will become apparent below. In contrast to this, explicit semantic relationships between populations contribute formally to semantic expressiveness and, hence, enrich the data model. More precisely, semantic relations between populations (which, in fact, amount to instances of semantic relationships between population members, or individuals) may link either two different populations (inter-population linkage) or individuals of one and the same population (intra-population linkage). Rather contrary to relationship modelling in traditional (relational) database theory, however, the semantic relationships between statistical populations are relationships between statistical *sample spaces* whence the semantics of relationships and the constraints resulting thereof deviate considerably from those in familiar data models.

Accepting that there are disjoint "homogeneous" classes, or populations, of empirically observable objects, each population is assigned a taxonomy used to split the collective of population members comprehensively and exhaustively into a set of disjoint sub-populations, or *traits*. From a linguistic point of view, this taxonomy is expressed by means of a system of *descriptive* features applying particularly to

.2 THE DATA MODEL

the members of the population under consideration (contrary to *discriminative* features used, formally or informally, to classify individuals with respect to population membership, that is: to define populations). Although the subdivision of objects into populations is taken for granted, it is by no means evident or unequivocal how to arrange such a system of descriptive features since this depends very much on the intended type(s) of discourse. Taking a closer look, taxonomic systems typically consist of a couple of descriptive dimensions with respect to each of which an individual assumes some particular value. Basically, this amounts to a structuring of sub-populations, or traits, such that, in principle, each discerned trait corresponds to a specific value combination taken over all descriptive dimensions taken into account at all; formally, a trait becomes represented as a point in a *property space* defined as a multi-dimensional lattice spanned by descriptive dimensions as depicted in **Fig. 2.2** (based on a graphical idea of [Lenz, 1993b]). This factoring of traits into products of elementary components, or observable characteristics, introduces a *formal data language* which can be used algebraically to denote intensionally, in terms of set-theoretical expressions, any sub-population that can be stated as a set of traits within the taxonomy. Evidently, it is this factoring of traits which enables the rich semantic structure presupposed by any reasonable analytical discourse (cf., for instance, the notion of *factorial tables* in discrete multivariate analysis [Bishop *et al.*, 1975]). Less clear, in general, is the process of actually constructing a property space which involves two stages: first, there must be some decision about, or agreement upon, the descriptive dimensions to be concerned and included in the data model; secondly, to each dimension a definite set of values to be distinguished must be assigned. Although rather easy for isolated, one-shot data models, the definition of multi-dimensional taxonomic data languages turns out as a tedious (not to say: insurmountable) task since the resulting language should, at the same time, be of utmost generality and parsimony (cf. [Lee and Hotaka, 1989] with respect to "primitiviness" of attributes). In particular, such a language is subject to (semantic) compatibility with already existing taxonomic systems to be integrated by the common data model and, ideally, it should rather neutrally admit the formation of *all* conceivable sub-populations (that is, sets of traits) that might be of interest at one time or another, permitting, additionally, a wide range of breakdowns (cross-classifications) of investigated sub-populations. Moreover, in view of the persistence of statistical data bodies the language ought to be as stable as possible over time in order not to preclude a temporal integration of data for even quite extended observation periods. Conversely, description dimensions should be *elementary* in order to

- achieve a certain granularity of the induced property spaces (which pertains to the economy of measuring: an increasing number of discerned terms provides more information – at least in terms of information theory – at possibly identical measuring cost) and
- avoid the (involuntary) introduction of redundancy incurred implicitly by partially overlapping description criteria (note that despite there is, in formal terms, no such overlap it resides deeply buried in extra-symbolic meanings).

Furthermore, for obvious reasons, it is mandatory to exclude descriptive dimensions from a property space if they depend *functionally* [Ullman, 1982] on others.

Practically, formal data languages are established in a more or less pragmatically oriented negotiation procedure taking into account criteria of necessity, utility, economy, and political consensus. With respect to data integration, this amounts to an "organic" derivation of taxonomic systems by taking existing systems as starting point and generalizing the found, bottom-level taxonomies such as to arrive at a *least common multiple* to be used as de facto standard for the common data model. An obvious and appealing advantage of this approach is the simple relationship between original taxonomies and the derived generalized taxonomy which, of course, facilitates actual data integration considerably. Contrary to this, however, generalized taxonomies may still suffer from peculiarities inherited from initial taxonomies and, hence, be judged somewhat less ideal from a subject-matter point of view which likely would favour *universal* taxonomies – that is, taxonomies not biased terminologically towards a specific kind of discourse or analysis. A particular deficiency of grown taxonomies is the frequently observable asymmetry between concept (that is, sub-population) *defining*, or *identifying*, and concept *cross-classifying* description dimensions in property spaces which often crucially hampers effective and versatile data exploitation. As a matter of fact, data integration could stimulate a thorough reconsideration of locally well-established taxonomies in view of the enhanced opportunities of multiple re-use of empirical data, including completely unprecedented data usage. All of this could lead up downright to a reconstruction of traditional taxonomies, or measuring systems, by first *deconstructing* concepts and subject-matter notions used hitherto, thus stripping off the historical burden of redundant, artificial, or unsuitable properties, before taxonomies are assembled anew in a concise, unbiased way. As a consequence of the taxonomy dilemma (and having in mind the long-lasting endeavours of, for instance, statistical classification efforts in standardization of nomenclatures), one might also prefer to adopt a "semi-pragmatic" approach yielding, in a sense, an ideal taxonomy though restricted to a subset of the ultimately conceivable universe of discourse and just general enough to include the relevant portions of considered bottom-level taxonomies to be integrated. Advantageously, METASTASYS's modest requirements in nomenclature standardization facilitate taxonomy formation because a *technically* unifying terminology is demanded only, without claiming a specific consent or even an explicit approval of its "appropriateness" or political desirability by involved negotiating parties.

In statistical terms, property spaces correspond to measure spaces, that is, each descriptive dimension is regarded as a *variable* (measurement axis) in a multivariate observation, or sample, space; the set of discerned values for each measurement axis is called a (measuring) *scale*. Thus, in METASTASYS, the formal data language associated with a population provides a measuring space as well such that a formal bottom-up derivation of statistical random variates is supported fully and transparently. In particular, traits can be considered consistently as sample points, or elementary events, amenable to axiomatic event algebraic processing, and the set of discernible traits spans an event space, Ω, on which probability distributions may be modelled for statistical or decision-theoretical analyses.

In addition to terminological and statistical functions, the data model fulfils a pivotal *operational* role, too. Basically, it

- provides the symbol structures to describe available data bodies formally,
- carries an algebraic language of names to designate statistical aggregates, and
- sets the stage for mechanized output derivation by supplying the pertinent – logical – data structures.

The formal description of available data bodies consists of a (metadata) representation of the symbolic re-expressions – in terms of mapping function images – of data languages used in the originating contexts of data bodies with respect to the established unifying terminology, thus highlighting the captured taxonomic differences between all data sources taken into account. The established unifying terminology, in turn, provides the basic vocabulary of a formal language used to designate data aggregates such that the structure of expressions itself discloses the information necessary to steer actual data aggregation or, more precisely, the formal deduction of designated aggregates from symbolic source data representations. The material data structures, finally, are linguistic devices denoting the physical operands actually conveying the statistical information of interest. Operationally, data structures are referenced and manipulated by expressions of a formal operand language the internal structures of which are described in subsequent paragraphs. In practice, of course, these symbol structures first have to be converted to definite database views before real data processing can take place.

2.2.1 Populations, Scales, and Qualitative Characteristics

As already mentioned, the totality of observable entities (in statistical terms: carriers of observable characteristics) is partitioned into a collection of classes called *populations*. Formally, this collection of discerned populations is conceptualized as a special METASTASYS set called POPSET:

Definition 2.2.1–1:
> POPSET is a BASET of distinguished entities, each representing a particular well-defined generic type of statistical observation, or sampling, unit. In particular, to each generic type of observation unit there must be associated a non-empty set of real-world instances such that, in principle at least, it is possible to enumerate the set of instances (that is, to create a population *register*) or to set up an effective (random) *sampling scheme*.

Although this POPSET may grow monotonically, its inception requires utmost care lest later augmentations with new entity types are not affected negatively or even inhibited by a thoughtless choice of previous elements.

The taxonomies assigned to the members of POPSET are created in a bottom-up fashion by first defining a set of generic scales which, subsequently, are adapted to the needs of specific taxonomic systems. Formally, a (measuring or observation) scale is an abstract device (function) returning, if applied to an individual instance of a population defined in POPSET, some empirical value chosen from a predefined scale domain. In METASTASYS, these notions are settled as follows:

Definition 2.2.1–2:
A MODality is a *distinct discrete* VALue (other than a SPECVALue) used to designate a particular possible outcome of a (theoretical) observation device.

Definition 2.2.1–3:
A SCALE is an OBJECT comprising a SCALEDOM which, in turn, is a BASET (that is, an *ordered* set) of MODs. A SCALE may be identified with a *label*, SCALAB, taken from a LABSET called SCALESET.

By definition, though different SCALEDOMains may share nominally identical VALues, they are considered *disjoint* throughout.

Some of the SCALEs come with already defined domains whence they are termed *special* scales:

Definition 2.2.1–4:
DICHOSCALE is a special SCALE with two MODs only, 'YES' and 'NO'.

Definition 2.2.1–5:
CARDSCALE is a special SCALE with MODs equal to non–negative integer numbers.

Definition 2.2.1–6:
NATSCALE is a special SCALE with MODs equal to positive integer numbers.

Definition 2.2.1–7:
INTSCALE is a special SCALE with MODs equal to integer numbers of either sign.

METASTASYS provides a predefined collection of SCALE types comprised in a special set called SCALETYPE such that each instance of a SCALE belongs to one of these predefined types:

Definition 2.2.1–8:
SCALETYPE is the DEFSET of discerned SCALE types comprising (at least) the VALues 'CATEGorical', 'DICHOtomous', 'BINary', 'ENUMerative', 'CARDinal', 'NATural', 'RADix', 'POSitive', 'NONNEGative', 'INTeger', 'SPATial', and 'TEMPoral'.

It should be remarked that SCALETYPE is not identical with the statistical scale *level*, although there is, of course, some interrelation. Apparently, DICHOSCALE, CARDSCALE, NATSCALE, and INTSCALE are of SCALETYPEs 'DICHOtomous', 'CARDinal', 'NATural', and 'INTeger', respectively.

.2 THE DATA MODEL

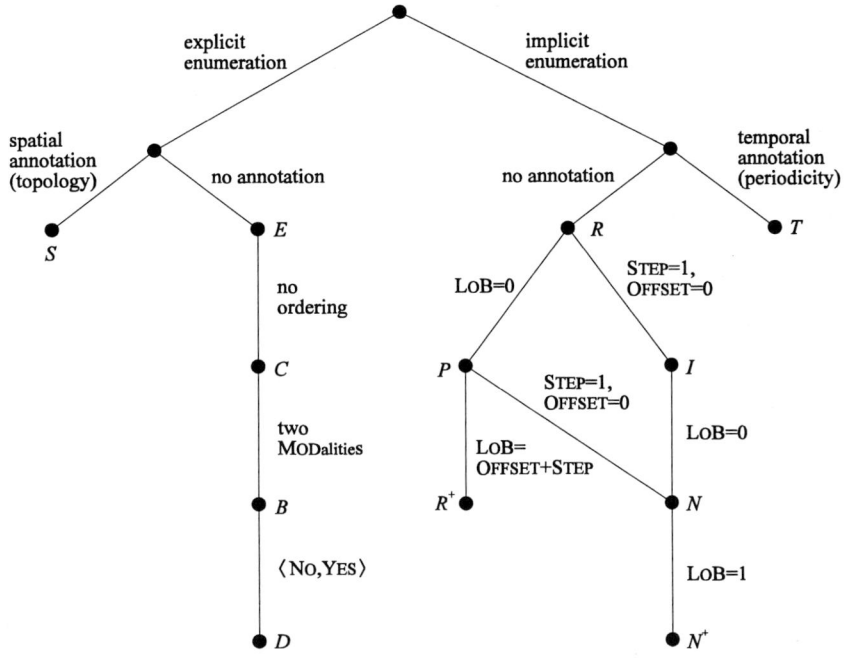

Figure 2.2.1: SCALETYPE Hierarchy

SCALETYPEs build a hierarchy as depicted in **Fig. 2.2.1**; **Tab. 2.2.1** (overleaf) summarizes the main features of SCALEs. Excepting special SCALEs, scale definition is achieved either by explicit enumeration of elements (where the defining sequence is interpreted as the intended element ordering of MODs) or by a parameterized domain specification. For non-ordered SCALEs, of course, the definite enumeration sequence of elements is immaterial. For implicitly specified SCALEs, the *range* of MODalities is delimited by lower and upper bounds, LOBound and HIBound, respectively, where LOB may assume the SPECVALue 'LINF' and HIB the SPECVALue 'HINF' denoting that the actual range of the SCALE extends as far as necessary to represent occurring observed or measured values (thus assuring finiteness of *empirical* domains). STEP indicates the step size of numerical SCALEs, viz. RAD, POS, and NONNEG, while OFFSET controls the relative (to "0") position of SCALE origins.

Default settings for SCALE parameters are as follows: HIB=HINF, LOB=LINF, STEP=1, and OFFSET=0, unless overridden by definition as shown in **Fig. 2.2.1**; only the parameters listed in **Tab. 2.2.1** can be assigned deviating values actually. Hence, under default conditions, CARD=NONNEG, NAT=POS, and RAD=INT.

A rather specific semantics is attached to the SPATial and TEMPoral scales. In addition to the enumeration of MODs, these scales may be annotated with topologi-

Table 2.2.1: SCALETYPE Summary

SCALETYPE	Symbol	Ordered	MODS	Specification of SCALEDOMain
DICHOtomous	D	no	⟨NO, YES⟩	–
BINary	B	no	explicitly	enumeration (two MODS)
CATEGorical	C	no	explicitly	enumeration
ENUMerative	E	yes	explicitly	(factored) enumeration
CARDinal	N	yes	implicitly	HIB
NATural	N^+	yes	implicitly	HIB
INTeger	I	yes	implicitly	LOB, HIB
RADix	R	yes	implicitly	LOB, HIB, STEP, OFFSET
POsitive	R^+	yes	implicitly	HIB, STEP, OFFSET
NONNEGative	P	yes	implicitly	HIB, STEP, OFFSET
SPATial	S	no	explicitly	(factored) enumeration, annotation
TEMPoral	T	yes	implicitly	(factored) enumeration, annotation

cal information in order to express relationships other than a linear ordering between MODS. In particular, each MOD of a SPATial scale may be associated with a node in a *neighbourhood graph*, the edges of which are indicating the topographic neighbourhood of regions or areas represented by the respective MODS. Furthermore, nodes can be furnished with a vector of attributes holding, for instance, the coordinate information (such as center of gravity), boundary polygons, etc., of MODS needed for thematic mapping in a geographical information system. With respect to the TEMPoral scale, a directed *periodicity graph* can be set up stringing together the MODS (time units) representing periodical repetitions, that is, periodically recurring ("stationary") times. Apparently, further SCALETYPES with specific annotation structures could be introduced on demand – this, in turn, implies that there are, for instance, statistical aggregation procedures and the respective extensions to the aggregate specification language (cf. Section 2.4) really making use of or presupposing this added semantics.

If SCALEDOMains can be expressed naturally by composite values, a factored enumeration by Cartesian products of sub-domains is convenient. In particular, TEMPoral scales can thus be composed out of years, days, hours, minutes, etc. For instance, a sample TEMP scale with hours as unit could be arranged simply by *hour.day.year* where *hour*=⟨0, 1, ..., 23⟩, *day*=⟨jan1, jan2, ..., dec31⟩, and *year*=⟨1950, ..., 2050⟩, inducing a SCALEDOMain of ⟨0.jan1.1950, 1.jan1.1950, ..., 22.dec31.2050, 23.dec31.2050⟩ (it must be kept in mind, however, that this definition implies invalid intercalary day dates such as '15.feb29.1995' as well). With respect to SPATial scales, *coordinate systems* could be expressed by factored enumeration rather elegantly either. The ordering of composite values within composite SCALEDOMains is derived lexicographically from the (explicit or implicit) enumeration ordering of sub-domains. Formally, any SCALEDOMain is eligible as a sub-domain for factored enumeration though it certainly is not advisable to make excessive use of these *supercodes*.

Operationally, SCALE objects respond to query messages such as **lab?**, **dom?**, and **type?**; where appropriate, scale parameters can be inquired by using parameter names as message selectors (such as, for instance, **offset?**). For SCALE definition, these messages can also be used *assertively* by replacing the *?*-sign with a semicolon followed by the specified argument value.

Specificity of Scales and Definition of Aggregation Levels

Used as basic building blocks of subsequent property space construction, SCALE-DOMains typically are prepared ahead of time. Some of them will serve rather specific purposes while others may be arranged in a fairly general way; for instance, for SPATial and TEMPoral scales a single encompassing SCALEDOMain, respectively, would be favourable in view of least restricted data integration. Accordingly, unique SCALEs to be used in each conceivable property space are termed *universal*, whereas SCALEs with restricted usability are termed *general*, and SCALEs adapted to a single property space will be called *(domain) specific*.

Quite frequently, the direct reference to MODalities leads to clumsy and tiresome formal expressions at least badly suited for human use. Hence, a means to arrange named *hierarchical groupings* of a SCALEDOMain's MODalities is provided explicitly. Technically, *aggregation levels* are introduced as GRUSETs on the SCALE-DOMain (cf. Subsection 2.1.2):

Definition 2.2.1–9:
With respect to a given SCALE, an AGGLEVel is a particular GRUSET introducing a grouping of either MODalities of the SCALEDOM or an already defined (lower-level) AGGLEV for this SCALE. Each AGGLEVel is given a label, AGGLAB. To the elements of an AGGLEVel's GRUSET names are attached defining a BASET called AGGDOM. AGGBASE is the GRUSET comprising an AGGLEVel's input, that is, the set of elements being (re-) grouped in AGGLEVel.

Since, functionally, AGGDOMains *are* SCALEDOMains, AGGDOMains may be specified implicitly as well provided that the AGGDOMain specification is consistent with both, the SCALEDOMain and the aggregation grouping structure of AGGLEVel. For the sake of uniformity, if AGGDOM is the SCALEDOMain itself, AGGLEV is defined as the BASPART of the SCALEDOM consisting of all singleton MODalities. As is evident from this definition, as many alternative AGGLEVels may be specified for a SCALE as suggested by practical requirements.

Technically, aggregation levels are sub-objects of SCALEs responding to messages like **lab?**, **dom?** and **base?** (which, by dropping the *?*-sign, can be used for AGGLEV definition as assignment messages). Within a SCALE, AGGLEVels can be accessed by the **agglev:** message with an AGGLAB argument.

Formal Denotation of Qualitative Characteristics

Now, for each population in POPSET, the associated taxonomic system is composed of a set of specifically instantiated SCALEs used as descriptive dimensions of the population's property space; these dimensions are called *qualitative characteristics*, or *qualities*, for short:

Definition 2.2.1–10:
Each QUality denotes a particular property space's descriptive dimension; formally, it comprises a DIMDOMain as well as a set of dimension attributes, DIMPROPS. For the sake of easy reference, each QUality is assigned a unique *label*, QULABel, taken from a LABSET called QUSET. Each QUality also has a type, QUTYPE, attached to it, as explained below.

Definition 2.2.1–11:
The DIMDOMain of a QUality is an *augmented* AGGDOMain, that is a (possibly grouped and restricted) SCALEDOMain extended by the SPECVALue 'MISSVAL' and, depending on property space structure (cf. Subsection 2.2.2), the SPECVALue 'NULL'.

Definition 2.2.1–12:
MISSVAL is a SPECVAL included in every DIMDOMain used to denote the case of *missing*, or not recorded, observation values because of refused answers or failures of measuring devices. With respect to the element ordering of a DIMDOMain, MISSVAL always comes last.

Definition 2.2.1–13:
NULL is another SPECVAL to be included in the DIMDOMains of those QUalities which do not apply generally but depending on particular *value combinations* with respect to other QUalities, that is, NULL is appended to the DIMDOMain of *dependent* QUalities. With respect to the element ordering of a DIMDOMain, NULL always *precedes* the MODalities.

It is important to distinguish semantically MISSVALues and NULLs because of their utterly different meaning in statistical information processing. For an adequate modelling of data dependencies, however, it is vital to discern situations where a necessarily present MODality of a QUality has not been recorded for some reason from situations where it is impossible for principal reasons to observe any MODality (of a given SCALE) at all; in a sense, NULLs are empirical values necessarily "missing" in particular circumstances (in traditional record-oriented database modelling, NULLs are dealt with either implicitly by introducing generalization hierarchies – cf. [Teorey, 1990] – or explicitly by stating "integrity constraints").

Symbolically, MISSVAL and NULL will be represented by symbols η and ϕ, respectively. As a mathematical shorthand notation, the SCALEDOMain of a QUality Q_k will be denoted by $D(Q_k) =_{def} \langle q_{k1},...,q_{kl_k} \rangle$ and the DIMDOMain of QUality Q_k by $\overline{D}(Q_k) =_{def} D_0(Q_k) \cup \{\eta\}$ where $D_0(Q_k) =_{def} D(Q_k) \cup \{\phi\}$ if Q_k is a *dependent* QUality, and $D_0(Q_k) = D(Q_k)$ otherwise. Furthermore, if a QUality Q_k's SCALE is of SCALETYPE ς and $\phi \in D_0(Q_k)$, $D_0(Q_k)$ will be abbreviated, generically, to ς_ϕ.

Definition 2.2.1–14:
The <u>DIMPROPS</u> of a QUality comprise several dimension specific description parameters such as DIMSCALE, AGGLEV, MEASLEV, MEASUNIT, MEASPREC, MEASTYPE, and ADDTYPE, as well as possibly further attributes not defined explicitly for the time being. <u>DIMSCALE</u> simply indicates the QUality's SCALETYPE; AGGLEV refers to the QUality's *aggregation level*; <u>MEASUNIT</u> denotes the physical *measurement unit* (if any) of the SCALE used; MEASLEV basically informs about the statistical scale *level* at chosen AGGLEvel (such as nominal, ordinal, interval, metric, etc.); <u>MEASTYPE</u> determines the functional role of the respective dimension from a *statistical modelling* point of view (factor, treatment, response, covariate, etc.); <u>MEASPREC</u> indicates the (numeric) *accuracy* of taken measurements (crisp, fuzzy, stochastic, etc.) as it may be relevant especially in the analysis of numerical data; ADDTYPE, finally, specifies the *additivity* behaviour of the QUality with respect to other property space dimensions and is defined succinctly below. Conceptually, the actual VALues of each element of DIMPROPS are comprehended in special sets (for instance, SCALETYPE for DIMSCALE). Some of the attributes are mandatory, such as DIMSCALE, AGGLEV, and ADDTYPE, while others depend on SCALETYPE or are optionally at all. Additionally, there may be a <u>DIMINFO</u> attribute providing further textual information about the QUality (for instance, a description of the actual measuring device).

Operationally, QUalities are OBJECTs responding to messages *lab?*, *dom?*, *dimdom?*, *dimprops?*, as well as one for each of the individual DIMPROPS (such as, for instance, *agglev?* or *addtype?*). The message *null?* will return TRUE if the asked QUality is dependent (that is, its DIMDOM includes the NULL value); the meaning of other messages is evident. All messages can also be used *assertively* for QUality definition by dropping the trailing *?*-sign or by replacing it with a semicolon if an argument follows; *null!* defines a QUality as a dependent one.

Among all QUalities, the *universal* QUalities *time* and *space* play a distinct role as temporal and spatial reference dimensions for observations. Although not being distinguished formally from other QUalities, they share a couple of specific properties: first, their labels are standardized (that is, *time* and *space* are always included in QUSET), and secondly, DIMSCALEs are predefined, viz. the DIMSCALE of *time* is a 'TEMPoral' SCALE, and the DIMSCALE of *space* is a 'SPATial' SCALE.

Additivity Constraints

Statistically, property spaces are arranged in order to aggregate data with respect to some of the spaces' dimensions. However, not all of the formally possible *summa-*

tions may be feasible in semantic terms whence this must be modelled explicitly at symbol level. Essentially, two types of observation, or measuring methods, must be distinguished with respect to a scale: measuring *over* a range or *at* a point. This distinction can be highlighted best by considering the temporal dimension: typically, cross-sectional observations on persistent objects are carried out *at* some point in time *over*, for instance, a spatial area as well as other observation dimensions. In case of *events* (that is, objects not extending in the time dimension), of course, the observation may range *over* time as well. Conversely, a particular dynamic phenomenon may be observed *at* different places *over* time (for instance, in an astronomical experiment involving several observatories). In each of these cases, apparently, summation is admissible only with respect to dimensions *over* which observation has been ranging; with respect to '*at*'-dimensions, of course, other forms of statistical data aggregation may apply nevertheless (such as taking – moving – averages etc.).

Thus, for the sake of easy specification, *defaults* are arranged for additivity constraints, viz. it is assumed that, most of the time, measurements or observations are taken on *persistent* individuals implying that measurement values are *additive* with respect to all dimensions *except* **time** unless stated otherwise. Additivity is described, for each dimension, by two special values as follows:

Definition 2.2.1–15:
  ADDTYPESET is a DEFSET comprising the SPECVALues 'FIXed' and 'FLUX'.
Definition 2.2.1–16:
  The ADDTYPE attribute of a QUality is a PLASET of composite VALues composed of a QUality (or QULAB) component and an ADDTYPESET component. For a particular QUality, its ADDTYPE defines the additivity status with respect to all *other* QUalities (of the same property space); if the status is FLUX then aggregations with respect to the indicated QUality is feasible, otherwise (status FIXed) it is not.

Thus, the default VALue for the temporal additivity constraint reads as '*time*.FIX' whereas, for any other QUality *Q*, the absence of additivity constraints is expressed as '*Q*.FLUX'. In view of these defaults, the ADDTYPE of a QUality needs only to state the VALues redefining additivity default settings which, in most cases, means that ADDTYPE is an *empty* SET. The ***addtype?*** and ***addtype:*** messages can be used to inquire and assign, respectively, the additivity constraints of a QUality.

Compressing QUality Definitions

Practical experience suggests to enhance the formal means to define property spaces. Although not really adding new expressiveness, some shortcut devices in quality specifications are desirable; in particular, quite frequently *clusters* of semantically similar description dimensions occur which, typically, are processed jointly in most statistical analyses. In this regard, METAMOD provides a *quality cluster* structure which comes in two strands: *solid* and *composite*. This gives rise to the following

Definition 2.2.1–17:
  A SOLQUality is an ordinary QUality as defined above without any particular semantic relation to other QUalities and, hence, with a *solid* (that is: unstructured) label QULAB. A COMQUality, however, is an element of an algebraic structure of *semantically related* QUalities where semantically related means that all COMQUalities belonging to the field, represented by a Cartesian product of QUFACTors, share *identical* DIMDOMains *and* DIMPROPS. In order to facilitate the efficient specification of such a field of QUalities, by convention QULABs of COMQUalities are expressed as *composite* VALues with components taken from the QUFACTors which are simple BASETs. A QUality's QUTYPE indicates whether it is a SOLQUality or a COMQUality by assuming either of the SPECVALues 'QSOLID' and 'QCOMPOSITE' of DEFSET QUTYPESET.

Conforming to general usage, the message *type?* sent to a QUality returns its type.

Definition 2.2.1–18:
  A QUCLUSTer is either a PLASET of SOLQUalities' QULABels sharing identical DIMDOMains and DIMPROPS or a BAXSET of QUFACTors generating the COMQUalities' QULABels of the Quality field (that is, COMVALs in dot notation). Formally, SOLQUalities are dealt with as 1-component QUFACTors. Each QUCLUSTer is assigned a *label*, QUCLUSTLABel, taken from LABSET QUCLUSTSET.

Definition 2.2.1–19:
  A DIMension of a property space is either a SOLDIMension (if it consists of a SOLQUality) or a CLUSTDIMension (if it comprises a QUCLUSTer). This distinction is represented by a DIMTYPE attribute assuming either of the SPECVALues 'DSOLID' or 'DCLUSTER' of DEFSET DIMTYPESET.

Thus, property spaces essentially are built of DIMensions which may be simple SOLQUalities, collections of SOLQUalities sharing scales and (most) attributes, or even structured arrays of COMQUalities. Especially in the latter case, property space definition is simplified considerably by saving the work of specifying over and over again semantically identical space dimensions. For the very same reason, DIMension OBJECTs respond to messages related to DIMPROPS just like QUality OBJECTs do; moreover, DIMension OBJECTs also accept messages like *type?*, *lab?*, and, if DIMTYPE equals 'DCLUSTER', *qufact?* (expectedly, these message selectors provide also an assertive mode for OBJECT definition).

  If CLUSTDIMensions are introduced which comprise non-additive QUalities, a shorthand notation is provided by replacing commas in the QUFACTors' BASETs with *vertical bars*; implicitly, this defines mutual additivity constraints between QUalities differing with respect to this QUFACTor's components.

In addition to the increase in efficiency gained by clustered specification, the clustering of QUalities, of course, adds yet more semantics to the data model in view of statistical analyses, as will become apparent later.

2.2.2 Property Spaces, Frames and Frame Linkages

After having defined measuring scales and populations, property spaces can be set up formally. In general, a property space comprises several DIMensions according to the multivariate observation structure of sample spaces; in any case, however, a property space definition must contain a *temporal* and a *spatial* axis since, evidently, any observation must happen somewhere at some time with respect to a spatio-temporal reference provided that discourse remains confined to *Newtonian* physics. Basically, an axis of a property space corresponds to a QUality and may, hence, use a particular aggregation level of the SCALE associated with an axis's QUality. Although SCALEs can be used ad lib, constrained *only* by the statistical properties of the measurement or observation procedure, the actual choice of SCALEs determines the subsequent possibilities of analyzing data from different property spaces jointly, as will be discussed shortly. Since, occasionally, populations may be heterogeneous in that not all of the observation dimensions included in a property space apply to each of the observable entities comprised in a population, some property space axes may become *dependent* on others which, in turn, calls for a specific treatment of these axes by, essentially, extending the DIMDOMains associated to dependent axes with a 'not applicable' value, viz. the NULL value introduced in **Def. 2.2.1–13**.

Using METAMODel, any kind of property space may be assembled provided that it comprises the universal QUalities *space* and *time* (it should be noted that a property space may contain several DIMensions with SPATial and/or TEMPoral SCALEs attached but those are distinguished formally from the universal QUalities *space* and *time* throughout). Each property space is now associated with a particular population contained in POPSET giving rise to the following definition:

Definition 2.2.2–1:
   A FRAME is an OBJECT consisting of a FRAMEUNIT, which is an element of POPSET, and a property space composed of at least three DIMensions, two of which are DIMensions holding the QUalities *space* and *time*, respectively. Each FRAME is given a *label*, FRAMELAB, taken from a LABSET called FRAMESET. The QUalities comprised in a FRAME's DIMensions are termed FRAMEQUalities.

By default, FRAMEs are labelled by their FRAMEUNITs' names as long as this results in a unique labelling of FRAMEs. Conceptually, a FRAME covers *all* observation DIMensions conceivably attached to the FRAMEUNIT; for practical purposes, however, such a *universal* FRAME's set of DIMensions may be *split* into a couple of smaller, possibly overlapping FRAMEs. For instance, the sample FRAMEs used in Chapters 3 and 4 for *persons* and (lodging) *enterprises*, respectively, could in fact

.2 THE DATA MODEL

be conceived of as excerpts of more comprehensive FRAMEs including several other DIMensions not considered in the present subject domains. However, the resulting set of split FRAMEs can be interrelated in statistical data aggregation and analysis only as far as the overlapping DIMs are sharing identical SCALEs (although the AGGLEVels – and, hence, statistical scale levels – may be different for respective axes in different FRAMEs).

For computational reasons, the DIMensions of a property space must come in some definite order settled during property space definition. Furthermore, within FRAMEs the DIMDOMs are represented in terms of AGGDOMains – that is, as BASETs – appropriately chosen: the MODalities of a FRAMEQuality are, in fact, names for GRUSETs of (SCALE) MODalities as defined in the respective SCALE-DOMain of the DIMension (in case that the chosen AGGDOM coincides with the underlying SCALEDOM – that is, the AGGDOM consists only of a subset of the *singleton* BASPARTition of this SCALEDOMain – the VALues of SCALEDOM are assumed as *names* for respective singletons in the AGGDOM, thus simply "lifting" the SCALEDOM MODalities to FRAME-level MODalities). Taken together, the FRAMEQUalities span the FRAME's (hypothetical) sample space represented as a factored (by QUalities) Cartesian product of AGGDOMs:

Definition 2.2.2–2:
The sample space, called <u>FRAMEDOMain</u>, of a FRAME is the set (sequence, computationally) of (FRAME-level) MODality combinations of the FRAMEQUs; these combinations are called <u>TRAITs</u>. These TRAITs, in fact, are represented intensionally in terms of an XGRID named <u>FRAMEGRID</u> which is a sequence of FRAMEQUalities actually defining, by their DIMDOMs, the Cartesian product of TRAITs. The *ordering* of FRAMEQUalities is represented by a <u>FRAMEPROFile</u> of cardinality <u>FRAMESIZE</u>. Both FRAMEGRID and FRAMEPROF are derived from the FRAMEARRangement used to set up a FRAME formally in terms of property space DIMensions.

Internally, the FRAMEPROFile is arranged as a PLASET of the FRAMEQUalities' QULABels. Letting denote $D(Q_k)$ the (FRAME-level) DIMDOM of FRAMEQUality Q_k, then $\mathbf{Q}_0 =_{def} \underset{k=1}{\overset{K}{\times}} D_0(Q_k)$, $\overline{\mathbf{Q}} =_{def} \underset{k=1}{\overset{K}{\times}} \overline{D}(Q_k)$, and, for completeness' sake, $\mathbf{Q} =_{def} \underset{k=1}{\overset{K}{\times}} D(Q_k)$, where K is the FRAMESIZE. $\mathbf{Q}_0$ comprises all TRAITs relevant in subject-matter terms whereas $\overline{\mathbf{Q}}$ is the set of *all* TRAITs of the FRAMEDOMain. Accordingly, a TRAIT $\mathbf{q} = (w_1, \ldots, w_K) \in \overline{\mathbf{Q}}$, $w_k \in \overline{D}(Q_k)$ for $1 \leq k \leq K$, is represented as XVALue $(\langle w_1 \rangle, \ldots, \langle w_K \rangle)$ of the FRAMEGRID; in subsequent usage, the term TRAIT may refer to either representation.

Access to the components of a FRAME is provided by a couple of messages with rather obvious meanings such as **lab?, unit?, arr?** (which, as usual, can also be used assertively by replacing the trailing question mark with a semicolon), as well as *prof?*, *grid?*, and *size?*.

In principle, each TRAIT represents one of the discerned sub-populations of FRAMEUNIT and, hence, FRAMEs provide a *conceptual* storage structure as follows:

Definition 2.2.2–3:
> A FRAMECELL is a pair of a TRAIT and an *instance storage register* holding the instances (that is, the population individuals' denotations) of the FRAMEUNIT's sub-population designated by the TRAIT. For each TRAIT defined by the FRAMEGRID, a FRAME comprises such a FRAMECELL. Symbolically, population instances are distinguished by unique INSTLABels taken from a LABSET called POPREGister; thus, FRAMECELLs in fact store INSTLABels.

Accordingly, TRAITs act as *keys* in the set of FRAMECELLs by which the contents of the instance storage registers are accessed. It must be kept in mind, however, that FRAMECELL merely is used as a conceptual device differing from actual symbol-level (and all the more: physical) storage organization. INSTLABels are arranged as COMVALues with at least *two* components the first one of which always refers to a FRAMEUNIT; the second component (and subsequent ones, if any) is then used to identify an individual within population FRAMEUNIT uniquely. By means of this convention, *population registers* are supported such that individuals can be traced throughout the modelled subject domain.

Structural Zeroes

From a semantic point of view, not every TRAIT representable formally in a FRAME is reasonable or possible at all. Thus, the portion of the Cartesian product of TRAITs defined in terms of a FRAMEGRID actually designating inadmissible (FRAME-level) MODality combinations (*structural zeroes*) need to be ruled out formally:

Definition 2.2.2–4:
> An inadmissible TRAIT, termed FRAMEDit, is represented as a BLOCK with a BASE equal to FRAMEGRID; since, frequently, FRAMEDits cover contiguous FRAME areas, they may be represented (non-uniquely, in general) in a condensed format by larger BLOCKs comprised in a then somewhat more parsimonious BLOCKCOLlection called FRAMEDITS. Each BLOCK in FRAMEDITS carries, in TAGS, a TAG (of TAGTYPE 'EDIT') with attribute ORIGIN set to the VALue 'frame'.

If a FRAME is free of FRAMEDs, FRAMEDITS is empty, $\langle\rangle$. Subsequently, the set of TRAITs in a FRAME remaining after removing all of the FRAMEDits is denoted as Q^*; accordingly, FRAMEDITS is a formal representation of $\overline{Q} - Q^*$. A FRAME's edits can be inquired by the *edits?* message.

.2 THE DATA MODEL

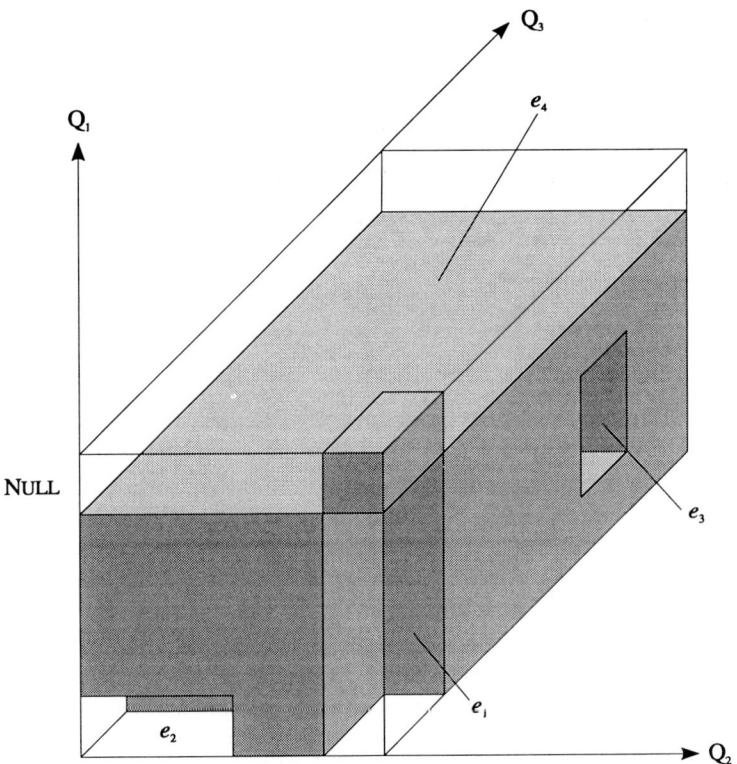

Figure 2.2.2–1: FRAMEDit Structures

If a FRAMEDit extends over the whole range of a FRAMEQuality, the DIMDOM of this *conditional* Quality *must* contain the NULL value; conversely, with respect to all other FRAMEQualities, FRAMEDITS will have to include FRAMEDs for all TRAITs having the NULL value in the conditional FRAMEQuality except for the portion of the FRAME where the former FRAMED extends over the whole range (except NULL) of the conditional Quality. For a simplified case with a 3-dimensional FRAME comprising one conditional (Q_1) and two independent FRAMEQualities (Q_2, Q_3), **Fig. 2.2.2–1** illustrates the resulting edits pattern by shading $\mathbf{Q}^*$ and leaving the FRAMEDits transparent. In this example, FRAMEDit e_1 covers the FRAME area where Q_1 does not apply, forcing Q_1 to assume the NULL value for all combinations of Q_2- and Q_3-MODalities comprised in BLOCK e_1 at the front-side right corner of the FRAME. FRAMEDit e_1, in turn, induces the *dual* "FRAMEDit" e_4 on top of the FRAME. Note that FRAMEDit e_2 in the lower front part of the FRAME does not extend over the whole range of Q_2 whence no NULL value needs to be introduced for this Quality. Likewise, the FRAMEDit e_3 shown in the rear part of the FRAME does not induce a further cut-out of the Q_1-NULL-FRAMEDit e_4. Of course, conditional QUalities may depend either jointly on several (independent) QUalities or recursively on other conditional QUalities (including mixed cases), with apparent consequences for the induction of further FRAMEDits.

Since each FRAMEQU's DIMDOMain comprises MISSVALue, the FRAMEDits induced by conditional QUalities extend to MISSVALues as well. In particular, a conditional QUality cannot take on any MODality except MISSVAL for the case of the conditioning QUality assuming MISSVAL already; this FRAMEDit will extend over the conditional QU's NULL value, too (since it remains undecidable if the conditional QUality had applied, would the conditioning QUality's MODality have been known).

FRAME Linkage

An instance of the METAMODel may be populated by several FRAMEs, either of identical or different FRAMEUNITs. Frequently, discerning several – related – statistical populations helps to better reflect the semantic structure of modelled reality and, additionally, may capture explicitly some salient features of population relationships that would induce rather contrived representations if squeezed into a single FRAME. FRAMEs sharing identical FRAMEUNITs are semantically related for obvious reasons; this implicit relationship is termed *unit linkage*. In case of FRAMEs with different FRAMEUNITs, however, semantic relations between FRAMEs, if any, need to be modelled explicitly in terms of *instance linkages*. In the latter case, two types of *instance linkage* are possible: *inter*-population and *intra*-population relationships between population instances (entities). METAMOD takes into account either of them, but remains restricted to *binary* relationships of *1:1*- and *1:n*-type, respectively (in case of inter-population relations, both *1:n*- and *m:1*-relationships between two FRAMEs are admissible simultaneously). Even in view of these restrictions, in case of intra-population linkages questions of relationship *consistency* may arise (for instance, symmetric kinship relations in a *person* FRAME) which, however, will not be treated here. It is mandatory for obvious reasons, that (explicitly) linked FRAMEs share identical SCALEs as far as there is an overlap in FRAMEQUalities. Furthermore, in order to maintain consistency, shared FRAMEDITS must coincide formally as well.

In relationship modelling, METAMOD borrows from ER-conventions (cf., for instance, [Chen, 1976; Teorey, 1990]) adapting them as appropriate to fit specifically the representation of relationships *between statistical populations*. In particular, METAMOD also distinguishes between (strong) entity types (= populations) and *weak* entity types, where weak entity types are understood to be *dependent* in their existence on the (strong) entity type they are linked to. For instance, with respect to the tourism example discussed in Chapter 4, *room* and *arrival* FRAMEs are weak since depending on the existence of the *enterprise* FRAME (cf. **Fig. 4.3.2–1**). Apparently, any weak FRAME is always related to exactly *one* strong FRAME. Adopting more or less common graphical usage of symbols, the frame (instance) linkages discerned in METAMOD are shown in **Fig. 2.2.2–2**. Single-line boxes represent

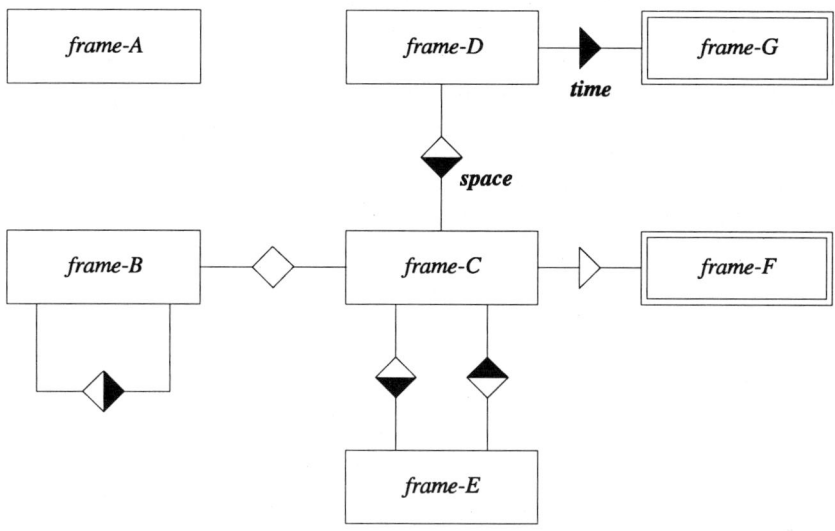

Figure 2.2.2–2: FRAME Linkage Types

strong entity FRAMEs, double-lined boxes weak entity FRAMEs. *1:1*-linkages are represented by blank squares whereas black triangles indicate the 'many'-side partner of a *1:n*-relationship. In "strong-weak" linkages the 'one'-side always points to the strong *parent* FRAME of the linkage; moreover, in this case relationships are symbolized as triangles instead of squares. In **Fig. 2.2.2–2**, *frame-A* is not related explicitly to any other FRAME (that is, only quality linkage may apply); *frame-B* has one *1:n intra*-population linkage as well as a 1:1 *inter*-population linkage attached to it. FRAMEs *frame-C* and *frame-E* are, in fact, in a *m:n*-relationship; *frame-F* and *frame-G* are both weak, depending on *frame-C* and *frame-D*, respectively.

A specific feature of *1:n*-instance linkages regards the "coupling" of universal QUalities, *space* and *time*, which applies if the spatial (temporal) relationship between related instances is fixed with respect to – observation – time (space); in other words, with respect to coupled QUalities repeated observation is disallowed or assumed infeasible for subject-matter reasons. Thus, as an example, in a spatial coupling in the linked FRAME for any *time* MODality *at most one* occurrence of an instance linked to a parent FRAME instance is admissible. Apparently, *spatial* couplings predominate but, for instance, in an astronomical observation setting it may happen that a particular phenomenon is observed *at the same time in different places* thus giving rise to a *1:n, temporally coupled* relationship between, say, a (strong) *phenomenon* FRAME and a (weak) *observation* FRAME. In case of weak FRAMEs, a definite coupling mode *must* be indicated for which, then, the respective universal QUality may be even completely *inherited* (that is, taken as is) from their parent FRAME. By definition, weak FRAMEs do not share any QUality except the universal ones with their parents; moreover, the aggregation level of a universal QUality not inherited by a weak FRAME *must not* be higher than the aggregation level of the respective QUality in the (strong) parent FRAME. Furthermore, the arrangement of such a linkage implies *additivity* with respect to the involved QUalities. In **Fig. 2.2.2–2**, *frame-C* is coupled with *frame-D* with respect to **space** whereas *frame-G* is ***time***–coupled to *frame-D*. If a pair of FRAMEs is linked *1:n* in either direction the *same* coupling mode must apply in both linkages for obvious reasons. Considering the tourism example again, a *room* of an *enterprise* obviously has, at any time, the very same location of the enterprise it is part of; hence, the (weak) *room* FRAME inherits the *space* axis from the *enterprise* FRAME.

Technically, explicit frame linkages are arranged as *frame association tables*:

Definition 2.2.2–5:
> A FRASSTAB is a binary relation (formally, an ordered sequence of pairs) the first component of which contains a single *key* label (taken from some POPREGister) whereas the second component is a PLASET of associated *reference* labels (taken from either the same POPREGister as the key labels or from a different POPREGister). Each FRASSTAB is identified by a two-component COMVALue label, FRASSTABLAB, such that the first component states the key FRAMELABel, and the second component the reference FRAMELABel.

FRASSTABs carry two attributes, FRASSKEY and FRASSREF, referring to the FRAMEUNITs attached to the key and reference parts, respectively; these attributes can be inquired by ***keyunit?*** and ***refunit?*** messages; ***frasstablab?*** returns an FRASSTAB's label. In order to retrieve the reference part (that is, the PLASET of labels for FRASSREF entities), a FRASSTAB is sent the keyword message ***key:*** followed by an appropriate argument key denoting a FRASSKEY instance.

FRASSTABs associate the linked FRAMEs' INSTLABels by adjoining the *key* part to the 'one'-side and the *reference* part to the 'many'-side of a linkage (thus, due to the symmetry of *1:1*-linkages, formally two FRASSTABs need to be arranged, one for each direction of interpretation). Accordingly, FRASSKEY will be set to the 'one'-side FRAMEUNIT and FRASSREF to the 'many'-side FRAMEUNIT.

In a particular FRAME linkage, instance linkages may be *infeasible* for a subset of TRAITs, that is population instances belonging to particular TRAITs of a parent FRAME are excluded from participating in the instance relation with the linked FRAME:

Definition 2.2.2–6:
> LINKEDITS is a BLOCKCOLLection (with BASE FRAMEGRID), associated with the parent FRAME of a FRAME linkage, denoting the set of TRAITs excluded from the instance linkage with a particular linked FRAME. The LINKEDits (BLOCKs) contained in LINKEDITS are TAGged exactly according to FRAMEDITS.

Formally, LINKEDITS put consistency constraints on FRAME linkages. As a matter of fact, LINKEDITS affect FRAMEs *as a whole*; it is impossible, apparently, to enforce linkage restrictions involving combinations of individual TRAITs in both FRAMEs linked (this can be achieved only by merging the linked FRAMEs into a single FRAME).

For simplified access, FRAME linkage information is attached directly to the involved FRAMEs; to this end, each FRAME carries a further description component:

Definition 2.2.2–7:
   The LINKPROPS of a FRAME comprise two attribute groups documenting FRAME linkages, viz. KEYLINKS and REFLINKS. The former attribute group comprises attributes indicating, for each of the linkages the FRAME participates in (i) a FRASSTABLAB, (ii) a LINKTYPE ('strong', 'weak') indicator, (iii) a LINKMODE (*space*, *time*) indicator (if a coupling mode is defined for the linkage), (iv) a LINKDEGree indicator ('*1*' or '*n*'), (v) LINKEDITS (except defaults), and (vi), if LINKDEGree is *n*, optionally a set of named (by AGGLABs) LINKSCALEs which, formally, are AGGLEVels for CARDSCALE to be used for *adjoined* QUalities (cf. Section 2.3). REFLINKS comprises only the first triplet of attributes (i) – (iii) indicating the linkages where the FRAME participates in the FRASSREF role.

These attributes are accessed, naturally, by **keylinks?** and **reflinks?** messages. Explicit linkages are stored by FRASSTABs only, if both FRASSKEY and FRASSREF are strong entity type FRAMEs; in case of a weak entity type participating in a frame linkage, the weak entities simply carry the INSTLABs of the strong entities they are linked to; for this reason, the FRASSTABLABs in KEYLINKS and REFLINKS are replaced by FRAMELABs accordingly.

Whenever more than two FRAMEs are linked (either implicitly or explicitly) in a METAMODel, the representation of *inter*-FRAME *additivity types* cannot be added simply as another attribute to KEYLINKS or REFLINKS structures; quite the contrary, this information pertains directly to (parent) FRAMEs. Basically, for all QUalities except the universal ones, unless the additivity *default* settings apply, an explicit ADDTYPE specification for each pair of QUalities (or DIMensions) from *different* FRAMEs is mandatory. The *intra*-FRAME additivity modes carry over without change from linked FRAMEs to parent FRAMEs. By extending **Def. 2.2.2–1**, the additivity information is stored as yet another FRAME component, LINKADDTYPES, comprising an association table of (pairs of) QUalities of linked FRAMEs and attached additivity types.

2.2.3 Domain Concepts

FRAMEs are the pivotal information structure of the METAMODel in that they provide the *conceptual frame* (hence the name), that is, a taxonomic language, partitioning a (statistical) population of observation (or sampling) units into an algebraic field of homogeneous sub-populations (TRAITs). This *theoretical data language*, in turn, supplies the terminological *handles* to statistical model building as well as to algebraic database operations, that is, statistical models and data manipulations will refer (only) to object sets representable in terms of algebraic expressions built out of the elementary vocabulary provided by the theoretical FRAME data languages.

From a subject-matter point of view, statistical analysis always concerns *quantitative statements about a specific population*; more precisely, with respect to populations as introduced formally in the METAMODel, statistical statements concern some selected set of TRAITs of a particular FRAME. The composition of such a set of TRAITs, of course, is determined by subject-matter, or *conceptual*, considerations which suggests the following definition:

Definition 2.2.3–1:
A domain concept definition, CONDEF, is a subset of TRAITs expressed in terms of a FRAMEGRID such that this subset does not comprise any FRAMEDits.

In symbol notation, a domain concept is an element $\mathbf{c} \subseteq \mathbf{Q}^*$ of a FRAME's *domain concept space* $\mathbf{C_Q}$, that is $\mathbf{c} \in \mathbf{C_Q} = 2^{\mathbf{Q}^*}$. Given that there are elementary (set) operations inducing the Boolean lattices over $2^{\overline{D}(Q_k)}$ for each of the FRAMEQUalities' K DIMDOMains $\overline{D}(Q_k)$, it is easily seen that all CONDEFs definable in a FRAME can be expressed in terms of (PLA)XGRIDs (with BASE FRAMEGRID); cf. Subsection 2.4.1. Formally, however, a different kind of representation, called *box* representation, is chosen for computational reasons:

Definition 2.2.3–2:
Each CONDEF is transformed into a PLABOX called CONBOX, the TEXTure of which equals the WRAPPING (cf. **Def. 2.1.2–20**) of the BLOCKCOLLection representing all TRAITs included in the CONDEF. Since the TEXTure of a CONBOX, termed CONGRID, is an XGRID, the difference between this CONGRID and the BLOCKCOLL of TRAITs it wraps is expressed as another BLOCKCOLL termed CONEDITS.

It must be noted that the transformation of CONDEFs into CONBOXes (facilitated by a *box!* message) introduces a second type of edits in addition to FRAMEDits both of which need to be distinguished semantically. Furthermore, FRAMEDITS and CONEDITS may overlap. Since all of the resulting edits are gathered in CONEDITS, its BLOCKs are equipped with TAGs of TAGTYPE 'EDIT' such that any FRAMEDit inherited by a CONBOX (that is, as far as it extends in the CONBOX) keeps the TAG with the ORIGIN attribute set to 'frame'; otherwise CONEDITS' BLOCKs get a TAG with ORIGIN set to 'condef'. The typical situations arising in this transformation stage are illustrated in **Fig. 2.2.3–1**: in each diagram, the shaded area indicates the CONDEF (which may be tessellated arbitrarily into BLOCKs); the rectangles represent FRAMEDits in a simple two-dimensional FRAME where it is assumed that Quality Q_2 depends on Quality Q_1. The thick-lined box in each diagram shows the resulting CONGRID whereas the added dashed blobs depict those BLOCKs of CONEDITS introduced by the transformation of CONDEFs into box representation.

.2 THE DATA MODEL

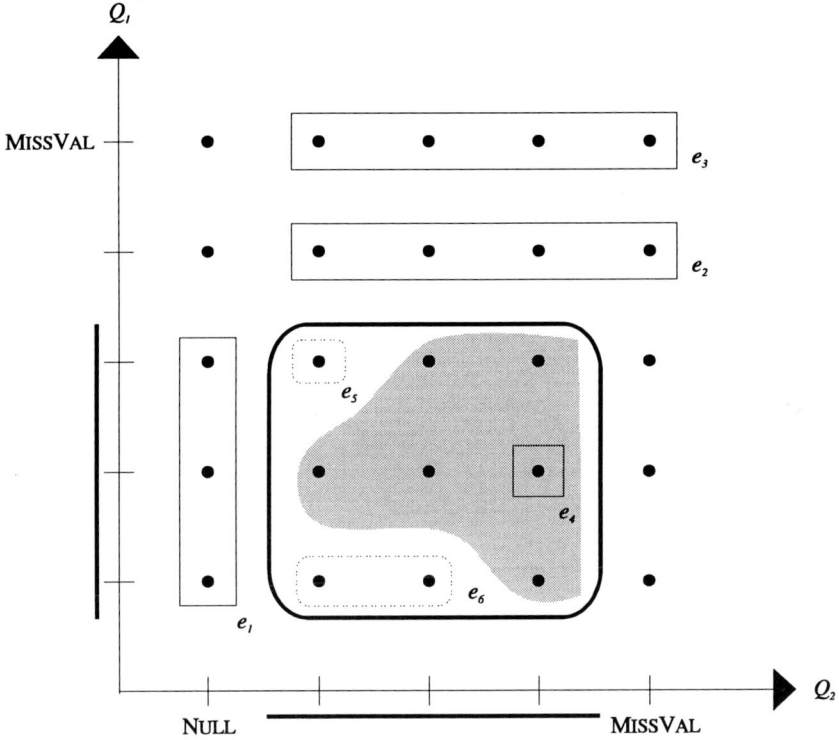

Figure 2.2.3–1 (a): CONBOX Representation with CONEDITS

Thus, in case **(a)** two BLOCKs, e_5 and e_6, are added to CONEDITS (each tagged with 'condef') while FRAMEDit e_4 is inherited from FRAMEDITS. Case **(b)**, next page, shows a simple case, where the resulting CONBOX will have an empty CONEDITS component. Case **(c)**, shown overleaf, is a little more complicated because CONDEF hits a FRAMEDit (e_2) extending over its whole range with respect to Q_2; therefore, the CONGRID must adjoin the NULL value of Q_2, thus inheriting the FRAMEDit e_1, too. Note also that the MISSVALue-TRAIT of Q_2 must not be included in the resulting CONGRID (due to standard box representation criteria), and that the remaining part of e_2 is inherited to CONEDITS as a BLOCK tagged with 'frame'.

Axis Rectification and Concept Profiles

Sometimes, the definition of a domain concept implies the specification of *constraints* involving several of a FRAME's QUalities simultaneously. Essentially, this means that a CONDEF is composed of non-orthogonal arrangements of TRAITs. Since, typically, there is a specific meaning attached to such kind of non-orthogonal domain concepts, it is certainly advisable to provide a formal structure reflecting

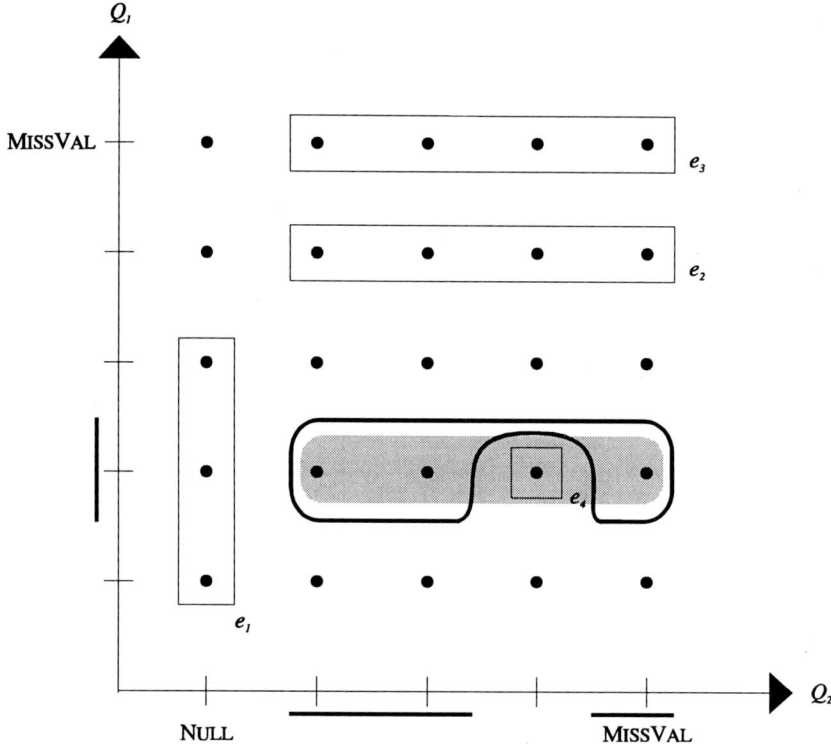

Figure 2.2.3–1 (b): CONBOX Representation without CONEDits

those concepts' semantics instead of restricting representations to ordinary box structures. In particular, it seems reasonable to support the representation of numerical constraints, termed *simplices*, which, basically, restrict the ranges of MODalities of numerical SCALEs by setting bounds to sums of VALues over several QUalities, that is, letting $Q_p,\ldots,Q_r$ be the QUalities involved in a simplex constraint, $\sum_{k=p}^{r} m_k \Theta n$ where $m_k \in D(Q_k)$, for all k, and $\Theta \in \{=,\leq,<,\geq,>,\neq\}$.

The QUalities participating in a simplex are required to be of *same* SCALE S; as a consequence, simplices may be defined either within a CLUSTDIMension of a FRAME or, otherwise, over a couple of DIMensions with identical DIMDOMains. The bounding VALue, n, must be chosen from the SCALEDOM of a *constraint* SCALE S' algebraically closed with respect to $(r-p)$-fold addition over S (note that if S is a CARDSCALE, S' is a CARDSCALE either; in other cases it is less easy to infer the constraint SCALE).

.2 THE DATA MODEL

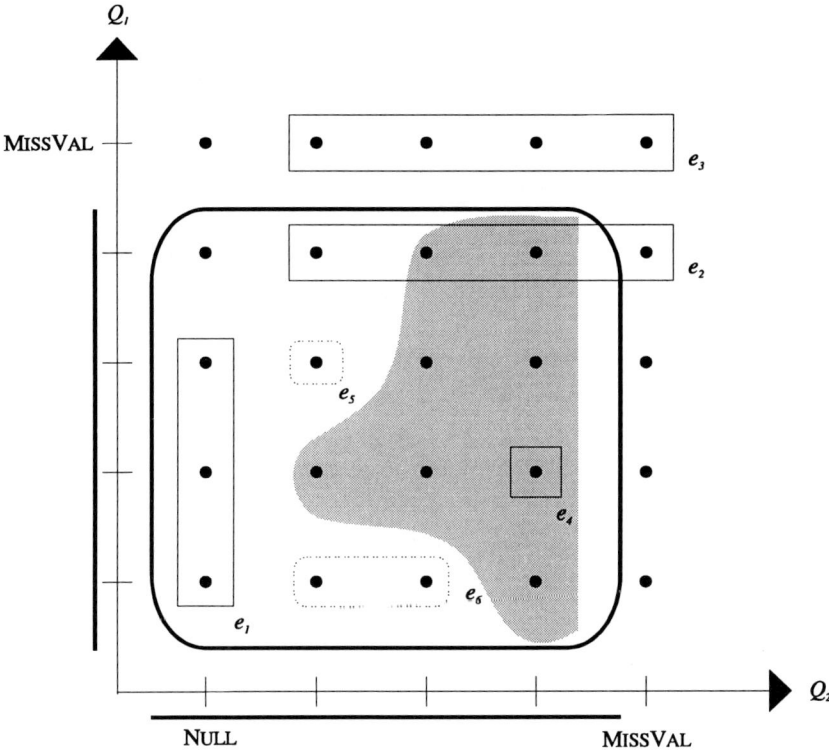

Figure 2.2.3–1 (c): CONBOX Representation Including NULL

Apparently, there is a further couple of structurally similar constraints (such as absolute value function, quadratic forms, etc.) which could be treated in like fashion but will not be discussed in detail subsequently. As a matter of fact, constraints may have little or even no structure at all; in such cases, of course, constraints need to be coded explicitly. **Fig. 2.2.3–2** (overleaf) illustrates both, unstructured and simplex, cases in a simple two-dimensional layout highlighting also the general principle followed in dealing with constraints: formally, FRAME-QUalities involved in a constraint are re-expressed in terms of a mapping to an intermediary DIMension, or *concept axis*. In the unstructured constraint case **(a)**, for each collection of MODality combinations a new MODality for the concept axis must be introduced whereas in case of structured constraints such as simplices (diagram **(b)** overleaf) the association of MODality combinations with the MODalities of the concept axis is arranged implicitly. For obvious reasons, unstructured constraints will be used in exceptional cases only, viz. if there is practically no desire to apply any breakdowns to the concept with respect to QUalities involved in such a constraint. Semantically, the MODalities of intermediary concept axes re-

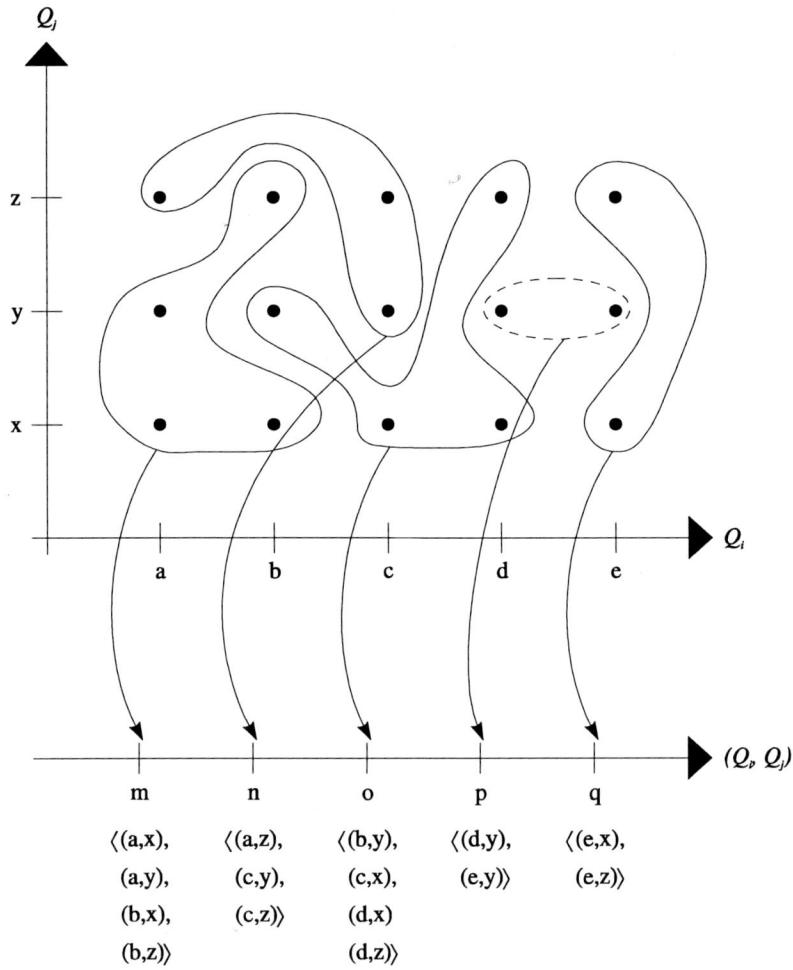

Figure 2.2.3–2 (a): AxRectification – General Case

present *supercodes*, viz. compositions of more elementary codes (defined at FRAME-MODality level).

In order to cope with constraints in formal domain concept definitions, the representation of CONBOXes is augmented with a *concept profile* which, in turn, refers to *axis rectifications* defined as follows:

Definition 2.2.3–3:
An AxRectification defines a *mapping* of a set of FRAMEQUalities to an Axis such that each XVALue in the RECTDOMain, that is the XGRID defined by the sequence of FRAMEQUalities called AxRECTPROFile, is assigned to *at most* one

.2 THE DATA MODEL

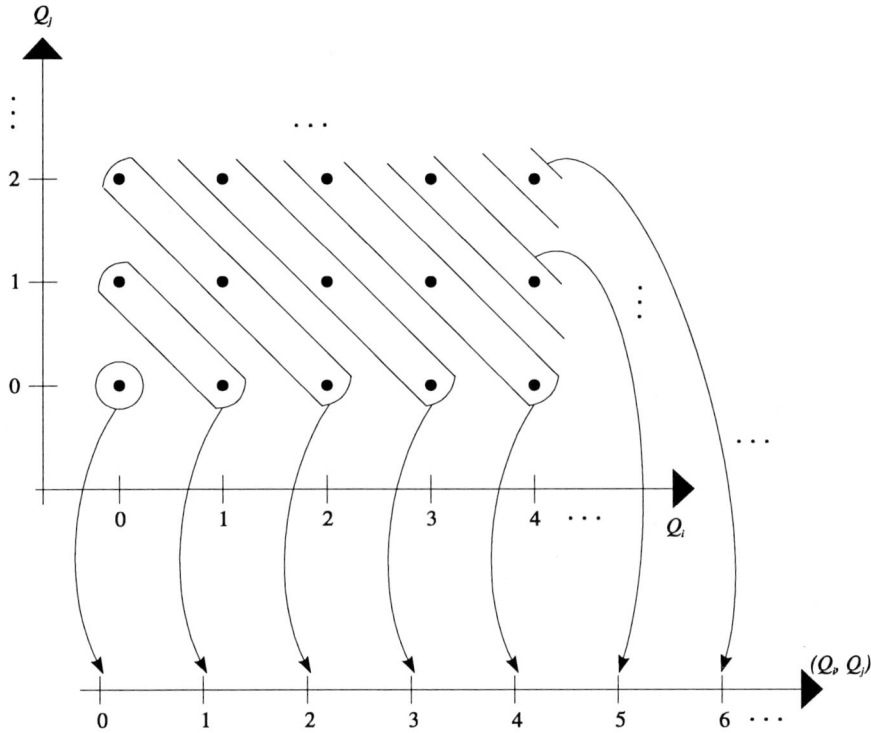

Figure 2.2.3–2 (b): AxRectification – Simplex Case

MOdality of the AxDoMain. Formally, an Axis is yet another QUality, the DIMDoMain – now dubbed <u>AxDoM</u>ain – of which ought to be chosen suitably to fit the intended RECTification mapping. In particular, the AxDoMains may include SPECVALues as specified for DIMDoMains. The actual definition of the mapping, <u>RECTDEF</u>, may be achieved either by setting up an explicit <u>RECT-TAB</u>le, associating each AxDoMain MOdality (dubbed <u>AxVAL</u>ue) with the set of RECTDoMain XVALues mapping to the MOdality, or by stating a symbolic <u>RECTFORM</u>ula implicitly characterizing the associated XVALue collection.

Care is in place when *conditional* FRAMEQUalities (that is, QUalities which only apply depending on specific MOdalities of other QUalities) participate in an AxRECTPROFile. In these cases, the resulting AxDoMain always comprises the NULL value except that the conditional FRAMEQUalities participating in an Ax-RECTPROFile depend *exclusively* on FRAMEQUalities *within* this very same AxRECTPROFile: then, of course, all FRAMEDITS involving the FRAMEQUalities participating in the considered AxRECTPROFile are vanishing at Axis level and, accordingly, the AxDoMain will *not* include NULL. **Fig. 2.2.3–3** (overleaf) illus-

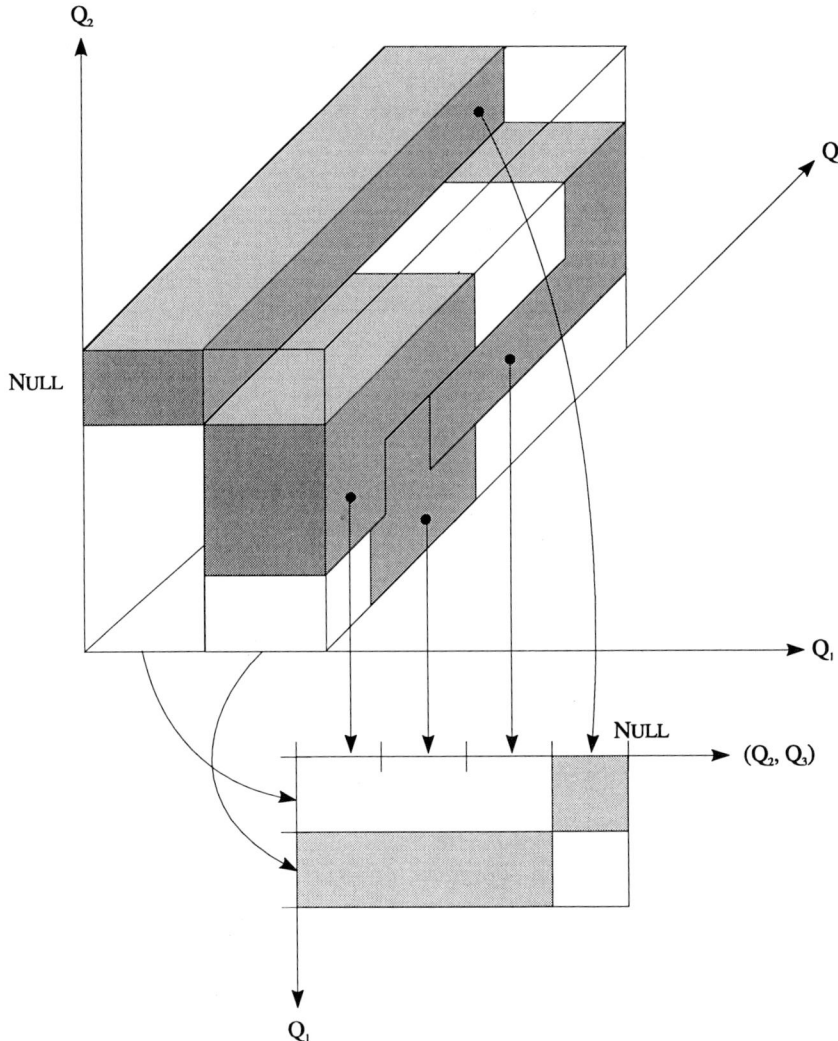

Figure 2.2.3–3: FRAMEDits in Axis Rectifications

trates, by a simple example, the transformation of FRAMEDits caused by an Ax-RECTification. The example assumes that Q_2 is dependent on Q_1 whereas Q_2 and Q_3 are rectified.

Normally, Axes are semantically identical to FRAMEQualities (that is, when no constraint applies); the AxRECTification amounts to the *identity* mapping then. Otherwise, the AxRECTification associates an appropriate Quality with the Ax-DOMain. Conceptually, replacing RECTDOMs with AxDOMs *rectifies* concept de-

.2 THE DATA MODEL

finitions at CONBOX level in order to facilitate standard CONCEPT processing (cf. Section 2.5). Having in mind the two-stage procedure of rectification, it may be reasonable for concept definition first to introduce rather general AXRECTifications the AXes of which then are used to arrange CONGRIDs just as if there had not been any rectification.

In case of special mapping functions used in rectification, syntactical patterns for coding the RECTFORMulae must be designed such as '*id*' for the identity mapping or

$$\Sigma_{Q_k \in C} Q_k \; \Theta \; n$$

for simple mappings on numerical SCALEs; in the latter case the symbolic denotation, in fact, has to be transformed, for numerical computations, into

$$\{q \mid \Sigma_{Q_k \in C} q[Q_k] \; \Theta \; n\},$$

where $q[Q_k]$ stands for the projection of TRAIT q over FRAMEQUality Q_k (thus returning the MOdality of a q with respect to Q_k) and C denotes the set of FRAMEQUalities involved in the simplex constraint. Note also that, in both of these cases, the AXDOM is set equal to the DIMDOMain of input FRAMEQUalities and, hence, any SPECVALues present in the (identical) DIMDOMains of such Qualities involved in a constraint are *inherited* to the AXDOMain.

In terms of rectification, the general CONCEPT concept of METASTASYS is now introduced:

Definition 2.2.3–4:
A CONCEPT is an OBJECT consisting of a CONPROFile and a CONBOX. The CONPROFile of a CONCEPT is a sequence of *concept axes* where each concept axis, CONAX, is defined by an AXRECTification such that, for reasons of consistency, the union of AXRECTPROFiles in these AXRECTifications covers *all* FRAMEQUalities of the FRAME the CONCEPT is based upon, and the AXRECT-PROFiles are *mutually disjoint* for all pairs of occurring AXRECTs. The CON-GRID of the CONCEPT's CONBOX, in fact, refers to the BAXGRID spanned by the CONAXes's AXDOMains (comprising CONVALues). Like Qualities, CONAxes are *labelled* by CONAXLABels which are QULABels. Each CONCEPT is *labelled* by a CONLAB taken from a LABSET called CONSET.

In view of CONCEPTs, CONDEFs are most naturally expressed based upon their AXDOMains. This, in turn, amounts to a re-expression of FRAMEDITS, as far as relevant anymore, at CONGRID level before the CONDEF can be transformed into the corresponding CONBOX. This, however, is easily achieved since the required re-expression of FRAMEDITS amounts to intersecting each FRAMEDit with the CONGRID in FRAMEGRID representation (viz. as CONDEF obtained by applying the

AxRect mappings *inversely*) and translating the outcome back to ConGrid representation (that is, using AxDomain MODalities instead of Frame-level MODalities).

Since Concepts are fairly complex Objects there is quite a lot of messages defined for Concept management only part of which will be presented here. Basically, *prof?* and *box?* retrieve the respective Concept components; *lab?* returns a Concept's label and *frame?* the associated Frame. Concepts also understand a *base?* message returning the BaxSet (BaxGrid, to be precise) the ConGrid is based upon which, due to intermediary rectifications, may differ from the one returned by the *grid?* message sent to a Concept's Frame. The ConProfile of a Concept is set up explicitly by instantiating a Concept with a *prof:* message the argument of which must be a consistent sequence of AxRectifications. Occasionally, Concepts with *identity* AxRectifications *only* will be called *normal* Concepts.

2.2.4 Data

Eventually, Concepts are used to represent *statistical figures*. In the Metastasys set-up, the "smallest" Concepts that can be expressed at all are Traits of a Frame, that is: sub-populations, and, hence, within MetaMod, data always means *summary* information. More specifically, FrameCells have been introduced (cf. Subsection 2.2.2) associating each Trait with an instance storage register destined to hold the InstLabels (elements of FrameUnit type) sharing Trait characteristics. According to the definition given in Subsection 2.2.2, those collections of members of sub-populations are termed *mesodata*, that is, observation records not yet subjected to a numerical aggregation procedure. Now, this conceptual data model is extended to a logical data structure. As a first step, the notion of *data box* is introduced:

Definition 2.2.4–1:
A <u>DatBox</u> is an Object identical to a ConBox except that (i) its TEXTure, called <u>DatGrid</u>, is *grouped* (that is, a DatBox is a GruBox), and (ii) <u>DatEdits</u> is a BlockCollection such that none of its <u>DatEdits</u> (Blocks) is partitioned further by the XGrouping induced by the DatGrid. The elements of the XGrouping induced by DatGrid are termed *concept traits*, <u>ConTraits</u>.

DatBoxes are the actual metadata structure for *meso*data: essentially, ConTraits can be justly viewed as a DatBox's *summary attribute* (in common terminology, cf. Section 1.3.1) although being actually modelled in Metastasys as an Object carrying several further attributes describing the represented aggregate value with respect to statistical, confidentiality, etc. characteristics and providing, if need be, specific (textual) annotations of relevance in data interpretation. The actual data values, viz. mesodata and macrodata, are kept in *data cells*:

Definition 2.2.4–2:
A DATCELL is a pair of a CONTRAIT and a DATREGister where DATREGister is a *data storage register* holding either a collection of INSTLABels (mesodatum) or the numerical value of a statistical *summary* function (macrodatum).

Note that, in the most extreme case, when the XGROUPing induced by DATGRID consists of a single CONTRAIT, only *one* DATCELL is attached to the DATBOX indicating that there is, in fact, no breakdown of the CONCEPT as defined by the underlying CONGRID. Conversely, an XGROUPing represents a cross-classification of the CONCEPT covered by DATGRID resulting in a corresponding number of DATCELLs according to the number of lattice points induced by the cross-classification expressed in terms of the XGROUPing. In many practical cases, DATCELLs may in fact represent generically defined *functions* (such as, for instance, statistical model equations) returning the contents of associated DATREGisters upon evaluation at respective CONTRAIT arguments.

The collection of DATCELLs attached to a DATBOX is comprised in a *file* structure:

Definition 2.2.4–3:
A FILE is an OBJECT comprising a DATBOX and a DATFILE where DATFILE is a sequence of DATCELLs such that each CONTRAIT induced by the DATBOX's DATGRID is associated with a DATCELL. Furthermore, the DATFILE carries an attribute, SUMTYPE, describing the type of summary value held in the DATFILE's DATREGisters. The admissible SPECVALues of the SUMTYPE attribute are gathered in a DEFSET called SUMTYPESET, among the SPECVALues of which there are, for instance, 'MESO', 'COUNT', 'SUM', etc. Depending on the SUMTYPE of the DATFILE there may be further attributes adding elements of summary type descriptions; for instance, in case of SUMTYPE 'SUM' there will be a SUMARGument attribute holding the CONAXLABs of the concept axes over which summation has been carried out. Each FILE is labelled by a FILELABel taken from a LABSET called FILESET.

Additionally, FILEs carry several other attributes as well, particularly those describing collectively the *(inferential-)statistical* properties of the data represented by the FILE. However, in the present set-up concentrating on the *algebraic* transformation of domain concepts in terms of data management operations, *inferential-statistical metadata* (cf. Subsection 1.5.2) will not be considered further and, hence, any reference to the corresponding formal representation structures are omitted. Likewise, the physical data organization will remain unspecified (except some considerations concerning the distribution of data; see Section 2.6). From a functional point of view, it is rather immaterial how the access to DATREGisters is actually brought about although, in practice, this issue may turn out less trivial and certainly needs a thoughtful solution especially for FILEs with a huge number of CONTRAITs (many of which may in fact have empty DATREGisters associated).

FILEs can be processed by messages like *box?*, *grid?*, *edits?*, *lab?*, *frame?*, *base?*, *prof?*, *sumtype?*, *sumarg?* etc. attributes accessing FILE components or components of sub-OBJECTs of FILEs; additionally, the *at:* message followed by a suitable CONTRAIT returns the contents of the addressed FILE's DATFILE *at* the indicated location (DATREGister). As usual, FILE messages also provide an assertive mode for FILE construction.

As a matter of fact, FILEs could be extended by attaching more than one DATFILE to a DATBOX such that these DATFILEs are of different SUMTYPEs (for instance, COUNT and SUM) to be combined arithmetically in subsequent computations. Conceptually, of course, there is no difference between this extended representation and a couple of separated FILEs with identical DATBOXes except that in the extended FILE representation a specific access mechanism would be required for addressing the individual DATFILEs it comprises.

2.3 Data Source Mapping

The METAMODel as described in the previous section provides a generic framework (or "shell", to use a term favoured in computer science) for representing statistical data aggregates. To become useful at all, this framework, of course, must be filled with actual data obtained from *data sources* (roughly, data sources can be conceived of as – physical – repositories of statistical observation data at any stage of processing). More specifically, in the present context this means, first of all, the arrangement of *metadata* describing structure and contents of the data sources which should so become integrated. In particular, existing as well as prospective data sources have to be captured *formally* such that – by means of these formal representations of data sources' contributions – the envisaged semantic integration of source data is achieved effectively.

Conforming to established usage (cf. [Sheth and Larson, 1990]), the term 'mapping' will be introduced to denote the formal "correlation" between semantics and source and semantics at the *federated* METAMODel level (but note that these mappings are defined in terms of statistical property spaces rather than database schemas, as "schema integration" usually amounts only to a subordinate *technical* task of statistical data integration and data harmonization). Referring (loosely) to Sheth and Larson's [1990] diagrammatic representation, METASTASYS's META-MODel instances are obtained by first converting local data models (LDM) – "local schemas" in database terminology – of participating source data holdings into so-called *component data models* (CDM) in a pre-processing step whereupon a formal mapping from component data models to the global *federated data model* (FDM), expressed as a METAMODel instance, takes place as depicted in **Fig. 2.3–1** (overleaf). Functionally, local component data models could be viewed as a specific kind of local "export schema" enriched with semantics required for arranging effectively the subsequent mapping(s) to the FDM set up; to this end, CDMs re-

.3 Data Source Mapping

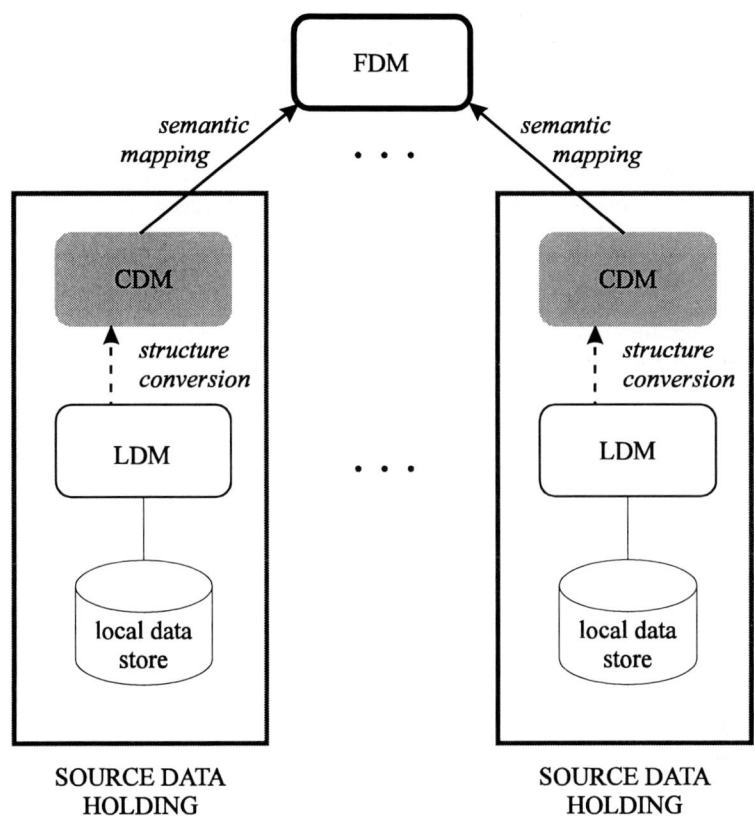

Figure 2.3-1: Source Data Holding Integration Architecture

express the LDMs involved using a predefined representation standard, that is, a uniform-structure description language. As a matter of fact, CDMs may exist *virtually* only provided that CDM structure can be evoked behaviourally at access time by pre-processing functions.

In technical terms, interoperability in the federated system of contributing source data holdings is facilitated by splitting local data model (or "schema", loosely speaking) translation and integration tasks in (i) a structure conversion step, migrating LDMs to CDMs, and (ii) an "ontology" conversion, or *semantic mapping*, step defining the subject-matter relationship between local and federated semantics. As the latter is captured in an OO-*common data model* (cf. [Pitoura *et al.*, 1995]) dealing with, essentially, a single data structure – generalized statistical tables – structure conversion basically has to bring about METAMODel-compliant interface behaviour; METASTASYS does not demand any specific representation requirements for

CDMs except that source data exported to the FDM must share the prescribed *OO view* (note that this is quite in contrast to the common concepts of schema integration where semantic integration and structure conversion are not separated that strictly).

The subsequent paragraphs of this section will outline how this *semantic mapping* of data sources is facilitated in METASTASYS and which preparatory measures have to be taken, in general, to enable the integration of data sources into a particular METAMODel instance. Throughout, the crucial relevance of the data source mapping step should be kept in mind: the virtual semantic integration of source data is determined *exclusively* by these mappings; all semantics dropped or omitted in arranging the formal mappings is lost inevitably for all data management and statistical operations taking place further on.

Despite the formal nature of data source mappings, their arrangement is obtainable mainly in a tedious manual way; economically, in a sense, the effort invested into the creation of mappings represents the added value of making data become (better) integrated semantically. The pivotal task in the development of formal data source mappings consists of achieving a terminology rich and powerful enough to express, within rather broad domain areas, the subject-matter concepts of interest and particularly, to account for a formal explication of semantic differences between the terminologies actually used in pertinent data sources. In restricting the focus to terminological aspects (and, thus, leaving out, for instance, the equally important pragmatic issue of the originally intended usage of data provided by a data source) a *theoretical data language* needs to be set up such that, for each population comprised in the language, the taxonomies of actual data sources (henceforth called *empirical data languages*) can be associated in formal terms with the respective elements of the theoretical data language. Plainly speaking, for each population a theoretical (that is, empirically possibly even fictitious) sample space is constructed such that the *images* of empirical sample spaces can be represented in terms of the theoretical sample space's data language. The arrangement of the basic vocabulary providing all notions necessary to fully establish the theoretical data language is a matter of skill and domain competence, though pragmatically guided by the terminological requirements for expressing the *differential semantics* of incorporated empirical data languages: apparently, semantic data integration addresses *relative* semantics only, that is, the theoretical data language merely states explicitly the relative difference between related notions of empirical languages, and never needs to go beyond this horizon.

A natural starting point for the inherently heuristic approach to designing a theoretical data language is an empirical analysis of a domain's *populations* (that is, the sampling units used), *sample spaces*, and *measurement scales*. As a first step, quite easy to achieve most of the time, the relevant domain populations can be singled out. Thus, the task is readily reduced to the creation of basic vocabularies for – possibly interrelated – populations. At data source side, sample space terminology, in general, is determined by observation procedures, survey questionnaires, or previous aggregations of sample or census data whence the compilation of a suitable theoretical *frame* language amounts to establishing a FRAME the QUalities of

.3 DATA SOURCE MAPPING 147

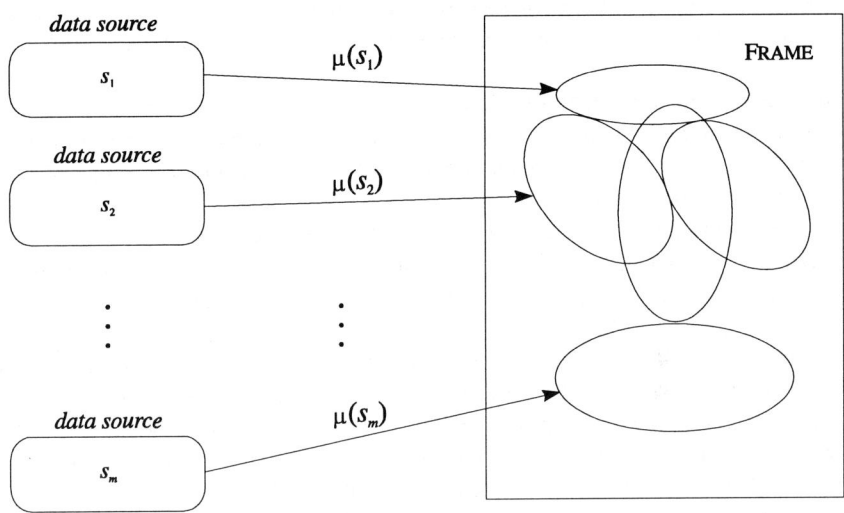

Figure 2.3-2: Data Source Mappings' Semantic Loci

which cater for re-expressing all of the really observable values in terms of the FRAMEQUalities' MODalities. Thus, basically, with respect to a data source contributing to a particular FRAMEUNIT, the main elements of a *source description* comprise

- the observation unit/population,
- a general source description providing various taxonomic, inferential-statistical, and technical metadata items (cf. Subsection 1.5.2), and
- the axis descriptions of the source's sample space, especially the measuring scales (observable values, measurement units) used.

Essentially, given that FRAMEUNIT and population, or sampling, units indeed match, the mapping of a data source – its component data model, in fact – amounts to the specification of a *mapping function*, μ, associating both, source axes with FRAMEQUalities, and observation values with MODalities. Diagrammatically, the situation is shown in **Fig. 2.3–2** where the "blobs" within 'FRAME' refer to the areas (collections of TRAITs), or *semantic loci*, covered by the data sources' images. From a technical point of view, the μ functions are obtained by first scrutinizing the involved data sources' empirical languages and deriving an appropriate theoretical (FRAME) data language; thereupon the empirical data languages are formally mapped, one by one, to the derived theoretical data language. This procedure may be rather straightforward if the structures of data sources and FRAMEs coincide to a large extent, but it may even turn out dauntingly awkward especially if the relationship between observation axes and QUalities is an intricate one (for instance, if the respective relationships are quite different between data sources), making it difficult to design FRAMEs such that source data can still be mapped without undue loss of information. Furthermore, depending on the type of sampling unit and the nature of observation axes (associated QUalities, respectively), the FRAMEs' additivity constraints are determined accordingly. Additivity may be constrained especially with respect to time, for instance in case of repeated surveys (so-called panels) where sample data are gathered several times from the same set of – persistent – sampling units.

In addition to these terminological mappings, *frame linkages* may be established (at mesodata level) wherever suitable. A particular difficulty in the formation of FRAMEs or, more precisely, the *domain concept spaces* established on top of FRAMEs, results from the typically rather disparate *source edits* (infeasible combinations of source values in data sources). In fact, there may be two kinds of source edits, viz. edits induced by a systematic *dependency* of an observation axis from another (such as, for instance, in case of conditional questions in a survey questionnaire), or *sporadic* edits denoting combinations of observation values which are impossible for other reasons. In contrast to sporadic edits which can be processed manually only, dependency edits are amenable to a somewhat more formal treatment; cf. Appx. B with respect to an automated edit derivation procedure. Despite this, however, it always must be decided by subject-matter criteria which edits are established as *global edits* at FRAME level and thus, by definition, as source edits; for reasons of consistency, of course, no data source can contribute soundly data to a global FRAMEDit.

Data Source Typology

Before a formal language for stating the μ functions can be developed, the various *types* of data sources occurring have to be surveyed. Having in mind the goal of being able to integrate virtually any kind of data source, at least four source types deserve closer examination (cf. Subsection 1.3.1):

- the classical *observation matrix* (usually dubbed "case-by-variate" matrix in view of the rectangular structure of observation values where each line of the matrix represents an observation case holding the values, column by column, of a multivariate observation) including its extension to multi-way data;
- *censuses* and *surveys*;
- *repeated surveys* (that is, sequences of structurally identical surveys repeated with respect to either the same, or partially changed, or totally different samples/(sub-) populations) including panels (longitudinal surveys), and
- already *aggregated observation data* (particularly, time series).

Tab. 2.3 summarizes these source types indicating – by asterisks – the prevalent data level (cf. Subsection 1.3.1) at which these source types usually provide data (asterisks are parenthesized for less frequent cases).

.3 DATA SOURCE MAPPING

Table 2.3: Source Types and Data Levels

Source type	micro	pre-macro	lower-level macro	higher-level macro
observation matrix	*			
survey	*	(*)	*	(*)
replicated survey	(*)	(*)	*	*
aggregate			(*)	*

Observation matrices and (one-shot) surveys, in general, are analyzed by the researcher(s) or institution(s) in charge of planning and carrying out the complete chain of data producing and analyzing steps; therefore, if available at all, microdata can be provided. However, with respect to surveys, aggregate breakdowns may be published only and access to microdata be inhibited for reasons of data privacy, for instance. Replicated surveys and other data collected periodically (for instance, on account of legal regulations) typically are available only as aggregates – that is: macrodata – mainly due to the bulky nature of microdata. Frequently, surveys and replicated surveys generate clusters of related macrodata at related aggregation levels such that, especially, the resulting macrodata share common marginals. It must be borne in mind, however, that ensuring non-redundancy of mapped data sources is of utmost importance since it is hardly possible to infer, or keep track of, the semantic relationships between data sources as well as between individual datasets at a later point in time except for those relationships created internally. As a further issue to consider, macrodata sources often provide arithmetically or functionally transformed source data; in those cases, it may become necessary to *reconstruct* the original property spaces generating the data before undergoing any transformation since, from METASTASYS's point of view, the empirical data languages of these original property spaces in fact have to be mapped. Hardly surprising, data sources do not yet provide data at mesodata level.

METASTASYS stipulates a data input interface at least at mesodata level; thus, microdata sources generally have to be converted to mesodata sources which involves, of course, a rather trivial pre-processing step (possibly keeping track of frame linkages as well) by no means affecting the terminological mapping of data sources. On the other hand, microdata excepted, all other source types essentially deliver data at levels readily fitting the METAMOD FILE structure since, naturally, numerical aggregate values are associated to combinations of source values, or *source traits*, to be mapped to (sets of) FRAME TRAITs which, in turn, are represented formally in terms of DATBOXes.

2.3.1 Source Views

In preparing the formal component data model underlying the data source mappings, *source* will denote henceforth a particular empirical sample space conforming to the (informal) definition of property spaces in Section 2.2. As a consequence, a data source in fact comprises more than one *source* in most cases. Now, mirroring the arrangement of FRAMEs, *source views* (of data sources) are captured in terms of METAMOD *source frames* and *source boxes* defined as follows:

Def. 2.3.1–1:
> SOFRAME is a conceptual property space used to describe a (view of a) data source formally; a SOFRAME's axes are termed SOVARiables, and to each SOVARiable a set of observation values, called SOVALues, is attached. A SOVARiable's SOVALues constitute a SCALEDOMain according to previous definitions which is extended by the SPECVALue 'MISSVAL' and, possibly (if the respective SOVARiable is *dependent* on other ones), the SPECVALue 'NULL'; the extended set of SOVALues is termed SOVARDOMain. To each SOFRAME, a FRAMEUNIT from POPSET is associated. SOFRAMELABel, taken from a LABSET called SOFRAMESET, denotes the unique *label* of a SOFRAME. Likewise, SOVARiables are *labelled* with unique labels, SOVARLABels, taken from a LABSET called SOVARSET.

Definition 2.3.1–2:
> The sequence of SOVARiables spanning a SOFRAME is called SOFRAMEPROfile the cardinality of which is SOFRAMESIZE. SOFRAMEDOMain denotes the set of tuples, called SOTRAITs, implicitly characterized by the SOFRAMEGRID, that is the XGRID composed of the SOFRAMEPROfile SOVARiables' SOVARDOMains. The set of *inadmissible* SOTRAITs is represented as a BLOCKCOLLection called SOFRAMEEDITS; each BLOCK in SOFRAMEEDITS carries a TAG (of TAGTYPE 'EDIT') with attribute ORIGIN set to VALue 'soframe'. The set of admissible SOTRAITs, expressing the "difference" between *all* SOTRAITs defined by SOFRAMEGRID and SOFRAMEEDITS, is termed *source concept*.

Definition 2.3.1–3:
> SOBOX is a PLABOX the TEXTure, called SOGRID, of which is a BLOCK with BASE SOFRAMEGRID, and the EDITS, SOEDITS, of which are a BLOCKCOLLection (of SOTRAIT BLOCKs), also with BASE SOFRAMEGRID, of *source edits*, SOEDits, suitably defined. Apparently, each BLOCK of SOFRAMEEDITS hit by SOGRID must be included in SOEDITS. Other BLOCKs in SOEDITS not inherited from SOFRAMEEDITS are TAGged with ORIGIN 'source'.

Practically, SOEDits may be represented by arbitrary subdivisions; in particular there is no need to distinguish between *explicit* and *implicit* edits as defined by Felligi and Holt [1976] since implicit edits are automatically covered by explicit edits (the distinction is relevant, however, for determining suitable imputations in

.3 DATA SOURCE MAPPING

preparatory data processing stages). Since METAMOD does not take care of or aim at any *canonical* representation of edits, further analyses of the edits' structure can be abandoned without harm.

Finally, a (conceptual) file of source data is represented in a *section* structure comprising, as sub-structure, a sequence of *source cells*:

Definition 2.3.1–4:
A source cell, SOCELL is a pair of a SOTRAIT, represented as a tuple (XVALue) of a SOFRAMEDOMain, and either a DATREGister as defined in **Def. 2.2.4–2** or a data-transforming *function* (evaluable code such as a database retrieval clause expressed in a suitable data manipulation language) returning the contents of the DATREGister upon runtime evaluation.

Definition 2.3.1–5:
SOSECTion is an OBJECT consisting of a SOBOX and a SOFILE where SOFILE is a sequence of SOCELLs such that each SOTRAIT within the SOBOX's SOGRID (which does not coincide with a BLOCK in SOEDITS) is associated with a SO-CELL. Each SOSECTion is identified by a *label*, SOSECTLABel, taken from a LABSET called SOSECTSET; SOSECTLABs are COMVALues composed of several (two, at least) components, the first of which always equals the SOFRAMELABel of the SOFRAME the SOSECTion contributes to.

With respect to a given SOFRAME, source datasets are represented in terms of SOSECTions. Like DATFILEs, SOFILEs are furnished with a SUMTYPE and, if need be, a SUMARG attribute; in case of mesodata sources, the SUMTYPE of the SOSECTion is 'MESO', in case of macrodata sources the SUMTYPE usually is 'COUNT' (but may be 'SUM' as well) accordingly. Note that it is not advisable to map source data of SUMTYPEs other than *elementary* ('MESO', 'COUNT', and 'SUM'; cf. Subsection 2.4.2) directly. Hence, to be eligible for data source mapping, source data aggregates of non-elementary summary types are better transformed into elementary SUMTYPE formats; if a SOSECTion should provide macrodata of a non-elementary summary type, it must comprise a further (trace) component containing a description *as if* the SOFILE were derived algebraically from operand macrodata of elementary summary type(s) according to META-STASYS's table calculus (see, particularly, Subsection 2.4.4). Occasionally, a data source – in its traditional information processing context – may comprise several SUMTYPEs at once; in such a case, of course, data sources (files) must be decomposed in a pre-processing step into partitions of homogeneous SUMTYPEs each of which, formally, has to be mapped individually (to the same FRAME, in general); see below.

Depending on the state of statistical processing of source data, various attributes attached to SOFRAMEs or, if varying between SOSECTions, to SOSECTions are used to record the information necessary to describe this processing state for subsequent computing, statistical analysis, and interpretation tasks (for instance, information about weighting schemes, applied anonymization procedures, description of sampling schemes, textual term definitions, annotations, etc.).

SOVARiables, SOFRAMEs, SOBoxes, and SOSECTions are processed by messages completely analogous to QUalities, FRAMEs, BOxes, and FiLEs, respectively, using the very same message selectors as far as applicable.

Pre-processing Steps

Datasets to be integrated into a federated METASTASYS data model are assumed to be *pre-processed* – or, in Sheth and Larson's [1990] terms, *data-transformed* – in several respects such that genuine local data semantics becomes re-expressed and perhaps augmented in order to conform to a "map-able" component data model as outlined above. First of all, clean and edited (that is, logically consistent data) are expected. In case of datasets sensible with respect to confidentiality, it is also expected that INSTLABels, if any, are changed technically to unique labels which prevent disclosure of the identity of the population instance being recorded; furthermore, (non-meso) input data may be modified suitably to increase confidentiality by applying anonymization techniques. If, at mesodata level, population instances are linked at all, these instance linkages must be recorded in AssTAB format (analogous to FRAssTABs; cf. **Def. 2.2.2–5**) for further processing.

In addition to this, the transformation of source data (as is) into SOSECTions implies, in general, a couple of further pre-processing operations. For one thing, SOFRAMEs must be arranged such that each SOVARiable gets associated with *one or more* FRAMEQUalities (cf. next subsection with respect to structural details of these associations). This, in turn, may require either of the following steps:

- *Nested* observation dimensions (for instance, questions in a survey questionnaire) have to be flattened out if the nested (conditional) dimension represents a refinement of response values of the dimension it depends upon, and both dimensions shall become associated with the *same* FRAMEQUality. In this case, letting $V^{(d)}$ denote the conditional variable with domain $D(V^{(d)})$, and $V^{(c)}$ the conditioning variable with domain $D(V^{(c)})$, such that $V^{(d)}$ is observed only if the response to $V^{(c)}$ has been $v^{(c)}$, the *condition* value, a SOVARiable V is arranged with domain

$$D(V) = \left(D(V^{(c)}) \setminus \{v^{(c)}\}\right) \cup D(V^{(d)}),$$

provided that, of course,

$$\left(D(V^{(c)}) \setminus \{v^{(c)}\}\right) \cap D(V^{(d)}) = \{\}.$$

- In case of *non-nested conditional* variables, the SOVARDOMains are augmented with the NULL value, ϕ.

.3 DATA SOURCE MAPPING 153

- *Implicit* variables of a source view are turned into explicit SOVARiables; in particular, spatial and temporal references of a source dataset have to be introduced as explicit SOVARiables (for becoming mapped to the universal QUalities, *space* and *time*) if not already present as source variable.
- Similarly, *lurking* dimensions need to be introduced as explicit SOVARiables. A source dataset's dimension is lurking if there is no variable becoming associated with any of the FRAMEQUalities the dataset is mapped to. In this case, a fictitious SOVARiable is arranged with a sole (arbitrary) SOVALue.
- All SOVARDOMains are extended to include MISSVAL.
- If it is not possible to achieve an association of source variables with FRAMEQUalities directly, the response values of several (related) source variables must be combined such that a single SOVARiable is obtained eventually the SOVARDOMain of which will, in fact, comprise *supercodes* of the combined source variables as its SOVALues.
- Conversely, if a SOVARiable gets associated with a couple of FRAMEQUalities such that the SOVALues can be decomposed into *a factored product* of these FRAMEQUalities' MODalities, it is advisable to arrange intermediary SOVARiables for each of the factors; the factor-SOVARiables are then associated *1:1* with the corresponding FRAMEQUalities (thus avoiding rather complicated SOVARDOMain mappings subsequently).
- Numerical variables, in general, receive a specific pre-processing due to the particular treatment of numerical SCALES in METASTASYS. Basically, numerical variables are represented as numerical SOVARiables but, since source data frequently comprise *macrodata* already, this fact needs to be represented explicitly. As a consequence, if numerical source values indeed represent aggregates (such as sums), the underlying SCALEDOMain (that is, the SCALE against which the population instances had – or may have – been measured against originally) must be introduced as SOVARDOMain and the type of aggregation actually applied must be stated accordingly by the SOFILE's SUMTYPE and SUMARG attributes. Very often, naturally, the same SCALEDOMain may apply to the SOVARDOMain and the FRAMEQUality becoming associated in the data source mapping.

In order to illustrate the last point, **Fig. 2.3.1–1** (overleaf) exhibits the set of SOTRAITs implicitly created by a simple time series recording, say, a sum value, s_t, for a sequence of time periods (for instance, years). Omitting the mandatory *space* axis for the sake of brevity, the corresponding SOFRAME will include a SOVARiable, called *value* in the figure, with a SCALEDOM of, perhaps, an N or P SCALE-TYPE. Thus, for each time period, t, the SOTRAIT extends over the whole SCALE-DOMain since

$$s_t = \sum_{i=1}^{I} n_i^{(t)} v_i^{(t)}$$

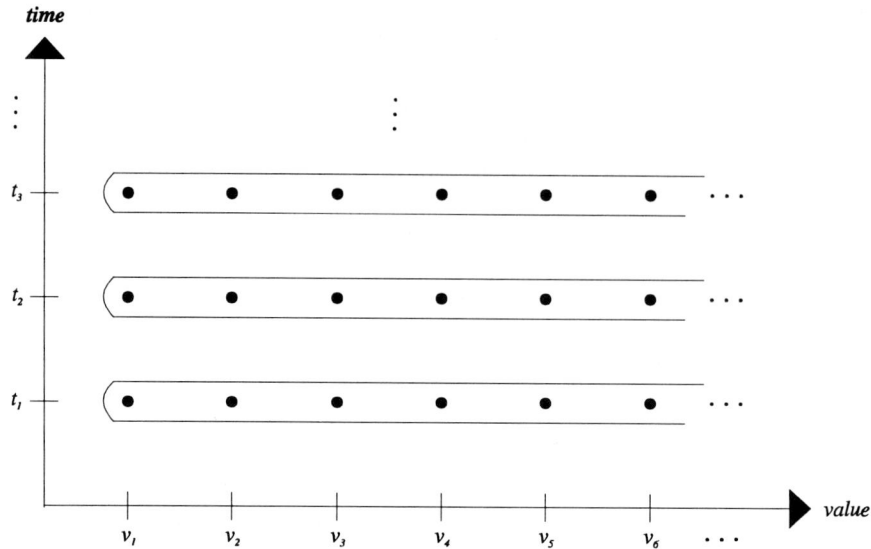

Figure 2.3.1–1: Time Series SoTraits Representation

where the $n_i^{(t)}$ are the not (any longer) known numbers (frequencies, in fact) of population instances (sampling units) with SoValue $v_i^{(t)}$ for the *value* dimension at time *t*; *I* denotes an upper bound chosen at least as large to cover the range of empirically occurring $v_i^{(t)}$'s. The s_t values, of course, are kept in DatRegisters associated with the SoTraits as depicted in **Fig. 2.3.1–1**. The SoFile's SumType, apparently, will be 'Sum' with SumArgument '*value*' indicating that the data values in fact represent sums taken over the *value* dimension of the SoFrame. By the way, remember that numerical Scales are always viewed as explicit property space axes in MetaMod which is somewhat opposite to the usual arrangement of macrodata where numerical values are associated directly with combinations of non-numerical values (that is, of categorical variables) rather than represented as additional axes; the representation, however, complies fully with standard *relational* microdata representation (especially observation matrices).

Data Source Roles, Facets, Sections, and Versions

With respect to the practical data organization, the source view as defined above benefits from a terminological refinement as follows. Traditionally, data production, maintenance, and dissemination centers around a few dedicated institutions being in charge of statistical data processing; in most cases, these institutions are statistical offices operating on a legal basis. Practically, with a network-oriented integration of data and, hence, data sources, in mind (cf. Section 2.6) a data source may fairly well contribute not only to a single FRAME but to several FRAMEs at the same time. The contribution of a data source to a particular FRAME will be called its *role* henceforth. Furthermore, within a role, a data source may contribute in different ways to the very same FRAME as will be explained in the next subsection; these different mappings will be called *facets* of a data source. Thus, a source view, essentially, amounts to the specification of a data source facet. However, facets determine only *how* a data source contributes data to a FRAME but does not tell which data it contributes. Actual data contributions are termed *sections*, and represented formally as SOSECTions (Sundgren [1992] prefers to call this the *occurrence layer*; cf. Subsection 1.3.1).

In practice, SOSECTions cover only part of the (intensionally) defined facet such that SOSECTions incrementally fill the source view with actual – physical – datasets (in Sundgren's [1992] terminology, facets roughly correspond to the *series layer*). As suggested by the name, sections ideally are *disjoint*, that is, a combination of sections (to larger ones) can be achieved simply by concatenating datasets since a homogeneous semantics is warranted by the corresponding facet definitions. Finally, the situation must be faced that sections may see several *updates* during their lifetime: updates mean that the SOSECTions' SOBOXes remain constant but the data held in attached SOFILEs is modified for whatever reason. For instance, in official statistics, datasets are updated every now and then when there is a pressing need to supply information quickly, but final numbers are published with delay after several rounds of imputations, corrections and adjustments. Although these updates may have little impact statistically, it is advisable to support keeping track of them (at least for the sake of deterministic reproduction) by distinguishing explicitly the *versions* of a section. Formally, versions are represented appropriately as SOSECTion attributes one of which being a TIMESTAMP indicating a version's *delivery* time. **Fig. 2.3.1–2** (overleaf) depicts some of the possible relations between a data source and several FRAMEs. Note that, if a data source contributes to *linked* FRAMEs, the association between linked FRAMEUNIT instances is inherited by *each* version of respective sections.

2.3.2 Mapping Definitions

Once a SOFRAME is associated with a FRAME on account of their shared FRAME-UNIT, the specification of μ amounts to the definition of axis mappings, that is, the taxonomic part of μ denoting the formal transformation of a data source facet to a FRAME is in fact *decomposed* into a set of mappings denoting the formal transformation of SOVARiables to FRAMEQUalities (disregarding, for the sake of simplicity, the inferential-statistical part of μ which, again, will consist of both, a general mapping of statistical data source characteristics as well as specific mappings for each of the involved axes). Of course, the decomposition of μ into component mappings cuts down complexity by an order of magnitude and, at the same time, increases the

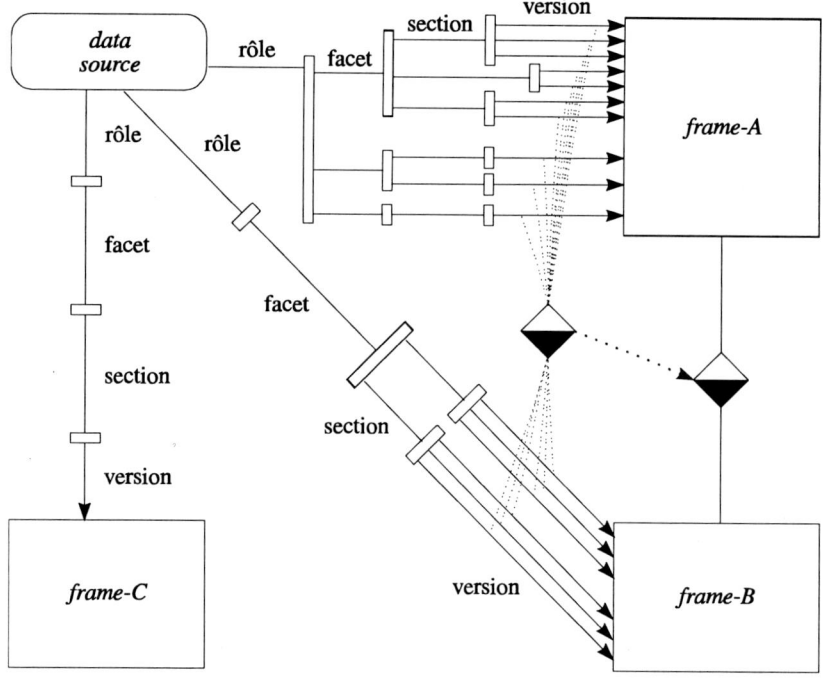

Figure 2.3.1–2: Relation Patterns Between Data Sources and FRAMEs

efficiency of mapping representations; moreover, it fits perfectly the standard representation of statistical aggregates chosen as fundamental data structure in METASTASYS, viz. the concept of BOXes.

Before expounding the formal set-up of data source mappings, it needs to be recalled that the relationships of SOFRAMEs to FRAMEs – as arranged in the METAMODel – on account of shared FRAMEUNITs may, as a matter of fact, involve *linked* FRAMEs as well. However, SOFRAMEs and FRAMEs may be linked differently; especially when data sources contribute *aggregate* data (that is, pre-macro- or macrodata), these aggregates may include information actually pertaining to *weak* FRAMEs attached to the aggregate's (strong) FRAMEUNIT but hide this relation in the attained aggregation level. As a consequence of this, *virtual* FRAMEs bear two types of QUalities, viz. those defined in their own FRAMEPROfiles as well as those "inherited" from linked FRAMEs; the latter QUalities are dubbed *adjoined* QUalities in contrast to the former ones, called *proper* QUalities. In particular, to a weak FRAME's proper QUalities all QUalities of its parent FRAME are adjoined. Conversely, a strong FRAME assumes all QUalities of a *1:1*-linked weak FRAME as adjoined ones whereas in case of *1:n*-linked weak FRAMEs, the weak FRAME's QUalities are arranged in a CLUSTDIMension of SCALETYPE N such that the weak

.3 DATA SOURCE MAPPING

FRAME's FRAMEPROfile, excluding universal QUalities, is converted to a set of QUFACTors and the AGGDOMains attached the FRAMEQUalities are turned into the QUFACTors' BASETs (basically, disregarding *space* and *time*, each TRAIT of the weak FRAME now corresponds to a distinct adjoined COMQUality in the parent FRAME). Thus, in what follows, both proper and adjoined QUalities of a FRAME are not discerned with respect to data source mappings. It must be noted, however, that the adjunction of QUalities from linked FRAMEs may introduce implicitly several further FRAMEDits, viz.

- the FRAMEDITS of the linked FRAME(s) as far as they relate to adjoined QUalities, and
- the LINKEDITS as defined for the involved FRAME linkages (cf. **Def. 2.2.2–7**);

in either case, these FRAMEDits extend over the other FRAME's whole FRAMEDOMain, except for shared FRAMEQUs.

Now, essentially, the image of a SOFRAME, under μ, is obtained by factoring out μ's component mappings: generically, for a SOFRAME $\mathbf{V}$ associated to FRAME $\mathbf{Q}$, $\mu: \mathbf{V}^* \to \mathbf{C}_\mathbf{Q}$, with $\mathbf{V}^*$ defined analogous to $\mathbf{Q}^*$, such that

$$\mu(\mathbf{v}) = \mu\big((v_1, \ldots, v_m, \ldots, v_M)\big), \text{ for } \mathbf{v} \in \mathbf{V}^*$$

is, in fact, replaced by

$$\mu(\mathbf{v}) = \underset{m=1}{\overset{M}{\times}} \mu_{lm}(v_m)$$

where μ_{lm} denotes the *m*-th component mapping. Subsequent paragraphs describe in detail the structure of component mappings and how these mappings can be arranged at all.

A source view (facet) mapping consists, basically, of two components: the mapping of SOFRAME and the mapping of SOFRAMEDITS. However, after the SOFRAME mapping has been achieved, the SOFRAMEDITS can be subjected already to this mapping in order to yield the *image* of SOFRAMEDITS. Hence, μ concerns primarily the mapping of SOFRAMEs.

As a first step of defining a component mapping, each SOFRAME's SOVARiable must be associated with the corresponding set of FRAMEQUalities by means of a *variable mapping*. In simple cases, a SOVARiable maps to a single FRAMEQUality but it may map to any combination of FRAMEQUalities as well. In view of the ordering of FRAMEQUalities (as induced by the FRAMEARRangement), the variable mappings are *sequence-valued* (that is, PLASETs) having as range the powerset of FRAMEQUalities. More specifically, the union of images of a SOFRAME's SOVARiables must result in a (proper) partitioning of the FRAMEPROfile. Formally, letting $\mu_V^{(s)}$ denote a variable mapping for a source s and letting $\langle V_1^{(s)}, \ldots, V_M^{(s)} \rangle$ denote the SOFRAMEPROfile of s with SOFRAMESIZE M, then

$$\bigcup_{m=1}^{M} \mu_V^{(s)}\left(V_m^{(s)}\right) = \hat{Q}_F$$

where $\hat{Q}_F = \left\langle Q_1^{(f)}, \ldots, Q_K^{(f)} \right\rangle$, the FRAMEPROFile of the associated FRAME

$$F = \mu_V(s)$$

such that *F unit?* = *s unit?*. Jointly, the variable mappings of a source view define an *image profile* as follows:

Definition 2.3.2–1:

The IMPROFile of SOFRAME s is the sequence $\left\langle \mu_V^{(s)}\left(V_{(1)}^{(s)}\right), \ldots, \mu_V^{(s)}\left(V_{(M)}^{(s)}\right) \right\rangle$ of IMAXPROFiles $\mu_V^{(s)}\left(V_m^{(s)}\right)$, $1 \leq m \leq M$, generating a GRUSET of FRAMEQUs ordered with respect to the QUality ordering induced by the FRAMEPROFile of FRAME $\mu_V(s)$ such that $\mu_V^{(s)}\left(V_{(i)}^{(s)}\right) \triangleleft \mu_V^{(s)}\left(V_{(j)}^{(s)}\right)$ for $1 \leq i < j \leq M$.

In preparing component mappings, each element of an IMPROFile, called *image axis*, is assigned an *axis domain* defined in terms of a rectification:

Definition 2.3.2–2:

The image axis domain, IMAXDOMain, of each image axis, IMAXis, is determined by an AXRECTification associating each IMVALue (that is, IMAXDOMain MODality) with a set of (one- or higher-dimensional) XVALues of the RECTDOMain induced by the respective variable mappings $\mu_V^{(s)}$; the RECTDEFinition may be defined in terms of a RECTTABle or a RECTFORMula (cf. **Def. 2.2.3–3**).

In the standard case, of course, IMAXis and FRAMEQUalities will be in *1:1* correspondence, although, in general, IMAXDOMains may not encompass the associated FRAMEQUality's entire DIMDOMain.

Now, for each SOVARiable involved in a SOFRAME mapping, the *value* mapping formally associates each SOVALue of the SOVARDOMain $\overline{D}$ (defined analogous to DIMDOMains of QUalities; cf. Subsection 2.2.1) with a set of IMVALues such that the union of the SOVALues' images results in a GRUSET with BASE IMAXDOMain:

Definition 2.3.2–3:

For each IMAXis $\mu_V^{(s)}\left(V_m^{(s)}\right)$, letting $\mathcal{H}_V^{(s)}$ denote a suitable BASPARTition of IMAXDOMain, the GRUSET of IMVALues is defined by the value mappings

.3 DATA SOURCE MAPPING

$\mu_{lm}(v)$, for all $v \in \overline{D}(V_m^{(s)})$, such that $\mu_{lm}(v) \in \mathcal{H}_V^{(s)}$ and $\mu_{lm}(v') \neq \mu_{lm}(v'')$, and hence disjoint, for $v' \neq v''$ ($v', v'' \in \overline{D}(V_m^{(s)})$).

Basically, IMVALues represent the *effective measurement resolution* (granularity) of observation as compared to the maximal (theoretically) resolution defined by FRAMEQUs. Thus, evidently, the DIMDOMains of FRAMEQUs determine the common language for re-expressing the observable measurement values (empirical languages of SOVARDOMains) as distinguished in different data sources.

The interaction between the specification of IMAXes and value mappings is illustrated in **Fig. 2.3.2** (overleaf); first, for a fictitious FRAME with QUalities Q_1 and Q_2 an IMAXis with six MODalities $\langle a,b,c,d,e,f \rangle$ in its IMAXDOMain is arranged whereupon SOVARiable V is mapped to this IMAXis by defining

$$\mu_{lV}(x) = \langle a,d \rangle,$$
$$\mu_{lV}(y) = \langle c \rangle,$$
$$\mu_{lV}(z) = \langle e,f \rangle.$$

In this sample mapping, IMVALue b is not used; it should be noted, however, that other sources may contribute to the arranged IMAXis in a different way (unless using other IMAXes definitions at all) employing another possible BASPARTition.

Most of the time, variable mappings will favourably gain singleton images, viz. SOVARiables are associated *one-to-one* with FRAMEQUalities. In such cases, of course, IMAXDOMains and the corresponding FRAMEQUalities' DIMDOMains will simply coincide.

Although the definition of value mappings as given in **Def. 2.3.2–3** is, in principle, formally complete it does not highlight the particular cases of mapping the SPECVALues MISSVAL and NULL. Trivially, MISSVAL always maps to MISSVAL, that is, for any SOVARiable V, $\mu_{lV}(\eta) = \langle \eta \rangle$. A little more intricate is the mapping of NULL values because of its dependence on whether the associated IMAXis is defined as *conditional* or not. Basically, if a SOVARiable were defined conditional while the associated IMAXis were not (the inverse case is inconsistent), two interpretations would be conceivable: where it does not apply (at data source side), the conditional SOVARiable may be considered as supplying either (i) missing data, or (ii) no data at all. Despite the seeming arbitrariness of this distinction, however, it turns out that adoption of interpretation (i) is inhibited by the definition of value mappings since, in that case, $\mu_{lV}(\eta) = \mu_{lV}(\phi) = \langle \eta \rangle$. As a matter of fact, admitting such an ambiguous mapping would blur the clear distinction between structural and empirical non-responders (where the latter, in contrast to the former, *could* have responded in principle). Conversely, interpretation (ii) leads to a *splitting* of the property space within the SOVARDOMain of the conditioning SOVARiable into both

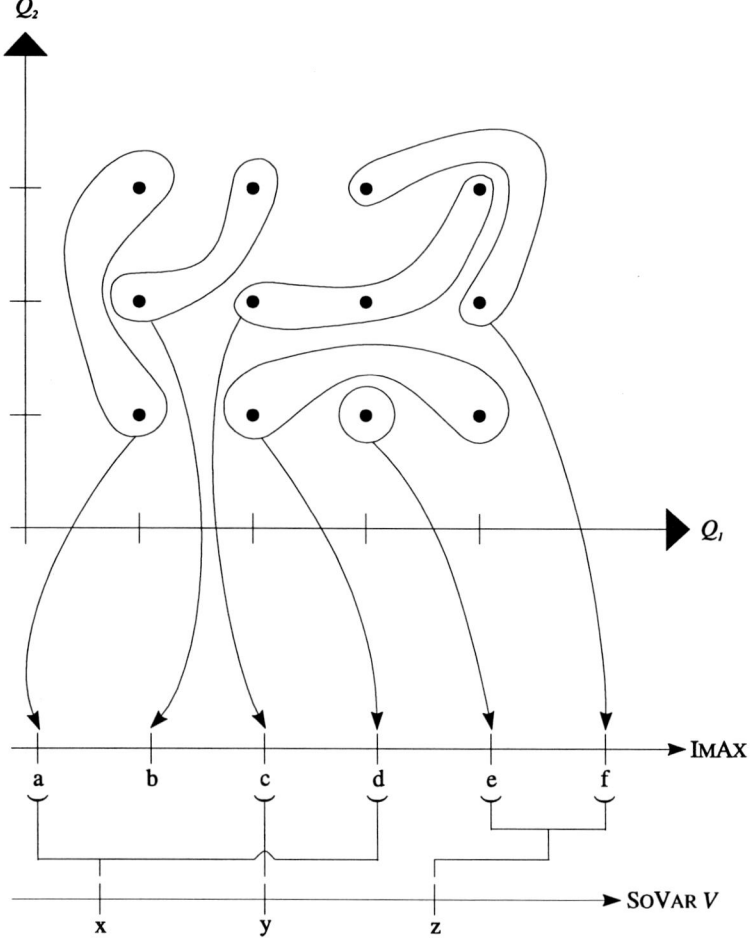

Figure 2.3.2: A Sample SoVar Mapping Using an ImAxis

alternative SoFrames, viz. where the conditional SoVariable does apply and where it does not. In the latter *facet*, the conditional SoVariable becomes a lurking one, being mapped to the sole ImValue comprising all Modalities of the corresponding ImAxDomain. In the other facet (SoFrame) so derived, the formerly conditional SoVariable is no longer conditional and thus Null can be dropped from its SoVarDomain. Otherwise, if *both* the SoVariable and the associated ImAxis are defined conditional, $\mu_{IV}(\phi) = \langle \phi \rangle$.

2.3.3   Source Images

Within a data source's roles the mapping of a facet s is represented formally in terms of variable and value mappings, denoted by functions μ_V and $\mu_{lm}^{(s)}$, respectively; furthermore, a facet is associated with a FRAME by the facet mapping $\mu_V^{(s)}$. In view of the derivation of statistical aggregates (cf. Section 2.5) both, these functions' images and the formal designation of the mapping functions' definitions, have significant importance within the devised table calculus. While mapping images are used to decide the derivability of statistical aggregates, function definitions control the actual translation of data access and data transformation steps expressed in image (that is: FRAME) terms into source (that is: SOFRAME) terms. In contrast to this, the further translation of retrieval and data management operations from SOFRAME terms into the implementation terms of actual data holdings is outside METASTASYS's scope.

The fundamental data structure used for representing data sources, called *image*, is defined as follows:

Definition 2.3.3–1:
 An IMAGE is a FILE OBJECT comprising an IMBOX and an IMFILE. IMBOX is a GRUBOX, the TEXTure, called IMGRID, of which is determined by the value mappings (IMVALues) for a SOFRAME's variable mappings rendering the IMAGE's IMPROFile. The IMEDITS of an IMBOX is a BLOCKCOLLection of IM-EDits (that is, BLOCKs), expressed in terms of BASE IMGRID, designating the *image traits*, IMTRAITs, not belonging to the mapped source concept or being ruled out otherwise in the SOBOX, that is those elements of the XGROUPing induced by IMGRID corresponding to SOEDits. IMFILE is a DATFILE such that each IM-TRAIT induced by IMGRID, except IMEDits, is associated with a DATCELL. Each IMAGE is assigned a unique *label*, IMLABel, taken from a LABSET called IMSET; IMLABs are COMVALues consisting of (at least) two components, the first of which always being the FRAMELAB of the FRAME the IMAGE contributes to, and the second component referring to the SOSECTion generating the IMAGE.

Formally, IMFILEs are transformed into SOFILEs simply by replacing the images of variable and value mappings with corresponding SOVARiables and SOVALues, respectively. For obvious reasons, the *versions* of a SOSECTion induce IMAGE versions as well which, again, can be represented by IMAGE attributes (moreover, versions could be referenced also by a *third* IMLAB component, for instance). The IMEDITS of an IMAGE can be derived mechanically by mapping the SOEDits one by one to the IMGRID; the mapped SOEDits keep their TAGS unchanged.

Being a FILE, essentially, IMAGEs can be processed by standard FILE messages as defined in Subsection 2.2.4.

Basic Mapping Components

The representation of μ is split into the components *facet*, *variable*, and *value* mapping each of which is stored by means of a *function definition*:

Definition 2.3.3–2:
A function definition, FUNDEF, is a sequence of pairs associating an ARGument (assuming a particular OBTYPE) with a MAPVALue (also assuming a particular OBTYPE).

FUNDEF OBJECTs are accessed either by **arg:** or **mapval:** messages; the latter, of course, responds deterministically only if FUNDEF encodes a *bijective* mapping. A FUNDEF is compiled by **arg: =:** messages establishing new (or replacing previous) ARG/MAPVAL-pairs. The OBTYPEs of a FUNDEF's ARGument and MAPVALue are inquired by **argtype?** and **mapvaltype?** messages, respectively. These messages are provided in an assertive mode as well.

Definition 2.3.3–3:
FACMAP is a FUNDEF associating SOFRAMEs with FRAMEs; technically, SOFRAMEs' and FRAMEs' *labels* are associated, that is, a FACMAP's ARG type is SOFRAMELAB and its MAPVAL type is FRAMELAB. Thus, FACMAP encodes the function μ_V.

Definition 2.3.3–4:
VARMAP is a FUNDEF associating SOVARiables with IMAXPROFiles (that is, PLASETs of FRAMEQUalities). For the sake of simplified handling, VARMAPs are assigned labels, VARMAPLABs, taken from a LABSET called VARMAPSET; strictly speaking, such a subdivision of this FUNDEF is not necessary since SOVARiables bear unique SOVARLABels. Thus, VARMAP encodes the function $\mu_V^{(s)}$ using *s* as VARMAPLABel. The ARG type of a VARMAP is SOVARLAB while its MAPVAL type is IMAXPROFile, that is a PLASET of QULAB (semantically, furthermore, the set of admissible QULABs is restricted to the FRAMEPROFile of FRAME $\mu_V(s)$).

Definition 2.3.3–5:
VALMAP is a FUNDEF associating SOVALs with IMVALs (that is, PLASETs of IMAXDOMain MODalities). VALMAPs are *labelled* by VALMAPLABs taken from a LABSET called VALMAPSET. Again, it is advisable to use two-component COMVALues as VALMAPLABels such that the first component refers (by label) to the mapped SOVARiable and the second component to the mapped SOFRAME. Thus, VALMAP encodes the functions $\mu_{lm}^{(s)}$ using *m* as first and *s* as second component of VALMAPLABel. The ARG type of a VALMAP is SOVALue while the MAPVAL type is IMVALue (constrained to the range of $\mu_V^{(s)}$).

.3 DATA SOURCE MAPPING
163

In addition to the ordinary FUNDEF messages, FACMAP, VARMAP, and VALMAP respond to a *lab?* message (returning the FUNDEF's label) which is used in assertive mode (that is, *lab:*) to assign a label to a FUNDEF.

Further Mapping Components

The outlined components of the mapping function μ concern the taxonomic metadata layer, essentially. Metadata elements of inferential-statistical and technical type (cf. Subsection 1.5.2) associated with source data have not been considered explicitly since these aspects – excepting summary type management – are of minor importance in the *algebraic* view of METASTASYS's table calculus. Nevertheless, it is conceivable to handle those types of metadata uniformly by attaching yet another component to FILEs (or to the BOXes containing the FILE structure description) holding a BLOCKCOLLection of TAGged BLOCKs such that pertinent metadata is carried by TAGs associated with BLOCKs extending over the relevant portion of the respective FILE (BOX). This kind of representation of source description elements offers the advantages

- of using information structures and algorithmic processing functions already available in the general set-up and
- of propagating the represented information through all operations applied to FILEs.

For instance, a particular annotation or definition reference associated with a SOSECTion will be conveyed through all table derivation stages although only part of the original SOSECTion may be kept at all; conversely, information being attached to dropped parts of a SOSECTion will be cut off simultaneously. However, due to the disjointness property of BLOCKCOLLs, metadata attached to BLOCKs may become subdivided (and thus replicated) in a somewhat unnatural way (unless, say, the BLOCK refers to a single IMTRAIT like a *footnote* to a table cell).

Every now and then, the extensions of SOSECTions may change; in particular, in case of periodically generated data (such as annual time series, etc.) the number of SOSECTions may grow beyond reasonable size. Provided that SOSECTions can be concatenated simply with respect to a particular SOVARiable (such as *time* in the periodically extended time series example) these updates are achieved rather easily: first, the SOBOX has to be adapted accordingly by extending the range (SOVAR-DOMain) of the SOVARiable in terms of which the SOSECTions are strung together; secondly, the SOFILEs must be merged to fit the resulting SOBOX. It should be noted, however, that from a formal point of view of subsequent table derivation there is absolutely no difference as to how SOSECTions are arranged effectively. Hence, the rearrangement of SOSECTions and, correspondingly, their IMAGEs, is driven primarily by physical data maintenance measures.

2.4 Data View Specification

The efficacy of statistical data retrieval depends upon two prerequisites, viz. knowledge about where to find pertinent data and knowledge about the *structure* of pertinent data. While the latter is indispensable for assessing if a record of data really contributes information to a concerned question or problem and certainly cannot be dealt with in a fully mechanized way, the former prerequisite clearly belongs to the genuine tasks of any information system deserving its name. In a sense, however, both prerequisites do not differ pragmatically; particularly, there is no point in looking for data that eventually will turn out useless with respect to the problem at hand. In other words, much could be gained if as much structure as possible entered the data search stage of the retrieval process such that only "promising" records of data are spotted and reported. Facing the rather impressively huge storage of statistical and scientific/empirical data gathered for generations and, at even increasing rates of volume, presently and in future in a distributed and asynchronous effort of society, the structural aspect of data queries is likely to be the only well-determined starting point of statistical data retrieval in most circumstances. Thus, although *context-free* support in locating data still may prove worthwhile, responses could very well be overwhelming as regards returned data volumes and at the same time be void of relevant data. With respect to this mournful outlook, the efficiency of "data hunting" ought to be improved by admitting a *structural specification* of the data sought in order to draw attention to rather qualified remaining data. In statistical terms, structural data specification amounts to describe

- *about what* an empirical statement,
- *which type* of statement, and
- *how* this statement should be made or is demanded.

Evidently, the *what* part refers to some sort of carrier of observable characteristics (statistical population); the *type* of statement determines the desired kind and level of statistical data aggregation of empirical data; the *how* part decides a specific presentation format of the derived statistical aggregate(s). In short, structural data specification should be limited consciously to this indispensable and, in logical terms, purely denotational statement; in particular, the data request should *not* involve prescriptive, or algorithmic, elements controlling the search or data aggregation process. Ideally, the structurally specified data request is set against a virtual statistical database (cf. Section 2.1) possibly – or rather, hopefully – providing pertinent data and, particularly, *structurally compatible* data. Moreover, in the latter case, retrieved data should and could be, if necessary at all, subjected to the statistical data aggregation demanded, thus generating explicitly statistical *data views* on the virtual statistical database.

.4 Data View Specification

METASTASYS's approach to data integration and, in particular, the data structures of its METAMODel have been devised to create a *uniform virtual statistical database* on top of existing distributed and heterogeneous statistical data holdings (cf. Section 2.6). In abstracting from both, the physical allocation of data and the structural heterogeneity of data holdings, the concept of data integration by means of a virtual database borrows from the relational database model [Codd, 1970] which strictly discerns a physical and a logical ("conceptual") database layer. Only the latter needs to be concerned in data retrieval, admitting conceptual data views irrespective of how data is actually stored at the physical data record level. Amongst several advantages of the relational model (for instance, cf. [Ullman, 1982]), the most important one is its proper emphasis on the logical structure of data represented in a *semantic* data model, that is real-world ontology in the modelled domain is cast into entities and relationships between entities, both expressed and represented in terms of attributes, or descriptive dimensions called "(random) variables" in statistics.

Despite this initial concordance between relational and statistical database systems, however, there is virtually little more that both types of system have in common. Essentially, traditional database theory (cf. [Ullman, 1982] as a primary source of reference in what follows) is focused on the design and implementation of transaction-oriented data stores such that each individual piece of data can be retrieved and logically combined with virtually any other piece of data stored in the database system (completeness property of retrieval languages). Apparently, this aim contrasts sharply with statistical data retrievals typically implying some aggregation applied to *collectively* retrieved (instance) data. While this is acknowledged very well in database research (for instance, cf. [Shoshani, 1982]) the superficial similarity between the relation format as used in relational databases and statistical observation data (notably, case-by-variate data matrices) has distracted many from the obvious fact that relation semantics and statistical aggregate semantics are rather discordant; for one thing, relational database theory concerns primarily the *column structure* of relations (functional and multi-valued dependencies in view of normalization of relations to prevent retrieval redundancies and update anomalies) whereas statistical data retrieval naturally has in mind *concept structures* (cf. Section 2.2), that is elements of concept (event) algebras defined on statistical property (sample) spaces. Although it is apparent that the category spaces of statistical aggregates (called *statistical entity spaces* in Rafanelli and Ricci's [1993] terminology; cf. Section 1.3.2) may be represented in terms of relations with a functionally dependent summary attribute appended and the category spaces can, in principle, be dealt with in terms of relational operators, this approach mistakes formal with semantic concordance: standard relational algebra does not reflect the specific structure of statistical aggregates and, hence, needs to be augmented with a lot of structure-preserving constraints eventually disturbing the advantages of formal coincidences with the relational model. More specifically, since statistical data retrievals do not retrieve (category space) tuples for their own sake but for denoting statistical (sub-) populations carrying some aggregate (summary) feature, operations applied to statistical aggregates, in general, transform systematically both, an aggregate's category space and its summary attribute(s) concurrently (cf. [Rafanelli and Ricci, 1991]). In order to support this conceptual view of statistical database operations adequately, the FRAME model has been introduced in METASTASYS where *domain concepts* designate the taxonomic scope (category space, loosely speaking) of statistical aggregates with respect to some FRAME and FILEs are the data representation structures attached to domain concepts. As a matter of consequence, this problem-driven conceptual view of structural statistical data retrieval also leads to internal data representations and operations quite different from those used in, for instance, relational algebra (the relational model, of course, may still be used favourably to record, for instance, technical metadata such as role-facet, facet-section, or section-version relationships).

The natural entities (instances) of statistical data models are observation units, that is individual members of statistical populations. With respect to the virtual statistical database envisaged, the generalized property space arranged for each such population of interest (cf. Section 2.2) resembles – though in a rather specific sense – a *universal* relation (*universal category relation* in Malvestuto's [1991] terminology). Recalling that, in relational database theory, the universal relation is a device to free the database user from memorizing the relation structure of a database's data model while still preserving the (join) dependency structure of involved relations and, thus, attains logical data independence, in METAMOD the role of universal relations is taken by *universal* FRAMEs in the sense that – with respect to a particular statistical population – each descriptive dimension (basically, a variable) fed or contributed by a data source considered is attached to this population's one and only *super*-property (sample) space. Operationally, a FRAME acts as a taxonomic platform (cf. Section 2.3) to integrate different data sources each of which possibly contributes only marginal distributions (*marginals*, for short) to the FRAME's full property space which, in relational terminology, amounts to say that a data source then consists only of a subset of FRAME "columns" or, in METAMOD terminology, a data source's variables actually map to a subset of FRAMEQUalities (letting all other FRAMEQUalities mutate into *lurking* source dimensions). Hence, relative to a particular universal FRAME, the structural specification of a statistical data retrieval abstracts from the structures of contributing data sources just as universal relation interfaces abstract from the underlying relation structures they are composed out of. Apparently, this type of logical data independence is a key concept on the way to achieving semantic data integration, that is in effectively establishing the virtual statistical database. Formally, as defined in Subsection 2.2.2, (universal) FRAMEs are represented by

- a FRAMEDOMain expressed in terms of TRAITs generated by a Cartesian product, $\mathbf{Q} = \times_{k=1}^{K} D(Q_k)$, of FRAMEQUalities Q_k, $1 \le k \le K$, and
- a BLOCKCOLLection of FRAMEDits to be subtracted from $\mathbf{Q}$.

In relational terms, letting $\pi_Q(r)$ denote the projection of relation r over component Q, trivially $\mathbf{Q} = \times_{k=1}^{K} \pi_{Q_k}(\mathbf{Q})$; although $\mathbf{Q}^*$, the FRAMEDOMain with FRAMEDits removed, could be composed by a natural join of suitably arranged partial

.4 DATA VIEW SPECIFICATION

FRAMEs, this is not advisable in view of sporadic FRAMEDIts inhibiting an effective decomposition (moreover, decomposed "FRAME relations" lead to rather clumsy representations) whence the *dual representation* of FRAMEs by $(Q, Q - Q^*)$ is in fact favourable. Roughly speaking, this dual representation, denoting Cartesian products $C = \times_{k=1}^{K} \pi_{Q_k}(C)$, such that $C \subseteq Q$, by $(\pi_{Q_1}(C), \ldots, \pi_{Q_K}(C))$, and using FILE structures, that is *mesodata* format as defined in Subsection 2.2.4, to align taxonomic concepts with actual data thus provides a simple, natural, powerful and in fact sufficient representation of statistical aggregates henceforth termed *table normal form* (TNF, for short). In view of TNF, database operations such as selections, joins, and projections are redefined appropriately (cf. Section 2.5) to model a complete and consistent algebra of statistical database operations allowing the formal algebraic transformation of mapped source data (*images* in TNF) into structurally specified statistical aggregates (*targets* in TNF either).

As already pointed out, each TRAIT $q \in Q^*$ denotes, as an atomic element of a population's property space, a particular (feasible) sub-population enclosed in a FRAME. Now, statistical aggregates always refer to domain concepts **c** representing non-empty sets of TRAITs $\{q|\psi(q)\}$, thus giving rise to a concept specification language resembling a tuple relational calculus language. Again, however, in contrast to standard relational calculus languages, ψ advantageously exploits FRAME structure, that is ψ is expressed favourably in terms of TNF either. This straightway suggests a rather specific query language as developed in the following subsections. Noting further that, quite frequently, arrays of taxonomically related statistical aggregates – disjoint unions of type $\bigcup_j \{q|\psi_j(q)\}$ – are requested instead of "stand-alone" aggregates and that, in fact, some kind of statistical aggregation $\theta(\bigcup_j \{q|\psi_j(q)\})$ applied to the sets of TRAITs rather than the sets of TRAITs themselves is of interest, the basic components of a statistical aggregate database language can be summarized (cf. Subsection 1.3.1) to comprise

1) a concept specification (*about what*),
2) (optionally) a cross-classification (*how*),
3) a summary type specification (*which type*).

Concept specification and cross-classification refer directly to the property space taxonomies of FRAMEs (that is, QUalities and SCALEDOMains or AGGDOMains defined on SCALEDOMains, respectively); sometimes also the summary type specification implies a reference to such a taxonomy. Thus, in ontological terms, these taxonomies fulfil the pivotal function in structural statistical aggregate specification by providing *universes of discourse* – (formal) languages to meaningfully and unambiguously denote statistical aggregates with respect to agreed upon taxonomic FRAMEs of reference.

If there is no cross-classification component in an aggregate specification, a *statistical indicator* [Olenski, 1992] (viz. a single non-structured aggregate value) is denoted; otherwise, cross-classification of a concept induces an *array* of (mutually disjoint) statistical indicators indexed by one or more common breakdown criteria expressed in terms of property space dimensions of the underlying FRAME, thus reflecting the algebraic relationship(s) between the domain concepts designating the taxonomic scopes of involved array elements, viz. scalar statistical indicators. Bearing in mind that, according to this definition, traditional statistical tables constitute a subclass of arrays of statistical indicators and that an array may degenerate to a scalar, any kind of statistical aggregate is subsumed formally to the notion of a *generalized statistical table* called META henceforth; it will be understood throughout that METAs are represented in TNF. In turn, the devised statistical database query language reduces to a generalized statistical table specification language, METASPEL, for expressing structural statistical data views vis-à-vis the virtual statistical database.

In a more comprehensive perspective, of course, generalized statistical tables are but the primitive operands of subsequent table operations either composing several – taxonomically related – METAs to ensembles (layout structures) or combining them arithmetically. This gives rise to several extensions of the core METASPEL structures which, however, will not be discussed in full depth. Moreover, the following exposition of METASPEL will refer to *internal* tables only, that is no attention is paid to the conversion of METAs into visible layouts (*external* tables); may it suffice here to state that this conversion, in general, is by no means trivial.

2.4.1 Target Concepts

Domain concepts as introduced in Subsection 2.2.3 basically are subsets of TRAITs of some FRAME; formally, for any concept c, $c \in C_Q$ if C_Q denotes the powerset 2^{Q^*} inducing the *concept algebra* over Q^* for some FRAME Q (letting Q represent, pars pro toto, a FRAME in what follows). In view of TNF, however, the direct specification of domain concepts in terms of TRAIT sets is replaced with a synthetic method of definition in terms of *hypervals* which are, if Q's FRAMESIZE is K, K-dimensional intervals representable as PLAXGRIDs according to **Def. 2.1.2–16** relative to the FRAMEGRID generated by Q's K AGGDOMains. Since any $c \in C_Q$ is representable as a (finite) set of hypervals ("measurability" of concepts) and hypervals, in turn, are expressed by *regular concepts* obtained from *primitive concepts*, the elementary foundation of concept specification is a language for primitive and regular concepts.

Relative to a FRAME Q, the simplest structure to describe verbally is a *restriction* of Q with respect to a single dimension. In this case, it obviously is sufficient to state the name of the restricted dimension and the actual range of this dimension. In METAMOD, dimension names are QULABels; the ranges may be any subsets of the DIMDOMain (excluding NULL) associated to this QULABel's FRAMEQUality.

.4 DATA VIEW SPECIFICATION

Formally, letting $D_+^{(k)} = \overline{D}(Q_k) - \langle \phi \rangle$ and $W_k \subseteq D_+^{(k)}$, for $1 \le k \le K$, primitive concepts are defined as follows:

Definition 2.4.1–1:
With respect to a FRAME **Q**, a primitive concept, PRIMCON, is a set of TRAITS

$$\{q \mid q \in \overline{D}(Q_1) \times \cdots \times \overline{D}(Q_{k-1}) \times W_k \times \overline{D}(Q_{k+1}) \times \cdots \times \overline{D}(Q_K)\},$$

for some $k \in \{1, \ldots, K\}$, denoted as PLAXGRID

$$\mathcal{C}_{k;W_k}^P =_{def} \left(\overline{D}(Q_1), \ldots, \overline{D}(Q_{k-1}), W_k, \overline{D}(Q_{k+1}), \ldots, \overline{D}(Q_K)\right)$$

with BASE

$$\mathcal{C}_{\text{BASE}}^P =_{def} \left(\overline{D}(Q_1), \ldots, \overline{D}(Q_K)\right).$$

Thus, to determine a particular PRIMCON it is sufficient to indicate k and W_k. A specific PRIMCON is $\mathcal{C}_Q =_{def} \mathcal{C}_{\text{BASE}}^P$, the concept representation of **Q** itself. Note that the definition of a PRIMCON may include implicitly one or several FRAMEDits in its scope; this does not cause any harm, however, since FRAMEDITS are taken account of indirectly by transforming domain concept representations into TNF as shown shortly. Given the definition of primitive concepts, *regular* concepts (called *orthogonal categories* by Chen et al. [1989]) can now be expressed simply as intersection of PRIMCONS:

Definition 2.4.1–2:
With respect to a FRAME **Q**, a regular concept, REGCON, is a set of TRAITS

$$\{q \mid q \in \overline{D}(Q_1) \times \cdots \times W_{k'} \times \cdots \times W_{k''} \times \cdots \times \overline{D}(Q_K)\},$$

such that

$$\mathcal{C}_{k';W_{k'}}^P \otimes \mathcal{C}_{k'';W_{k''}}^P =_{def} \left(\overline{D}(Q_1), \ldots, \overline{D}(Q_{k'}) \cap W_{k'}, \ldots, \overline{D}(Q_{k''}) \cap W_{k''}, \ldots, \overline{D}(Q_K)\right),$$

letting $\otimes$ denote the (primitive) *concept product* operation; for $r \ge 1$, a REGCON is properly stated as PLAXSET

$$\mathcal{C}_{k_1,\ldots,k_r;W^{[1]},\ldots,W^{[r]}}^R =_{def} \bigotimes_{j=1}^{r} \mathcal{C}^P\left(W^{(k_j)}\right),$$

letting $\mathcal{C}^P\left(W^{(k_j)}\right) =_{def} \mathcal{C}^P_{k_j;W^{[j]}}$ and $W^{[j]} =_{def} W_{k_j}$ for $1 \le j \le r$, with BASE

$$\mathcal{C}^R_{\text{BASE}} =_{def} \left(\overline{D}(Q_1), \ldots, \overline{D}(Q_K)\right) = \mathcal{C}_Q.$$

Evidently, the concept product is both, associative and commutative in its arguments. Furthermore, setting $k' = k''$, the space of REGCONs obviously subsumes the space of PRIMCONs. In terms of a formal concept specification language, a simple grammar of regular expressions will do (using BNF with Arden's replication star and printing *meta-grammar constants* boldface):

```
<rc-expr>   ::=   <pc-expr> [ & <pc-expr> ]*
<pc-expr>   ::=   <QuLAB> :  = <set-of-MoD>
```

where '&' reads as *and* to become translated into '⊗' operationally (of course, this language can be embellished considerably as discussed in Subsection 2.4.3).

Apparently, any hyperval expressible in FRAME terminology (excepting NULLs) can be represented as a REGCON. In order to represent any domain concept $\mathbf{c} \in \mathbf{C}_Q$ of FRAME **Q**, a concept sum is needed additionally:

Definition 2.4.1–3:
Given a FRAME **Q**, the (finite) domain concept space $\overline{\mathbf{C}}_Q$ comprises the set of all sums of REGCONs defined inductively as

$$\overline{\mathbf{C}}_Q =_{def} \bigcup_{n>0} \left\{ \bigoplus_{j=1}^n \mathcal{C}^R_{T(j)} \,\middle|\, T(j) \in \left[j \mapsto k_1, \ldots, k_r; W_{k_1}, \ldots, W_{k_r}\right] \right\}$$

where $\left[j \mapsto k_1, \ldots, k_r; W_{k_1}, \ldots, W_{k_r}\right]$ denotes the class of index mappings for all possible values $k_1, \ldots, k_r$ and $W_{k_1}, \ldots, W_{k_r}$, $r \ge 1$, and the *concept sum* simply denotes a union of regular concepts (with identical BASE),

$$\mathcal{C}^R_{T(j_1)} \oplus \mathcal{C}^R_{T(j_2)} =_{def} \left\langle \mathcal{C}^R_{(T(j_1))}, \mathcal{C}^R_{(T(j_2))} \right\rangle$$

such that the ordering of the union set (sequence) is determined lexicographically by the regular concepts' parameter sets, that is $\mathcal{C}^R_{(T(j_1))} \triangleleft \mathcal{C}^R_{(T(j_2))}$ (cf. Subsection 2.1.2), and the BASE of $\mathcal{C}^R_{T(j_1)} \oplus \mathcal{C}^R_{T(j_2)}$ is $\mathcal{C}_Q$.

.4 DATA VIEW SPECIFICATION

Like $\otimes$, the concept sum $\oplus$ is associative and commutative in its arguments. By extending the concept grammar with the rewriting scheme

$$\text{<dc-expr>} \quad ::= \quad \text{<rc-expr> [| <rc-expr>]*}$$

where '|' reads as *or* and translates into $\oplus$ operationally, domain concepts can be stated formally in terms of primitive concepts (as atomic structures), concept product ('&') and concept sum ('|') operations, resulting in expressions resembling logical *disjunctive normal form* used for representing the *characteristic concept function*. With respect to data structures, CONDEFs (cf. **Def. 2.2.3–1**) are naturally stored as BLOCKCOLLections (with BASE $\mathcal{C}_Q$) such that each BLOCK represents one REGCON of a CONDEF.

It is fairly easy to see that this CONDEF language is *complete*; obviously, $\mathcal{C}_Q \in \overline{\mathbf{C}}_Q$ and, letting $\mathbf{q}[Q_k] =_{def} \pi_{Q_k}(\langle \mathbf{q} \rangle)$, since, for some $w'_k \in \overline{D}(Q_k)$, $\mathbf{q}[Q_k] = \langle w'_k \rangle = W_k$ for any $\mathbf{q} \in \mathbf{Q}^*$ and, hence, there is a regular concept $(\langle w'_1 \rangle, \ldots, \langle w'_K \rangle) \in \overline{\mathbf{C}}_Q$ (the XVALue representation of a TRAIT), each domain concept $\mathbf{c} = \{\mathbf{q} | \psi(\mathbf{q})\} \in \mathbf{C}_Q$ in fact is representable (note, however, that $\mathbf{q}$'s with NULL components cannot be accessed directly). By extending the definition of the domain function D to concept representations $\mathcal{C}_c$ of concepts $\mathbf{c} \in \mathbf{C}_Q$,

$$D(\mathcal{C}_c) = D\left(\bigoplus_{j=1}^{J} \mathcal{C}^R_{T(j)}\right) =_{def} \bigcup_{j=1}^{J} D\left(\mathcal{C}^R_{T(j)}\right) = \mathbf{c}$$

where, for $1 \leq j \leq J$, $D\left(\mathcal{C}^R_{T(j)}\right) = D\left(W^{(j)}_1, \ldots, W^{(j)}_K\right) = \underset{k=1}{\overset{K}{\times}} W^{(j)}_k$ with $W^{(j)}_k \subseteq \overline{D}(Q_k)$, $1 \leq k \leq K$, suitably chosen. Thus, the domain concept space $\overline{\mathbf{C}}_Q$ provides indeed a language for $\mathbf{C}_Q$ in terms of a Boolean algebra $(\mathcal{C}_Q, \otimes, \oplus)$ with elements already represented in a format "close to" TNF (that is, as XSETs with BASE $\mathcal{C}_Q$).

The specification of domain concepts turns out as a rather tedious task if forced to obey explicitly all FRAMEDits; hence, as it has been tacitly assumed in previous definitions, FRAMEDITS are better dealt with implicitly: disregarding possibly interfering FRAMEDits reduces, in general, the number of regular concept components (*J*) of domain concepts considerably and thus supports a subject-matter view on concepts. Formally, this means that domain concepts are expressed in terms of $\overline{\mathbf{Q}}$ rather than $\mathbf{Q}^*$ and that, for the sake of simplicity, ψ is replaced with ψ' such that, with respect to the dual representation of FRAMEs, $\mathbf{c} = \mathbf{C} \cap (\overline{\mathbf{Q}} - \mathbf{Q}^*)$, letting $\mathbf{C} = \{\mathbf{q} | \psi'(\mathbf{q})\} \in \overline{\mathbf{Q}}$, for a concept $\mathbf{c} = \{\mathbf{q} | \psi(\mathbf{q})\} \subseteq \mathbf{Q}^*$. In other words, arranging a

CONDEF reduces to stating **C** instead of **c** since the conversion of **C** into **c** is achieved automatically.

As a further simplification CONDEFs can be assigned (unique) names (akin to the *def* statement used in the concept definition language of D'Angiolini [1992]; cf. Subsection 1.3.2) to be used in place of the fully stated CONDEFs; moreover, in conjunction with named CONDEFs concept *hierarchies* are established easily by applying the distributive law of $\otimes$ with respect to $\oplus$ letting, for any concept $\mathcal{C} = \mathcal{C}^R_{T(1)} \oplus \cdots \oplus \mathcal{C}^R_{T(u)} \in \overline{\mathbf{C}}_\mathbf{Q}$ and some primitive concept $\mathcal{C}^P_{k;W_k} \in \overline{\mathbf{C}}_\mathbf{Q}$,

$$\mathcal{C} \otimes \mathcal{C}^P_{k;W_k} = \left(\mathcal{C}^R_{T(1)} \otimes \mathcal{C}^P_{k;W_k}\right) \oplus \cdots \oplus \left(\mathcal{C}^R_{T(u)} \otimes \mathcal{C}^P_{k;W_k}\right).$$

This is accounted for simply by adding the rewriting schemes

```
<dc-expr>   ::=   <dc-expr> [ & <pc-expr> ]*
<dc-expr>   ::=   <CONLAB>  [ & <pc-expr> ]*
```

to the CONDEF specification grammar where <CONLAB> (cf. **Def. 2.2.3–4**) is the name assigned to a <dc-expr>.

Thus far, the formal concept language is capable to express the *about what* component of structural statistical aggregate specification. Additionally needed, however, is a method of determining the *how* component as well. At the taxonomic level, in particular, this component comprises the array structure of aggregates – so-called *cross-classifications* of (generalized) statistical tables. For obvious reasons, any cross-classification applies to *all* TRAITs included in a concept; more specifically, it regards all regular concepts a CONDEF is composed out of. This suggests an alternative kind of concept representation which also complies ideally with the dual representation of FRAMEs: as indicated in Subsection 2.2.3, a CONDEF $\mathcal{C}_c$ is converted to *box representation* by

- "wrapping up" $\mathcal{C}_c$ with a *regular* concept $\mathcal{B}_c \in \overline{\mathbf{C}}_\mathbf{Q}$ and
- representing the formal difference between $\mathcal{C}_c$ and $\mathcal{B}_c$ in terms of another CONDEF $\mathcal{E}_c \in \overline{\mathbf{C}}_\mathbf{Q}$;

thus, $\mathcal{C}_c$ is represented equivalently as a pair $\mathbf{B}_c = (\mathcal{B}_c, \mathcal{E}_c)$ such that $D(\mathcal{C}_c) = D(\mathbf{B}_c) =_{def} D(\mathcal{B}_c) - D(\mathcal{E}_c)$. Disregarding further components of BOXes for the time being, $\mathbf{B}_c$ is equivalent to the CONBOX representation of the domain concept **c**. By the way, $\mathbf{Q}^*$ transforms into $\mathbf{B}_\mathbf{Q} = (\mathcal{C}_\mathbf{Q}, \mathcal{E}_\mathbf{Q})$, the "ur-concept" of FRAME **Q**, letting $\mathcal{E}_\mathbf{Q}$ state the FRAMEDITS $\overline{\mathbf{Q}} - \mathbf{Q}^*$, since

$$D(\mathbf{B}_\mathbf{Q}) = D(\mathcal{C}_\mathbf{Q}) - D(\mathcal{E}_\mathbf{Q}) = \mathbf{Q}^*.$$

.4 DATA VIEW SPECIFICATION

In view of this unifying concept representation, cross-classifications now need only to be applied to $\mathcal{B}_c$ irrespective of the internal CONDEF structure of the CONBOX (cf. **Def. 2.2.3–2**). Formally, a cross-classification is *modulated upon* $\mathcal{B}_c$ by inducing a *grouping* structure on $\mathcal{B}_c$'s axis components. This grouping structure, of course, applies to a subset of FRAMEQUalities only; those FRAMEQUalities in terms of which a concept is actually cross-classified will be called *cross-classifying* dimensions accordingly whereas the remaining FRAMEQUalities are termed *reference* dimensions. In technical terms, grouping structures are expressed as suitably adapted BASPARTitions (**Def. 2.1.2–13**) as follows:

Definition 2.4.1–4:
Let $\mathcal{H}(Q_k)$ denote the set of all BASPARTitions of Q_k's defining AGGDOMain; then the set of DIMPARTitions Q_k, $\overline{\mathcal{H}}(Q_k)$ is defined as

$$\overline{\mathcal{H}}(Q_k) =_{def} \{h'_k \cup \langle\langle\eta\rangle\rangle \,|\, h_k \in \mathcal{H}(Q_k)\}$$

where $h'_k = \begin{cases} h_k \cup \langle\langle\phi\rangle\rangle & \text{if } \phi \in D_0(Q_k) \\ h_k & \text{otherwise} \end{cases}$.

Now, $\mathcal{B}_c - \left(W_1^{(c)},\ldots,W_K^{(c)}\right)$, with $W_k^{(c)} \in \overline{D}(Q_k)$, is replaced with

$\mathcal{B}_c^G = \left(G_1^{(c)},\ldots,G_K^{(c)}\right)$ where $G_k^{(c)} \in \mathcal{G}_k = \left\{g \cap W_k^{(c)} \,|\, g \in \overline{\mathcal{H}}(Q_k)\right\}$, for $1 \leq k \leq K$,

such that $\gamma(\mathcal{B}_c^G) = \gamma\left(G_1^{(c)},\ldots,G_K^{(c)}\right) =_{def} \left(\gamma\left(G_1^{(c)}\right),\ldots,\gamma\left(G_K^{(c)}\right)\right) = \left(W_1^{(c)},\ldots,W_K^{(c)}\right)$

with $\gamma\left(G_k^{(c)}\right) =_{def} \bigcup_{g \in G_k^{(c)}} g$; introducing a grouping structure does not, of course, change a concept's taxonomic scope, that is $\mathcal{B}_c^G\,plain! = \mathcal{B}_c$. More specifically, for each *reference* dimension k, $W_k^{(c)}$ is replaced with $g_0^{(k)} \in \mathcal{G}_k$ such that $\left(W_k^{(c)} \cap D(Q_k)\right) \in g_0^{(k)}$ and, if $\eta \in W_k^{(c)}$, also $\langle\eta\rangle \in g_0^{(k)}$. For each *cross-classifying* dimension k, a $G_k^{(c)} \neq g_0^{(k)}$ is substituted; the latter DIMPARTitions are specified explicitly in terms of AGGLEvels as expounded in Subsection 2.4.3 below. With respect to FRAMEQUality Q_k the GRUSET $G_k^{(c)}$ determines the grouping of (AGGDOM-) MODalities into groups $g \in G_k^{(c)}$, that is the MODalities comprised within a group g are subjected to the statistical aggregation function $\theta(\mathcal{C}_c)$. Hence, $\mathcal{B}_c^G$ induces an XGROUPing (**Def. 2.1.2–19**) of $\mathcal{C}_c$'s TRAITs such that each element of the XGROUPing represents a single statistical indicator. Apparently, if each $G_k^{(c)}$ comprises a *single* group $g_k^{(c)} \subseteq D(Q_k)$ only (except $\langle \eta \rangle$ and $\langle \phi \rangle$) there effectively is no cross-classification – although there may have been specified one – and, hence, the statistical aggregate is a (single) statistical indicator. At the taxonomic level, a concept's TNF can now be defined:

Definition 2.4.1–5:
With respect to a FRAME **Q**, a domain concept $c \in C_Q$ is in *table normal form* (TNF) if and only if **c** is represented as a CONBOX with (i) a *grouped* CONGRID $\mathcal{B}_c^G$ and (ii) a BLOCKCOLLection (with BASE $\mathcal{C}_Q$) of CONEDITS equivalent to $\mathcal{E}_c$ partitioned such that each BLOCK in CONEDITS is completely contained in an element of the XGROUPing induced by $\mathcal{B}_c^G$. Furthermore, TNF implies that

$$D(\mathcal{E}_Q) \cap D(\gamma(\mathcal{B}_c^G)) \subseteq D(\mathcal{E}_c).$$

Symbolically, a domain concept in TNF is represented as a GRUBOX where the grouping pattern of the GRUBOX's TEXTure (which, formally, is a GRUXGRID) carries the *array* structure of the statistical table associated to the concept. In view of **Def. 2.2.4–1**, TNF conforms to the DATBOX structure, that is each element of the XGROUPing induced by $\mathcal{B}_c^G$ is a CONTRAIT to which a DATREGister (**Def. 2.2.4–2**) becomes associated eventually holding the value of the statistical aggregate computed for the respective element.

From a practical point of view, domain concepts (without a grouping structure defined) are more persistent than definite cross-classifications. Hence, it appears reasonable to provide a repository of *ungrouped* domain concepts $\{\mathbf{B}_{c_r}\}_r$ (for instance, identifiable by CONLABs) such that cross-classifications become determined at retrieval time actually. Furthermore, a storage of domain concepts in CONDEF format, $\{\mathcal{C}_{c_r}\}_r$, instead of the internal CONBOX structure is preferable for the sake of (easier) mental tractability; in particular, CONDEFs may be stored in terms of ψ' for the users' convenience.

Axis Rectification, Functional Transformations, and Frame Linkages

In practice, the structural specification of statistical aggregates as well as cross-classifications is not restricted to single FRAMEs. First, concept definitions may become rather awkward if only the basic terminology of QUalities and MODalities is available, particularly in case of concepts with non-orthogonal shape. To this end, the framework must be extended, as defined in Subsection 2.2.3, to comprise *axis rectifications*, thus replacing QUalities and MODalities by the more general constructs of AXes and AXVALues. These replacements imply rather straightforward adaptations of the previous definitions only and, hence, deserve no further

.4 DATA VIEW SPECIFICATION

discussion (the major departures, in fact, concern the concept definition language; cf. Subsection 2.4.3).

This amendment, however, is not sufficient yet. Another minor addition concerns the treatment of *functional transformations* of AXes; this, apparently, does not require further formal prerequisites since any such function is either bijective or it groups domain values to image values, both of which are representable easily in terms of AGGLEvels. Thus, for instance, taking the *log* of an *R*-SCALE amounts to arranging a new AGGDOMain comprising the log-values of the original *R*-SCALE's AGGBASE VALues. Of course, a special definition device for setting up functionally defined AGGDOMs would be desirable to avoid the rather tedious manual specification of involved value mappings. Obviously, a functional transformation may affect several QUalities at once; in such a case, the transformation has to be split in an AXis definition followed by a suitable AGGLEvel definition. Note also that these functional transformations pertain to SCALEs only (and not to observations) whence many seemingly "functional" scale transformations are excluded by definition; for instance, computing the *duration*, or *starting* and *ending* points of, say, the presence of observed phenomena from recorded *time* values is a *statistical* (that is: a model-based) transformation not affecting the *time* QUality at all.

A more profound extension of the structural specification of statistical aggregates results from established FRAME *linkages.* In other words, concept specifications with respect to a single FRAME as considered hitherto are but a special case. For this reason, the notion of domain concept is extended to the notion of *target concept* and, accordingly, FRAMEs are extended to *target* FRAMEs, TAFRAMEs for short.

Analogous to the data source mapping definitions (Subsection 2.3.2) a (parent) FRAME's *proper* QUalities are augmented with *adjoined* QUalities of linked FRAMEs; additionally, the FRAME linkage *type* determines the usage of adjoined QUalities in concept specification. Now, conceptually, a TAFRAME is defined in the following

Definition 2.4.1–6:
A <u>TAFRAME</u> is an OBJECT consisting of a FRAMEUNIT (taken from POPSET) and comprising the FRAMEQUalities of a FRAME *F* (with same FRAMEUNIT) as well as a (possibly empty) set of *adjoined* FRAMEQUalities of FRAMEs linked to FRAME *F*. Adjoined QUalities are adjoined *virtually*, that is, only as far as mentioned in a definite target concept specification. Basically, any referenced QUality of the FRAMEUNIT's universal FRAME is included in a TAFRAME (in case of split FRAMEs); furthermore, any such QUality in a FRAME linked to *F* by LINKDEGree *1* is adjoined; finally, QUalities of *1:n*-linked FRAMEs (whether LINKMODE is *weak* or not) are adjoined as CLUSTDIMensions (of SCALETYPE *N*) such that the linked weak FRAMEs' FRAMEPROFiles, excluding universal QUalities, are converted to sets of QUFACTors and the respective AGGDOMains attached to each FRAMEPROFile's QUalities are turned into the QUFACTors' defining BASETs. In the resulting <u>TAFRAMEPROF</u>ile, the DIMensions are ordered according to the POPSET ordering.

Thus, formally, a TAFRAME conforms to the FRAME structure of **Def. 2.2.2–1** albeit with adjoined DIMensions as indispensable for the specification of a particular (target) concept. The adjoined N-SCALE DIMensions from 1:n-linkages, of course, represent intermediately aggregated (counted) instances of linked FRAMEs (this, by the way, provides – though indirectly – a means to express the *some, atmost, atleast*, etc. relations of D'Angiolini's [1992] concept definition language; cf. Subsection 1.3.2). In case of *intra*-population 1:n-linkages, the respective FRAME is conceptually duplicated before the adjoined CLUSTDIMensions of QUFACTors are formed.

Specific caution is in place in matching universal QUalities, particularly in weak FRAME linkages not inheriting universal QUalities: since, at the linked FRAME's side, the resolution (AGGLEvel) may be lower than at the parent FRAME's side, the linked FRAME first has to be "aggregated" such that universal QUalities are identical to the parent FRAME's AGGLEvels of *space* and *time*. This, in turn, implies *additivity* (cf. Subsection 2.4.2) of the linked FRAME's QUalities with respect to those universal QUalities involved. Technically, such an aggregation amounts to introduce a grouping of MODalities for the linked FRAME's QUalities in terms of an AGGLEvel the AGGDOM of which is in subset relationship to the *defining* AGGDOM of the parent FRAME's respective QUality. Apparently, if these AGGLEvels or ADDMODEs do not comply, FRAMEQUalities *cannot be adjoined* and, hence, the TAFRAME not composed (but note that in case of *weak* linkages these constraints are enforced by definition). These elementary SCALE aggregations, if needed at all, can be carried out in the background completely hidden from the actual target concept specification.

As in data source mapping, the adjunction of QUalities from linked FRAMEs could give rise to new FRAMEdits induced by the LINKEDITS defined for a FRAME linkage. Again, LINKEDITS are interpreted as *dependency edits*, that is the linked FRAME's QUalities do not apply for TRAITs enclosed in LINKEDITS and, hence, the corresponding DIMensions get a NULL value appended when becoming adjoined. As with implicit adaptation of AGGLEvels for universal QUalities, necessary adaptations of FRAMEDITS for a TAFRAME are achieved internally without further explicit specification. Likewise, population instances not assuming any TRAIT in a linked FRAME are assigned MISSVALUes for the linked FRAME's adjoined QUalities (DIMensions) implicitly.

Target Concepts

Just as domain concepts have been defined in terms of a FRAME, *target concepts* can now be defined in terms of TAFRAMES:

Definition 2.4.1–7:
Each target concept TACONDEF defined, analogous to a CONDEF, with respect to TAFRAME is transformed into a PLABOX called TACONBOX, the TEXTure of which equals the WRAPPING of the BLOCKCOLLection representing all TRAITs included in TACONDEF. Since the TEXTure of a TACONBOX, termed TACON-

.4 Data View Specification

GRID, is an XGRID, the difference between this TACONGRID and the CONDEF it wraps is expressed as a BLOCKCOLL termed TACONEDITS.

Definition 2.4.1–8:
A TACONCEPT is an OBJECT consisting of a TACONPROFile and a TACONBOX (cf. **Def. 2.2.3–4**). The TACONPROFile of a TACONCEPT is a sequence of *target concept axes* where each target concept axis, TACONAX, is defined by an AXRECTification such that, for reasons of consistency, the union of AXRECTPROFiles in these AXRECTifications covers *all* FRAMEQUalities of the TAFRAME the TACONCEPT is based upon, and the AXRECTPROFiles are *mutually disjoint* for all pairs of occurring AXRECTs. The TACONGRID of the TACONBOX, in fact, refers to the BAXGRID spanned by the TACONAXes' AXDOMains (comprising TACONVALues). Like QUalities, TACONAXes are *labelled* by TACONAXLABels which are QULABels; in case of adjoined QUalities the QULABels are COMVALues the first component of which is indicating the FRAMELABel the QUality originates from. Each TACONCEPT is *labelled* by a CONLAB taken from a LABSET called CONSET (cf. **Def. 2.2.3–4**).

Now, letting $\left(\mathcal{C}_{Q_t}, \mathcal{E}_{Q_t}\right)$ denote the box representation of a TAFRAME Q_t such that $Q_t^* = D\left(\mathcal{C}_{Q_t}\right) - D\left(\mathcal{E}_{Q_t}\right)$, a TARGET T_t with respect to Q_t is defined in

Definition 2.4.1–9:
Letting T_U denote the (universal) *target concept space* of a FRAME Q (defined analogous to the concept space C_Q of a FRAME Q), T_t is a TARGET if and only if $t \in T_Q$ is a TACONCEPT on Q_t and T_t is in TNF with TACONPROFile $\wp_t$, the *grouped* TACONGRID $\mathcal{B}_t^G$, and the TACONEDITS $\mathcal{E}_t$, such that

$$D(\mathcal{E}_{Q_t}) \cap D(\gamma(\mathcal{B}_t^G)) \subseteq D(\mathcal{E}_t).$$

In general, the target concept space T_Q will be exhausted only partially since, for a particular target T_t, Q_t adjoins only a (small) subset of QUalities available for adjunction; thus, the effective target concept space T_{Q_t} induced by Q_t^* is a subspace of T_Q, that is $t \in T_{Q_t} \subseteq T_Q$. Practically, $T_{Q_t} = D(B_t^G)$, letting $D(B_t^G) =_{def} D(\gamma(\mathcal{B}_c^G)) - D(\mathcal{E}_c)$, is determined implicitly by augmenting the FRAMEPROFile of Q with just those adjoined QUalities necessary to state t.

2.4.2 Summary Types

Concept structures describe the taxonomic scope of statistical aggregates; in contrast to this, *summary types* characterize the meaning of a statistical aggregate's real data contained in the DATREGisters associated with CONTRAITs (cf. **Def. 2.2.4–2**). In terms of data requests (database queries), summary types specify the *kind of aggregation* the data (numbers, in general) in the intended statistical table should have undergone.

By definition, for any specific CONCEPT the associated FILE comprises DATREGisters of the *same* summary type only since CONCEPTs represent statistical aggregates in terms of (generalized) statistical tables. The summary types defined in the following paragraphs and comprised formally in a DEFSET called SUMTYPESET, are understood as specific *data types* in view of the statistical operations (basically, aggregations) that can be applied to the data in a CONCEPT's FILE. Thus, in particular, these summary types do not convey any information about the history or creation context of data (for instance, if numbers are observed or estimated, etc.); any such information has to be stored and provided in a different way, for instance in terms of *statistical* metadata.

Primarily, the devised virtual statistical database is conceived as a source of more or less elementary statistical observation data, that is empirical data at rather low levels of aggregation or data even not yet aggregated. Since, essentially, the aim of statistics is data *aggregation*, the cardinal task of a statistical database system is an aggregation-oriented retrieval of observation data. Moreover, quite frequently the output of statistical database retrieval becomes the input of subsequent "higher-level" statistical analysis (such as some model-based inference etc.), whence it is advisable to provide outputs in a suitably pre-processed format which, in most cases, implies some intermediary data aggregation as well.

Now, in view of the taxonomic structure of property (sample) spaces captured in terms of FRAMEs, data aggregation first of all amounts to retrieve unions of a property space's sub-populations (that is, TRAITs) such that the empirical observations belonging to those TRAITs are counted and, thereafter, the obtained counts possibly are processed further arithmetically. Arguing measure-theoretically, data aggregation is operationalized by an *additive set function* establishing a homomorphism between the concept algebra defined on the Boolean lattice (of disjoint sets) over C_Q, or T_Q, respectively, and the additive group over the set of cardinals (natural numbers including zero), $(N,+)$. From this point of view, letting a TACONCEPT $t_* = \bigcup_j t_j \in T_Q$ where each $t_j \subseteq Q_{t_*}^*$ is a statistical indicator's (sub-)TACONCEPT and letting further θ denote a statistical aggregation function, $\theta(t_*) = \{\theta(t_j)\}_j$ such that, eventually, each $\theta(t_j)$ reduces to the additive set function counting the

.4 DATA VIEW SPECIFICATION

FRAMEUNIT instances comprised in t_j and, particularly, a FRAME Q (more specifically, a TAFRAME $Q_{t.}$) can be viewed as a *universal summary table* such that each TRAIT is associated with the (absolute) *frequency* of FRAMEUNIT instances as the elementary statistical aggregate. Hence, (disjoint) union of TRAITs can be replaced with numerical addition of set cardinalities. As a consequence of this, once the instance set-denoting TARGETs can be converted automatically into additive statistical aggregates, virtually all statistical aggregates of interest become retrievable because, in most cases, statistical aggregates either are additive by their very nature or can be derived arithmetically – that is, *computed* – from additive aggregates. As it turns out, however, the concept of statistical aggregation goes beyond simple additivity because – under specific circumstances – aggregating non-additive TRAITs is meaningful statistically and, hence, admissible.

In this extended view and recurring to terminology introduced previously, statistical aggregation – expressed in terms of deriving generalized statistical tables (factorial arrays of statistical indicators) – amounts to a particular kind of *projection* of universal summary tables onto specified subsets of their universal category relations $Q_{t.}^*$ such that the summary data for the "projected" category relation is obtained by a statistical aggregation function θ (aggregating source data along the "projecting" directions) irrespective of θ being the standard additive set function or some other (non-additive) aggregation operator.

Taking a closer look at additivity and non-additive aggregation, unfortunately, blurs the clear distinction in yet another way as pointed out by Klensin [1991]: even "simple" additivity may in fact be not as simple as appearing initially after concerning additional information typically associated with observation data, such as measurement error distributions, imputations applied to the data during a preceding data cleaning stage, etc. Thus, adopting the simplified view of additivity is justified mainly by

- the generally low availability of this type of knowledge about data, and
- the methodological difficulty of taking this additional information into account properly in subsequent statistical data processing (the latter, of course, is a particularly weak excuse).

Whatsoever, METAMOD is prepared, in principle, to accommodate for pertinent data model extensions, for instance by adding BLOCKCOLLections of TAGged BLOCKs associated with IMTRAITs carrying the additional data description as required (certainly more critical is the arrangement of suitable aggregation operators exploiting this added knowledge about empirical data).

Instance Retrieval

Although the elementary *summary* type of METASTASYS is the *number of* FRAMEUNIT *instances* (either assigned theoretically to or observed empirically for some (CON)TRAIT), METAMOD takes yet one step back conceptually and admits a setlevel representation of observation data in terms of so-called mesodata (cf. Subsection 1.5.1).

Def. 2.4.2–1:
In METAMOD, a FILE is of SUMTYPE 'MESO' if and only if each of the DATFILE's DATREGisters contains a *set of* instance *labels* attached to the instances of the FRAMEUNIT the FILE's DATBOX refers to.

Strictly speaking, the SUMTYPE 'MESO' is not at all a proper summery type; however, there are two major arguments in favour of including the MESO summary type formally among METAMOD's summary types:

- in case of data sources providing microdata (that is, observations at individual record level) observation records may be identified by instance labels and, thus, become related – via FRASSTABs (cf. Subsection 2.2.2) or implicitly in case of weak linkages – to other observation records associated with a linked FRAME;
- as to elementary aggregation there is no difference logically whether sets of instances or the cardinalities of instance sets are stored but, once instance sets are replaced with counts, it is no longer possible to trace individual instances (over *time*, for example, if instances are persistent) and, thus, some information from the original microdata is lost apparently.

Albeit METAMOD does support a lossless representation of microdata, there is limited use of MESO data beyond that outlined above. For instance, despite the possibility to compute, for a given pair of *time* points, a *transition matrix* this data structure is not representable formally inside METAMOD (since this would correspond to a FRAME of type $Q^* \times Q^*$ and, thus, call for a specific FRAME calculus generating FRAME structures like $(Q^*)^n$, $n > 1$).

Because of the stipulated disjointness of TRAITs within FRAMEs, the MESO summary type is additive as far as additivity of FRAMEQualities (with respect to other FRAMEQualities) is defined in a particular METAMODel instance; by definition, taking set unions always amounts to take disjoint unions. For additive FRAMEQualities in some FRAME Q, letting $\xi(q)$ denote the contents of the DATREGister associated with TRAIT q, obviously

$$\xi(c) =_{def} \bigcup_{q \in c} \xi(q)$$

for any $c \in C_Q$. At MESO level, data aggregation – that is, taking unions of instance sets – is infeasible in case of non-additivity (although, of course, data aggregations with respect to summary types other than MESO are by no means ruled out automatically). In view of data source mappings (cf. Section 2.3) it should be noted that even if data sources provide microdata, SOTRAITs typically correspond to non-singleton concepts $\mu(v) \in T_Q$, that is *sets* of (IM)TRAITs, for which reason source

.4 Data View Specification

data come in terms of $\xi(\mu(\mathbf{v}))$ instead of $\xi(\mathbf{q})$. In MESO representation the SO-CELLS' DATREGisters contain either explicit unique instance labels as assigned to instances in *registers* or simply observation numbers as dummy labels. The latter case, of course, is rather unimportant unless a source dataset comprises repeated measurements; tracing sampling units across single source datasets is possible unambiguously only if instance labels from established POPREGisters are recorded. In case of SOEDits, the associated SOCELLs, if represented explicitly at all, must not contain any instance labels in their DATREGisters, that is DATREGisters constantly hold the empty instance set.

Instance Counting

Switching from summary type 'MESO' to summary type 'COUNT' implies the application of the "measure" function

$$n(\mathbf{q}) =_{def} |\xi(\mathbf{q})|$$

to the contents of DATREGisters. This extends immediately, for $\mathbf{c} \in \mathbf{C}_\mathbf{Q}$, to

$$n(\mathbf{c}) =_{def} \sum_{\mathbf{q} \in \mathbf{c}} n(\mathbf{q}) = \sum_{\mathbf{q} \in \mathbf{c}} |\xi(\mathbf{q})|.$$

For completeness' sake, with respect to $\mathbf{Q}, n(\mathbf{q}) = 0$ by definition if $\mathbf{q}$ coincides with a FRAMEDit; likewise, $n(\mathbf{c}) = 0$ if $\mathbf{c}$ denotes the *empty* concept $\emptyset$ (which may comprise any non-negative number of FRAMEDits). Thus, the *counting* function $n(\mathbf{c})$ – hereafter abbreviated to $|\mathbf{c}|$ for simplicity – is the elementary additive statistical function homomorphic from the set disjoint union of concepts (over $\mathbf{C}_\mathbf{Q}$) to the sum operator on natural numbers (including zero): for any disjoint pair $\mathbf{c}', \mathbf{c}''$, $|\mathbf{c}' \cup \mathbf{c}''| = |\mathbf{c}'| + |\mathbf{c}''|$. Analogously, counting aggregation applies to $\mathbf{T}_\mathbf{Q}$. In terms of METAMOD's SCALETYPEs, the range of $|\cdot|$ equals CARDSCALE.

Def. 2.4.2–2:
In METAMOD, a FILE is of SUMTYPE 'COUNT' if and only if each of the DATFILE's DATREGisters contains a CARDinal MODality.

Note that, theoretically, letting $n_\mathbf{Q} = |\mathbf{Q}^*|$, $f_\mathbf{Q}(\mathbf{q}) = n(\mathbf{q})/n_\mathbf{Q}$ for $\mathbf{q} \in \mathbf{Q}^*$, defines, for a FRAME $\mathbf{Q}$, a probability function on the (fictitious) *representative* table [Malvestuto, 1991] $\mathbf{Q}^*$, whence, for $\mathbf{c} \in \mathbf{C}_\mathbf{Q}$, $f_\mathbf{Q}(\mathbf{c}) = |\mathbf{c}|/n_\mathbf{Q}$. Hence, $f_\mathbf{Q}$ complies with the standard *contingency table* model for the universal summary table.

Numerically, $|\cdot|$ is the starting point for further statistical computations. In case of a microdata IMAGE s, the instance sets of SOTRAITs **v** in MESO type are replaced with IMTRAIT COUNTs $|\mu^{(s)}(\mathbf{v})|$ (if **v** is a SOEDIT, $|\mu^{(s)}(\mathbf{v})| = 0$, of course); this, in turn, induces an empirical distribution on $\mu^{(s)}(\mathbf{V}_s^*) \subseteq \mathbf{Q}^*$ such that

$$f_s(\mu^{(s)}(\mathbf{v})) = |\mu^{(s)}(\mathbf{v})| / n_{\mathbf{V}}^{(s)}$$ where $n_{\mathbf{V}}^{(s)} = \sum_{\mathbf{v} \in \mathbf{V}_s^*} |\mu^{(s)}(\mathbf{v})|$ and $\mathbf{v} \in \mathbf{V}_s^*$. Statistically, with respect to $\mathbf{Q}^*$, f_s defines a *conditional* (on $\mu^{(s)}(\mathbf{V}_s^*)$) probability distribution (except in the – rare – case $\mu^{(s)}(\mathbf{V}_s^*) = \mathbf{Q}^*$). In computational terms, the $|\mu^{(s)}(\mathbf{v})|$ represent *pre-macro* aggregate values, that is the cardinalities of observed instance sets (observation records) not yet subjected to further model-based statistical operators (such as weighting or estimation functions to scale up sample figures to population figures etc.).

As to additivity, the COUNT summary type behaves similar to the MESO summary type. Additionally, however, aggregations may be possible even in case of non-additivity of TRAITs with respect to certain FRAMEQualities. Specifically, repeated observations taken on persistent sampling units are non-additive temporally but could nevertheless be aggregated over **time** by means of *time series models*. Similarly, methods of *spatial statistics* may apply to observations which are non-additive spatially. In these special cases, the standard summation operator is replaced with a specific non-additive statistical aggregation operator θ such that $n(\mathbf{t}) = \theta_{\mathbf{q} \in \mathbf{t}} n(\mathbf{q})$ instead of $n(\mathbf{t}) = \sum_{\mathbf{q} \in \mathbf{t}} n(\mathbf{q})$; the operator θ, of course, may depend on further (model) parameters. Obviously, the θ operator (along with mandatory model parameters) must be included explicitly in the specification of statistical aggregates.

Numerical Scales and Summation

Practically, data sources provide datasets either in microdata (MESO) or in COUNT format most of the time. Nevertheless, a SUMTYPESET comprising 'MESO' and 'COUNT' types only is sufficient neither in view of the integration of existing data aimed at nor internally in view of derivable statistical aggregates. Especially with respect to *numerical* SCALEs a further elementary summary type is indispensable: the SUM summary type combining (absolute) frequencies with MODalities of numerical SCALEs to obtain numerical *feature sums* in terms of weighted sums. Simply speaking, the 'SUM' type bridges the gap between "categorical" (that is, discrete-value based) and numerical statistics accentuated somewhat artificially in METAMOD for the sake of achieving uniform FRAME structures. Formally, *summation* always refers to a summation dimension S comprising either a single FRAMEQuality or a subset of a CLUSTDIMension's FRAMEQualities (in order to assure

.4 DATA VIEW SPECIFICATION

identical SCALEs for all QUalities involved in a summation). Now, letting the SUMARG S denote a (non-empty) summation dimension,

$$\Sigma_S(\mathbf{t}) =_{def} \Sigma_{\mathbf{q} \in \mathbf{t}}\left(\Sigma_{s \in S}\mathbf{q}[s]\right)n(\mathbf{q})$$

is the feature sum of $\mathbf{t} \in \mathbf{T_Q}$ with respect to S. Implicitly, the SCALE of Σ_S is $\overline{R}$ where, disregarding MOdality ordering, $\overline{R} =_{def} R \cup \{o\}$ for a suitably chosen (in fact, determined by S) radix SCALE R and the additional group axiom $(\forall x \in R).x + o = o + x = x$ (cf. Chen et al. [1989]); MOdality 'o' is used to denote the feature sum of the empty concept, $\Sigma_S(\emptyset) = o$.

Apparently, summation over S implies *full* additivity between all QUalities within summation dimension S. Contrary to counting, summation changes FRAME-GRID semantics in that a FRAME $\mathbf{Q_t}$'s original FRAMEPROfile $\wp_\mathbf{t}$ is projected onto $\wp_\mathbf{t} - S$ such that to each TRAIT of the "projected" FRAMEGRID the feature sum (over S) is attached. Since summations imply a change of summary type from 'COUNT' to 'SUM', they *cannot* be applied sequentially (such as counting); however, counting still applies after summations (now, of course, summing SUMs instead of COUNTs). Moreover, since there is only a single summation feasible in any aggregation and $\Sigma_S(\mathbf{t})$ as defined above can be restated, by commutativity of addition, as

$$\Sigma_S(\mathbf{t}) = \sum_{w=1}^{W}\left|\{\mathbf{q} \in \mathbf{t} | \Sigma_{s \in S}\mathbf{q}[s] = m'_w\}\right|m'_w,$$

letting $\langle m'_1,\ldots,m'_W\rangle$ denote the (implicit) SCALEDOMain closed with respect to $(|S|-1)$-fold addition over $D(Q) = \langle m_1,\ldots,m_U\rangle$, the numerical SCALEDOMain of QUalities $Q \in S$, summation can be applied *after* counting throughout. For instance, in a two-dimensional sample FRAME as depicted in **Fig. 2.4.2** (overleaf), a TARGET $\mathbf{T_{t_*}}$ with $\mathcal{B}^G_{\mathbf{t_*}} = \left(\langle D(Q_1)\rangle,\langle\langle a\rangle,\langle b,c\rangle\rangle\right)$, assuming that $\mathcal{E}^G_{\mathbf{t_*}} = \langle\rangle$ and $\wp_{\mathbf{t_*}} = \langle\langle Q_1\rangle,\langle Q_2\rangle\rangle$, such that $\mathbf{t_*} = \mathbf{t_1} \cup \mathbf{t_2}$, where $\mathbf{t_1} = \langle\langle m_1,a\rangle,\ldots,\langle m_U,a\rangle\rangle$ and $\mathbf{t_2} = \langle\langle m_1,b\rangle,\ldots,\langle m_U,b\rangle,\langle m_1,c\rangle,\ldots,\langle m_U,c\rangle\rangle$, for the requested summation $\Sigma_{\{Q_1\}}(\mathbf{t_*}) = \{\Sigma_{\{Q_1\}}(\mathbf{t_1}),\Sigma_{\{Q_1\}}(\mathbf{t_2})\}$ resolves into computing

$$\Sigma_{\{Q_1\}}(\mathbf{t_1}) = \Sigma_u\left|\langle\mathbf{q}_{m_u a}\rangle\right|m_u = \Sigma_u n(\mathbf{q}_{m_u a})m_u$$

and

$$\Sigma_{\{Q_1\}}(\mathbf{t_2}) = \Sigma_u\left|\langle\mathbf{q}_{m_u b},\mathbf{q}_{m_u c}\rangle\right|m_u = \Sigma_u\left(n(\mathbf{q}_{m_u b}) + n(\mathbf{q}_{m_u c})\right)m_u,$$

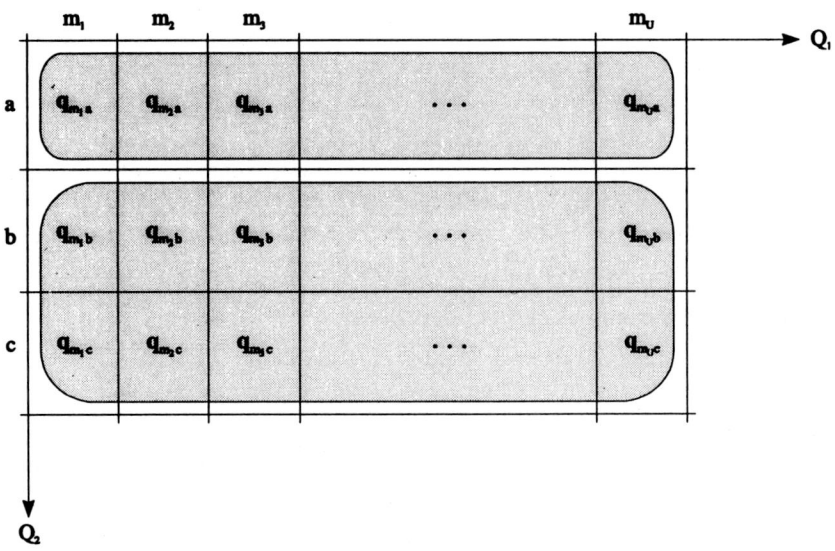

Figure 2.4.2: Sample FRAME Summation

respectively. While the last summation formula is useful for summation in the presence of constraints, it is always possible to *cascade* summations by exploiting the distributivity of multiplication over addition, that is

$$\Sigma_S(t) = \sum_{s \in S} \left(\Sigma_{\{s\}}(t) \right)$$

and, hence, $\Sigma_S(t)$ can be obtained by adding the (intermediary) summations for any partitioning of the summation dimension S.

Summation of Classified Data

In defining summation it has been assumed tacitly that the COUNTs going to be summed originate from ungrouped MODalities. This, of course, must by no means hold since data source mappings may contribute instances or counts for MODality *groupings* only, that is effective measurement resolution at source may be lower than accounted for by the associated FRAMEQUalities' theoretical SCALEDOMain. As usual in statistics, in such cases a suitable *class representative* $\tilde{m}$ for each MODality grouping (interval) $\langle m_l, \ldots, m_u \rangle$ replaces $q[s]$ in the summation formula.

Strictly speaking, the choice of $\tilde{m}$ depends on statistical model considerations and, thus should be controlled explicitly. Although this is not included in the present set-up of METASTASYS's data source mappings, it might prove necessary to do so

.4 Data View Specification

unless these choices are deferred until retrieval time and, hence, become a part of aggregate specification (but note that, conversely, if data sources contribute SUM data already, regaining the cell frequencies depends on $\tilde{m}$ as well which speaks in favour of augmenting the mapping documentation rather than the aggregate specification language).

Generalized Summation

Despite being somewhat less general than counting, summation provides the indispensable interface to numerical statistical processing of aggregates. Furthermore, summation extends naturally from sums of *scalars* to sums of *products* used to compute higher and mixed moments of statistical distributions. Apparently, letting

$$\Sigma_S^{(d)}(\mathbf{t}) =_{def} \sum_{w=1}^{W} \left| \{ \mathbf{q} \in \mathbf{t} | \Sigma_{s \in S} \mathbf{q}[s] = m'_w \} \right| (m'_w)^d$$

computes the d-th, $d > 0$, (non-central, of course) moment; in particular, $\Sigma_S^{(1)} \equiv \Sigma_S$. Setting $d = 0$, $\Sigma_S^{(0)}$ becomes a special version of counting (inducing implicitly an AxRECTification in terms of the FRAMEQualities comprised in S). Mixed moments, as required, for instance, to compute (partial) correlations, are obtained from product sums

$$\Sigma_{S_1 \ldots S_H}^{(d_1, \ldots, d_H)}(\mathbf{t}) =_{def} \sum_{\mathbf{q} \in \mathbf{t}} \left(\prod_{h=1}^{H} \left(\Sigma_{s_h \in S_h} \mathbf{q}[s_h] \right)^{d_h} \right) n(\mathbf{q}).$$

For instance, setting $H = 2$, $d_1 = d_2 = 1$, and $S_1 = \{Q_a\}$ and $S_2 = \{Q_b\}$, respectively, $\Sigma_{S_1, S_2}^{(1,1)}(\mathbf{t}) = \sum_{\mathbf{q} \in \mathbf{t}} \mathbf{q}[Q_a] \mathbf{q}[Q_b] n(\mathbf{q})$. Of course, $\Sigma_{\{Q_a\},\{Q_a\}}^{(1,1)}(\mathbf{t}) = \Sigma_{\{Q_a\}}^{(2)}(\mathbf{t})$ in case of $Q_a = Q_b$. Thus, in terms of $|\cdot|$, Σ_S, $\Sigma_S^{(2)}$, $\Sigma_{S_1, S_2}^{(1,1)}$, etc., means, variances, covariances, correlations, as well as other univariate, bivariate or multivariate measures can be computed in terms of elementary SUMTYPEs.

Def. 2.4.2–3:
In METAMOD, a FILE is of SUMTYPE 'SUM' if and only if each of the DATFILE's DATREGisters contains an $\overline{R}$ MODality either obtained by applying a summation (to a FILE of COUNTs) or obtained from source data. SUM data are accompanied by further attributes (description elements), viz. a SUMARGument attribute comprising the sequence of summation dimensions $S_1, \ldots, S_H$, a SUMDEGree attribute comprising the sequence of summation degrees $d_1, \ldots, d_H$, and a SUMARITY attribute holding the VALue H. By default, $H = d_1 = 1$.

The summary types 'MESO', 'COUNT', and 'SUM' exhaust METASTASYS's elementary SUMTYPES. For convenience, this terse basic *repertory* could be extended by several summary types derivable from the elementary ones by applying various arithmetical functions. From a statistical point of view, natural candidates of non-elementary summary types are *average, rate, percentage, index, variance, standard deviation, skewness*, etc. Furthermore, with respect to complete summations over numerical DIMensions, *quantiles, mode, range, min, max, variation coefficient*, etc. are conceivable SUMTYPES. In subsequent considerations, however, only the SUMTYPES of *rate* and *percent* will be referred to as illustrating examples of non-elementary summary types.

Obviously, a data source can provide data of *any* SUMTYPE defined in METAMOD although data sources contributing elementary summary types are preferable in view of data aggregation and, particularly, data integration.

As a remark at the margin, in a generalized perspective *statistical models* (and their parameters) represent special kinds of summary types, too. For instance, the centroids of a cluster analysis or the (main and interaction) effects of a variance analysis model could be modelled as further SUMTYPES. This possibility, however, is just indicated here, particularly because in many cases the standard BOX representation of METAMOD does not match all requirements of the extended SUMTYPESET (ANOVA, for example, would require a layered BLOCKCOLLection of – specially TAGged – BLOCKs representing the scopes of model effects – a single encompassing BLOCK for the overall mean, a set of BLOCKs for each main effect, and so on). Representable straightforwardly, as a matter of fact, would be *estimated* distributions (in terms of COUNT and SUM DATFILEs) as well as *residuals*. In the longer run, it may be very advantageous to include such model structures in METASTASYS especially with respect to data propagation procedures (cf. Subsection 2.5.2) needed to extrapolate empirical distributions for "dataless" FRAME areas.

Summary Type Accommodation

In METAMODel instances comprising several linked FRAMEs, the adjunction of FRAMEQUalities to obtain TAFRAMEs requires an implicit *accommodation* of differing SUMTYPES. How this accommodation is accomplished depends, essentially, on the availability of explicit instance linkages: as long as the instances of parent and linked FRAMEs are associated by explicit (key and reference, respectively) instance labels – that is, data are of 'MESO' type – ordinary *algebraic* adjunction of FRAMEQUalities applies albeit aggregations ranging over TRAITs in the linked FRAME induce corresponding groupings on the (transformed) adjoined QUalities in the parent FRAME's (TA)FRAMEPROfile. Note that for $1{:}n$-linkages QUality adjunction always implies counting over the n replications of associated instances in the linked FRAME. Otherwise, FRAMEs can be linked *statistically* by aggregating marginal distributions only (cf. Subsection 2.5.2).

2.4.3 METASPEL

The formal language used to specify structurally data views as described above needs to meet two essential criteria: on the one hand, it ought to be conceptually simple (which, from the user's point of view, just means *expressive power*) and, on the other hand, formal language expressions should, to the highest degree possible, be transformable mechanically into database retrieval and subsequent data aggregation operations necessary to derive the specified data view effectively – if it is derivable at all. In logical terms, such a formal language is generated by a *grammar* defining recursively the syntactical structure of formulae denoting statistical aggregates or, more specifically, generalized statistical tables such that the formal structure of a denotation alone determines the derivation (that is, the computational process) of the denoted aggregate. In other words, the set of aggregate expressions (data views) induced by the formal data view grammar coincides, ideally, with a space of *nameable* statistical aggregates (generalized tables); practically, of course, only a subset of nameable aggregates may be derivable (cf. [Sato, 1989]) since source data typically to not support a derivation of all theoretically conceivable tables; this could be termed *empirical incompleteness* in contrast to the desired theoretical completeness of the data view language.

The grammar of the proposed aggregate specification language, METASPEL (META Specification Language), comprises two major components:

- a fixed syntactical structure part which, evidently, represents a choice from several more or less equivalent alternatives;
- a collection of parameter sets provided by METAMODel taxonomies, that is, specifically, the BASETs of a definite METAMODel instance (domain data model).

In principle, there is a bijective relation between METASPEL sentences (that is, formulae generated by the grammar) and internal META OBJECTs; practically, however, METASPEL will be used to denote REQUESTs – TARGETs augmented with a summary type specification – only; in general, there is no point in re-translating METAs into METASPEL expressions although META derivations can be viewed, at the metadata level, as formal proofs obtained from *rewriting* the METASPEL descriptions of data sources (taking the role of axioms denoting IMAGEs) into REQUEST METASPEL expressions, and META derivations, in turn, are conceptualized as the transitive closure of a binary relation over METASPEL expressions defined intensionally by a set of rewriting rules (cf. Section 2.5). As a matter of fact, from a procedural point of view, the definition of METASPEL is rather obsolete since the OBJECT structures corresponding to METASPEL expressions could be composed directly as well (though in an unreasonably clumsy way – dropping METASPEL surely would violate the "expressive power" premise). Moreover, METASPEL itself will be hardly attractive in any real implementation of METASTASYS if compared to the superior features of, for instance, *graphical user interfaces* admitting a combination of verbal and non-verbal modes of REQUEST specification (cf. Section 2.7). In spite of this, however, any useful statistical data view specification language certainly must provide expression structures similar to those of METASPEL and hence, in its essence, paraphrase METASPEL.

METASPEL Sentence Structure

In view of the restriction in using METASPEL only as a REQUEST specification language, a straightforward language design has been chosen. Furthermore, the following paragraphs give an abbreviated, semi-formal account of METASPEL instead of an in-depth presentation – a full description certainly would become somewhat lengthy and obscure the language's formal structure rather than expound it. For simplicity, the presentation also excludes explicit layout specification details; practically, it may be advantageous to defer the specification of external tables (layouts) until the actual derivation of internal tables for several reasons: the derivation of an internal table may fail, the internal table derived eventually may depart from the requested internal table (calling for an adapted layout), and – last but not least – an internal table, in general, admits several layouts to be decided rather independent from internal table specification.

The structure part of METASPEL provides a *sentence scheme* such that each METASPEL sentence comprises a sequence of *clauses* concatenated by *semicola*. Although clauses have a uniform outer structure, consisting of a unique *key phrase* (typed in boldface) followed by a *parameter phrase*, the clauses contained in a sentence to be well-formed depend (partly, at least) on various parameter relationships. **Tab. 2.4.3** (overleaf) lists the keywords naming the "native" METASPEL clauses the presence or absence of which, in a sentence, does not depend on each other; this overview indicates the status of each clause, lists the default values (if any), and provides a short clause description. Mandatory clauses need to appear explicitly in a sentence only if there is no default value assigned. The **ON** clause is exceptional since, of course, there is no point in including it into a sentence in case that a domain data model consists of a single FRAME only; however, for the sake of better readability, either a **COMPUTE**, a **FOR**, or an **ON** clause should be contained in any METASPEL sentence. Basically, there is no specific clause ordering to be obeyed; subsequently, however, clauses will be ordered as listed in **Tab. 2.4.3** throughout. To be valid, a METASPEL sentence must not contain repeated occurrences of clauses with identical key phrases.

The parameter phrase of a clause (which follows its key phrase) provides the actual clause parameter(s). In some cases, parameter phrases are simple keywords either but, in general, parameter phrases have their own syntax as expounded subsequently in connection with a short discussion of each of the elementary METASPEL clauses.

.4 DATA VIEW SPECIFICATION

Table 2.4.3: Native METASPEL Clauses

Key phrase	Status	Default value	Description
COMPUTE	mandatory	table	specifies the *type of output*
OF	mandatory	n	specifies the aggregate's *summary type*
ON	mandatory	the sole FRAME's FRAMELAB in a single-frame model; none otherwise	specifies the domain FRAME of TARGET
FOR	mandatory	#	specifies the *generic target concept* of the aggregate
WITH	optional	–	*restricts* the target concept
BROKEN DOWN BY	optional	–	modulates a *cross-classification* onto the generic target concept
OVER	mandatory	–	adds the *space* restriction to the generic target concept
IN	mandatory	–	adds the *time* restriction to the generic target concept

With respect to the limited formal development of METASPEL the **COMPUTE** clause is, strictly speaking, superfluous; internal tables are always numerical aggregates corresponding to the standard image of statistical tables (tabular arrangements of summary numbers). Thus, a

COMPUTE table

clause creates an ordinary statistical table (except that the internal table generated may still comprise a high number of breakdown dimensions needing a specific "flattened" layout arrangement). In the longer run, however, the parameter value 'table' may be substituted by different aggregate types such as, for instance, graphical representations of summary data or more complex statistical analyses; particularly in the latter case, of course, the parameter phrase of the clause would have to comprise further elements (say, model parameters, or the like) which, alternatively, could be supplied by yet another corresponding clause.

The ON Clause

In domain data models comprising several FRAMES, one of these FRAMES must be chosen as the statistical population of FRAMEUNITs which the statistical aggregate will make a quantitative statement about. To this end, the **ON** clause's parameter phrase states the FRAMELABel associated with the specific FRAME of interest. If a domain data model comprises a single FRAME only, the **ON** clause can, of course, be left out by virtue of the default definition.

The OF Clause

The **OF** clause of a METASPEL sentence defines the internal table's summary type and belongs to the more intricate clause species. In simple cases, **OF**'s parameter phrase reduces to 'MESO' or 'COUNT' (which may be abbreviated symbolically to '#' and 'n', respectively). In practice, MESOdata retrieval will happen seldom whereas summary type COUNT is standard and, hence, chosen as default value (assumed if there is no **OF** clause in a METASPEL sentence). In case of summations, the parameter phrase starts with a 'Σ' symbol followed by a summation dimension argument SUMARG,

$$\Sigma \ \text{<arg>}$$

where <arg> must refer to a numerical QUality associated with the (TA)FRAME specified in the same sentence's **ON** clause; that is, <arg> coincides either with a QULAB of a *numerical* QUality of the **ON**-FRAME or with a QULAB of an *adjoined* numerical QUality of a FRAME linked to the **ON**-FRAME. In the latter case, the QULAB needs to be prefixed with the linked FRAME's FRAMELAB; syntactically, letting L_F denote the linked FRAME's FRAMELAB and L_Q denote the involved QUality's QULAB, such a *frame prefix* is written as

$$L_F \blacklozenge L_Q$$

In case of a *1:n* FRAME linkage, the simplicity of this expressions conceals entirely that, in fact, the L_Q-part becomes a QUFACTor by implicit conversion of the linked FRAME's QUalities to QUFACTors of the parent (TA)FRAME whence summation over SUMARG L_Q implies aggregation over all other QUFACTors first before summation can take place. Furthermore, <arg> may consist of a whole CLUSTDIMension of numerical (COM)QUalities and, additionally, QUality subsets of CLUSTDIMensions. QUality subsets may be either specified explicitly, by putting the respective QUalities' QULABs – separated by commas – into set braces, or implicitly by stating factored QULABs in dot notation and inserting a '#' (*catch-all* sign) for components to be summed over as well as the respective QUFACTors VALues to be matched/selected (examples are given amply in Section 4.4). Obviously, dot notation is useful only as far as COMQUalities are still referenced uniquely. Set brace and dot notation with catch-all signs may be used concurrently, assuming a set union of dot notation-specified QUality sets included in the summation argument. For instance, assuming that a QUCLUSTer is generated by three QUFACTors, F_A, F_B, F_C, and m_1 as well as m_2 are among F_B's component VALues; then

$$\Sigma \ \{ \ \#.m_1.\# \ , \ \#.m_2.\# \ \}$$

.4 DATA VIEW SPECIFICATION

denotes a summation over all COMQUalities with second component identical to either m_1 or m_2. Even more fanciful notation patterns are, of course, conceivable. It must be noted that, implicitly, summations over QUality sets induce AXRECTifications (with AXDOMains inferred internally). The syntax of **OF** parameter phrases can be enriched considerably as will be discussed later on.

The FOR Clause

METASPEL's **FOR** clauses are used to specify target concepts. More precisely, these *clause concepts* provide a skeletal TACONCEPT specification only since the TACONCEPT specification, in general, is affected by parameter information from several clauses of a METASPEL sentence. In particular, a clause concept usually denotes a *generic* predefined – ungrouped – domain TACONDEF (possibly omitting *space* and *time* specifications) which, by means of a formally assigned CONLAB, is retrieved by name from a repository of domain concepts (only in case of very simple-structured target concepts a definition is reasonable *in situ*). Basically, clause concepts are defined in terms of the simple grammar of the concept specification language expounded in Subsection 2.4.1; for convenience, this grammar is extended slightly by replacing the `<pc-expr>` rule with the more comfortable version

```
<pcx-expr>   ::=   <ax-ref> [ @ <agglev> ]:   <val-expr>
```

where `<ax-ref>` either refers directly, by AxLABel, to one of the **ON**-FRAME's QUalities or Axes defined in terms of its QUalities, or `<ax-ref>` is an expression of type '$L_F \blacklozenge L_A$' as defined above, that is an AxLABel prefixed by a linked-FRAME reference (FRAMELABel L_F) which, by definition, reads implicitly as '$\Sigma\ L_F \blacklozenge L_A$' in case of L_F labelling a *1:n*-linked FRAME (in this particular case, letting L_A=# is also admissible understanding that '#' encompasses all QUFACTors generated by QUality adjunction). As long as no `<agglev>` is stated, Axes may be defined *in situ* by introducing constraints over numerical QUalities as expounded in Subsection 2.2.3. As an enhancement to restriction specification, `<val-expr>` may refer to an AGGLEVel – by substituting its AGGLAB for `<agglev>` – defined for the CONAXis `<ax-ref>`. Admissible substitutions for `<val-expr>` are

(i) sets of MODalities either in terms of explicit enumerations of VALues enclosed in set braces and separated by commas or implicit enumerations by using *interval notation*,
(ii) single MODalities prefixed by a '='-sign, or
(iii) the keyword '**except**' followed by a term of either shape (i) or (ii).

Additionally, type (i) terms may be prefixed with a '='-sign throughout; all MODalities referenced in a `<val-expr>` are understood to refer to the `<agglev>` stated. In interval notation, the set of denoted MODalities is specified, with respect to the MODalities' orderings, by an expression

<low-bound> .. <hi-bound>

where the MODality substituted for `<low-bound>` *must precede* the MODality substituted for `<hi-bound>` in the referenced MODality ordering. It is admissible, however, to omit (at most) one of the bounding MODalities implying that the denoted range extends to the most extreme MODality occurring empirically.

In case of specific selections of a linked FRAME's instances it may be more convenient to refer directly to a domain concept defined in that FRAME's terms. This, in turn, implies a further extension of `<pcx-expr>`-syntax by arranging rewriting rules like

<pcx-expr>   ::=   <FRAMELAB> ♦ <rc-expr>

or

<pcx-expr>   ::=   <FRAMELAB> ♦ <CONLAB>

respectively, where `<rc-expr>` and `<CONLAB>`, of course, are restricted to terminology and definitions attached to the linked FRAME labelled `<FRAMELAB>`.

Quite frequently, the clause concept refers to the ur-concept of the eventual TACONCEPT; in this case, the symbol '#' is inserted as a shorthand in **FOR**'s parameter phrase. If a METASPEL sentence does not contain a **FOR** clause at all, it is assumed to be '**FOR #**' by default.

The WITH Clause

In many cases, generic clause concepts are restrained just for the purpose of deriving the aggregate of current interest. This is achieved by a **WITH** clause the parameter phrase of which conforms syntactically to

<with-expr>   ::=   <pcx-expr> [& <pcx-expr>]*

The restrictions specified in **WITH** clauses denote PRIMCONcepts multiplied with the clause concept CONDEF as defined in Subsection 2.4.1. Evidently, **WITH** clauses are optional.

Universal QUalities: The IN and OVER Clauses

The specification of concept restrictions with respect to the *universal* QUalities *space* and *time* is separated from other clauses because of its distinctive importance. First, as has been pointed out already, predefined clause concepts are often generic in the sense that they do not involve a spatial and/or temporal restriction of their range of validity (or, if they do, this range of validity may be quite general as well) within the referenced FRAME. Secondly, the specification of temporal and spatial coverage of a statistical aggregate needs utmost care in order not to produce, rather involuntarily, (a huge amount of) statistical indicators of little interest at considerable computational costs. For both reasons, two special clauses are included in the METASPEL language to handle spatial and temporal TACONCEPT restrictions; additionally, both clauses admit, simultaneously, a specification of temporal and spatial breakdowns.

With respect to *space*, restrictions are specified in the **OVER** clause, the parameter phrase of which may assume one of the formats supported by the following syntax rule:

```
<st-expr>    ::=
        [ <agglev1>: ] [=] <val-expr1>
        [ with <agglev2>: [=] <val-expr2> ]
```

where `<agglev1>` rewrites to an AGGLAB denoting one of the AGGLEVels defined for the *space* QUality of the **ON**-FRAME and `<val-expr1>` is a term either

(i) enumerating a set of `<agglev1>`-MODalities explicitly (enclosed in set braces and separated by commas if there is more than one element; otherwise, set braces can be omitted), or

(ii) enumerating such a set implicitly either in interval notation as defined above or by inserting a "catch-all" sign ('#') if the whole range is meant, or

(iii) if the referenced `<agglev1>`-MODalities are COMVALues, enumerating this set in dot notation such that each of the components may be of either type (i) or (ii).

Identically, `<val-expr2>` is composed though referring to `<agglev2>`-MODalities where `<agglev2>` must rewrite to an AGGLAB denoting an AGGLEVel strictly higher than `<agglev1>`. If the clause contains a '**with**'-keyword only those `<agglev1>`-MODalities are included in the TACONCEPT which are contained fully in the range of `<agglev2>`-MODalities. Prefixing the `<val-expr1>` term by '=' means that all MODalities go in one group, that is, *space* becomes a *reference dimension* not receiving a breakdown; otherwise, the `<val-expr1>` elements are used as breakdown values, or spatial categories, of the final table.

Temporal restriction is treated completely analogous to spatial restriction except that, for the sake of distinction, the clause's key phrase changes to **IN** and the sub-expressions of the parameter phrase refer to the *time* DIMension of the **ON**-FRAME.

The BROKEN DOWN BY Clause

In most cases, statistical aggregates comprise an array of statistical indicators; these arrays are induced by breakdowns, or cross-classifications. In METASPEL, cross-classifications with respect to DIMensions other than *space* and *time* are specified by means of the **BROKEN DOWN BY** clause. In its basic format, the parameter phrase of this clause conforms to the syntax rules

```
<x-expr>     ::=   <xdim-expr> [ * <xdim-expr> ]*
```

```
<xdim-expr>    ::=    <ax-ref> [ @ <agglev> ]
```

where, defining `<ax-ref>` as previously, each of the terms `<xdim-expr>` introduces one particular breakdown dimension such that the resulting breakdown discerns exactly the VALues of `<agglev>`; thus, AGGLEVels are used to induce the intended grouping structure of AXVALues. For the sake of abbreviation, `<ax-ref>` may also be substituted by a CLUSTDIMension, particularly if the same `<agglev>` is to be used for breaking down each of the QUalities in this CLUSTDIMension. A specific case arises for *summations* over CLUSTDIMensions of COMQUalities: because of the factorial structure of included COMQUalities, breakdowns may be expressed in terms of QUFACTors; hence, in the latter case the definition of `<x-expr>` changes to

```
<x-expr>    ::=    <qufact> [ * <qufact> ]*
```

dropping '`[ @ <agglev> ]`' which is unnatural in this context; all occurring `<qufact>`-expressions, of course, must belong to the very same CLUSTDIMension the summation ranges over. Clearly, the breakdowns of all `<xdim-expr>` or `<qufact>` terms in an `<x-expr>` generate collectively the actual cross-classification of the (restricted) clause concept; AXVALues no longer contained in the TACONCEPT are truncated (this, in fact, may *restrict* the TACONCEPT even further if concept borders cut through MODality groups defined implicitly by `<agglev>` MODalities). TACONAXes (except *space* and *time* which receive special treatment) not occurring in `<xdim-expr>` belong to the *reference dimensions* of the TARGET and, hence, do not appear as *cross-classifying* dimensions in the final table. Like **OF** clause parameter phrases, **BROKEN DOWN BY** clause grammar could be enhanced considerably; however, only a few convenient extensions will be discussed shortly below.

METASPEL Compared to Database Query Languages

In summarizing, the TARGET and, thus, the virtual TAFRAME of a REQUEST is determined collectively by the parameter phrases of **FOR, WITH, BROKEN DOWN BY, OVER, IN**, and, in case of summations, possibly by **OF** clauses. Very faintly, METASPEL resembles database query language of the SQL genre [Chamberlin *et al.*, 1976; Ullman, 1982]; roughly, compared with SQL's *select-from-where* structure the **FOR** clause corresponds to *select*, the **ON** clause to *from*, and the **WITH, OVER**, and **IN** clauses to *where*. Quite on the contrary, however, it must not be overlooked that the couple of clauses presented may not be sufficient to facilitate the derivation (cf. Section 2.5) of the specified table: especially interspersed aggregation or disaggregation steps may require a couple of further specifications (such as statistical model structures accompanied with suitable parameter data) to be supplied in terms of additional – not yet developed, in fact – clauses, too (alternatively, of course, if the necessity to supply this information is not foreseen

.4 DATA VIEW SPECIFICATION

at REQUEST specification time it may be supplied interactively during the table's derivation stage as demanded). For instance, because (dis-)aggregations over temporally non-additive instances are very frequent, a special clause type for the determination of temporal averaging (including, say, a data smoothing or a seasonal adjustment model) could be introduced (cf. Subsection 2.4.2).

From Simple to Compound Tables

Making use of the METASPEL grammar as defined up till now enables the specification of a rather broad class of *simple* generalized statistical tables if tables comprising no more than *one* array of statistical indicators are called simple. Practically, simple tables are extended very often to *compound* tables such that families of semantically related internal simple tables are combined either for the sake of joint presentation in external tables (for instance, in printed publications) or for further arithmetical processing. As to the joint presentation of simple tables in a unified layout (cf. [Fortunato *et al.*, 1987]), three major cases can be distinguished:

- The compound table may be a combination of a single simple *kernel* table with all other simple tables directly derived from it by computing *margin* tables. Apparently, alignment of margin tables is possible only if the kernel table is cross-classified; margin tables are obtained from the kernel table by dropping one or more of its cross-classifying dimensions (that is, by changing it into a reference dimension).
- Occasionally, a compound table consists of hierarchically related target concepts; such a decomposition of tables into sub-tables corresponding to split target concepts is called *dissection*.
- Compound tables also result from *mounting* simple tables with formally unrelated target concepts provided that all simple sub-tables are subjected to the same common cross-classification (if any).

In view of the rather high degree of overlap in specifying the simple sub-tables of compound tables it is certainly advisable to augment the basic structure of METASPEL to account for at least the most frequent compound table formats. In addition to simple tables, however, an eye must be kept on the formal structures of compound tables admitted in order not to sacrifice the convertibility of internal compound tables into visible tabular and graphical layouts.

The most important of the mentioned cases regards the alignment of kernel and margin tables. Since margin tables share target concepts and summary types with associated kernel tables except some collapsed cross-classifying dimension(s), a syntactical extension of parameter phrases of the **BROKEN DOWN BY, OVER,** and **IN** clauses suggests itself. Basically, in **BROKEN DOWN BY** clauses any `<xdim-expr>` or `<qufact>` term can be either prefixed with 'Σ|' or suffixed with '|Σ' (with apparent meaning); furthermore, sub-expressions of several (prefixed/suffixed or not) `<xdim-expr>` or `<qufact>` terms can be put in parentheses and, in turn, prefixed or suffixed *jointly* in the same way (thus, of course, creating *nested* structures). For instance, letting A_1, A_2, and A_3 denote some arbitrary Axes, expressions like

$$(\Sigma | A_1 * A_2) | \Sigma * A_3 | \Sigma$$

are well-formed. In this example, the kernel table would, ceteris paribus, comprise a cross-classification with respect to $A_1 * A_2 * A_3$; the margin tables induced would have cross-classifications $A_1 * A_2$, $A_2 * A_3$, A_2, A_3, and a sum total without any cross-classification. Similarly, the expression

$$A_1 | \Sigma * A_2 | \Sigma * A_3 | \Sigma$$

would induce margin tables for *all* factor patterns possible, that is $A_1 * A_2 * A_3$, $A_1 * A_2$, $A_1 * A_3$, $A_2 * A_3$, A_1, A_2, A_3, and, again, a sum total without any cross-classification. Computationally, there is no difference between prefix and suffix "marginalization" but this position information may be used for layout determination later on. In an analogous fashion, these prefixes and suffixes can be used to augment the <val-expr1> term in the parameter phrases of **OVER** and **IN** clauses, respectively. Furthermore, in case of COMVALues, prefixes and suffixes may be included into the component sub-expressions generating the respective ranges (cf. Section 4.4 with respect to pertinent examples). Alternative to the use of prefixes and suffixes, marginalization may be specified by a *depth indicator* as well, if AGGLEVels are compiled hierarchically (as it happens, typically, with regional classifications in *space* DIMensions). The depth indicator specifies the number of requested aggregation levels ("batch sums") downwards, implying the inclusion of all intermediary-level sums. Syntactically, the depth indicator could be distinguished by prefixing the integer *depth* number by an underscore symbol; by definition, the depth indicator always comes after the specification term of the respective cross-classifying dimension. In parameter phrases, prefixes, suffixes, and depth indicators may be used side by side provided that for each cross-classifying dimension at most one marginalization operator is specified.

If a table should exhibit statistical figures not only for some particular target concept but for one or more of its sub-concepts, this can be achieved simply by extending the **FOR** clause grammar: in addition to naming a single clause concept the rewriting rule is extended to

```
<for-expr>   ::=
     <ConLab> [ >> <sub-ConLab> [ wrt <pcc-expr> ] ]*
```

where <sub-ConLab> must refer to a domain concept defined formally as sub-concept of <ConLab>, '**wrt**' reads as "dissected further with respect to" and <pcc-expr> is identical to the <pcx-expr> term of **WITH** clauses except that prefixing the <val-expr> term within this <pcx-expr> term by a '='-sign is not admitted. Semantically, the <ConLab> concepts gets accompanied by one or more of its direct or indirect sub-concepts which, optionally, may become dissected further by VALues comprised in <val-expr> of the <pcc-expr> term. It is by no means required that the sub-concepts entered in a **FOR** clause be disjoint (this, however, would hold automatically in case of dissections for apparent reasons). In principle, also a hierarchical nesting of sub-concepts is conceivable but not supported by the term definition given above.

Sometimes, related simple tables may share all but a few specification clauses. To support economic table specification, METASPEL is augmented with a syntactical *folding* mechanism. Basically, this folding mechanism multiplies out multiple parameter phrases where the latter simply are repeated parameter phrases enclosed in square brackets and separated by commas. Thus, letting Γ_h and Γ_t denote the (possibly empty) starting and ending clause sequences, respectively, of a METASPEL sentence $\Gamma_h \kappa [\alpha_1,...,\alpha_K] \Gamma_t$ such that $\kappa[\alpha_1,...,\alpha_K]$ is a clause with key phrase κ and multiple parameter phrase $[\alpha_1,...,\alpha_K]$, this sentence is formally rewritten to $[\Gamma_h \kappa \alpha_1 \Gamma_t,...,\Gamma_h \kappa \alpha_K \Gamma_t]$ (due to the full commutativity of folding, inner brackets can be stripped off throughout). Despite its simplicity, this folding of specifications is a rather powerful principle which even can be extended further, for instance, to take into account particular cross-classification structures by replacing, in **BROKEN DOWN BY** clauses, the separating commas by '+' symbols; combined with proper bracketing and the distributive law of '*' with respect to '+' provides an elegant way of fairly complex breakdown specifications within individual tables. For instance, letting again A_1, A_2, and A_3 denote some arbitrary AXES,

$$(A_1 + A_2) * A_3 \equiv A_1 * A_3 + A_2 * A_3$$

which, in turn, converts to $[A_1 * A_2, A_1 * A_3]$ before becoming subjected to the folding mechanism (additionally, for later table composition and, particularly, layout preparation juxtaposition information will have to be stored separately as well).

Quite often, statistical computations imply an arithmetical combination of (simple) internal tables with identical taxonomic scope. For instance, with respect to the elementary summary types defined in METASTASYS, computing *averages* amounts to derive two intermediary simple tables first, viz. a table of SUM as well as a table of COUNT, to get the quotients. This type of table arithmetics is accounted for in METASPEL by augmenting the standard folding mechanism with *functional* folding accordingly. For apparent reasons, the envisaged table arithmetics is restricted to the **OF** clauses of METASPEL sentences such that, formally, the separating commas between multiple parameter phrases in **OF** clauses are replaced by arithmetical operator symbols or, viewed differently and more generally, multiple parameter phrases become arguments of arithmetical expressions. Thus, symbolically, letting $\alpha_1,...,\alpha_K$ denote the arithmetical expression f's argument parameter phrases, **OF** $f(\alpha_1,...,\alpha_K)$ rewrites to

$$f(\mathbf{OF}\,\alpha_1,...,\mathbf{OF}\,\alpha_K)$$

and, using Γ_h and Γ_t as defined above, a sentence $\Gamma_h\,\mathbf{OF}\,f(\alpha_1,\ldots,\alpha_K)\Gamma_t$ rewrites to

$$f(\Gamma_h\,\mathbf{OF}\,\alpha_1\,\Gamma_t,\ldots,\Gamma_h\,\mathbf{OF}\,\alpha_K\,\Gamma_t)\,.$$

Functional folding and standard folding are not commutative; either functional foldings are deferred until other standard foldings no longer apply, or "inner" foldings are removed by yet another rewriting rule transforming

$$f(\ldots,\Gamma_h\,\kappa[\alpha_1,\ldots,\alpha_K]\Gamma_t,\ldots)$$

into

$$[f(\ldots,\Gamma_h\,\kappa\alpha_1\,\Gamma_t,\ldots),\ldots,f(\ldots,\Gamma_h\,\kappa\alpha_K\,\Gamma_t,\ldots)]\,.$$

For convenience, of course, several arithmetical standard expressions of statistical computing (such as means and variances) could be defined as special parameter phrase templates provided that SUMTYPESET is extended to include the corresponding summary types. Although the arithmetical combination of tables with different TARGETs (but possibly identical summary types) looks deceptively similar to the arithmetical combination of tables with identical TARGETs differing as to summary types only, there is amazingly little formal coincidence between both types of table processing: simple tables and compound tables consisting of simple tables specified by single METASPEL sentences are but a subclass of a yet more general class of statistical tables, the META class of OBJECTs, the elements of which can be combined in a specially devised table algebra discussed in Subsection 2.4.4.

The collection of simple (internal) tables specified in a single METASPEL sentence has to be arranged in a joint uniform (external) table layout eventually. In general, internal tables still admit several logically – but perhaps not visually ! – equivalent graphical layouts; for this reason, further specification parameters – *table view* definitions including the arrangement of cross-classifying dimensions in (at most) three-dimensional space (row, column, layer), table header, table stub, footnote design, etc. – are required to arrive at a unique table presentation. Since the present exposition abstracts from table layout issues, only one remark is added noting that compound table specifications usually introduce more or less strict constraints on feasible table layouts. Thus, in case of compound tables, at least some table composition information must be conveyed from the lexical analysis of METASPEL sentences to the layout specification stage. Essentially, three types of spatial relationships between the sub-tables of a compound table occur: alignment, interposition, and nesting. *Alignment* is a general kind of table juxtaposition applicable whenever two sub-tables share the same set of cross-classifying dimensions. *Interposition* is more specific in that it is used to align kernel and associated margin tables by interposing, in a hierarchical fashion, the table entries of kernel and margin tables. *Nesting*, finally, is a special variant of interposition admitting changing breakdowns at the lower level of interposed tables. Nesting is induced either by

FOR clauses specifying dissections (all of which become subjected to the same cross-classification pattern) or by *aligned* breakdowns specified in **BROKEN DOWN BY** clauses using the '+'-symbol. Interposition is the standard layout structure for hierarchically arranged multiple breakdowns. Formally, this juxtaposition information is recorded in a *metatable* expression extending the representation of METASPEL sentences obtained by functional folding (that is, moving nestings and breakdown alignments "outwards").

2.4.4 METAAlgebra

How powerful and comprehensive a data view specification language like METASPEL ever may be, there is always room for yet more specific operations and aggregate structures. Apparently, beyond a certain point it becomes unjustifiably difficult to understand and use a language integrating virtually all features in a rather closed formal language design. In this respect, METASPEL tries to strike a balance between simplicity and generality, leaving the more involved operations on statistical aggregates to a *table algebra* being discussed in its rudiments in this subsection. From the user's point of view, operations in this table algebra are carried out in a stepwise fashion, in contrast to the "all at once"-specification of aggregates by METASPEL sentences. Formally, making such a clear distinction between METASPEL and table algebra is motivated mainly by the fact that, in general, the operations of the table algebra lead to operands of a structure more complex than the one representable in terms of native METASPEL clauses (that is, summary type management is no longer sufficient exclusively to capture the semantics of the operations applied); specifically, the arithmetical combination of METAs specified by METASPEL sentences induces a calculus such that, at the TARGET level, the structure of the computations needs to be recorded explicitly in addition to the operand structure descriptions. Thus, in full generality, META descriptions must comprise a *trace* component recording the transformations applied to the input arguments (which are METAs as well) in order to assure a complete protocol of any META's derivation chain; the input terminals of each derivation chain, of course, are METAs originating from METASPEL sentences by definition. Morphologically, the trace component of METAs is a symbolic representation of the formula generating the output META it is attached to which also determines a META's "data type" (that is the summary type of a META's data layer DATFILE). Note that the representation of functional foldings, as introduced in the previous subsection, can be regarded as a particularly simple instance of such a trace component.

The operational tasks of the META algebra can be grouped into four major areas (cf. the PC-AXIS system [Nordbäck, 1992]): table editing, table arithmetics, table layout design, and table export formatting. These groups comprise varying numbers of operations, and none of it will be described exhaustively. Despite their practical relevance, layouting (for instance, to shape tables according to regulated standards such as [DIN 55 301, 1978; Ö-Norm A 6195, 1989]) and export formatting operations (in accordance to interface standards of, for instance, statistical analysis packages or EDI standards) are omitted completely from subsequent considerations mainly because their output is not used any further *within* the META algebra. Table editing operations enable various kinds of table "surgery" (such as – possibly conditional – selection, deletion, or replacement of table entries or even whole partitions of METAs) and table rearrangement (such as sorting by categorical or numerical criteria). There would have to be also special editing functions affecting the metadata layer of METAs only (for instance, adding or changing footnotes, entering table annotations, etc.)

The most important group of table operations in the present context, of course, is table arithmetics. First of all, once computed from a METASPEL sentence, a META may be marginalized or de-marginalized supplementarily. In computational terms, marginalizing a META is facilitated by sending it a Σ|: or a |Σ: message stating the name (AXLAB) of the cross-classifying dimension to be aggregated over; practically, marginalizing amounts to either really aggregate (if the corresponding margin table is not among the addressed META's sub-tables) or to drop the META's kernel table (otherwise). Note that, implicitly, *all* sub-tables of a META are affected by a marginalizing message sent – if the particular margin table alone were of interest, it would have to be selected, if need be, by a separate table editing operation. Just like marginalizing a META, it can be de-marginalized sending it a Σ^{-1}: message specifying a suitable breakdown in terms of an <xdim-expr> as message argument. However, if a re-computation of the META including the additional cross-classifying dimension from the (virtual) database turns out impossible, a ***data:*** message providing disaggregation rates must be supplied in conjunction with the Σ^{-1}: message.

Given the possibility to modify aggregates as specified by METASPEL sentences at some later point in time, it may turn out useful to provide a store of METASPEL sentence *templates* such that individual templates can be retrieved and, after initializing template parameters (if any), submitted to the evaluation process.

Within METASPEL table arithmetics remains fairly restricted; many relevant statistical tables can be obtained only by extending the set of admissible table operations considerably. In fact, the modest capabilities of **OF** clause arithmetics needs to be expanded to embrace a whole domain of "structurally compatible" tables. In view of table arithmetics, this structural compatibility of operands splits into three branches:

- The *grouping compatibility* requires all operand tables to be combined belonging to the same "mother" FRAME, having identical TACONPROFiles, and, with respect to shared cross-classifying dimensions, having identical MODality groupings (though, formally, AGGLEvels may be different).
- The *dimension compatibility* requires that the cross-classifying dimensions of all operand tables are (proper or improper) subsets of the "maximal" operand table (that is the operand table with the largest number of cross-classifying dimensions). This precondition provided, the "margin operands" can be expanded to the size of the maximal one by repeating their data parts along the collapsed

cross-classifying dimensions as often as necessary (this principle of dimensionality adjustment occasionally has been called $E \times E$ compatibility; cf. [Edlefsen and Jones, 1986]).
- The *summary type compatibility* requires that, by means of established type conversion rules, the operand tables' summary types comply and determine the output table's summary type uniquely.

Jointly, grouping, dimension, and summary type compatibility determine structural compatibility of an expression's operand tables. On account of these compatibility checking rules, a formal arithmetical META algebra can, in principle, be introduced comprising the standard operations of basic arithmetics, functions, and comparison operations (returning BOOLEan METAs) at the table entry (DATCELL) level viewing the data component of METAs as *multi-dimensional matrices* processed element-by-element. However, it must be noted that, in some cases at least, semantics of META algebra deviates from standard semantics of statistical table management; for instance, adding two structurally compatible METAs, in general, will produce different results if addition is done with respect to a DIMension declared non-additive within the underlying FRAME (because summation would then be inhibited at the level of internal aggregation).

Percentages and Rates

As a simple example of the usefulness and necessity of the arithmetical META algebra, tables of summary type *percentage* are considered. Percentages are not among the elementary summary types because a quotient of two tables of summary type COUNT needs to be computed, and a summary type is, by definition, not elementary if it depends on an elementary one. Furthermore, to obtain percentages as familiar, quotients must be multiplied by a factor of 100. Formally, the derivation of a percentage table implies the specification (i) of an enumerator table E determining the eventual taxonomic scope of the result table, and (ii) of a denominator table D which is a margin table with respect to E. Both tables are, of course, structurally compatible since D can be expanded to the size of E simply by repeating its contents with respect to the collapsed cross-classifying dimensions of E. Likewise, "multiplication" by a *constant table* '100' is achievable (apparently, constant tables are structurally compatible with *any* table). The resulting META comprises a DATBOX identical with the one of E, descriptions of (or references to) E and D, and a trace component representing the formula '$100*(E \div D)$'.

With respect to summary types of computed METAs two approaches are possible: if a computing formula is occurring quite often (such as the formulae for rates and percentages) and, additionally, data sources contribute macrodata conforming to this formula's output frequently, it can and should be included as a *non-elementary summary type* in SUMTYPESET by introducing a unique SUMTYPE name – say, '<u>RATE</u>' for rates, '<u>PERCENT</u>' for percentages, etc.; otherwise, a default summary type (say, 'COMPUTED') is assigned which, obviously, does not provide any structural information about a META's actual summary type (except it is not elementary).

In order to simplify the frequent specification of selected non-elementary summary type aggregates further, METASPEL syntax may be amended as well. For instance, in case of percentages, letting '%' denote the summary type 'PERCENT', the parameter phrase of **OF** clauses is extended to comprise patterns like '%(n)' and '%(Σ <arg>)'; correspondingly, another, *non-native* METASPEL clause defining the percentage's denominator table is introduced: since enumerator and denominator tables differ with respect to cross-classifying dimensions (only), either the denominator table's cross-classifying dimension(s) or the collapsed cross-classifying dimension(s) of the enumerator table must be stated explicitly (preferring the more economic method, respectively). The denominator table's cross-classification is specified by the **BASED ON** clause whereas collapsed breakdowns are specified by the **DROPPING** clause; in both cases, standard parameter phrase syntax of **BROKEN DOWN BY** clauses applies except that, additionally, *space* and *time* are valid <ax-ref> substitutions as well.

Even simpler is the handling of RATEs; in this case, the arguments E and D may be any structurally compatible tables. As a convenient METASPEL notation, an '**OF RATE**'-clause could be defined which, then, must be accompanied by a **FOR**-clause of shape

$$\text{FOR } E \div D$$

such that each corresponding METASPEL sentence – note, that there may be several of them, if the folding mechanism applied – is decomposed in a special trailing unfolding operation rewriting

$$\Gamma_h \text{ FOR } E \div D \, \Gamma_t$$

to

$$100 * \left(\Gamma_h \text{ FOR } E \, \Gamma_t \div \Gamma_h \text{ FOR } D \, \Gamma_t \right).$$

In quite a similar way, indices could be dealt with by introducing, for instance, an

$$\text{OF INDEX}(t)$$

clause where t refers to a *time*-MODality of the same AGGLEVel as the one used in the respective METASPEL sentence's **IN**-clause.

Temporal Comparisons

Similarly, other non-elementary summary types can be handled. For instance, since comparisons of aggregates for changing points of time are requested frequently, **OF** clause syntax could be extended to include a 'Δt' parameter phrase denoting the difference between two time points (MODalities of a FRAME's *time* AXis); accordingly, METASPEL's native **IN** clause would have to be replaced with, say, a non-native **BETWEEN** clause having parameter phrases conforming to the syntax

`<st-expr>` **and** `<st-expr>`

meaning that all differences between pairs of left-hand and right-hand side time points are thus requested.

By means of these amendments quite complex tables can be specified (and, by the way, stored as templates for later and repeated usage) in terms of METASPEL sentences although it must be remembered that, in fact, METASPEL sentences containing non-native clauses are pre-processed into META algebra expressions first before operand tables are generated and inserted in the META algebraic expressions eventually returning the specified aggregate(s).

META Variables

METASPEL is a conceptual language for the specification of statistical aggregates which can be combined to even more complex ones in META algebra expressions. Adding the concept of META variables, intermediary METAs can be stored temporarily and referenced by variable names later on in other META algebraic expressions. In METASTASYS, the assignment operation is denoted by the binary '←' operator; submitting a META variable name to evaluation is completely equivalent to submitting the corresponding META specification. Using META variables, METAs can be processed interactively (that is in a stepwise, conditional fashion) at the interface level. Although not envisaged presently, the META algebra could be developed into a complete interactive table programming language – possibly even including elements of direct manipulation interfaces such as those reviewed in Subsection 1.3.2 – mimicking statistical programming language systems like S-PLUS [StatSci, 1993; Becker, 1994] or LISP-STAT [Tierney, 1990].

2.5 META *Derivation*

When a REQUEST is submitted it faces the bunch of IMAGES available. More specifically, since IMAGES comprise both description and data, the REQUESTS are confronted with these IMAGES' IMBOXes and SUMTYPEs. Symbolically, letting $\mathfrak{I}_Q$ denote the set of all IMAGES contributing to T_Q, T_* a TARGET based on some $t_* \in T_Q$, s_* a SUMTYPE, and $r_* =_{def} \langle T_*, s_* \rangle$ a particular REQUEST, the derivability relation can be expressed simply as

$$\mathfrak{I}_Q \vdash_Q M_*$$

such that, as a side effect, each derivable r_* is supplemented with the DATFILE D_* comprising the numerical information corresponding to r_*; jointly, T_* and the supplementary DATFILE D_* of SUMTYPE s_* determine a FILE structure (cf. **Def. 2.2.4–3**) representing the core component of the generated output META M_*. In what follows it is understood implicitly that only those IMAGEs belonging, or contributing, to the same FRAME also T_* refers to are considered; apparently, this IMAGE selection is accomplished easily and, hence, there is nothing more to be said in this respect.

The IMAGEs, in general, are divided into two classes, viz. those with elementary SUMTYPEs and those with non-elementary SUMTYPEs. In view of a derivation calculus, the former class is of considerably more interest because derivations can be broken down into sequences of elementary "proof" steps such that the additivity property of the data objects is preserved; in the latter class only a rather limited set of transformations can be applied feasibly and, thus, very little remains derivable from IMAGEs with non-elementary SUMTYPEs. However, for completeness' sake, "non-elementary" IMAGEs will also be included in the following exposition.

In terms of man-machine interaction, a REQUEST states a specific kind of *question* to be answered against the pool of IMAGEs connected to a domain META-MODEL. The internal question-answering mechanism must now, in some effective way, transform the stored positive evidence (coded in empirical observation data or already statistically processed observation data comprised in IMAGEs) into the data views expressed as REQUESTs formally. Computationally, this is tantamount to the design of a set of algebraic operators which can be combined to generate, by enumeration, the set (the transitive closure, in fact) of all formal statements derivable from IMAGEs for checking the set membership of a REQUEST. Practically (and abstracting from minor details), this derivation procedure splits into three layers: to derive an answer to a REQUEST r_*

- (a subset of) IMBOXes need to be transformed into the TARGET T_*,
- the (involved) IMAGEs' SUMTYPEs need to be transformed into s_*,
- the (involved) IMFILEs need to be transformed arithmetically into D_* corresponding to the IMBOX and SUMTYPE transformations of the metadata layer.

As a major consequence, METASTASYS's table derivation is separated into a metadata layer calculus and a data layer calculus. While the former investigates the derivability of a REQUEST from the available IMAGEs by means of symbolic reasoning (*feasibility verification* in Basili and Meo-Evoli's terms [1992]), the latter commences – by executing a *resolution plan* [Basili and Meo-Evoli, 1992] composed during the metadata layer table derivation stage – with actual database retrieval and arithmetical data transforming operations only *after* a REQUEST's derivability has been assured. In a sense, this two-stage approach systematically expands the notion of "query optimization" well-known from the database domain: before computing a query's extension the query expression itself is scrutinized and pre-processed in various respects; only then "deferred evaluation" may take place. Formally, at the metadata level, the derivation, or table, calculus strings together operators *rewriting* the IMAGEs' representations $d_i =_{def} \langle I_i, s_i \rangle$, where – for rea-

.5 META DERIVATION

sons of formal symmetry – $\mathbb{I}_i$ denotes the i-th IMAGE's IMBOX $\mathbf{B}_i = \left(\mathcal{B}_i^G, \mathcal{E}_i \right)$ in TNF with IMPROfile $\wp_i$ (cf. **Def. 2.3.3–4**), and s_i denotes its SUMTYPE, such that eventually a given r_* is obtained if possible at all. Thus, in fact, letting $\mathfrak{I}_Q^D$ denote the set of description parts (which, strictly speaking, comprise several other metadata components in addition to d_i playing vital roles in META derivations) of the IMAGEs comprised in $\mathfrak{I}_Q$, the derivability relation reduces to

$$\mathfrak{I}_Q^D \vdash_Q r_*.$$

From a problem solving view each operator is used to reduce the *semantic distance* between REQUEST and IMAGE representations; a REQUEST is derivable formally, if there is an operator sequence removing this semantic distance. Moreover, instead of brute enumeration of derivable REQUESTs, the formal structure of semantic distances can be exploited beneficially to diminish the search complexity of the table derivation procedure. After deriving a REQUEST successfully, translating a completed derivation's operators from the metadata layer into the corresponding data layer operations yields, more or less straightforwardly, a definite resolution plan (query evaluable at the database level) generating the REQUEST's DATFILE D_*, that is the actual table entries.

At any time, a submitted REQUEST r_* meets a fixed set of IMAGEs and, hence, IMAGE descriptions $\mathfrak{I}_Q^D$. Accounting for practical demands and emphasizing the problem solving perspective, METASTASYS extends the strict derivability relation such that, if r_* cannot be **Q**-derived from $\mathfrak{I}_Q^D$, r_* is "relaxed" to some derivable $\tilde{r}_*$ by admitting (meta-)data-driven adaptations of r_*. In turn, the eventual response to a query r_* may be $\tilde{M}_*$ instead of the intended but not derivable M_*. Pragmatically, $\tilde{M}_*$ may be considered as a semantic approximation to M_* in view of actual empirical evidence (data availability). It goes without saying that $\tilde{M}_*$ is represented consistently by replacing T_* and s_* with $\tilde{T}_*$ and $\tilde{s}_*$, respectively, as necessary. At present, however, it is not completely clear which types of REQUEST relaxations should be allowed and carried out automatically.

In METASTASYS's semantic distance reduction engine, METAGEN, the operators used for deriving METAs are classified into three groups in view of their different consequences for (automatic) summary type management (cf. [Meo-Evoli et al., 1992]). Moreover, this distinction contributes to a problem decomposition insofar as operators particularly of the first two classes frequently concern a single IMAGE or table at a time and thus, evidently, do not interlace.

- *Algebraic concept transformators* (ACTs) affect the BOX structure (taxonomic scope and grouping structure) of IMAGE descriptions such that the corresponding table operations at the data layer (called *table management* by Meo-Evoli et al.

[1992]) do not refer to any statistical model, that is if they change a table's data layer at all the involved data transformations remain restricted to rather basic arithmetics such as taking (weighted) sums over grouped MOdalities etc. (both, "simple" and "semantic" reducibility of Basili and Meo-Evoli's [1992] 'StEM' SDB query processor belong to the ACTs class). In particular, ACTs never rely on additional (parameter) information actually controlling data transformations as apparently indispensable, for instance, in weighting/estimation or disaggregation operations. As an immediate consequence of these restrictions the calculation algorithms achieving the corresponding data layer effect of changes in the IMAGE descriptions is determined transparently from the operands' metadata and, hence, need not be stated explicitly or by default. Thus, as long as only ACTs are involved in a REQUEST derivation, data processing is specified exhaustively by $\langle T_*, s_* \rangle$. Obviously, transparent SUMTYPE management is of particular importance in establishing TAFRAMEs internally in case of *1:n* FRAME linkages when MESO data need to be converted to COUNTs implicitly for adjoined QUalities.

- *Statistical concept transformators* (SCTs) affect the data layer of tables only (that is, the DATFILE entries) and do *not* change the metadata layer description of tables except, of course, with respect to statistical metadata recording model-based changes of the data layer. Typical examples of SCTs are estimation procedures (grossing up, for example, sample figures to population figures) or data smoothers applied to cross-classified tables (arrays). Meo-Evoli *et al.* [1992] assigned the name *data analysis* to the SCT class of operators including *any* type of operation changing summary types (which implies a more general usage of this term than admitted in METASTASYS).

- *Mixed concept transformators* (MCTs), finally, comprise the residual cases of operators involving both algebraic and statistical components. MCTs either extend the taxonomic scope of tables or rearrange the internal table structure (cross-classifying dimensions) on account of statistical models and thus imply algebraic as well as statistical concept transformations. The class of MCTs comprises operations for imputations, disaggregations (including de-marginalizing), interpolations, and data propagation methods (including forecasts and other types of extrapolations). Apparently, MCTs are more complex operations than ACTs and SCTs; like SCTs, MCTs always have to be fed with further parameter data. In contrast to ACTs and SCTs, MCTs typically combine information of several IMAGES or tables concurrently and, in doing so, obstruct the decomposition of REQUEST processing into independent sub-derivations; rather conversely, MCTs may launch sub-queries on their own for providing required parameter data internally.

Simply speaking, ACTs are the direct counterparts to elementary database operations (such as selections, restrictions, projections, merges, etc.) albeit operating at the CONCEPT level. Occasionally, MCTs are rather "close" to ACTs in that it is often possible to introduce default assumptions supplying tacitly the parameters of MCTs. As far as such default settings are accepted (that is, not changed actively) these

.5 METADERIVATION

MCTs behave like ACTs from a user's point of view. This mechanism is particularly useful in standard situations like aggregations over *time* or *space* in presence of non-additivity or in case of de-marginalizing ("aggregation" of marginals) where averaging over contiguous regions (in terms of SCALE semantics as defined by neighbourhood or periodicity graphs; cf. Subsection 2.2.1) or stochastic independence [Malvestuto, 1989], respectively, are rather natural default assumptions. Other defaults may be less popular (or harder to justify) and, in some cases, there may be no reasonable default settings at all. Anyway, defaults must be amenable to replacement (for instance, by suitable extensions of the METASPEL syntax; cf. Subsection 2.4.3) with formal assumptions introduced explicitly (cf. the bygone *Algos* project; [Graves and Manor, 1986]) and, irrespective of where the controlling parameters originate from, they must be protocolled automatically in a META's *statistical* metadata component to keep track appropriately of all operators applied.

To be an algebra, the involved operations must be closed with respect to the algebra's carrier set of objects. In METAGEN, algebraic closedness of operations is established simply by enforcing all operands to conform to IMAGE descriptions, that is, essentially, *pairs* of grouped CONDEFs in TNF (**Def. 2.4.1–5**) referring to the same FRAME **Q** and SUMTYPEs; by definition, REQUESTs fit this format, too. For the sake of easy reference, this description format will be referred to as <u>PROMETA</u> (proto-META). In parentheses, however, it should be remarked that, for practical purposes, PROMETAs have to comprise some more metadata elements (such as statistical metadata) which are simply omitted from the present exposition but need, of course, a similarly rigorous formal treatment in the longer run; the outlined PROMETA structure would have to be augmented accordingly. Given this abbreviated object representation, METAGEN operators can be stated as term rewriting rules transforming PROMETAs. Practically, the operators are condensed to *parameterized rewriting schemata* the parameters of which are inferred by "measuring" the semantic distance between given input(s) and desired output, viz. – rewritten – IMAGE descriptions and REQUESTs, respectively. A further criterion of an algebra's set of operations is minimality, or non-redundancy. METAGEN uses a sparse set of operations each of which fulfils a different purpose (cf. Subsection 2.5.1) whence these operations are non-redundant for apparent reasons. Finally, the table calculus ought to be complete and, thus, has to have a sufficient set of operations either. Despite the evident meaning and relevance of this claim it is difficult to make the notion of completeness semantically concise because completeness implies the presence of a definite and commonly agreed upon yardstick (for instance, Meo-Evoli *et al.* [1992] expound a "complete" calculus for a restricted class of table management operations). In principle, of course, given the expressivity of a particular domain METAMODel language, a definite framing of all metadata components of METAs not concerned presently, and a fixed set of statistical methods for data processing, a *model* can be set up providing an effective yardstick for determining the completeness of any axiomatized data theory. In practice, conversely, it will suffice to demonstrate compellingly that the devised calculus is indeed capable to derive all but a few of the "really important" statistical aggregates on the grounds of common judgement. Unfortunately, a trustworthy assessment may even be harder to achieve than a formal proof of completeness.

In logical terms, **Q**-derivability is decidable since, at any time, (i) the set of available IMAGES $\mathfrak{I}_Q$ is finite, and (ii) there is only a finite set (basically: a constant set) of operators the parameters of which are fed with values from finite domains, viz. the BASETs defining the theoretical data language of FRAME **Q**. Hence, trivially, each derivation is finite (there is no room for König's [1950] *infinity lemma* to apply) and, thus, the set of derivable METAs – though quite big, in general – is finitely enumerable (either there are no more term rewritings applicable or they become idempotent). Conversely, however, to prove a particular r_* from $\mathfrak{I}_Q^D$ involves, in general, to consider, one by one, each of $\mathfrak{I}_Q^D$'s non-empty subsets and check if r_*, or a related $\tilde{r}_*$, is derivable from it; this apparently cannot be achieved in polynomially bounded time and – though any formal proof is omitted here – **Q**-derivability is *NP*-complete (equivalent to the SAT problem; cf. [Garey and Johnson, 1979]). Acknowledging both, finiteness and intractability of enumerating all possible **Q**-derivations, there is ample room for problem-tailored heuristics employing various divide-and-conquer strategies as expounded in Subsection 2.5.3.

Although the following presentation gives a rather scanty account of METAGEN (mindful of the vast scope of details to be considered eventually) it should be remembered that "standard" table derivations can be dealt with even in this provisional set-up; more distinctly, METAGEN readily catches up in power with virtually all statistical database retrieval approaches proposed hitherto (cf. Subsection 1.3.2) while being far from worked out fully.

2.5.1 Basic METAGEN Operations

The functionality of METAGEN is achieved by a couple of operators which can be separated naturally into three groups with respect to the overall META derivation scheme (cf. Subsection 2.5.3) consisting of a *search* stage and a *transformation* stage which, in turn, may be divided further into an *integration* stage and an *extraction* stage. In this triplet, the search stage is responsible for collecting IMAGES actually contributing information (data) to a given REQUEST ("what is available ?"); the integration stage investigates the possibilities to merge (some of) the IMAGES collected in the search stage ("how can it be consolidated ?"); finally, the extraction stage retrieves and shapes the relevant portion of combined IMAGES in view of the TARGET specifications ("what can be gained from it ?") of a processed REQUEST. For apparent reasons, each stage comprises rather specific operations – conceptualized as formal operators – which, in turn, are composed out of various lower-level operations. Hence, for the sake of a condensed account, these lower-level operations are discussed first such that higher-level operations and METAGEN operators are developed synthetically proceeding from elementary to increasingly complex functions. Generally, the following exposition is limited to the operators' algebraic

.5 META DERIVATION

aspects – in other words, only ACTs and MCTs fully controlled by defaults are considered whereas SCTs will be omitted from the presentation. In symbolic terms, METAGEN's operators are conceptualized as unary, parameterized functions (though with a set-valued argument wherever necessary); at calculus level, only these operators are visible (in contrast to the operators' internal sub-operations). According to the present set-up, operator arguments always conform to IMAGE descriptions d_i or structurally equivalent operands obtained by applying META-GEN operators to IMAGE descriptions; in the longer run, of course, these metadata descriptions must be augmented particularly with statistical metadata required for controlling properly the application of SCTs as well as the more intricate MCTs. Formally, the operators embody METAGEN's rewriting schemata provided with specific parameter information obtained from "measuring" the semantic distance between an operator's input operands and the desired operator output. Thus, an operator's parameters are instantiated typically by binary *parameter functions* investigating specific aspects of (transformed) IMAGEs and a given REQUEST. In a few cases, however, parameter data may be contributed dynamically or obtained from a search process. Contrary to the a priori parameters computed by parameter functions, a posteriori parameters are added to the operand descriptions during operator execution (however, after termination of operator execution one cannot tell formally whether a parameter had been supplied a priori or not). In particular, a posteriori parameters can be used as a protocolling mechanism in case of choice decisions made interactively, thus admitting semi-automatic, user-driven META derivations in a strict consistency-preserving way.

In their most elementary form, METAGEN's parameter functions refer to individual BOX dimensions since operations affecting BOXes can, in most cases, be decomposed straightforwardly into operations affecting single DIMensions (or AXes) only. With respect to operand descriptions, almost any operation refers to BOXes for which reason an outline of METAGEN's operations soundly starts with dimension-wise parameter functions.

Whether originating from IMAGEs or REQUESTs, BOX dimensions ("factors") are GRUSETs (cf. Subsection 2.1.2) formally. The following function definitions are easily converted into FOOL expressions as listed in Appx. A.

Auxiliary Parameter Functions

A rather trivial, auxiliary parameter function computing group intersection is defined in

Definition 2.5.1–1:
Letting G' and G'' be two GRUSETs of same BASE, the *group intersection* of G' and G'' is a GRUSET $\lambda(G', G'')$ obtained by intersecting elements $g' \in G'$ and $g'' \in G''$, respectively, computed by functions

and

$$\lambda(G',G'') = \begin{cases} \lambda_0(g',G'') \cup \lambda(G'-g',G'') & \text{if } G' \neq [] \\ [] & \text{otherwise} \end{cases}$$

$$\lambda_0(g',G'') = \begin{cases} \lambda_0(g',G''-g'') + (g' \cap g'') & \text{if } G'' \neq [] \\ [] & \text{otherwise} \end{cases};$$

'+' and '−' denote element addition and subtraction, respectively, as defined in Appx. A.5.

For convenience, $G' \cap G''$ may be used as a notation variant equivalent to $\lambda(G',G'')$ in view of FOOL's type polymorphism (cf. Appx. A.5). Apparently, group intersection is both commutative and associative. Group intersection extends easily to more than two arguments as stated in

Definition 2.5.1–2:
Letting $\mathbf{G} = \{G_1,...,G_J\}$, $J > 1$, with all elements $G_j \in \mathbf{G}$, $1 \leq j \leq J$, having the same BASE and letting further $G \in \mathbf{G}$,

$$\lambda^*(\mathbf{G}) = \begin{cases} G & \text{if } \mathbf{G} = \{G\} \\ \lambda(G, \lambda^*(\mathbf{G} \setminus \{G\})) & \text{otherwise} \end{cases}$$

where '\' denotes (ordinary) set difference.

Apparently, $\lambda^*(\{G',G''\}) = \lambda(G',G'')$. A generalized version of group intersection is group tiling which appends the groups (elements) of two GRUSETs G' and G'' not included in $\lambda(G',G'')$ as defined in

Definition 2.5.1–3:
Letting G' and G'' denote two GRUSETs of same BASE, and letting further denote $\mathbf{C}G$ the SET complement of $\gamma(G)$ with respect to the BASE of G (expressed as a singleton GRUSET), the *group tiling* of G' and G'' is a set (a GRUSET, in fact)

$$\tau(G',G'') =_{def} \lambda(\mathbf{C}G',G'') \cup \lambda(G',G'') \cup \lambda(G',\mathbf{C}G'')$$

which is computed accordingly.

.5 META DERIVATION

In FOOL notation, $\tau(G',G'')$ is replaced by $G'\#G''$ (cf. Appx. A.5). Evidently, group tiling is both commutative and associative. Like group intersection, group tiling extends outright to $\tau^*(\mathbf{G})$ defined analogous to $\lambda^*(\mathbf{G})$.

Another simple parameter function is group restriction as defined in

Definition 2.5.1–4:
Letting, again, G' and G'' be two GRUSETs of same BASE, the *group restriction* of G' and G'' is a set (a GRUSET, in fact)

$$\rho(G',G'') =_{def} \{g' \in G' \mid (\exists g'' \in G'').g' \cap g'' \neq \langle\rangle\}$$

computed by functions

$$\rho(G',G'') = \begin{cases} \rho_0(G',g'') \cup \rho(G',G''-g'') & \text{if } G'' \neq [] \\ [] & \text{otherwise} \end{cases}$$

and

$$\rho_0(G',g'') = \begin{cases} \rho_0(G'-g',g'') + g' & \text{if } (G' \neq []) \wedge (g' \cap g'' \neq \langle\rangle) \\ [] & \text{otherwise} \end{cases}$$

where $g' \in G'$ and $g'' \in G''$, respectively.

In general, group restriction is neither commutative nor associative; fortunately, however, $\rho(\rho(G',G''),G''') = \rho(\rho(G',G'''),G'')$. To see this, think of the extreme case of $\gamma(G'') \cap \gamma(G''') = \langle\rangle$ – where γ is the "un-grouping" function introduced in Subsection 2.4.1 replacing a GRUSET with the union of its PLASETs – but $\rho(G',G'') \neq []$ as well as $\rho(G',G''') \neq []$; then, of course,

$$\rho(G',\rho(G'',G''')) = [] \neq \rho(\rho(G',G''),G''')$$

is still possible but the ordering of restrictions applied to G' is immaterial throughout because, evidently, $\rho(\rho(G',G''),G''') \equiv \lambda(\rho(G',G''),\rho(G',G'''))$, and λ is commutative. In FOOL, $\rho(G',G'')$ is denoted as $G' \bullet G''$ (cf. Appx. A.5).

Now, in terms of group intersections and group restrictions some useful group relationships can be established formally. The first of these is group subsumption as defined in

Definition 2.5.1–5:
Letting G' and G'' denote two non-empty GRUSETs of same BASE, G' *subsumes* G'' – $G' \prec_G G''$ symbolically – if and only if

$(\forall g'' \in G'').(\exists H \subseteq G').g'' = \bigcup_{g \in H} g$; computationally, group subsumption is the Boolean function

$$G' \prec_G G'' = \begin{cases} (\gamma(\rho_0(G',g'')) = g'') \wedge (G' \prec_G (G'' - g'')) & \text{if } G'' \neq [] \\ \daleth & \text{otherwise} \end{cases}$$

where $g'' \in G''$.

Note that, by this definition, $G' \prec_G [] = \daleth$ for the sake of the function's well-foundedness; letting $G'' = []$ otherwise lacks any proper meaning, however. Note also that $G' \prec_G G''$ implies $\gamma(\rho(G',G'')) = \gamma(G'')$. Obviously, $\prec_G$ defines a partial ordering on the BASPARTitions (cf. **Def. 2.1.2–13**) of a given BASET since group subsumption is reflexive, antisymmetric (identitive), and transitive, that is

- $G \prec_G G$,
- $(G' \prec_G G'') \wedge (G'' \prec_G G') \Rightarrow (G' = G'')$, and
- $(G' \prec_G G'') \wedge (G'' \prec_G G''') \Rightarrow (G' \prec_G G''')$.

Occasionally, for a given pair of G' and G'' neither $G' \prec_G G''$ nor $G'' \prec_G G'$ may hold but G' and G'' still have a common "overlap" with crisp boundaries; more specifically, group match is defined in

Definition 2.5.1–6:
Letting G' and G'' denote two non-empty GRUSETs of same BASE, G' matches G'' – $G' \approx_G G''$ symbolically – if and only if

$$(\exists G).(\lambda(G',G'') \subseteq G) \wedge ((G \prec_G G') \wedge (G \prec_G G''))$$

such that G **base?** = G' **base?** = G'' **base?**; computationally, group match is the Boolean function

$$G' \approx_G G'' = \begin{cases} \daleth & \text{if } \gamma(\rho(G',\lambda(G',G''))) = \gamma(\rho(G'',\lambda(G',G''))) \\ \mathsf{L} & \text{otherwise} \end{cases}.$$

Thus, G conforms to $\lambda(G',G'')$ in the overlap area $\gamma(G') \cap \gamma(G'')$ inheriting the remaining groups from either of the GRUSETs G' and G''. Trivially, $(\lambda(G',G'') = []) \Rightarrow (G' \approx_G G'')$. Group match induces an equivalence relation since, apparently, $\approx_G$ is reflexive, symmetric, and transitive. Moreover, group

.5 META DERIVATION

subsumption implies group match; $(G' \prec_G G'') \Rightarrow (G' \approx_G G'')$ where, of course, $G' = G$ for G as defined in **Def. 2.5.1–6**. The inverse implication, though, does not hold. The notion of group conformation, in turn, is specified in

Definition 2.5.1–7:
Letting G' and G'' denote two non-empty GRUSETs of same BASE, G' *conforms* to G'' – $G' \cong_G G''$ symbolically – if and only if $(\exists G).(G' \subseteq G) \wedge (G'' \subseteq G)$ such that G ***base?*** $= G'$ ***base?*** $= G''$ ***base?***; computationally, group conformation is the Boolean function

$$G' \cong_G G'' = \begin{cases} 1 & \text{if } \rho(G',G'') = \rho(G'',G') \\ L & \text{otherwise} \end{cases}.$$

As trivial consequences, $G' \cong_G G''$ implies that $G' \approx_G G''$ as well as $\lambda(G',G'') = \rho(G',G'') = \rho(G'',G')$; obviously, $(\lambda(G',G'') = []) \Rightarrow (G' \cong_G G'')$, too. Like $\approx_G$, also $\cong_G$ induces an equivalence relation (as a sub-relation of $\approx_G$). Several further miscellaneous relations for GRUSETs (such as $\subset_G, \subseteq_G$, etc.) are defined in Appx. A.5.

A straight group intersection may not be appropriate all the time; an asymmetric kind of re-grouping a GRUSET's elements is group division as defined in

Definition 2.5.1–8:
Letting G' and G'' denote two non-empty GRUSETs of same BASE, $\delta(G',G'')$ is the GRUSET obtained from *dividing* G' by G''; computationally,

$$\delta(G',G'') = \begin{cases} \lambda_0(\gamma(G'),G'') & \text{if } G' \prec_G \rho(G'',\lambda_0(\gamma(G'),G'')) \\ \lambda_0(\gamma(G'),\tau(G',G'')) & \text{otherwise} \end{cases};$$

that is, a *tiling* of G' and G'', restricted to G', is computed unless G' subsumes G'' where both GRUSETs actually overlap.

Note that in case of $G' \prec_G G''$, $\rho(G'',\lambda_0(\gamma(G'),G'')) = \rho(G'',G'') = G''$; conversely, $\lambda_0(\gamma(G'),G'') = G''$ for $G' \approx_G G''$ as well and, hence, provided that $G' \prec_G \lambda_0(\gamma(G'),G'')$ at least, it is sufficient to truncate G'' appropriately. In algebraic terms, δ is a somewhat odd function since its properties depend on the group relationship between G' and G''; in general, this function is neither commutative nor associative. On the other hand, given that $G' \prec_G G''$ as well as $G' \prec_G G'''$ (but neither $G'' \prec_G G'''$ nor $G''' \prec_G G''$),

$$\delta(\delta(G',G''),G''') =$$
$$= \delta(G'',G''') = \lambda(G'',G''') = \lambda(G''',G'') = \rho(G''',G'') =$$
$$= \rho(\rho(G',G'''),G'').$$

Thus, in this special case, group divisions can be re-ordered safely.

A standard operation of statistical databases is aggregation. However, aggregations generally must adapt to the grouping structure present in real datasets. In METAGEN the basic operation of adaptive group aggregation – which, by the way, may be viewed as a generalized kind of *set union* – is defined in

Definition 2.5.1–9:

Letting, again, G' and G'' be two (non-empty) GRUSETs of same BASE, the *group aggregation* of G' and G'' is a set (a GRUSET, in fact) returned by the function

$$\alpha(G',G'') = \begin{cases} \alpha\big(G'-g',\langle\alpha_+(g',G'')\rangle \cup \alpha_0(g',G'')\big) & \text{if } G' \neq [] \\ G'' & \text{otherwise} \end{cases}$$

where

$$\alpha_+(g',G'') = \begin{cases} \alpha_+(g',G''-g'') \cup g'' & \text{if } (G'' \neq []) \wedge (g' \cap g'' \neq \langle\rangle) \\ \alpha_+(g',G''-g'') & \text{if } (G'' \neq []) \wedge (g' \cap g'' = \langle\rangle) \\ g' & \text{otherwise} \end{cases}$$

and

$$\alpha_0(g',G'') = \begin{cases} \alpha_0(g',G''-g'') + g'' & \text{if } (G'' \neq []) \wedge (g' \cap g'' = \langle\rangle) \\ \alpha_0(g',G''-g'') & \text{if } (G'' \neq []) \wedge (g' \cap g'' \neq \langle\rangle) \\ [] & \text{otherwise} \end{cases}$$

with $g' \in G'$ and $g'' \in G''$, respectively.

For given g' and G'', $\alpha_+(g',G'')$ and $\alpha_0(g',G'')$ are complementary to each other. Since, essentially, $\alpha(G',G'')$ is composed of SET unions it is fairly easy to see that this function is both commutative and associative. Group aggregation affects neither group conformation nor group match, that is

$$(G' \equiv_G G'') \Rightarrow (G' \equiv_G \alpha(G',G'') \equiv_G G'')$$

and

$$(G' \approx_G G'') \Rightarrow (G' \approx_G \alpha(G',G'') \approx_G G''),$$

respectively. In addition to this, $(G' \prec_G G'') \Rightarrow (\alpha(G',G'') \prec_G G'')$ and, more interestingly, $(G' \prec_G G'') \Rightarrow (\lambda(\alpha(G',G''),G'') = G'')$: in case of group subsumption, group aggregation removes the differences in groupings in the overlapping area. Furthermore, $(G' \approx_G G'') \Rightarrow (\rho(G',G'') \prec_G \alpha(G',G''))$ which, by symmetry of $\approx_G$, holds for G'' as well. Making use of FOOL's type polymorphism, $G' \cup G''$ may be written equivalently to $\alpha(G',G'')$; cf. to Appx. A.5. Analogous to group intersection (**Def. 2.5.1–2**), $\alpha(G',G'')$ can be extended to $\alpha^*(\mathbf{G})$.

In specific cases, group restriction is combined favourably with group aggregation as defined in

Definition 2.5.1–10:
Letting G' and G'' denote two non-empty GRUSETs of same BASE, *group selection* from G' by G'' is a set (a GRUSET, in fact) computed by function

$$\sigma(G',G'') = \rho(\lambda(\alpha(G',G''),\gamma(G')),G'').$$

Thus, in contrast to group restriction, group selection extracts group-aggregated elements of G'. In general, group selection, like group restriction, is neither commutative nor associative. Furthermore, unless $\rho(G',G'') = \rho(G',G''')$, $\sigma(\sigma(G',G''),G''') \neq \sigma(\sigma(G',G'''),G'')$ for any triplet G',G'',G''' of GRUSETs sharing the same BASE. However, group selection induces group subsumption, $G' \prec_G \sigma(G',G'')$ because $\alpha(G',G'')$ remains confined to $\gamma(G')$. FOOL notation abbreviates $\sigma(G',G'')$ to $G' * G''$.

A pivotal parameter function of METAGEN is group unification used for the determination of a joint grouping pattern for a *set* of (at least two) GRUSETs. Group unification, however, is quite different from other parameter functions in that its output usually comprises a set of candidate solutions instead of a single well-determined function value. Because of its complexity, group unification is presented in terms of an algorithm instead of a closed function definition; moreover, this algorithm relies on a special data structure as defined in the auxiliary

Definition 2.5.1–11:
Given a set of $J \geq 1$ GRUSETs $\breve{\mathbf{G}} = \{\breve{G}_1,\ldots,\breve{G}_J\}$ all of same BASE; then a *group unification graph* $\Gamma(P,E)$ is constructed with node set $P = \bigcup_{j=1}^{J} \breve{G}_j$ using ordinary set union and edge set $E = \varepsilon(P)$ computed by the functions

$$\varepsilon(P) = \begin{cases} \{\} & \text{if } P = \{\} \\ \varepsilon_+(p, P\setminus\{p\}) \cup \varepsilon(P\setminus\{p\}) & \text{otherwise} \end{cases}$$

and

$$\varepsilon_+(p, P) = \begin{cases} \{\} & \text{if } P = \{\} \\ \varepsilon_+(p, P\setminus\{p'\}) & \text{if } (P \neq \{\}) \wedge (p \cap p' = \langle\rangle) \\ \{\langle p, p' \rangle\} \cup \varepsilon_+(p, P\setminus\{p'\}) & \text{otherwise} \end{cases}$$

where $p, p' \in P$.

The group unification graph encodes the information of mutually overlapping groups which necessarily have to go into different (subsets of) BASPARTitions and, hence, candidate partitions. These candidate partitions are obtained by looping over the edges of the group unification graph and (intermediary) partitions as follows:

[initialize] set $G := \{P\}$;
[outer loop over edges] *for each* $e = \langle g', g'' \rangle \in E$ of $\Gamma(P, E)$
  [inner loop over partition set] *for each* $p \in G$
    [updating of partition set] *if* $\{g', g''\} \subset p$ *then*
$$G := (G\setminus\{p\}) \cup \{p\setminus\{g'\}, p\setminus\{g''\}\};$$

Upon termination, set G comprises GRUSETs derived by recursive splitting of P; each GRUSET is a candidate partition. In the worst case, group unification returns $\breve{G}$ itself, that is $G = \breve{G}$. In graph-theoretical terms, letting $\Gamma^c(P, E)$ denote the complementary graph of $\Gamma(P, E)$ with respect to the complete graph K_P [Harary, 1969] (Γ^c joins nodes of P not joined in Γ and vice versa), set G collects the maximal complete sub-graphs of $\Gamma^c(P, E)$.

Using the group unification graph structure, group unification can now be defined formally; it must be kept in mind, however, that group unification generally depends on group decompositions implying the availability of pertinent disaggregation data.

Definition 2.5.1–12:
Letting $\mathbf{G} = \{G_1, \ldots, G_J\}$ denote a set of $J \geq 1$ GRUSETs of same BASE, *group unification* is a three-step procedure:
1) Find the maximal effective grouping $\breve{G}_j$ for each $G_j \in \mathbf{G}$, $1 \leq j \leq J$. Apparently, $\tau^*(\mathbf{G}) \prec_\mathbf{G} \breve{G}_j \prec_\mathbf{G} G_j$ (thus, in general, implying intermediary

.5 META DERIVATION

disaggregation of G_j to $\breve{G}_j$ which, due to lacking disaggregation data, may fail to coincide with $\tau^*(\mathbf{G})$); ideally, of course, $\rho(\tau^*(\mathbf{G}), G_j) = \breve{G}_j$ for all j simultaneously. In the worst case, $\breve{G}_j = G_j$; that is, the maximal effective grouping of G_j is identical to G_j itself.

2) Having determined $\breve{\mathbf{G}}$, initialize the group unification graph $\Gamma(P,E)$ as defined in **Def. 2.5.1–11**.

3) Compute the function $\omega(E,G)$ with $G := \{P\}$ defined as

$$\omega(E,G) = \begin{cases} G & \text{if } E = \{\} \\ \omega(E \setminus \{e\}, \omega_e(e,G)) & \text{otherwise} \end{cases}$$

where $e \in E$,

$$\omega_e(e,G) = \begin{cases} \{\} & \text{if } G = \{\} \\ \omega_0(\omega_e(e, G \setminus \{p\}) \cup \omega_+(e,p)) & \text{otherwise} \end{cases}$$

where $p \in G$, and

$$\omega_+(\langle g', g'' \rangle, p) = \begin{cases} \{p - g', p - g''\} & \text{if } \{g', g''\} \subseteq p \\ \{p\} & \text{otherwise} \end{cases}.$$

Dominated solutions are ruled out by applying the functions

$$\omega_0(G) = \begin{cases} \{\} & \text{if } G = \{\} \\ \omega_x(p, \omega_0(G \setminus \{p\})) & \text{otherwise} \end{cases}$$

where $p \in G$,

$$\omega_x(p,G) = \begin{cases} G & \text{if } \omega_?(p,G) \\ G \cup \{p\} & \text{otherwise} \end{cases}$$

and

$$\omega_?(p,G) = \begin{cases} \top & \text{if } (\omega_?(p, G \setminus \{p'\}) \vee (p \subseteq p')) \\ \bot & \text{otherwise} \end{cases}$$

where $p' \in G$.

To give a simple illustration of group unification, assume that $P := \{g_1, \ldots, g_6\}$ is a set of groups (originating from some couple of GRUSETs $\mathbf{G} = \{G_1, \ldots, G_J\}$) and **Fig. 2.5.1–1** (overleaf) **(a)** exhibits the corresponding group unification graph $\Gamma(P,E)$. In this case, $\omega(\varepsilon(P), \{P\}) = \{\langle g_1, g_2, g_5 \rangle, \langle g_1, g_3, g_5 \rangle, \langle g_3, g_6 \rangle, \langle g_4, g_6 \rangle\}$ as the set of non-dominated candidate partitions of P which can be verified easily by inspecting the complementary graph $\Gamma^c(P,E)$ in **Fig. 2.5.1–1 (b)**. To simplify notation, $\omega(\mathbf{G}) \equiv \omega(\varepsilon(P), \{P\})$ will be used henceforth.

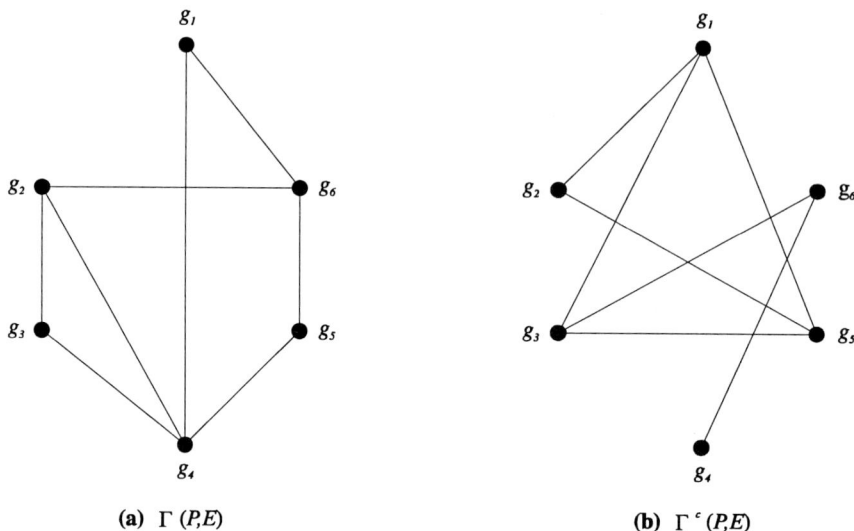

Figure 2.5.1–1: Sample Unification Graph

XSET-level Parameter Functions

By and large, these parameter functions are transferable to XSETs without further adaptations. In particular, group intersection (λ, λ^*), tiling (τ, τ^*), restriction (ρ), division (δ), aggregation (α, α^*), and selection (σ) are accomplished at XSET level simply by applying the respective operation to each of the XSET's K dimensions concurrently. In case of group relationships ($\prec_G, \approx_G, \cong_G$; but also $\subset_G, \subseteq_G$, etc.), these relationships hold at XSET level if and only if the respective group relationships hold for *each* of the K dimensions. Somewhat specific, however, is group unification because re-assembling the MODality groupings of different GRUSETs with respect to all K dimensions is successful only on condition that the selected candidate partitions taken jointly over the K dimensions result in a *feasible* GRU-XSET (cf. **Def. 2.1.2–17**): this, of course, imposes a constraint on the choice of groups since a GRUXSET is feasible if and only if the partitions in each of its K dimensions are composed out of the same set of input GRUXSETs. More formally, GRUXSET feasibility is stated in

.5 META DERIVATION

Definition 2.5.1–13:

Let $\mathbf{G}_k = \{G_k^{(1)}, \ldots, G_k^{(J)}\}$ a set of k-components of $J > 1$ GRUXSETs $\mathcal{B}_1^G, \ldots, \mathcal{B}_J^G$; furthermore, letting $\omega(\mathbf{G}_k) \equiv \omega_k(E_k, G_k)$ denote the set of candidate partitions obtained by group unification with respect to dimension k ($1 \leq k \leq K$), enumerating the elements of $\omega_k(E_k, G_k) = \{\omega_k^{(l_k)}\}_{l_k}$ for $1 \leq l_k \leq L_k = card(\omega_k(E_k, G_k))$, and associating each $\omega_k^{(l_k)}$ with an index set I_{kl_k} comprising the identifying indices of the k-th dimension's input partitions contributing at least one group to $\omega_k^{(l_k)}$, the *grid unification* $\omega(\mathcal{B}_1^G, \ldots, \mathcal{B}_J^G)$ returning the set of feasible GRUXSETs obtained from $\mathcal{B}_1^G, \ldots, \mathcal{B}_J^G$ is characterized plainly as

$$\omega(\mathcal{B}_1^G, \ldots, \mathcal{B}_J^G) =_{def} \left\{ \bigcap_{k=1}^{K} I_{kl'_k} \,\middle|\, \bigwedge_{k=1}^{K} (1 \leq l'_k \leq L_k) \wedge \left(\bigcap_{k=1}^{K} I_{kl'_k} \neq \{\} \right) \right\}.$$

Note that even if group unification fails completely, $\omega(\mathcal{B}_1^G, \ldots, \mathcal{B}_J^G) \neq \{\}$ for then each of the K dimensions supplies its original GRUSET as input partition, that is for each of the $J > 1$ input partitions there is a singleton $I_{kl'_k}$ and, thus, $\bigcap_{k=1}^{K} I_{kl'_k} \neq \{\}$. For each solution element in $\omega(\mathcal{B}_1^G, \ldots, \mathcal{B}_J^G)$, the vector $(l'_1, \ldots, l'_k, \ldots, l'_K)$ records the (reference indices to) candidate partitions for a feasible GRUXSET. Apparently, always $\prod_{k=1}^{K} L_k \geq 1$ but, in general, the number of index set intersections to be tested is considerably lower than J^K (although, theoretically, it may even exceed this value because, by sharing groups appropriately, a set of partitions may generate further candidate partitions implicitly even though all of the non-shared groups are mutually disjoint). Testing intersections stepwise by dimension helps to speed up the procedure because many combinations will die out sooner or later.

Furthermore, METAGEN makes use of an operation in a sense "dual" to group unification at XSET level which does not have a counterpart at GRUSET level at all: given again a set of $J > 1$ GRUXSETs $\mathcal{B}_1^G, \ldots, \mathcal{B}_J^G$, the function $\varpi(\mathcal{B}_1^G, \ldots, \mathcal{B}_J^G)$ computes a partition of $\{1, \ldots, J\}$ such that the K-dimensional $\mathcal{B}_j^G$ are divided into mutually disjoint GRUXSETs clusters (equivalence classes) as defined in

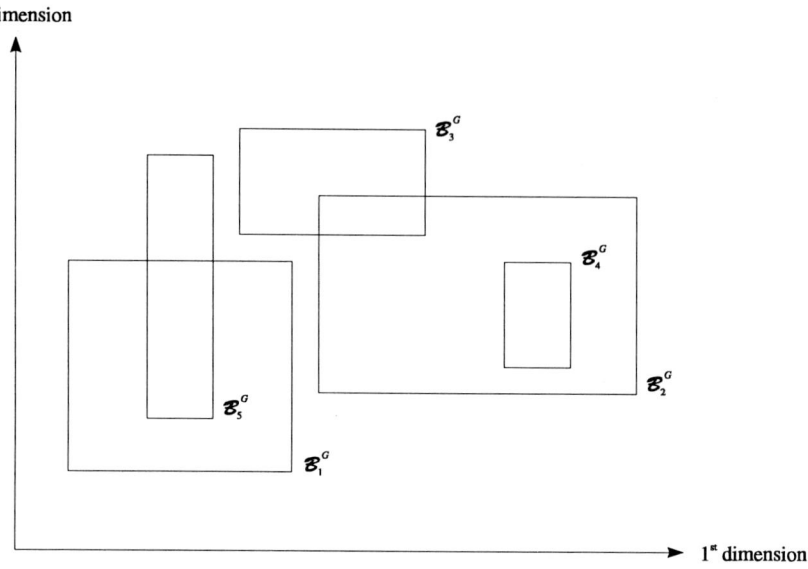

Figure 2.5.1–2 (a): Sample TEXTures for Clustering

Definition 2.5.1–14:

Letting $\Gamma_\varpi(\mathbf{G}_k, E_k)$ denote a graph with node set $\mathbf{G}_k = \{G_1^{(k)},...,G_J^{(k)}\}$ where $G_j^{(k)}$, $1 \leq j \leq J$, represents the k-th component, for $1 \leq k \leq K$, of $\mathcal{B}_j^G$. Let further E_k, the edge set of Γ_ϖ, contain an edge $\langle j', j'' \rangle$ for each pair of GRUSETs $G_{j'}^{(k)}$ and $G_{j''}^{(k)}$, $j' \neq j''$, if and only if $\lambda(G_{j'}^{(k)}, G_{j''}^{(k)}) \neq []$. Now, an XSET *clustering* $\varpi(\mathcal{B}_1^G,...,\mathcal{B}_J^G)$ is obtained simply by

1) computing $E_\varpi =_{def} \bigcap_{k=1}^{K} E_k$;
2) constructing the cluster graph $\Gamma_\varpi^*(\{\mathcal{B}_1^G,...,\mathcal{B}_J^G\}, E_\varpi)$ by re-labelling $\mathbf{G}_k$ in Γ_ϖ and replacing the edge set with E_ϖ;
3) determining the set of maximal connected components of Γ_ϖ^* partitioning $\{\mathcal{B}_1^G,...,\mathcal{B}_J^G\}$.

.5 META DERIVATION

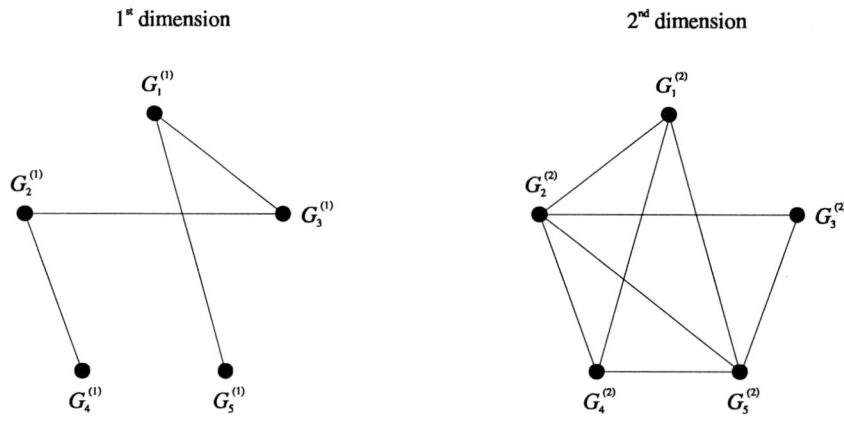

Figure 2.5.1–2 (b): GRUSET Intersections by Dimension

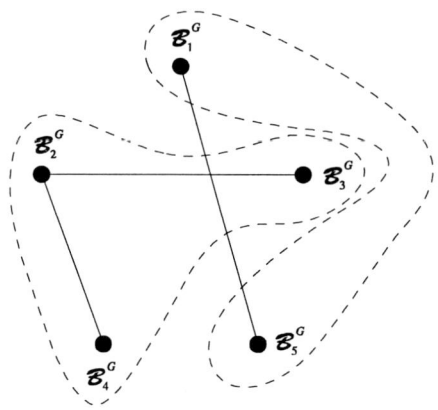

Figure 2.5.1–2 (c): TEXTure Clusters

Apparently, $\varpi\left(\mathcal{B}_1^G,\ldots,\mathcal{B}_J^G\right)$ comprises $\leq J$ clusters of GRUXSETs such that, for any pair $\mathcal{B}_{j'}^G$ and $\mathcal{B}_{j''}^G$ taken from different clusters, $\lambda\left(\mathcal{B}_{j'}^G,\mathcal{B}_{j''}^G\right) = \emptyset$ (within clusters GRUXSETs may or may not be mutually disjoint). Fig. 2.5.1–2 illustrates clustering by means of a simple example in two dimensions; **(a)** exhibits the semantic loci of five sample GRUXSETs (TEXTures), **(b)** shows $\Gamma_\varpi(G_k, E_k)$ for $k = 1,2$, and **(c)**, finally, presents Γ_ϖ^* with the resulting (two) components surrounded by dashed lines.

BOX Operations

In view of METAGEN operators, parameter functions are applied rarely to XSETs in the first place; rather are BOXes the operands of primary concern. Fortunately, however, most of the functions apply straightway to BOXes as well insofar as usually the TEXTure (which, by definition, is an XSET or XGRID) of BOXes is affected only. The major exception to this general rule are group intersection and tiling operations which may involve a rearrangement of the BOX's EDITs part in order to maintain TNF representation (cf. **Def. 2.4.1–5**): compartmentalizing existing groups will, in general, also cut through EDits whence split EDits need to be removed from EDITS and re-introduced as BLOCKs according to the new XGROUPing resulting from intersecting or tiling a BOX. This rearrangement of EDITS extends, of course, to the TAGS attached to individual EDits (basically, the TAGs of a split EDit are passed on simply to each of its successors). It goes without saying that real BOX intersection (that is, an intersection beyond intermediary intersection of TEXTures) implies a *union* of EDITS of the intersected BOXes (in the intersection area) whence intersection and union operations at the BLOCK and BLOCKCOLLection layer, respectively, have to be among the set of elementary METAGEN operations. Basically, inserting a BLOCK into a BLOCKCOLLection amounts to a simple addition if the BLOCK is disjoint with each BLOCK of the BLOCKCOLLection; otherwise, BLOCKs are split along the cutpoints induced by their common intersection (note that in K-dimensional space the intersection of K-hypervals results in another K-hyperval while the union of K-hypervals returns, in general, a *set* of K-hypervals or, in METASTASYS's terms, a BLOCKCOLLection). Apparently, intersections and unions of BLOCKCOLLections reduce to corresponding BLOCK-wise operations.

2.5.2 METAGEN Operators

As discussed above, PROMETAs, the – at present only partially specified – operands of METAGEN, carry four major metadata components: PROFiles, GRIDs, EDITs, and SUMTYPEs. A minimal requirement of any joint processing of PROMETAs is PROFile uniformity lest virtually all structural properties of the FRAME-based table calculus and the computational advantages drawn thereof are lost. Likewise, to achieve a final output PROMETA, differing input SUMTYPEs have to become unified somewhere during the derivation stage. Both types of operations are accomplished by rather specific METAGEN operators affecting only PROFiles and SUMTYPE components of PROMETA arguments, respectively, whereas all of the other operators deal with PROMETAs' BOX components. Irrespective of these distinctions, METAGEN operators are denoted uniformly by introducing an "operator bracket" notation

$$\lfloor \chi_{op} \rfloor^{op} D$$

.5 META DERIVATION

where **op** identifies the particular type of operator and χ_{op} supplies **op**'s (a priori) parameter data usually expressed in terms of *parameter functions*; both, the admissible formats and purpose, of χ_{op} are determined by **op**. In the restricted account of METAGEN given here, **op** may assume one of the values **c, s, u, d, m, p, r, a,** and **t**, each representing a particular operator type as described in detail in the following paragraphs. Instead of presenting these operator types neutrally, the descriptions already reflect the operators' later usage to make for a smoother explanation of METAGEN. In general, several METAGEN operators are applied in sequence to yield a self-contained derivation. Sequential operator application is represented formally by stringing together operator bracket expressions; hence, a sequence

$$\mathbf{op}_1 \circ \mathbf{op}_2 \circ \cdots \circ \mathbf{op}_n(D)$$

is denoted simply as

$$\lfloor \chi_{op_n} \rfloor^{op_n} \cdots \lfloor \chi_{op_2} \rfloor^{op_2} \lfloor \chi_{op_1} \rfloor^{op_1} D$$

implying that, stage by stage, each evaluated expression $\lfloor \chi_{op_i} \rfloor^{op_i} \cdots \lfloor \chi_{op_1} \rfloor^{op_1} D$ provides the argument to $\mathbf{op}_{i+1}$. Sometimes, the elements $d_i \in D$ are subjected to METAGEN operators; in that case, set notation

$$\left\{ \lfloor \chi_{op_i} \rfloor^{op_i} d_i \right\}_i .D$$

is used where it is understood that the set of evaluated element expressions is again a well-formed METAGEN operand. Whenever an operator spawns a sub-REQUEST to feed χ_{op}'s parameters, this initializes a derivation branch and so transforms derivation sequences into (directed) derivation trees. Apparently, tree-shaped derivations require careful construction in order to prevent adverse statistical interferences which may be introduced all too easily by insufficient tracking of IMAGEs used procedures applied in different derivation branches.

The *C*oerce Operator

A fundamental, though subordinate METAGEN operator is the *coerce* operator used to unify PROFiles of PROMETAs. Given an input argument D, this operator is stated formally as

$$\lfloor \chi_c \rfloor^c D$$

where χ_c supplies the PROFile the PROMETAs in D have to assume on exit. More specifically, χ_c consists of (i) a switch which may take on either one of the states 'align' or 'adapt' and (ii) a parameter function governing the processing of D depending on the switch state.

In the 'align' mode, letting $D = \{d_1,\ldots,d_n\}$ a set of $n \geq 1$ IMAGE descriptions and r_* a TARGET, *coerce* is controlled by the parameter function $\alpha^*(\{\wp_1,\ldots,\wp_n,\wp_*\})$ if $\wp_i$ denotes d_i's PROFile and $\wp_*$ denotes r_*'s PROFile, respectively. Hence, internally, each $\wp_i$, $1 \leq i \leq n$, is transformed into $\alpha^*(\{\wp_1,\ldots,\wp_n,\wp_*\})$ which, for obvious reasons, implies a rearrangement of the IMAGES' BOXes $\mathbf{B}_i$ (unless, trivially, $\wp_i = \alpha^*(\{\wp_1,\ldots,\wp_n,\wp_*\})$): for each IMAGE involved, for any set of its AXes becoming embraced in a single group – representing a "unifying" AXis – of the resulting joint, or coerced, PROFile, those AXes' AXVALues are first combined in terms of a Cartesian product (returning, in a sense, supercodes); in a second step, for each of the obtained AXVALue combinations the RECTDOMain elements associated with these AXVALues are composed (again in terms of Cartesian products), thus "puffing up" the original AXVALues to combined XVALues generating the RECTDOMain of the "unifying" AXis; finally, $\wp_i$ can be replaced with and $\mathbf{B}_i$ re-expressed in terms of the arranged supercodes relative to $\alpha^*(\{\wp_1,\ldots,\wp_n,\wp_*\})$. Re-expressing $\mathbf{B}_i$, of course, affects both IMGRID $\mathcal{B}_i^G$ and IMEDITs $\mathcal{E}_i$ components. As with AXRECTifications, the re-expressions of both components may interfere, viz. whenever an individual IMEDit becomes enclosed fully in the "unifying" AXis's RECTDOMain this IMEDit has to be dropped from re-expressed IMEDITs and the respective RECTDOMain must be truncated accordingly. Note that, because $\wp_* \prec_G \alpha^*(\{\wp_1,\ldots,\wp_n,\wp_*\})$, each d_i is re-expressed in terms of $\mathbf{T}_*$'s AXVALues (as stated in $\mathbf{T}_*$'s generating METASPEL sentence), or supercodes thereof. In general, of course, also $\mathbf{T}_*$ needs to be re-expressed analogous to the $\mathbf{I}_i$. Note also that reshaping the IMPROFiles does not exert any effect on the arrangement of attached IMFILEs (only the DATCELLs' access keys are changing).

Ideally, $\wp_1 = \cdots = \wp_i = \cdots = \wp_n = \wp_*$ or, at least, $\wp_1 = \cdots = \wp_i = \cdots = \wp_n$ and $\wp_i \prec_G \wp_*$; then, in either case, $\wp_* = \alpha^*(\{\wp_1,\ldots,\wp_n,\wp_*\})$ and, obviously, in the former case *coerce* does not change PROFiles at all. In the worst case, unfortunately, all AXes go into a *single* "unifying" AXis, thus effectively converting the factored representation of IMAGEs and CONCEPTs in terms of XSETs into a representation in terms of XDOMains (cf. **Def. 2.1.2–23**) which, for obvious reasons, is rather clumsy to deal with computationally (in fact, this extreme case conforms to operating at the extensional database level). Since $\alpha^*(\{\wp_1,\ldots,\wp_n,\wp_*\})$ is likely to grow high-dimensional, excessive collapsing of AXes may quickly lead to prohibi-

.5 META DERIVATION

tively large RECTDOMains suggesting to abort operator execution (formally, *coerce* would then return UNDEF, $\bot$).

In the 'adapt' mode, *coerce* goes one step further: in addition to finding suitable joint group aggregations of several PROFiles, the operator searches for minimally aggregated common PROFiles by investigating the possibilities to *decompose* the IMAGES' PROFiles in view of $\wp_*$. Compared to the 'align' mode, the 'adapt' mode is considerably more expensive and risky because despite a careful screening of candidates no (useful) decompositions may be obtainable after all; furthermore, different decompositions of IMAGE PROFiles may give rise to varying patterns of coerced PROFiles difficult to bring into a preference ordering mechanically. Technically, in *coerce*'s 'adapt' mode $\tau^*(\{\wp_1,\ldots,\wp_n,\wp_*\})$ replaces the former group-aggregating α^* function; internally, the operator spawns a set of sub-REQUESTs r_i, one for each $d_i \in D$, asking for disaggregation tables (of RATEs) with PROFiles "broken down" with respect to $\delta(\wp_i, \tau^*(\{\wp_1,\ldots,\wp_n,\wp_*\}))$. Instead of responding with the requested PROFile, however, r_i may deliver some $\breve{\wp}_i$ such that

$$\tau^*(\{\wp_1,\ldots,\wp_n,\wp_*\}) \prec_G \breve{\wp}_i \prec_G \wp_i$$

only, and in the worst case even $\breve{\wp}_i = \wp_i$ may be obtained. As a consequence, *coerce* will eventually achieve the joint PROFile $\alpha^*(\{\breve{\wp}_1,\ldots,\breve{\wp}_n,\wp_*\})$ rather than the "optimal" $\tau^*(\{\wp_1,\ldots,\wp_n,\wp_*\})$. Letting $\breve{M}_i$, with PROFile $\breve{\wp}_i$, denote the (PRO-) META of disaggregation RATEs derived for r_i, the operation

$$\lfloor \text{adapt}; \alpha^*(\{\breve{\wp}_1,\ldots,\breve{\wp}_n,\wp_*\}) \rfloor^c D$$

rewrites to

$$\lfloor \text{align}; \alpha^*(\{\breve{\wp}_1,\ldots,\breve{\wp}_n,\wp_*\}) \rfloor^c \{\lfloor \text{prof}; \breve{\wp}_i; \breve{M}_i \rfloor^d d_i\}_{i=1,\ldots,n} . D$$

where $\lfloor \chi_d \rfloor^d$ denotes a METAGEN *disaggregate* operator discussed below; basically, each $d_i \in D$ is first subjected to a disaggregation before becoming coerced. If either $\wp_1 = \cdots = \wp_i = \cdots = \wp_n = \wp_*$, or $\wp_1 = \cdots = \wp_i = \cdots = \wp_n$ and $\wp_i \prec_G \wp_*$, then apparently $\lfloor \text{adapt}; \alpha^*(\{\wp_1,\ldots,\wp_n,\wp_*\}) \rfloor^c D$ is totally equivalent to $\lfloor \text{align}; \alpha^*(\{\wp_1,\ldots,\wp_n,\wp_*\}) \rfloor^c D = \lfloor \text{align}; \alpha(\{\wp_1,\wp_*\}) \rfloor^c D$.

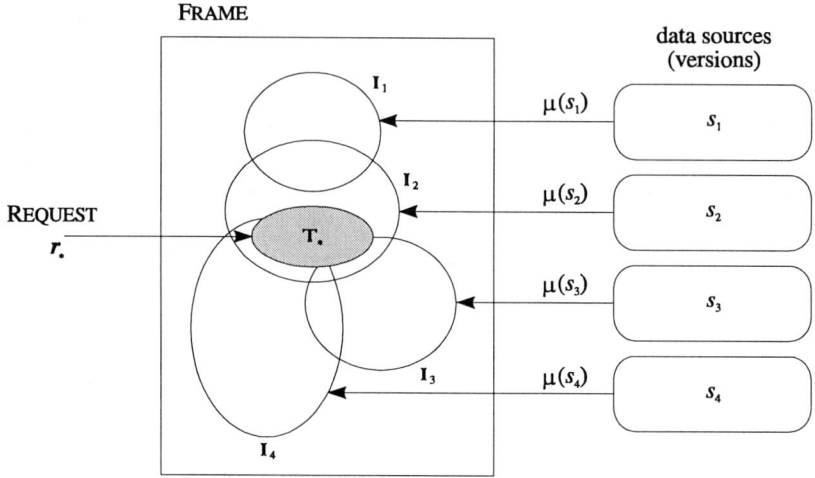

Figure 2.5.2–1: IMAGE Selection

Practically, adaptive coercing of PROFiles will be hardly successful except perhaps in cases where IMAGEs supply "aggregate" data to IMAXes representing (subsets of QUalities of) QUCLUSTers (cf. **Def. 2.2.1–18**). Otherwise, IMAXes would have been introduced because of orthogonally non-decomposable VALMAPs (cf. **Def. 2.3.3–5**) from SOVARs to FRAMEQUalities and, therefore, posterior decompositions usually are out of question. Thus, of the two modes of the *coerce* operator, the 'align' mode is the more relevant one. If, for one reason or another, only constant PROFiles can occur the *coerce* operator is rendered superfluous.

The *S*earch Operator

Given a REQUEST r_*, at the beginning of each derivation a search for relevant IMAGEs is conducted. An IMAGE, represented by PROMETA d, is relevant if it actually "contributes" to the TARGET T_* such that its SUMTYPE complies with s_*. Schematically, **Fig. 2.5.2–1** illustrates a case of four IMAGEs $d_1,...,d_4$, three of which – viz., d_2, d_3, d_4 – are contributing to T_*; $\mu(s_i)$ is a casual notation used to indicate source mappings (cf. Section 2.3) bringing forth, inter alia, the IMBOXes I_i. The target contribution of an IMAGE can be captured in terms of the "concept overlap" between the IMBOX I and T_*: apparently, if there is no such overlap, d does not contribute to r_* and can be excluded from further consideration; conversely, the "volume" of the overlap provides a yardstick to rank an IMAGE's con-

.5 META DERIVATION

tribution to r_*. Borderline cases excepted, it is, however, not possible to value properly the contribution of a single IMAGE in isolation because this is a joint function of all IMAGEs contributing to r_* and, hence, is highly interdependent with (statistical) integration properties of each individual IMAGE selection d participates in. Computationally, IMAGE selection amounts to a *generate-and-test* procedure (cf. Subsection 2.5.3) – in a first step, accordingly, candidate selections of IMAGEs are generated by screening available IMAGEs using fairly crude "context-free" eligibility criteria. The basic operation of this procedure consists of checking, for each available IMAGE, (i) target coverage and (ii) summary type compliance.

Target coverage is proposed as a numerical, standardized measure indicating how good an IMAGE's $\mathbb{I}$ fits a given TARGET T_*. Remembering that $\mathbb{I} = \left(\mathcal{B}_i^G, \mathcal{E}_i \right)$ and $T_* = \left(\mathcal{B}_*^G, \mathcal{E}_* \right)$ in BOX representation, and extending the definition of '∩' and '∪' operations to BLOCKs and BLOCKCOLLections straightforwardly, this measure is introduced formally in

Definition 2.5.1–15:
For given $\mathbb{I}$ and T_*, the *target coverage* $c(\mathbb{I}, T_*)$, $0 \leq c(\mathbb{I}, T_*) \leq 1$, is the product of box overlap $b(\mathbb{I}, T_*)$ and edits mismatch $e(\mathbb{I}, T_*)$, that is

$$c(\mathbb{I}, T_*) =_{def} \begin{cases} b(\mathbb{I}, T_*)(1 - e(\mathbb{I}, T_*)) & \text{if } b(\mathbb{I}, T_*) > 0 \\ 0 & \text{otherwise} \end{cases}.$$

The *box overlap*, in turn, expresses the ratio of T_*'s BOX volume covered by $\mathbb{I}$;

$$b(\mathbb{I}, T_*) =_{def} \frac{\left| D\left(\gamma\left(\mathcal{B}_i^G \right) \cap \gamma\left(\mathcal{B}_*^G \right) \right) \right|}{\left| D\left(\gamma\left(\mathcal{B}_*^G \right) \right) \right|}.$$

Similarly, on condition that $b(\mathbb{I}, T_*) > 0$ the *edits mismatch* expresses the ratio of discordance in EDITs in the overlapping area;

$$e(\mathbb{I}, T_*) =_{def} \frac{\left| D\left(\left(\mathcal{E}_i \cap \gamma\left(\mathcal{B}_*^G \right) \right) \cup \left(\mathcal{E}_* \cap \gamma\left(\mathcal{B}_i^G \right) \right) \right) - D\left(\mathcal{E}_i \cap \mathcal{E}_* \right) \right|}{\left| D\left(\gamma\left(\mathcal{B}_i^G \right) \cap \gamma\left(\mathcal{B}_*^G \right) \right) \right|}.$$

Note that $D(.)$ converts XSETs and BLOCKCOLLections to XDOMains to which the cardinality function $|\cdot|$ applies as usual in set theory. Computationally, of course, it is advisable to interchange both operations since computing set cardinalities ("volumes") from XSETs is considerably cheaper than from XDOMs. As is apparent from this definition, $c(\mathrm{I},\mathrm{T}_*)=1$ only if $b(\mathrm{I},\mathrm{T}_*)=1$ and, at the same time, $e(\mathrm{I},\mathrm{T}_*)=0$, that is I covers T_* entirely and EDITS coincide fully. Conversely, $c(\mathrm{I},\mathrm{T}_*)=0$ whenever $e(\mathrm{I},\mathrm{T}_*)=1$: apparently, an IMAGE does not contribute anything if the overlapping area – irrespective of how large it may be – merely comprises EDITS of either I or T_*.

Now, an IMAGE I is eligible if $c(\mathrm{I},\mathrm{T}_*)>0$; more specifically, an IMAGE is rejected if either $b(\mathrm{I},\mathrm{T}_*)=0$ or $D(\gamma(\mathcal{B}_i^G) \cap \gamma(\mathcal{B}_*^G)) \subseteq D(\mathcal{E}_*)$ since, in the latter case, I would not contribute to T_* at all (though even $e(\mathrm{I},\mathrm{T}_*)=0$ could still happen). Fortunately, both criteria are quite efficient to check computationally – to find that $b(\mathrm{I},\mathrm{T}_*)=0$ it is sufficient to compute $\gamma(\mathcal{B}_i^G) \cap \gamma(\mathcal{B}_*^G) = \emptyset$ which is a simple test on BLOCK disjointness; likewise, the second criterion can be replaced with testing BLOCK membership, viz. $(\gamma(\mathcal{B}_i^G) \cap \gamma(\mathcal{B}_*^G)) \subseteq \mathcal{E}_*$.

For an IMAGE to be eligible, however, it must also be of compatible SUMTYPE in addition to contributing to T_*. Essentially, an IMAGE's SUMTYPE s_i complies with s_* if s_* can be obtained from s_i by "upward" computing in terms of increasing the data aggregation level. Henceforth, this will be called SUMTYPE *upgrading* (in contrast to SUMTYPE *downgrading* basically involving some kind of re-sampling procedure to regain lower-level SUMTYPEs; for instance, SUMs could be downgraded to COUNTs, COUNTs to MESO, etc.). Thus, for example, if s_* =COUNT, all IMAGEs with SUMTYPE $s_i \in$ {MESO,COUNT} comply with s_*; other IMAGEs are rejected except they can be downgraded to COUNT somehow. In case of SUMs, the summary type is determined jointly by the attributes SUMARG, SUMDEG, and SUMARITY (cf. **Def. 2.4.2–3**). Thus, as a matter of fact, SUMs comply only if these attributes match *all at once*; otherwise, SUMTYPEs are different and, as a consequence, to render IMAGEs of non-matching SUM types compliant, an intermediary downgrading (to COUNTs) must be achievable. Note that by virtue of the principles adopted for source mappings (cf. Subsection 2.3.1), SUMTYPE downgrading – contrary to other forms of disaggregation – cannot be accomplished in terms of statistically related IMAGEs but presupposes, in general, external re-sampling models instead. In view of its weak importance in statistical data retrieval, SUMTYPE downgrading will be exempted from further discussion in this presentation.

With respect to a REQUEST r_*, the screening of IMAGEs to sift out the contributing ones is facilitated by a *search* operator

$$\lfloor \chi_s \rfloor^s D$$

which scrutinizes each $d \in D$ as to target coverage and summary type compliance. Essentially, χ_s supplies the parameters T_* and s_* which, in general, may be ac-

.5 Meta Derivation

companied in the longer run by further parameters controlling this operator's mode; particularly in view of distributed processing (cf. Section 2.6) where $\lfloor \chi_s \rfloor^s \mathfrak{I}_Q^D$ is replaced with $\left\{ \lfloor \chi_s^{(r)} \rfloor^s D_r \right\}_r$ such that $\bigcup_r D_r = \mathfrak{I}_Q^D$, parameters for tuning/synchronizing search as well as for output arrangement prescriptions may have to be inserted. On exit, the default output structure of a *search* operation is a set $D_* \subseteq D$ of selected IMAGEs; additionally, the elements of D_* are ordered decreasingly with respect to $c(\mathbb{I}, T_*)$. Of course, $D_* = \{\}$ may happen. In that case, as a remedial (though costly) action a relaxation of s_* to a suitably "upgraded" $\tilde{s}_*$ may be tried such that, eventually, $D_* \neq \{\}$ (in particular, it seems strongly advisable to clarify beforehand if $\tilde{s}_*$ would be an acceptable alternative to s_* at all). Typically, D_* will be small compared to D but it must be remembered that $\mathfrak{I}_Q^D$ includes *all* IMAGEs of sources mapping to any FRAME linked to Q (in the most extreme case the linkage encompasses the FRAME's universal QUalities only). Note also that being selected does by no means change an IMAGE irreversibly within an on-going derivation. Inevitably, of course, for intermediary processing to take place, IMAGEs generally become, at least, coerced to $\wp_*$ (the PROFile of r_*) by applying the operator

$$\lfloor \text{align}; \alpha(\wp_i, \wp_*) \rfloor^c d_i$$

to each $d_i \in D$, thus replacing, in fact, $\lfloor r_* \rfloor^s D$ with

$$\lfloor r_* \rfloor^s \left\{ \lfloor \text{align}; \alpha(\wp_i, \wp_*) \rfloor^c d_i \right\}_i . D$$

The 𝒰pgrade Operator

In general, for a set of IMAGEs d_i selected by applying a *search* operator or derived otherwise, these IMAGEs' SUMTYPEs s_i – though complying with $\tilde{s}_*$ (replacing or upgrading the original s_*) – may not be identical yet. Hence, unless $s_1 = \cdots = s_i = \cdots = s_n = \tilde{s}_*$, the SUMTYPEs have to be equalized by an *upgrade* operation. The corresponding operator

$$\lfloor \chi_u \rfloor^u D$$

is used to map each $d_i \in D$ to a PROMETA $d_i = \langle \mathbb{I}_i^u, s_i^u \rangle$ such that $s_i^u = s_i$ if the operation does not affect d_i, or $s_i^u = s_u$ where s_u denotes the output SUMTYPE to be achieved by *upgrade*. Depending on s_u this operation may rearrange $\mathcal{B}_i^G$ (that is, in general, $\mathbb{I}_i \neq \mathbb{I}_i^u$); this will happen particularly when MESO and COUNT summary types become upgraded to SUMs. Conversely, *upgrade* never modifies d_i's PROFile $\wp_i$ (note that in case of SUMmation upgrading any implicit collapsing of the AXes over which a SUM is ranging is recorded in the IMFILE's SUMARG; cf. Subsection 2.4.2). In any case, however, the attached IMFILE (copy) is affected by the *upgrade* operator – obviously, even if $\mathbb{I}_i = \mathbb{I}_i^u$, the DATCELL entries (DAT-REGisters, cf. Def. 2.2.4–2) of IMFILEs must be replaced with the corresponding numerical s_u-VALues eventually (but note that data-level operations are deferred until some $\tilde{r}_*$ has been derived successfully at the PROMETA level). The parameter χ_u of the *upgrade* operator is instantiated basically to the SUMTYPE s_u aimed at (note that in case of SUMTYPE 'SUM' s_u always embraces its determining attributes, too). *Upgrade* is organized selectively in that only those $d_i \in D$ are transformed the SUMTYPE of which is really "below" s_u; in other words, after application of an *upgrade* operator, the lowest-level SUMTYPE occurring is s_u. Apparently, s_u =MESO is not a sound choice since this would leave things entirely unchanged.

The *disaggregate* Operator

A methodically outstanding operation of METAGEN is disaggregation. Naturally, aggregation – which, in formal terms, is a deductive operation drawing consequences from premises mechanically – is achievable much simpler than disaggregation which, in information-theoretical terms [Brillouin, 1962], increases disorder (entropy): disaggregating empirical data means to *add* uncertainty ("noise" or neg-information, so-to-say) a posteriori. Hence, to make disaggregation a formal deductive procedure like aggregation this missing information (the "aggregation loss") must be supplied in terms of further data and statistical model structures. Probably because of this adverse precondition of additional information required, disaggregation is a kind of operation considered rarely in the statistical database field (the proposals of Falcitelli *et al.* [1989], Rafanelli and Ricci [1993], etc., being notable exceptions); being rooted in formal logic, database theory has always focused on the derivation of information readily contained in data storage instead of inserting information in order to get data out. With statistical databases, however, this traditional view is challenged and ought to be modified thoroughly since otherwise statistical retrieval capacity of database systems suffers considerably – without introducing some kind of disaggregation operator there is, essentially, little left to be retrieved effectively at all (more specifically, only stored datasets or sub-

.5 META DERIVATION

sets thereof remain obtainable then). Furthermore, statistical data retrieval is followed by some kind of statistical data combination quite frequently whence it is a matter of convenience to have this statistical functionality included in one "rounded-off" information system. Apparently, in terms of the distinction of the operator classes introduced above, disaggregation belongs to the MCTs affecting both, data and metadata layers of data objects. It goes without saying that any application of disaggregation operators is fraught with difficulties offering little hope for success; for one thing, if – as claimed – only non-redundant data sources are taken into account the search for appropriate disaggregation data is always likely to fail unless, in the last resort, there is made provision for supplying the disaggregation data required interactively – via user interface – as a posteriori parameter data (supplied, for instance, by means of a ***data:*** message; cf. Subsection 2.4.4). The power of the *disaggregate* operator nevertheless determines the functionality of METASTASYS to a considerable degree since it is liable for

- bringing about various cross-classifications as demanded by METASPEL requests as well as
- enabling statistical data combination in METAGEN's *merge* operation (see below) by decomposing stored aggregates as needed for further REQUEST processing.

In view of the numerous and intricate questions pending, however, only a sketchy account of the *disaggregate* operator will be given.

Within METAGEN, the *disaggregate* operator

$$\lfloor \chi_d \rfloor^d D$$

is used either for de-marginalizing or for re-grouping. De-marginalizing increases the number of breakdown dimensions (of a table or PROMETA) by at least one (thus, formally, converting one or more reference dimensions into cross-classifying dimensions) whereas re-grouping reshapes the grouping patterns of a cross-classifying dimension; in both cases, disaggregating formally causes a replacement of BASPARTitions at AXis level. Alternatively, disaggregation applies to BOXes or PROFiles. Based on the fact that formal conditions impede PROFile disaggregations in all but a few specific exceptional cases being grossly different, in general, from grouping disaggregation cases it is assumed that, despite conceivable interactions, disaggregations referring to both PROFile and BOX of an operand can always be separated into a PROFile disaggregation and a grouping disaggregation. Correspondingly, two modes of disaggregation are discerned in METAGEN signalled by a mode switch assuming either of the states 'prof' or 'group'. This mode switch is a part of χ_d; its state determines type and meaning of further disaggregation parameters. By definition, D is restricted to singletons, that is *disaggregate* refers to single operands (IMAGEs) throughout (in accordance with suggested notation conventions, $\lfloor \chi_d \rfloor^d D \equiv \lfloor \chi_d \rfloor^d d$ since $D = \{d\}$).

For PROFile disaggregation, the PROFile $\wp_i$ of d ought to be decomposed into $\wp_\mathbf{d}$ where $\wp_\mathbf{d} \prec_G \wp_i$; given $\wp_*$, $\wp_\mathbf{d}$ is determined as minimal decomposition of $\wp_i$ required for re-assembling $\wp_*$, that is

$$\wp_\mathbf{d} = \delta(\wp_i, \wp_*)$$

noting that, virtually, for any FRAME $\mathbf{Q}$ the "saturated" set of FRAMEQUalities defining a universal FRAME including all QUs of FRAMEs linked to $\mathbf{Q}$ provides these PROFiles' common BASE. To this end, *disaggregate* launches a sub-REQUEST $r_\mathbf{d} = \langle T_\mathbf{d}, s_\mathbf{d} \rangle$ such that $T_\mathbf{d}$ has PROFile $\wp_\mathbf{d}$ and TEXTure $\mathcal{B}_\mathbf{d}^G$, and $s_\mathbf{d}$ =RATE. It is clear that d (as well as any higher-level "parent" operand to which $r_\mathbf{d}$, whether directly or indirectly, contributes) is excluded from $\mathfrak{S}_Q^D$ with respect to the derivation of $r_\mathbf{d}$. Letting $\mathcal{B}_i^G$ denote d's TEXTure, $\mathcal{B}_\mathbf{d}^G$ is identical to $\mathcal{B}_i^G$ regarding all AXes remaining unchanged; the other AXes of $\mathcal{B}_i^G$ are replaced with the corresponding AXes groups of $\wp_\mathbf{d}$ by splitting $\wp_i$'s groups of FRAMEQUs accordingly. The structure of $\mathcal{B}_\mathbf{d}^G$ and, in fact, $\mathbf{B_d} = (\mathcal{B}_\mathbf{d}^G, \mathcal{E}_\mathbf{d})$ is determined by first inverting the RECTDEFs (**Def. 2.2.3–3**) generating the AXes of $\wp_i$ (this may revive original FRAMEdits obscured by AXRECTs) and, thereupon, fusing the FRAMEQUs to $\wp_\mathbf{d}$'s AXes. As an immediate consequence, this rearrangement is restricted to cases where the resulting AXes' AXDOMains are derivable directly from the involved RECTDEFs – which, particularly, does *not* hold for AXRECTs establishing non-orthogonally decomposable SOVAR mappings unless such AXes are decomposed entirely into their constituent FRAMEQUalities (because the latter, by definition, induce the ubiquitously admissible XGROUPing at the TRAIT level of a FRAME). Statistically, AXis decomposition amounts to de-marginalizing d implying, of course, that a removal of this decomposition by the inverse aggregation restores the original d again. This reveals yet another crucial restriction to disaggregation since AXes are decomposable only subject to their constituents' (that is, FRAMEQUs) *additivity* (although, in special circumstances, non-additive decompositions may be possible technically; think of, for instance, temporal disaggregations using time series models "inversely").

Now, as far as AXis rearrangements are achievable at all, in view of these structural limitations the disaggregation of d is facilitated by decomposing each of d's attached DATCELL entries into a set of new DATCELL entries becoming associated with it in terms of the induced AXis rearrangement; numerically, the *sum* of this new DATCELLs' entries must equal the content of the original DATCELL entry they are associated to. This simple and apparent *invariance property* of disaggregation applies irrespective of how precise $\breve{\mathcal{B}}_\mathbf{d}^G$, the TEXTure of $\breve{M}_\mathbf{d}$ eventually arrived at as response to $r_\mathbf{d}$, matches the grouping structure of $\mathcal{B}_i^G$. Note that neither $\wp_\mathbf{d}$ nor

$\mathcal{B}_d^G$ may be obtainable from available data or ad hoc (= a posteriori) supplied disaggregation information and, hence, some $\breve{\wp}_d$ and $\breve{\mathcal{B}}_d^G$ with $\wp_d \prec_G \breve{\wp}_d$ and $G_k^{(d)} \prec_G \breve{G}_k^{(d)}$ for all Axes k shared identically by $\mathcal{B}_d^G$ and $\breve{\mathcal{B}}_d^G$, respectively, must do. Because of this subsumption conditions, d and $\breve{M}_d$ are structurally compatible such that, eventually, disaggregation is carried out at the data level by simple table arithmetics as outlined in Subsection 2.4.4, that is computing the "product" of d and $\breve{M}_d$ using the latter as a table of weighting factors. The better $\breve{G}_k^{(d)}$ matches $G_k^{(d)}$ for the portion of $\wp_i$ not affected by the disaggregation, the more "informative" in terms of *conditional* de-marginalizing is the disaggregation accomplished effectively.

On condition that d in fact can be disaggregated with respect to its PROFile yielding $\breve{\wp}_d$ with $\breve{M}_d$, the operator is denoted formally as

$$\lfloor \text{prof}; \breve{\wp}_d; \breve{M}_d \rfloor^d d$$

By definition, this PROFile disaggregation does not change the grouping patterns of those Axes of $\wp_i$ left untouched.

For *group* disaggregation, the major difference to PROFile disaggregation consists in replacing the PROFile parameter with a TEXTure parameter. Correspondingly, a sub-REQUEST $r_d = \langle T_d, s_d \rangle$ is launched such that, again, T_d has TEXTure $\mathcal{B}_d^G$ and s_d =RATE, but this time $\wp_d = \wp_i$; additionally, $\mathcal{B}_d^G \prec_G \mathcal{B}_i^G$ must be fulfilled. The TEXTure $\mathcal{B}_d^G$ is determined as the minimal re-grouping of $\mathcal{B}_i^G$ required to obtain $\mathcal{B}_*^G$, the TEXTure of T_*, that is

$$\mathcal{B}_d^G = \delta\left(\mathcal{B}_i^G, \mathcal{B}_*^G\right).$$

To be effective, the $\breve{\mathcal{B}}_d^G$ achieved eventually must at least comprise the de-marginalized or re-grouped Axes among the cross-classifying dimensions. Assuming that there are $J > 1$ such Axes $k_1, \ldots, k_J$ and $\breve{G}_{k_1}^{(d)}, \ldots, \breve{G}_{k_J}^{(d)}$ are the corresponding post-disaggregation Axes' groupings (GRUSETs), the operator is denoted as

$$\lfloor \text{group}; k_1, \ldots, k_J; \breve{G}_{k_1}^{(d)}, \ldots, \breve{G}_{k_J}^{(d)}; \breve{M}_d \rfloor^d d$$

Again, $\breve{M}_d$ represents the disaggregation (PRO)META derived for r_d such that $\breve{B}_d = (\breve{\mathcal{B}}_d^G, \breve{\mathcal{E}}_d)$, the BOX of $\breve{M}_d$ with PROFile $\wp_i$ still unchanged; $\breve{\mathcal{E}}_d$ equals $\mathcal{E}_i$ except that the refined XGROUPing $\breve{\mathcal{B}}_d^G$ (relative to $\mathcal{B}_i^G$) may cause some EDits to become split in order to preserve $\breve{r}_d$'s (and $\breve{M}_d$'s) TNF.

In either disaggregation mode, if the attempted disaggregation of an operand cannot be brought about, of course,

$$\lfloor \chi_d \rfloor^d d = d$$

holds implying that $\breve{\wp}_d = \wp_i$ and $\breve{\mathcal{B}}_d^G = \mathcal{B}_i^G$, respectively; in this degenerate case $\breve{M}_d$ collapses into a scalar of unit weight. It should be obvious from its function that the invariance condition for $\breve{M}_d$ must, in fact, hold for $D^+(\breve{M}_d)$ because for TRAITs belonging to $\breve{\mathcal{E}}_d$ all RATE VALues have to equal 0 constantly.

The *M*erge Operator

The functional core piece of METAGEN and, hence, of METASTASYS altogether is the *merge* operator used to combine IMAGEs both formally and statistically. It comes as no surprise that *merge* is somewhat more complex than any of the other METAGEN operators because, in order to fulfil its role properly, *merge* integrates most of the previously defined subordinate functions. Roughly, the *merge* operation consists of a three-stage procedure taking the output of a preceding *search* operation as its argument and returning a set of candidate (PRO)METAs for further processing towards r_* or $\tilde{r}_*$, respectively. More specifically, since the search stage of a derivation may deliver a *stack* of IMAGE sets (called *retrieval sets* and denoted as $\Re_\nu(r_*)$, $\nu = 1, 2, \dots$; cf. Subsection 2.5.3), *merge* pops the stack elements one at a time and commences with processing the topmost element just popped. Informally, the three processing stages of *merge* may be summarized as follows.

In a first and preparatory *clustering* stage, the (respective) input retrieval set $\Re_{\nu'}(r_*) = \{d_1^{(\nu')}, \dots, d_{n_{\nu'}}^{(\nu')}\}$, $n_{\nu'} \geq 1$, is partitioned into mutually disjoint clusters of IMAGEs; this, in turn, requires two steps:

(i) To enable further operator applications, *coerce* $\Re_{\nu'}(r_*)$'s IMAGEs by commencing a coercing operation with $\tau^*(\{\wp_1, \dots, \wp_{n_{\nu'}}, \wp_*\})$ yielding, eventually, the set of coerced IMAGEs

$$\left\lfloor \text{adapt}; \alpha^*(\{\breve{\wp}_1, \dots, \breve{\wp}_{n_{\nu'}}, \breve{\wp}_*\}) \right\rfloor^c \Re_{\nu'}(r_*).$$

(ii) Taking these coerced IMAGEs, obtain a clustering of $\Re_{v'}(r_*)$ by submitting the IMAGEs' (coerced) IMGRIDs to the *clustering* function ϖ defined in **Def. 2.5.1–14**, $\varpi(\breve{\mathcal{B}}_1^G,\ldots,\breve{\mathcal{B}}_{n_{v'}}^G)$, returning a set of $J_{v'} \geq 1$ IMAGE clusters $\{\Re_{v'}^{(1)}(r_*),\ldots,\Re_{v'}^{(J_{v'})}(r_*)\}$ which, whenever unambiguous, will be abbreviated to $\{\Re_1,\ldots,\Re_J\}$, $1 \leq J \leq n_{v'}$, henceforth.

Of course, in case that $card(\Re_{v'}(r_*)) = 1$ the clustering stage can be skipped safely provided that $\Re_{v'}(r_*)$ is redefined, for later usage, as cluster appropriately. Otherwise, IMAGEs are separated into as many clusters as possible. Note that, in this stage, *coerce* will never abort processing since it would have done so already during the search stage.

After finishing the clustering stage, the *combining* stage is entered next. This pivotal stage is responsible for obtaining the actual combination of empirical IMAGE data. It is in this stage that the involved IMAGEs' work copies are getting transformed persistently (at the metadata level at least) for the first time. However, given that $s_* =$MESO having survived the search stage unchanged (that is, all IMAGEs in $\Re_{v'}(r_*)$ are in fact of MESO type) the combining stage is skipped since, at best, disjoint unions of IMAGEs may apply whereas numerical data combinations are inhibited definitely. Otherwise, each of the clusters $\Re_j$, for $1 \leq j \leq J$, generated in the preceding stage is submitted to a four-step procedure:

(i) Initially, the (original) $n_j \geq 1$ IMAGEs comprised in $\Re_j$ are *coerced* locally by commencing a coercing operation with $\tau^*\left(\{\wp_1,\ldots,\wp_{n_j},\wp_*\}\right)$ yielding, eventually, some set of coerced K-dimensional IMAGEs

$$\left\lfloor adapt; \alpha^*\left(\{\breve{\wp}_1,\ldots,\breve{\wp}_{n_j},\breve{\wp}_*\}\right)\right\rfloor^c \Re_j.$$

(ii) Having got IMAGEs coerced, *grid unification* at TEXTure level takes place. Letting $\mathcal{B}_1^G,\ldots,\mathcal{B}_{n_j}^G$ denote the TEXTures of coerced IMAGEs $d_1,\ldots,d_{n_j}$ of $\Re_j$, this grid unification (cf. **Def. 2.5.1–12**) presupposes a disaggregation of each (coerced) IMAGE in 'group' mode initialized with disaggregation TEXTure

$$\mathcal{B}_d^G = \delta\left(\mathcal{B}_i^G, \tau^*\left(\{\mathcal{B}_1^G,\ldots,\mathcal{B}_{n_j}^G,\mathcal{B}_*^G\}\right)\right)$$

unless $\mathcal{B}_i^G \prec_G \tau^*\left(\left\{\mathcal{B}_1^G,\ldots,\mathcal{B}_{n_j}^G,\mathcal{B}_*^G\right\}\right)$. In any case, this step returns, for each $\mathcal{B}_i^G$, $1 \le i \le n_j$, some $\breve{\mathcal{B}}_i^G$.

(iii) As a final preparatory step to data combination, grid unification is now applied to the $\breve{\mathcal{B}}_1^G,\ldots,\breve{\mathcal{B}}_{n_j}^G$ obtained in step (ii); more specifically, if $n_j > 1$, $\omega\left(\breve{\mathcal{B}}_1^G,\ldots,\breve{\mathcal{B}}_{n_j}^G\right)$ as defined in **Def. 2.5.1–13** is computed returning the set of feasible combinations of (coerced) IMAGEs. If $n_j = 1$, of course, $\breve{\mathcal{B}}_1^G$ itself represents the only feasible combination.

(iv) Now, provided that $n_j > 1$, for each IMAGE combination derived in step (iii) the set of $1 \le L = card\left(\omega\left(\breve{\mathcal{B}}_1^G,\ldots,\breve{\mathcal{B}}_{n_j}^G\right)\right) \le n_j$ pieces $\{\mathcal{P}_1,\ldots,\mathcal{P}_L\}$ determined by the combined component GRUSETs $\omega(\mathbf{G}_k)$, for $1 \le k \le K$, is generated such that, for $1 \le l \le L$, $\mathcal{P}_l \subseteq_G \breve{\mathcal{B}}_i^G$ for some $i \in \{1,\ldots,n_j\}$. In particular, for any pair of pieces $\mathcal{P}_{l'},\mathcal{P}_{l''}$, $l' \ne l''$, from a piece set, $\mathcal{P}_{l'} \equiv_G \mathcal{P}_{l''}$, that is pieces are *conforming*. At this point, statistical data combination may commence by considering, in turn, all occurring non-empty intersections of two or more pieces and achieving, eventually, a *mounting* $\mathcal{M}$ of (statistically) combined data ("evidence"). Formally, $\mathcal{M}$ is a set of $L' \ge 1$ (modified) pieces $\{\hat{\mathcal{P}}_1,\ldots,\hat{\mathcal{P}}_{L'}\}$ returned by the data combination function $\Xi(\mathcal{P}_1,\ldots,\mathcal{P}_L)$ left unspecified here; of course, each $\hat{\mathcal{P}}_{l'}$, $1 \le l' \le L'$ still conforms to its predecessor pieces, that is $\hat{\mathcal{P}}_{l'} \equiv_G \mathcal{P}_l$ for $1 \le l \le L$. Again, in case of $n_j = 1$ there is nothing left to combine whence $\mathcal{M} := \{\breve{\mathcal{B}}_1^G\}$.

Several remarks are in place here. First, this four-step procedure may be run in parallel for each of the clusters $\mathfrak{R}_j$ because there is no interaction between these processing threads. Secondly, and prompting cause for concerning parallel cluster processing, each step of this procedure potentially generates a set of candidate solutions instead of a single one leading to an adverse combinatorial growth of possible continuation branches. In steps (i) and (ii) branching is due to different disaggregations that may be derivable whereas in step (iii) the grouping patterns derived before govern the generation of piece sets; in step (iv) the statistical preconditions of available IMAGE data as well as the SCTs available and applicable to actually merge overlapping pieces determine the range of derivable solutions. For various reasons it is dauntingly difficult to assess at intermediary steps which of the developed solution branches will admit satisfactory overall results later on, so probably the best what can be done is to generate and maintain a pool of candidate solutions as large as feasible and tractable.

.5 META DERIVATION

Obviously, a critical part of the combining stage is the statistical data combination in step (iv). A particularly simple approach in fact avoiding any statistical data combination would be the *selection* of pieces such that a maximum coverage is achieved with as few pieces as possible (sometimes, this is in fact the only approach possible – in particular, IMAGEs with non-elementary SUMTYPEs usually impede data combination). Of course, this still may not determine a mounting uniquely; even worse, this strategy, in general, rejects pieces "dominated" by others and thus wastes information and effort already invested. A real data integration based on statistical models, however, turns out rather intricate as well since standard methodology (such as the aggregation of marginals of one and only representative table of frequencies by the iterative proportional fitting algorithm [Ireland and Kullback, 1968] used to estimate contingency tables based on a maximum entropy, or stochastic independence, principle; cf. also [Malvestuto, 1989]) rarely applies – for one thing, the "marginals" to become combined typically originate from random samples and hence, their marginals may, at best, converge asymptotically to the joint distribution's marginals of the representative table (tantamount to the universal FRAME) but will hardly match nominally on shared marginals. As a consequence, data combination within overlapping pieces, in general, is rendered conditional on stochastic homogeneity properties of log-linear models geared suitably to the situations at hand following the general methodology of discrete multivariate analysis (cf., for instance, [Bishop *et al.*, 1975]). It must be kept in mind as well that the disaggregations carried out to prepare for data combination are, from a statistical point of view, data combinations themselves; therefore, occasionally, disaggregations and data combinations may interfere, particularly if the members of a selected IMAGE set are used to disaggregate each other. Moreover, some of the IMAGEs may not be of COUNT type adding yet further complexity to data combination (necessitating, in general, IMAGE *upgrades* at least). Apparently, all of this opens a fertile field of applied interdisciplinary statistical and database research not yet worked up thoroughly.

The third, or *pasting* stage of the merge operation again comprises a sequence of lower-level steps used to patch up the partial results of the combining stage – viz., the mountings – and to package the pasted partial solutions into (PRO)METAs ready for further processing. Although some of the candidate solutions of the previous stage may be dropped, the pasting stage in general creates even further variants out of the available partial solutions.

Basically, unless there is just a single cluster, the mountings $\left\{ \mathfrak{m}_l^{(j)} \right\}_l$ obtained for the different IMAGE clusters $\mathfrak{R}_j$ generated in *merge*'s cluster stage need to be joined. Thus, in general, as a first step in the pasting stage, solution *nuclei* are formed by taking, in turn, one candidate mounting $\mathfrak{m}_{l_j}^{(j)}$ from each cluster. Nucleus composition is a straightforward combinatorial procedure; though, in order to prevent an excessive generation of nuclei it is advisable to proceed by sorting the mountings within clusters in terms of semantic coverage (measured in terms of TRAIT volume as used in **Def. 2.5.1–15** for computing target coverage) first such that nuclei occur heuristically ordered decreasing with respect to joint semantic coverage – apparently, the "larger" the chosen pieces within each of the clusters, the larger the joint coverage of a nucleus. In the single cluster case, of course, each of this sole cluster's mountings is a legitimate nucleus.

In general, solution nuclei need to become *coerce*d before other operations apply formally. Assuming that a nucleus consists of $J > 1$ mountings $\mathcal{M}^{(1)}_{l_1}, \ldots, \mathcal{M}^{(J)}_{l_J}$ with PROFiles $\wp_1, \ldots, \wp_J$, the joint PROFile $\wp_w$ is achieved by submitting, for $1 \le j \le J$,

$$\lfloor \alpha^*(\wp_1, \ldots, \wp_J) \rfloor^c \mathcal{M}^{(j)}_{l_j}$$

returning the coerced mountings $\widetilde{\mathcal{M}}^{(1)}_{l_1}, \ldots, \widetilde{\mathcal{M}}^{(J)}_{l_J}$.

Because $\alpha^*(\wp_1, \ldots, \wp_J)$ may, occasionally, lead up to rather unfavourable $\wp_w$ s (agglomerating too many of the original Axes into few resulting ones) this particular nucleus could be dropped or, alternatively, one or more mountings of minor relevance (as to net contribution to the overall semantic coverage of the nucleus) might be omitted from the nucleus to see if this helps to improve the structure of $\wp_w$. Any such decision, of course, depends on the number of nuclei available for scrutiny as well as the trade-off between reduction of coverage and gains in cross-classifications still maintained. Obviously, a formal decision model is hard to arrive at and, hence, decisions usually must be provided interactively by the end user as a posteriori parameter data.

As a next step, the coerced mountings have to be put together in BOX format. This is a technical operation requiring, essentially, that all mountings share the same grouping pattern with respect to each of the remaining Axes. This joint grouping pattern, or *virtual resolution*, is achieved by aggregating suitably the partaking mountings' grouping patterns. Letting

$$\hbar := h(\mathcal{M}) = h\left(\{\hat{P}_1, \ldots, \hat{P}_{L_\mathcal{M}}\}\right) =_{def} \alpha^*\left(\{\hat{\mathcal{E}}^G_1, \ldots, \hat{\mathcal{E}}^G_{L_\mathcal{M}}\}\right)$$

denote the "convex hull" of mounting $\mathcal{M}$ where $\hat{\mathcal{E}}^G_l$ is the TEXTure of piece $\hat{P}_l$, $1 \le l \le L_\mathcal{M}$, the virtual resolution of a nucleus is obtained by submitting

$$\lfloor \alpha^*(\hbar_1, \ldots, \hbar_J) \rfloor^a \mathcal{M}^{(j)}_{l_j}$$

where $\hbar_j \equiv h(\mathcal{M}^{(j)}_{l_j})$, for $1 \le j \le J$, and $\lfloor . \rfloor^a$ denotes the *aggregate* operator described below. Here, again, $\alpha^*(\hbar_1, \ldots, \hbar_J)$ might return too coarse a grouping for the requested breakdowns of T_* forcing to reconsider the choice of the present

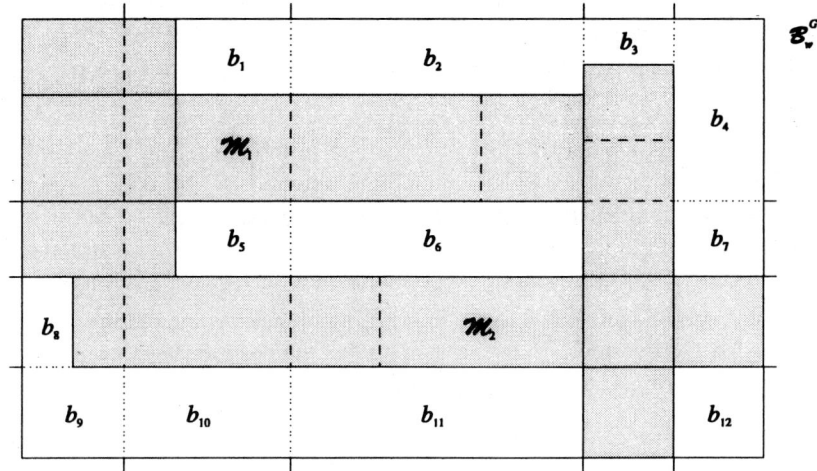

Figure 2.5.2–2: Patching up a Work Table

nucleus of an improvement of the virtual resolution by dropping one or more of the contributing mountings.

After coercing the mountings and achieving a joint XGROUPing on the nucleus concerned, the nucleus can now be converted into TNF. Note that this step is mandatory also for the single cluster case. The consolidation of a nucleus to a so-called *work table* requires three elementary steps: the derivation of the work table's TEXTure, its EDITS, and its SUMTYPE.

Letting a nucleus consist of $J > 1$ mountings $\mathcal{M}_{l_1}^{(1)}, \ldots, \mathcal{M}_{l_J}^{(J)}$, the TEXTure of work table w is $\mathcal{B}_w^G := \alpha^*(\hbar_1, \ldots, \hbar_J)$, as computed above. Of course, if $J = 1$ then, simply, $\mathcal{B}_w^G := \hbar_1$. In either case, however, $\bigcup_{j=1}^{J} D(\mathcal{M}_{l_j}^{(j)}) \subseteq D(\mathcal{B}_w^G)$; this means that the difference (if any) must be "filled in" with EDITs suitably. Formally, the BLOCK elements (CONTRAITS; cf. **Def. 2.2.4–1**) of the auxiliary XGROUPing induced by $\tau^*(\hbar_1, \ldots, \hbar_J)$ on $\mathcal{B}_w^G$ but not covered by any BLOCK $\ell \in \bigcup_{j=1}^{J} \mathcal{M}_{l_j}^{(j)}$ are entered as <u>BLANK</u>s into $\mathcal{E}_w$, the EDITS of w (note that, in terms of OBJECTs, the mountings can be converted easily into BLOCKCOLLections whence $\bigcup_{j=1}^{J} \mathcal{M}_{l_j}^{(j)}$ is a well-defined FOOL BLOCK union); each BLANK is TAGged by 'blank'. **Fig. 2.5.2–2** shows a (contrived) example joining two mountings $\mathcal{M}_1$ and $\mathcal{M}_2$ which results in a TEXTure $\mathcal{B}_w^G$ as indicated and twelve BLANKs $b_1, \ldots, b_{12}$. Recalling the notation conventions adopted, w denotes w's BOX $\mathbf{B}_w =_{def} (\mathcal{B}_w^G, \mathcal{E}_w)$ having PROFile $\wp_w$. Finally, to complete w's PROMETA representation, the SUMTYPE s_w is set equal to the highest-level SUMTYPE occurring among the contributing mountings; hence, in general, arranging s_w implies the intermittent application of *upgrade* operations accordingly. At this point, the work table is rendered in PROMETA format, that is $w = \langle \text{W}, s_w \rangle$.

Depending on the number of clusters discerned in the clustering stage and the nuclei obtainable from the candidate mountings returned by the data combining stage the pasting stage generates, in general, a more or less extensive set of work tables. Although work tables are obtained and processed in view of the structure of the initial REQUEST r_*, the resulting work tables may in fact depart considerably from requested specifications due to the intervening operations interacting in multifarious ways hard to overlook. Hence, assuming that the pasting stage has produced $n \gg 1$ work tables $w_1, \ldots, w_n$, this candidate set is sorted decreasingly in terms of *target coverage* $c(\text{W}_i, \text{T}_*)$ as defined in **Def. 2.5.2**. Having accomplished this, the sorted set (sequence) $\langle w_{(1)}, \ldots, w_{(n)} \rangle$ is handed back by the pasting stage as the final output of a *merge* operation.

Analogous to other operators, *merge* is encapsulated in a formal operator expression

$$\lfloor \chi_m \rfloor^m D$$

where D represents, in general, a pre-processed retrieval set and χ_m provides several parameters controlling the behaviour of the *merge* operation. Unlike other operator types' χ_{op}, χ_m mainly collects a posteriori parameters emanating from the various control decisions to be made (interactively) during the processing of D; it should be clear that χ_m only records parameter values committed to finally (and discards parameter values assumed tentatively while processing backtracks over failing trial solutions).

The *propagate* Operator

The specification of a REQUEST determines a *semantic locus* with respect to a particular FRAME but does so, in general, irrespective (or unaware) of available IMAGEs contributing to that FRAME. As a consequence, the IMAGEs selected and mounted to work tables may yield quite an unsatisfactory coverage of T_* full of gaps. Basically, there can be done little to remedy an apparent lack of data from the database point of view; in some cases, however, it may be possible to *propagate* statistical inferences already drawn from empirical data for one area of a FRAME to other, "dataless" areas resorting to statistical model structures supposed to hold theoretically at least. Roughly speaking, on condition of correlation properties reasonably to assume between qualified sub-populations (CONCEPTs) of a FRAME, verified – or, rather, not-yet-falsified – stochastic inferences regarding one sub-population may be transferred formally to other ones facilitating an *imputation* of (extrapolated) data. Clearly, model-based data imputation is a thorny issue most of the time and therefore requiring a careful and disciplined approach. Although METAGEN, in the longer run, could make good use of a sophisticated generalized *propagate* operator, here only a superficial description is outlined avoiding the intricacies encountered in a more detailed working out. For evident reasons, the propagate operator technically belongs to METAGEN's MCT class of operators.

The pivotal idea to realize data imputation mechanically rests on the notion of statistical *inheritance* [Rowe, 1991]: given that the statistical models fitted to, say, some CONCEPT $\mathcal{C}_1$ are captured explicitly in formal symbol structures, these inferences may be translated – "inherited"– to a semantically similar disjoint CONCEPT, $\mathcal{C}_2$, analogously. Referring to METAMOD and leaving aside semantic aspects of statistical relationships among fitted models and involved datasets, the semantic relationship between subject-matter CONCEPTs is encoded succinctly in CONCEPT respresentations using the respective FRAME's terminology, assigning a concise meaning to CONCEPT similarity. In statistical terms, the reasoning underlying the inheritance method is tantamount to stipulate the existence of a definite (but usually unknown) statistical super-model covering both $\mathcal{C}_1$ and $\mathcal{C}_2$ in terms of hierarchical factor inclusion, this way comprising several sub-models some of which – such as that of $\mathcal{C}_1$ – are known (to the degree of more or less convincing fits) while others are not. In its simplest appearance, the super-model may assume sub-model homogeneity (admitting "lateral" inheritance as termed by Rowe [1991]) but, of course, any kind of correlation or higher-level stochastic interaction (inducing "diagonal" inheritance in Rowe's [1991] terminology) may be considered legitimately provided that it is accounted for technically by multivariate statistical modelling.

Within the METASTASYS set-up, the availability of data propagations implies an expansion of IMAGE data $\mathfrak{I}_Q$ to $\overline{\mathfrak{I}_Q}$; contrary to the IMAGEs comprised in $\mathfrak{I}_Q$ and obtained by source mappings (cf. Section 2.3) supplying DATFILEs extensionally, PROPagations will contribute (observation-level) data merely intensionally unless evaluated definitely on demand ("anti-sampling"). Thus, only $\overline{\mathfrak{I}_Q}^D$ will in fact be represented formally. More specifically, since each PROP, in turn, depends on IMAGE data (that is, $\mathfrak{I}_Q$), the *propagate* operator can be stated as

$$\lfloor \chi_P \rfloor^P d$$

assuming that d is an IMAGE or work table with at least some EDits not belonging to FRAMEDits (naturally, if d is either free of EDits or comprises FRAMEDits exclusively, *propagate* will not exert any effect) and χ_P contributes

- a CONCEPT $\mathcal{C}_p$ selected among suitable and feasible EDits of d,
- some suitable *propagation table* $\tilde{M}_p$ used to derive a specific PROPagation for $\mathcal{C}_p$, as well as

- a statistical (propagation) model Θ_p used to compute the data extension of $\mathcal{C}_p$, that is deriving $\Theta_p(\check{M}_p)$ at the data level.

Alternatively, $\check{M}_p$ may be replaced with a reference to previously derived statistical model fits on condition that statistical analyses are recorded explicitly in METAMOD as indicated briefly at the end of Subsection 2.4.4. In general, the (PRO)META $\check{M}_p$ must be determined appropriately first by screening (*searching*) $\overline{\mathfrak{I}}_Q^D$ or, if this fails, by submitting a pertinent sub-REQUEST r_p returning some $\check{M}_p$ for a possibly relaxed $\tilde{r}_p$ only; in conjunction with $\mathcal{C}_p$ the obtained $\check{M}_p$ then determines the specific statistical propagation model parameter of χ_p. Advantageously, $\mathcal{C}_p$ is stated in TNF (but might be free of cross-classifying dimensions). In order to feed $\mathcal{C}_p$ properly, χ_p must assure two crucial syntactical conditions:

- the statistical propagation model Θ_p must transform $\check{M}_p$ into a work table $\Theta_p(\check{M}_p)$ with a PROFile identical to that of $\mathcal{C}_p$ (favourably, already $\check{M}_p$'s PROFile is identical to that of $\mathcal{C}_p$);
- the cross-classifying dimensions – termed *propagation* AXes – of $\mathcal{C}_p$ and $\Theta_p(\check{M}_p)$ must coincide fully *and* these AXes must be among the explanatory factors of Θ_p whereas, for all remaining *mute* AXes (if any), the only groupings must (should) be identical.

The more the mute AXes' groupings diverge, the more deteriorates the statistical plausibility of data propagation. If $\mathcal{C}_p$ and $\Theta_p(\check{M}_p)$ are disjoint (implying that groupings for at least one propagation AXis are disjoint) the PROPagation is in fact a pure extrapolation the feasibility of which, additionally, depends on topological properties determining particularly the taxonomic CONCEPT similarity – even if mute AXes coincide fully, $\Theta_p(\check{M}_p)$ should be as "close" to $\mathcal{C}_p$ as possible with respect to propagating AXes in order to avoid "remote" extrapolations (for instance, with respect to temporal and spatial vicinities, respectively, semantic distance can be measured effectively in terms of periodicity or neighbourhood graphs; cf. Subsection 2.2.1).

Data propagation cannot produce data by magic, of course, but it may be useful to fill annoying gaps in composite data aggregates whenever the choice of engaged propagation models is founded reasonably and the ratio of imputed to source data remains acceptable. Conversely, sometimes it is this data propagation being demanded for its own sake; for instance, the generation of forecasts entails a dataless area by definition (otherwise no predicting estimate would have to be asked for) which apparently calls for the application of data "propagation" models to historical data, though in a rather specific way using established time series modelling

.5 META DERIVATION

techniques instead of the log-linear modelling approaches perhaps preferable for data extrapolations with respect to other FRAME dimensions. In any case, of course, to be trackable transparently the derived PROpagations need to be declared as such, and must never be delivered without their statistical metadata documenting type and source of involved propagation data and models.

The *restrict* Operator

METAGEN's most conventional operator is the *restrict* operator used to select specified parts of an IMAGE or work table by stripping off those parts (groups) not touched by T_*. Formally, letting d denote the operand (PRO-) META to become *restrict*ed,

$$\left\lfloor \rho\!\left(\mathcal{B}_d^G, \mathcal{B}_*^G\right) \right\rfloor^r d$$

determines a restricted table d' with TEXTure $\mathcal{B}_{d'}^G = \chi_r = \rho\!\left(\mathcal{B}_d^G, \mathcal{B}_*^G\right)$ where $\mathcal{B}_d^G$ is d's and $\mathcal{B}_*^G$ is T_*'s TEXTure, respectively. By definition (**Def. 2.5.1–4**), $\mathcal{B}_{d'}^G \neq \mathcal{B}_*^G$ unless $\mathcal{B}_d^G \prec_G \mathcal{B}_*^G$; moreover, for the same reason, $\gamma\!\left(\mathcal{B}_{d'}^G\right) \neq \gamma\!\left(\mathcal{B}_*^G\right)$ unless $\mathcal{B}_d^G \approx_G \mathcal{B}_*^G$. Since $\gamma\!\left(\mathcal{B}_*^G\right) \subseteq \gamma\!\left(\mathcal{B}_d^G\right)$ can be assured neither by *merge* nor by *propagate* operations, *restrict* may in fact constrain T_* instead of d; this kind of TARGET relaxation is termed *clipping* because χ_r "clips" some part of $\mathcal{B}_*^G$, returning the relaxed TEXTure $\tilde{\mathcal{B}}_*^G$. Apparently, restriction crashes entirely if $\mathcal{B}_d^G \cap \mathcal{B}_*^G = \emptyset$. At the data level, a *restrict*ion simply removes those DATCELLs from d's DATFILE being attached to CONTRAITs stripped off. In this sense, *restrict* corresponds to ordinary database retrieval. With respect to the algebraic properties of group restriction ρ, *restrict* operators may be applied sequentially, in either order, to an operand d. Referring to Basili and Meo-Evoli [1992], the *restrict* operator achieves "simple reducibility" of macrodata (corresponding to MEFISTO*'s *restriction* operation; cf. [Falcitelli *et al.*, 1989]). The *restrict* operator, of course, is a pure ACT since absolutely no manipulation of data level entries is involved.

The *aggregate* Operator

The "upward" rearrangement of grouping patterns by collapsing MODality (Ax-VALue) groups is facilitated by METAGEN's aggregate operator. Drawing upon group aggregation as defined in **Def. 2.5.1–9**, an IMAGE or work table d is aggregated by

$$\left\lfloor \alpha\!\left(\mathcal{B}_d^G, \mathcal{B}_a^G\right) \right\rfloor^a d$$

where, again, $\mathcal{B}_d^G$ denotes d's TEXTure while $\mathcal{B}_a^G$ represents the aggregate XGROUPing aimed at such that $\mathcal{B}_d^G \prec_G \mathcal{B}_a^G$; if this condition is infringed, *aggregate* aborts processing and returns $\bot$. On condition of subsumption transitivity of TEXTures, aggregation may be applied several times consecutively to one operand d. At the data level, the table entries of collapsed CONTRAITs become combined arithmetically depending on the additivity type (cf. **Def. 2.2.1–15,16**) of aggregation; in case of plain ('FLUX'-type) additivity, of course, the arithmetical operation applying is ordinary summation (cf. Subsection 2.4.2). In the latter case, *aggregate* is a pure ACT (like *restrict*); otherwise, *aggregate* becomes a MCT unless a *default* non-FLUX aggregation method is applicable. Technically, aggregation can be framed up making use of (relational) *aggregation formation* and *aggregation-by-template* proposals of Klug [1982] and Özsoyoglu *et al.* [1987], respectively. Note that both, re-grouping and computing of margin sums (marginals), are accomplished by *aggregate* operations since, in METAGEN, marginalizing is just a borderline case of re-grouping gathering all AXVALues of an AXDOMain – except SPECVALues – in a single group (remember that, by definition, single-group AXes are reference dimensions; otherwise, AXes are cross-classifying dimensions). Basili and Meo-Evoli's [1992] StEM approach uses the term "semantic reducibility" for what *aggregate* does in METAGEN (in StEM, semantic reductions of macrodata are carried out by the MEFISTO* *reclassification* operation; cf. [Falcitelli *et al.*, 1989]).

The *trim* Operator

Quite frequently, *restrict* and *aggregate* operators are applied jointly suggesting the definition of yet another operator encompassing both operation types at once. The resulting *trim* operator uses group selection as defined in **Def. 2.5.1–10** to restrict and re-group an IMAGE or a work table d simultaneously,

$$\left\lfloor \sigma\!\left(\mathcal{B}_d^G, \mathcal{B}_t^G\right) \right\rfloor^t d$$

where $\mathcal{B}_t^G$ denotes the XGROUPing aimed at (that is, usually, $\mathcal{B}_t^G \equiv \mathcal{B}_*^G$), by "trimming" the TEXTure $\mathcal{B}_d^G$ of d. Although, in case of $\mathcal{B}_d^G \prec_G \mathcal{B}_t^G$,

$$\sigma\!\left(\mathcal{B}_d^G, \mathcal{B}_t^G\right) = \alpha\!\left(\rho\!\left(\mathcal{B}_d^G, \mathcal{B}_t^G\right), \mathcal{B}_t^G\right)$$

implies a straightforward rewriting of the *trim* operator to

$$\left\lfloor \alpha\!\left(\bar{\mathcal{B}}_d^G, \mathcal{B}_t^G\right) \right\rfloor^a \left\lfloor \rho\!\left(\mathcal{B}_d^G, \mathcal{B}_t^G\right) \right\rfloor^r d$$

.5 METADERIVATION

such that $\breve{\mathcal{B}}_d^G = \rho\!\left(\mathcal{B}_d^G, \mathcal{B}_t^G\right)$, the *trim* operator in fact applies irrespective of $\mathcal{B}_d^G \prec_G \mathcal{B}_t^G$ holding or not since group selection causes an internal clipping of $\mathcal{B}_t^G$ whenever necessary. Apparently, *trim*ming may be cascaded. At the data level, of course, *trim* induces the composite effect of restricting and aggregating.

2.5.3 METAGEN Derivation Schemata

The operators of METAGEN provide the basic building blocks to become synthesized to complete derivations; as pointed out in the introductory remarks to this section, for a given REQUEST r_* first a *resolution plan* composed of METAGEN operators is looked for such that the execution of this plan generates a META $\tilde{M}_*$ as output bearing the metadata layer description $\tilde{r}_*$ semantically "close" to (or, at best, even incident with) r_*. While, depending on r_*, the generated resolution plans will exhibit varying derivation (and, hence, operator) patterns it is important to note that, in fact, there are a few prototypical plan skeletons only. This fortunate circumstance simplifies the effective derivation of generalized tables considerably since, by and large, indeterminism of choices remains confined to the scope of single operators. More specifically, the inevitable – though computationally costly – search processes are focused towards the determination of the parameter data of operators located within partial precompiled operator sequences.

With respect to this stereotyped derivation structure, METAGEN paraphrases the "select–project–join" schema known from the relational database domain [Ullman, 1982], thus being also very much in line with earlier statistical database model proposals (cf., for instance, [Rafanelli and Ricci, 1991; Ghosh, 1991; Chen *et al.*, 1989]) which – mainly due to their relational database heritage – facilitate query processing in terms of a "select–project–aggregate" structure (the "aggregate" part usually refers to the conversion of microdata to macrodata); up till now, "join" has been neglected widely in the statistical database domain except for the rather specific case of aggregating marginals of an underlying (though unknown) representative – that is, joint distribution – table [Malvestuto, 1989, 1991]. From the statistical database point of view, however, the "select–project–aggregate" template enforces some severe restrictions compared to what is desirable from a general data-consumption oriented statistical information system: virtually each of these proposals presupposes that queries refer to a single operand table, viz. either a microdatum (say, a case-by-variates matrix structure) or a macrodatum (summary table comprising aggregated microdata), thus obviating the intricate "join" facet of database retrieval (to the author's knowledge, merely D'Atri and Ricci [1989] have undertaken an exceptional effort to design a statistical database system based on a universal relation interface implying an implicit join of observation case microdata). Apparently, any such approach precludes an effective system-controlled data *integration* from the very beginning – if any aggregation (and, if considered at all, disaggregation) remains restricted to a single operand, no derivation can bring about a fusion of several microdata sets or macrodata tables; on the contrary, even rather simple alignments of closely related datasets – such as the combination of repeated surveys to (multivariate) time series – challenges the derivation power of such a system.

Recalling the traditional approach to data integration which, essentially, uses metadata *descriptively* to create and maintain "surface representations" of semantic relationships between datasets facilitating a more efficient and convenient *browsing* mode to locate requested information in a database (cf. [Rafanelli, 1991] for an overview), the data integration task is in fact left to the end user submitting the database queries: it is up to him to find a sound way to integrate retrieved data by exploiting the available metadata, viz. the end user is burdened with the generation of a resolution plan in terms of a set of more or less high-level statistical database operators (such as those of, for instance, the MEFISTO* system; [Falictelli *et al.*, 1989; Rafanelli and Ricci, 1993; Rafanelli and Ricci, 1990; Meo-Evoli *et al.*, 1994]).

However, it can be anticipated all too easily that the more proficient and powerful statistical meta-information-based browsing systems will become the less will a human mind be able to catch up with the abundant information confronted with – that is to say, having to take into account all of the necessary details determining an admissible resolution plan will, in general, be demanding too much from the (end) users. Hence, METASTASYS attempts to re-deploy the efforts shared in statistical query processing among system and user by encoding taxonomic and statistical relationships between source datasets in terms of *operative* metadata to be used in a formal derivation of resolution plans shielding users from most of the minor details and calling for interaction only in case of substantial information missing to achieve a complete derivation. Predominantly, this additional information pertains to *statistical* data integration (that is, model-based "evidence combination" mostly excluded from the present discussion) whereas the taxonomic data integration is amenable to mechanized processing to a large degree (it remains to be explored to which extent the missing statistical model information can be captured as operative metadata – similar to taxonomic metadata – in order to permit a substitution of interactively interrogated parameter data by parameter data inferred mechanically from either the source datasets' metadata descriptions or suitable default assumptions). Thus, as far as METASTASYS fails to derive resolution plans automatically, it supports query processing by at least providing a succinct *work flow* management ensuring the internal consistency of derivations with regard to (statistical model) information supplied interactively. Although being still quite far from automating statistical database query processing fully, METAGEN's "select–merge–reshape" schema reconciles the "select–project–aggregate" schema of previous statistical database proposals with the "join" part of the standard "select–project–join" database query processing schema at the macrodata level by incorporating disaggregations (as far as possible statistically in view of supplied model information) and data integration operations whenever necessary: roughly speaking, "merge" re-

.5 META DERIVATION

places "join", whereas "reshape" generalizes and unifies both "project" and "aggregate".

The overall structure of a METAGEN derivation comprises two major stages, viz. (i) a *search* stage and (ii) a *transformation* stage. If a REQUEST is submitted to METAGEN, it is first of all analyzed lexically and parsed according to the META-SPEL grammar specifications (cf. Subsection 2.4.3). After this initial *query validation* step [Basili and Meo-Evoli, 1992] the structural information gathered from a valid METASPEL sentence is used to generate a *search stack* of retrieval sets; each stack element provides a possible selection of IMAGEs from which a response to the submitted REQUEST may be derivable in the subsequent transformation stage. In general, the search stage will generate several candidate retrieval sets (except in the default mode; cf. Subsection 2.5.2) but – depending on the search processing mode – will do so in a piecemeal fashion proceeding heuristically from most promising to less advantageous candidates (cf. Section 2.6). Anyway, upon termination of the search stage the transformation stage commences with the topmost element $\Re_0(r_*)$ of the search stack (which is removed from the stack). By scrutinizing the *semantic distance* between r_* and the current $\Re_0(r_*)$ selected from the search stack (i) a skeletal resolution plan is picked and (ii) the operators of it are instantiated appropriately such that, if a consistent instantiation of all operator parameters is achievable at all (*feasibility verification* in Basili and Meo-Evoli's [1992] terminology), a *derivation* (that is, a fully specified resolution plan) $\Phi(\Re_0(r_*))$ is obtained. Although the skeletal resolution plans are sequences of operators, a complete derivation may in fact be tree-shaped because, in general, some operators' parameters have to be fed (recursively) by nested sub-REQUESTs requiring completed resolution plans of their own (the composition of resolution plans touches on the issues of reasoning about plans and plan formation investigated in Artificial Intelligence research; cf. [Sacerdoti, 1977] as well as [Georgeff et al., 1985; Georgeff and Lansky, 1987] with respect to reasoning about processes and dynamic adaptation of plans to a changing "environment"). In case of operators depending on parameters obtained from sub-REQUESTs, the respective sub-resolution plans are inserted formally in those operators' χ_{op} s – hence, in general, $\Phi(\Re_0(r_*))$ assumes a nested structure similar to nested function composition. A feasible derivation, however, is *acceptable* only if $\Phi(\Re_0(r_*))$, when evaluated (at the metadata level), reduces to $r' \in U_\zeta(r_*)$ where

$$U_\zeta(r_*) =_{def} \{\tilde{r}_* | d(r_*, \tilde{r}_*) < \zeta\}$$

denotes the *acceptance region* of r_*; clearly, the concept of $U_\zeta(r_*)$ is used as a heuristic device only as long as the "distance measuring" function $d(r_*, \tilde{r}_*) \geq 0$ is left unspecified. Assuming that $U_\zeta(r_*)$ can be determined effectively, the acceptability of a derivation is controlled by the ζ parameter: setting $\zeta = 0$ would rule out any derivation generating a $\tilde{r}_* \neq r_*$ (since in this case obviously $U_\zeta(r_*) = \{r_*\}$) and may therefore be too strict a choice in many practical circumstances. Again, the definite choice of ζ depends on the preselected processing mode of the transformation stage: in a single-pass mode some threshold value $\zeta' > 0$ could be specified and a derivation $u^* \in U_{\zeta'}(r_*)$ chosen as final solution where

$$u^* = \min{}_{u \in U_{\zeta'}(r_*)}\{u\}$$

whereas in a multi-pass mode a monotonically increasing sequence $0 \leq \zeta_0 < \zeta_1 < \cdots < \zeta_n$ could be specified and, for $\Re_0(r_*)$ held constant, the transformation stage repeated until $U_{\zeta_i}(r_*) \neq \{\}$ the first time (that is, $U_{\zeta_0}(r_*) = \cdots = U_{\zeta_{i-1}}(r_*) = \{\}$). While the former approach stipulates a crisp distinction between acceptable and unacceptable solutions, the latter divides the (intensional) solution set into classes of practically "equivalent" solutions. If, irrespective of the transformation stage's processing mode, no generated solution is acceptable (which subsumes the case that no derivation could be found at all), control may be passed back to the search stage; if the search stack contains another retrieval set not yet investigated the transformation stage is re-entered, otherwise – depending on the search processing mode – the search stage could resume generating further stack entries. Thus, by and large, METAGEN's search and transformation stages are interlinked in terms of a *generate-and-test* procedure such that the transformation stage tests the candidate solutions brought forth by the search stage one by one and, as long as candidate solutions fail this test, backtracking continues in order to exhaust the space of candidate solutions.

Basically, METAGEN distinguishes three general processing "schemata" depending on s_* and the contents of the selected retrieval set:

- The *standard schema* applies if s_* belongs to the elementary SUMTYPES.

- If s_* is not an elementary SUMTYPE and $\tilde{M}_*$ is derivable directly from the retrieval set (thus avoiding the tedious composition of target tables out of intermediary operand tables), the *brief schema* derivation is conducted.

- If s_* is not an elementary SUMTYPE and $\tilde{M}_*$ is *not* derivable directly from the retrieval set the derivation forks into interdependent sub-derivations for each of the intermediary operand tables needed to obtain the requested s_* by table arithmetics (cf. Subsection 2.4.4); this will be called the *composite schema* of table derivation.

Roughly, each of these schemata corresponds to a skeletal resolution plan as explained below (but note that, since SCTs have been excluded deliberately from the

.5 META DERIVATION

present exposition, these skeletal plans omit statistical operators such as *weight* and *estimate* which are, for instance, indispensable for scaling up sample figures to population figures). The processing mode of skeletal plans, however, is controlled additionally by a couple of configuration settings which, for instance, will enable or disable disaggregations and/or propagations, determine the acceptance region of feasible derivations, and prescribe the structure and sequence of retrieval sets generated in the search stage (cf. Section 2.6).

Standard Derivation Schema

In *standard schema* processing, the transformation stage is decomposed into an *integration* stage and an *extraction* stage, thus replacing the standard two-stage derivation procedure conceptually by a three-stage procedure. First, the search stage comprises a *search* operator

$$\lfloor r_*; \chi'_s \rfloor^s \mathfrak{I}_Q^D(r_*)$$

where $\mathfrak{I}_Q^D(r_*) \subseteq \mathfrak{I}_Q^D$ denotes a possibly preselected IMAGE subset (assuming that r_* provides information helping to eliminate "pointless" IMAGEs from the outset). χ'_s, in turn, contributes the processing mode parameters controlling search stack generation; in particular, this processing mode determines the scope of IMAGEs considered for inclusion in retrieval sets, for instance, by excluding all IMAGEs requiring disaggregations. Furthermore, the processing mode is also responsible for the stack ordering of retrieval sets – retrieval sets with IMAGEs subsuming (parts of) T_* may favourably precede retrieval sets with IMAGEs matching T_* only (cf. Subsection 2.5.1), and so on.

Letting $\mathfrak{R}_0(r_*)$ again denote the topmost retrieval set on the search stack, the integration stage of the standard schema commences processing with applying an *upgrade* operator

$$\lfloor \text{COUNT} \rfloor^u \mathfrak{R}_0(r_*)$$

on condition that $s_* \neq$ MESO (otherwise, *upgrade* is suppressed). Supposing that $\mathfrak{R}_0^{(u)}(r_*)$ has been obtained this way, the integration stage proceeds to *merge* the upgraded IMAGEs by applying

$$\lfloor \chi_m \rfloor^m \mathfrak{R}_0^{(u)}(r_*)$$

which, contrary to the previous steps, will generally involve the interactive specification of (a posteriori) parameter data χ_m (χ_m may of course also comprise processing mode parameters specified a priori). Usually, *merge* generates a solution set $\mathfrak{R}_0^{(m)}(r_*)$ and, hence, marks a branching point of candidate solutions. Choosing (via χ_m, formally) some $w \in \mathfrak{R}_0^{(m)}(r_*)$, the integration stage concludes with a propagation step comprising, in general, a series of $L \geq 1$ non-interfering *propagate* operators

$$\lfloor \chi_p^{(1)} \rfloor^p \cdots \lfloor \chi_p^{(L)} \rfloor^p w$$

provided that w contains EDits different from FRAMEDITS that can be imputed effectively. Since, in general, different imputations may apply to w, the propagation step is yet another branching, or backtracking, point of candidate solutions.

Having chosen a work table w_p from the candidate set obtained at the end of the integration stage, the extraction stage of the standard schema commences by either applying a *restrict* or a *trim* operator depending on whether the data-induced or the target-induced grouping patterns should prevail. Usually, *trim*ming is preferred (converting data-induced groupings to cross-classifications specified in T_*) whence

$$\lfloor \chi_t \rfloor^t w_p$$

rewrites w_p to the trimmed table w'. Finally, it may be necessary to *upgrade* w' to match s_* by

$$\lfloor s_* \rfloor^u w'$$

achieving the derived $\tilde{M}_*$.

It certainly goes without saying that in derivations only *work copies* of IMAGE data are being processed; IMAGEs in permanent storage are in fact never changed unless updated explicitly in terms of new *versions* (cf. Subsection 2.3.1). Conversely, however, the reference to IMAGEs of mesodata by adjoined QUalities of a virtual FRAME will induce intermediary data accommodations implicitly in case of *1:n* FRAME linkages (causing a counting of linked FRAME instances; cf. Subsection 2.4.2).

As pointed out in the discussion of METAGEN operators, instead of a REQUEST r_* only a *relaxed* variant $\tilde{r}_*$ may be derivable from actual IMAGE data. This relaxation may refer either to r_*'s TARGET T_* or its SUMTYPE s_*. Because of automatic adaptations enforced by the operators applied, *over-upgrading* may lead to a SUMTYPE $\tilde{s}_*$ whereas target clipping and lack of data ("void" FRAME areas) alter T_* vigorously: particularly changes in CONCEPT coverage with respect to T_*'s reference dimensions might render the derived tables rather pointless (note that these changes imply a redefinition of or even blatant deviation from the CONCEPT specified in the REQUEST originally); certainly much easier to tolerate are changes

.5 META DERIVATION

with respect to T_*'s breakdowns unless they are upset entirely (more formally, $\mathcal{B}_*^G \neq \tilde{\mathcal{B}}_*^G$ might be rather uncritical as long as $\gamma(\mathcal{B}_*^G) = \gamma(\tilde{\mathcal{B}}_*^G)$ and $\mathcal{E}_* = \tilde{\mathcal{E}}_*$ hold, at least approximately). Apparently, these semantic subtleties ought to be captured formally in the specification of acceptance regions $U_\zeta(r_*)$ but possibly should, for the sake of improving computational efficiency of derivations by early cancelling of solution candidates unsuitable further on, be incorporated in the processing mode configuration already.

METAGEN's standard schema is a fairly general one encompassing a variety of more specific cases. Leaving aside a few exceptions, the standard schema covers virtually all types of derivations considered in previous proposals for statistical database systems. Moreover, since these special, though simple, cases make the lion's share of REQUESTs encountered practically, it is important to see how gracefully and efficiently they are dealt with by METAGEN. The simplest case conceivable at all assumes a singleton retrieval set $\Re_0$; if, furthermore, the sole IMAGE comprised in $\Re_0$ encloses T_* entirely, the integration stage can be omitted by and large (although coercing of PROFiles and group disaggregations may be involved). Otherwise, if T_* is not covered fully, a propagation might be tried which, in turn, may initiate a sub-derivation returning a sub-REQUESTed propagation table. If, finally, $\Re_0$ is *empty* at all, the standard schema has to be abandoned and replaced with an isolated propagation REQUEST alone (as it would be the case in a pure forecast REQUEST). More frequently, however, will perhaps occur retrieval sets comprising more than one IMAGE. In this case, the decisive criterion regards the disjointness of retrieved IMAGEs; obviously, *merge* operations are rather straightforward to carry out if involved IMAGEs are disjoint because both clustering and combining stages become obsolete (unless some of these IMAGEs' PROFiles and grouping patterns do not subsume $\wp_*$ or T_* and, hence, are in need of disaggregations). The remaining pasting stage, eventually, is particularly simple if all IMAGEs (=nuclei) share an identical PROFile since in this case only the common *virtual resolution* and the incurred BLANKs need to be determined. Apparently, the better the disjoint IMAGEs fit together with respect to the non-disjoint "AXes", the fewer BLANKs will have to be introduced. Optimally, conforming IMAGEs are either identical or disjoint to each other with respect to all of the AXes, enabling a simple alignment without creating any BLANK when pasting nuclei together.

A somewhat atypical – but perhaps no less important – case regards REQUESTs with SUMTYPE 'MESO'. Since in this case only *additive* set unions of instances are feasible, the standard schema is restricted to the merging of retrieval sets comprising, in the integration stage, *disjoint* IMAGEs only; accordingly, in the extraction stage only restriction and – provided that aggregations refer to additive dimensions exclusively – trimming operations apply. Despite these peculiarities it must be kept in mind, however, that 'MESO'-type REQUESTs accomplish METASTASYS's microdata retrieval, thus providing its interface to "classical" statistical methodology basing inferences on microdata rather than on macrodata (recalling that, for theoretical reasons, the latter are causing efficiency losses and, in general, biased estimates).

In addition to requiring rather delicate data combinations to achieve appropriate *mountings* of overlapping IMAGEs, REQUESTs may, in general, also call for fairly sophisticated interaction of individual processing steps within derivations. As an example, just think of a REQUEST asking for a (temporal) chaining of IMAGEs obtained, say, from some series of repeated surveys. Basically, the task consists of extrapolating one series of IMAGEs by re-weighting the data of another series such that the re-weighting information is drawn from a suitable data combination for the common overlap of both IMAGE series. Thus, the re-weighting data is obtained from a *merge* operation (possibly requiring some disaggregation of either IMAGE series to take place intermediately) supplying the propagation table for a *propagate* operation actually imputing the data for the IMAGE series to be chained. As highlighted by this example, any derivation being in need of sub-derivations contributing parameter data is facing serious challenges of internal consistency maintenance. Not only are sub-REQUESTs depending on the local contexts they are emerging from, they may also interact with each other *across* local contexts. As an immediate consequence of this, each derivation must be embedded in a "global" environment taking account of the relationships of local contexts. For instance, it is quite objectionable to use the same IMAGEs several times within a single derivation from a probabilistic point of view and, therefore, the place and reason for using an IMAGE needs to be recorded in explicit terms. This tracking of context interrelations is particularly important with respect to statistical model assumptions and, hence, is connected intimately to the statistical metadata layer. The maintenance of global derivation environments becomes even more important in view of the backtracking dynamics of the derivation process since, particularly with respect to statistical model assumptions introduced interactively more or less *ad hoc*, the mutual consistency of introduced assumptions can be safeguarded, if at all, by a mechanized means of protocolling and cross-checking. Likewise, in view of nested resolution plans the environment must prevent circularity and "deadlocks" (that is, mutually dependent sub-resolution plans) by recognizing self-referencing REQUESTs. How such a derivation environment is established and maintained best belongs to the more puzzling half of pending questions in statistical information system design (cf., for instance, [Cowley and Whiting, 1985; Froeschl, 1990]) and clearly relates to the development of *reason maintenance* systems (cf. [Stoyan, 1988] for an overview of basic considerations, and particularly [Dressler, 1988] with respect to assumption-based truth maintenance).

Brief Derivation Schema

It is common practice in statistical data processing (at least as far as official statistics is concerned) to prepare and maintain a stock of ready-made statistical aggregates – termed "standard tables" occasionally – used to respond quickly and efficiently to "frequently asked queries" instead of re-computing these aggregates over and over again from basic IMAGE data. Although, in general, the storage of pre-derived METAs is redundant logically (except in cases where the original IMAGE data is not available any more), the economic gains may nevertheless be considerable, particularly if access to this data storage is supported by additional semantic linkage structures (thesauri, for instance). Conversely, the opportunities to process pre-derived METAs further are limited basically to *restrict* operations making a selection of CONTRAITs because of the non-additivity of such METAs' non-elementary SUMTYPEs. However, it may still be possible to find propagations to extend the coverage of "standard" METAs stored. Correspondingly, the resulting *brief schema* for METAGEN derivations comprises either a *search–restrict* or a *search–propagate–restrict* operator sequence. As long as s_* refers to an elementary SUMTYPE and $\mathfrak{R}_0$ is a singleton, this brief schema complies fully with the standard schema of META derivation; if, on the other hand, s_* refers to a non-elementary SUMTYPE, brief schema derivation forces $\mathfrak{R}_0$ to be a singleton (although, in specific cases, *merge*s may still be achievable). In its most rudimentary instance, of course, the brief schema reduces to a simple search without any subsequent processing of $\mathfrak{R}_0$.

Composite Derivation Schema

In the most general case, a REQUEST r_* asking for a META with non-elementary SUMTYPE s_* might not be obtainable by a simple brief schema derivation but, instead, needs to be derived entirely from "scratch" or be composed recursively in patchwork fashion out of intermediary operand tables originating from subordinate standard or brief schema derivations. In particular, compound tables induced by functional folding (cf. Subsection 2.4.3) being computed from operand tables not derivable from – and, hence, hierarchically dependent on – a single kernel table require a kind of structurally "co-ordinated" derivation of the involved operand tables. To be useful, these operand tables must fulfil specifications rather tightly (in terms of "narrow" acceptance regions) in order to enable subsequent arithmetical composition successfully. This, in turn, puts yet higher demands on the global derivation environment now becoming responsible for enforcing a *concurrent* consistent control of *all* derivation threads confluent to the joint outcome. Compared to both standard and brief derivation schemata, this *composite schema* can be sketched at present rather vaguely only and it certainly requires quite a lot of thorough investigation before a tenable comprehensive solution comes into reach.

2.6 Distributed Request Processing

In order to accomplish a *structural* and, in particular, navigation-free access to statistical data (cf. Section 2.4), the design of METASPEL as well as the development of METAGEN have been guided by the well-founded database principles of physical and logical data independence. Practically, however, pertinent data is scattered inevitably over several computing sites interconnected typically by some kind of digital network infrastructure; in addition to being distributed physically, statistical data is stored rather differently on different local computer/software systems – using semantic data models, formal host languages, and hardware platforms almost as unequivocal as imaginable – constituting what nowadays is commonly called *heterogeneous federated databases* [Sheth and Larson, 1990; Hsiao et al., 1990]: task-driven combinations of pre-existing databases which, as a rule, have been designed and created independently of each other without anticipating the emerging requirements of data sharing and trans-system information linkage. Hence, in general, REQUEST processing amounts to a spatial decomposition into several interdependent sub-REQUESTs basically run in parallel and addressed – suitably adapted – to local database systems actually contributing empirical data to a submitted REQUEST.

Technically, the data integration approach envisaged in METASTASYS therefore implies a *network-based distributed derivation* of statistical data aggregates viewing the involved local database systems linked in the network conceptually as a single heterogeneous distributed (multi-) database system. Although distributed REQUEST processing adds yet another functionally intricate layer to META derivation it should be clear that there is virtually no way to do away with either physical data dispersion or data heterogeneity: whether maintained by statistical offices, governmental or private agencies, or multi-national companies and corporations, established data holdings usually have to support both local and non-local tasks and, hence, – if for no other than historical reasons – owe their tribute to local data production methodologies, data documentation standards, storage structures, and usage contexts. The very idea of an overall re-structuring of local data, being questionable from a logical point of view anyway, is certainly beyond economic reason – neither a (retrograde) duplication of bulk data prepared specifically to support data integration nor a re-engineering of customized local data usage provides a practically feasible solution in all but a few exceptional cases; moving (or copying) data to a single site to work around data distribution isn't even a thinkable option in reality since this would imply setting up artificial mega-databases both costly and clumsy to maintain. Quite on the contrary, in view of a distributed *demand* for statistical data the structural data integration approach is better backed up with a distributed processing strategy allocating the resources of local database systems dynamically as driven by the requirements of individual REQUESTs. In addition to its efficiency potentials (cf. [Sadreddini *et al.*, 1992]) such an approach supports the *subsidiary principle*, respects the local autonomy and responsibility of data suppliers, and acknowledges the low adaptation capabilities ("plasticity") of *legacy systems* to be bridged appropriately by situation-specific interfaces (cf. Section 2.3; in addition to the logical interfaces, of course, hardware as well as low-level communication interfaces are needed, too).

Assuming that there is a dedicated set of statistical data hosts (that is, local computing systems supplying statistical data) participating in an *interoperable* META derivation network, METASTASYS is designed as a distributed processing system in

.6 Distributed Request Processing

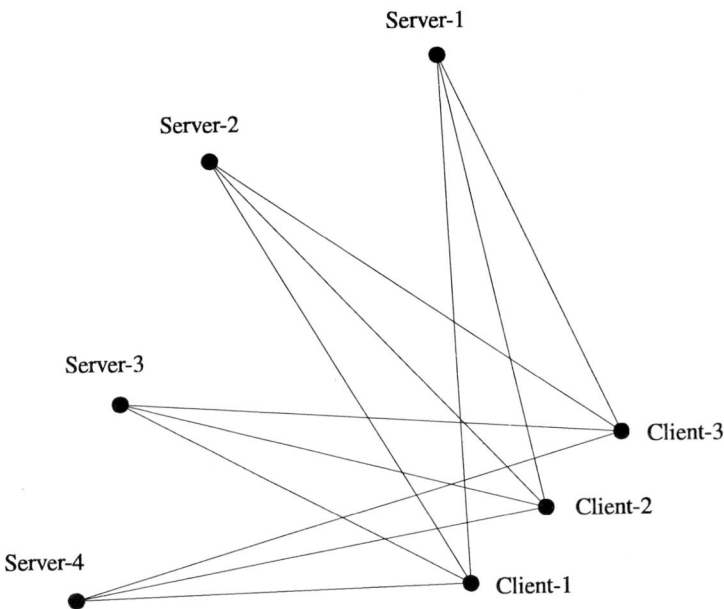

Figure 2.6–1: Virtual METANET Connectivity

terms of a specific client/multi-server architecture – called METANET – such that local data hosts act as *server* systems ("points-of-offer") and the network nodes submitting REQUESTS act as *client* systems ("points-of-demand"). In addition to the physical interconnection of data hosts by standard communication protocols, the statistical data supplied become interconnected logically at the metadata layer by

- arranging suitable METAMODel instances (cf. Section 2.2) comprising, first and foremost, FRAMEs and FRAME linkages, and
- defining mappings (cf. Section 2.3) of local data semantics to METAMODel instance semantics.

Despite the stipulated global scope and validity of METAMODels, individual clients need only possess a restricted window – or "external schema" in common database terminology – on those global data models (that is, they must have at least a *partial* METAMODel at their command to participate in the network); irrespective of a client METANET window's width, however, source data, original source data semantics (local terminology etc.), and attached local metadata – including specifically the mappings to METAMODels – reside at local METANET server sites exclusively and are visible in client windows mediated through METAMODels only. Letting clients specify and submit REQUESTs within their respective METAMODel windows, REQUEST processing – in a purely logical view – fans out on the METANET along the client-server links as indicated in **Fig. 2.6–1**, that is processing

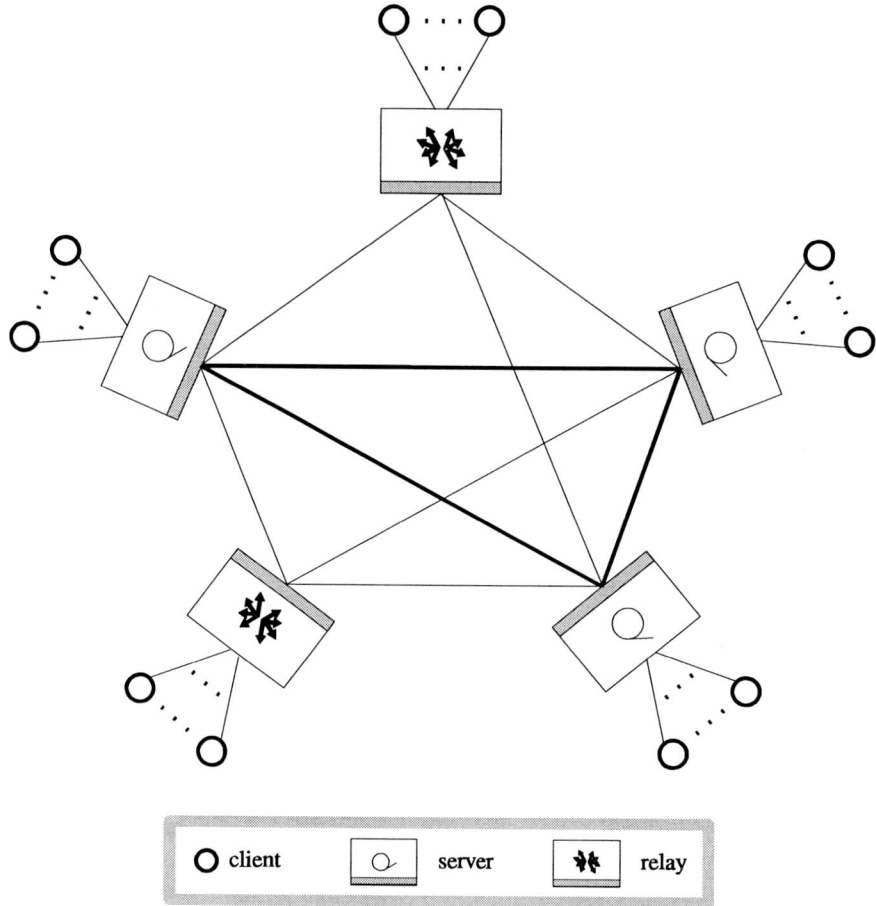

Figure 2.6–2: METANET Topology

makes use of a divide-and-conquer strategy shifting computations of compartmentalized sub-problems to server sites to avoid particularly mass data transfers (of microdata) over the network and to speed up responses by interoperable server processing controlled remotely from client sites (in a sense, the concentration of data is exchanged for a distribution of processing control).

The bipartite graph connection topology between server and client sites shown in **Fig. 2.6–1** is, however, merely a conceptual one; as is all too apparent, a clustering of clients – similar to the approach used in telephone exchanges – is advisable economically especially with respect to physical neighbourhood. The resulting connection topology, illustrated in **Fig. 2.6–2**, takes also into account that, in general, clients themselves will be connected to specific host systems which, in fact, may coincide with network servers. Hence, clients are linked physically to servers either

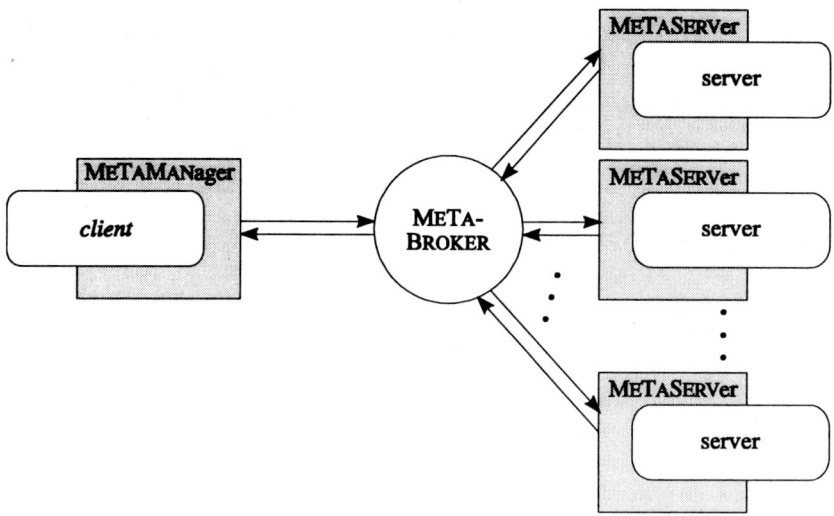

Figure 2.6–3: METANET Communication Flow

directly or by mediating relay nodes (in fact, carrying on the analogy with telephone switching systems, a relay node in **Fig. 2.6–2** may correspond to an embedded relay-sub-network). Now, to get access to METANET, a client must establish only a (local) link to a server or a relay otherwise, where it is assumed that servers fulfil the role of relays as well (note that, according to common terminology, this type of connection is termed client-server linkage quite in contrast to the higher-level functional client-server relationship view adopted in METANET).

For each established (or active) METANET client link, a client process is run communicating with the server processes using a specific METACOM protocol: since, essentially, integral OBJECTs such as REQUESTs and METAs are being communicated between client and server processes, plain message-passing protocols certainly must be augmented (or replaced ?) with OBJECT-passing facilities such as the *shared variable* communication paradigm proposed by Kühn [1994]; by virtue of METACOM, a client process faces the full stock of IMAGEs (including PROPagation models; cf. Subsection 2.5.2) supplied jointly by all servers comprised conceptually in METANET as a unique encompassing *virtual* OBJECT store, though restricted possibly to its individual METANET window. While METACOM facilitates the global access to the virtual METASTASYS database, physical and logical distribution of source data is dealt with by a METABROKER process mediating between clients and servers or, more precisely, between the METACOM interfaces of client and server processes, METAMANagers and METASERVers, respectively. Within this architecture as depicted schematically in **Fig. 2.6–3**, servers are viewed as IMAGE *pools* comprising (with respect to a particular METAMODel considered) source data and, associated to source data, the mapped source sample spaces expressed in METAMODel terminology (for the sake of efficiency, IMAGEs may in fact be backed up with derived METAs such as *standard tables* useful in answering "frequently asked questions" provided that non-IMAGE data sources are also integrated consistently in the respective METAMODels). Furthermore, servers provide for each FRAME, upon request by a METABROKER, a *pool containment* indicating, in METAMODel terminology either, the (rectangular) portion of the FRAME to which the server actually contributes source data (cf. **Fig. 2.5.2–1**); formally, pool containments are CONBOXes (cf. **Def. 2.2.3–2**) embracing, in the sense of least convex hulls, all (PRO)METAs (whether IMAGE or not) supplied by the enquired server to the FRAME considered, implying that a server definitely does *not* provide any empirical data outside its pool containments. It must be kept in mind, however, that pool containments often will comprise areas without any actual data ("blanks", so-to-say) which may or may not be filled by PROPagations later on (hence, despite initial appearance, a server may fail eventually to supply any data to a particular REQUEST).

Against the background of this functional METANET architecture, a REQUEST is understood as a process run under local client control vis-à-vis the virtual global METAMODel database. In particular, the *client* (process) initiates and supervises the derivation of a resolution plan and, in order to achieve this, decomposes REQUESTs logically into sub-REQUESTs (one for each operand table) in case of compound tables not reducible to a single substitute (such as a *kernel* table etc.; cf. Subsection 2.4.3); likewise, the client is liable for spawning and co-ordinating all sub-REQUESTs feeding (recursively) the parameters of the master resolution plan's operators, if need be. Apparently, controlling the derivation of the master resolution plan calling for co-ordinating all forked sub-REQUESTs implies that the client is responsible for maintaining the global derivation environment as well. However, the client – being always attached to one *broker* (METABROKER process) mediating the distributed data access to the *servers* – does not have to take care of the actual data distribution. In particular, this broker returns obtained (partial) results to the calling client to resume resolution plan generation or further processing of derived results. Expectedly, clients also run the front-end user interface of METASTASYS which, in addition to REQUEST formation and submission, plays a vital role in supplying a posteriori operator parameter data; moreover, it may support table layout, table algebraic, data export, etc. functions (cf. Subsection 2.4.4) necessary for convenient system handling.

A *broker*'s task consists of resolving the clients' data requests to spatially decomposed sub-requests submitted to specific METANET servers. Basically, brokers segment received requests – formally, (sub-) TARGETs of (sub-) REQUESTs – roughly by checking the intersection of (sub-) TARGETs with corresponding pool containments. To this end, brokers may utilize pre-compiled index structures speeding up this check procedure; for instance, some of the servers may – with respect to a considered FRAME – have entirely disjoint pool containments whereas other servers may be disjoint dynamically only (that is they are disjoint with respect to specific TARGETs and/or SUMTYPEs only). Furthermore, METAMODels could, in principle at least, provide other *semantic indices* helping to separate servers either in spatial, temporal, or subject-matter terms to improve the brokers' performance to sift out servers failing to contribute data to a request. Using these semantic indices, the search for contributing servers can be narrowed by issuing *informed request broadcasts* (instead of non-informed request broadcasts used as a default strategy) to preselected subsets of available servers (for instance, some REQUESTs may refer to IMAGEs supplied by a specific server only). It is also conceivable to introduce a *layering* of servers such that sub-nets of a METANET are condensed to *virtual* servers used to rule out larger sets of servers at once in terms of hierarchically nested pool containments. In the longer run, the TARGET-dependent selection of META-NET servers may call additionally for an economic ranking scheme based on (roughly estimated) transaction cost incurring from accessing and retrieving data from servers (noting that these costs are influenced by several factors like the volume of data to be transferred, physical client-server distance, access time, server-specific charging of transactions, data transmission cost, etc.). This information could then be passed on to the clients which – by adding, for instance, further cost estimates for various data transformations applied to retrieved data in a resolution plan derived eventually – can make a rational choice decision based on a more or less profound cost-benefit analysis; in particular, such a decision rule could be used in conjunction with acceptance regions (cf. Subsection 2.5.3) to determine Pareto-optimal solutions with respect to both criteria, cost and semantic deviation from r_*, of candidate resolution plans.

Heuristics for Ranking Retrieval Sets

In view of METAGEN's derivation schemata (cf. Subsection 2.5.3), a broker ought to provide a little more than a selected set of servers: since, in general, IMAGEs need to be combined to cover a REQUEST's TARGET, brokers in fact are responsible for creating the *search stacks* of retrieval sets used as solution candidates for resolution plan development. Taking a closer look, retrieval sets may include (i) IMAGEs from one or more servers and (ii) within servers, again, one or more IMAGEs. Apparently, IMAGEs may overlap *within* as well as *between* servers (the latter implying non-disjointness of pool containments, of course). In addition to this, both types of included IMAGEs may interfere since choosing more servers could reduce the number of IMAGEs participating in a retrieval set within servers and vice versa. An important aspect of this interaction is the frequent coincidence of spatial and subject-matter disjointness of pool containments and, hence, IMAGEs (which is mainly due to historical developments: for instance, statistical offices have statutory obligations to operate within a certain territory, etc.); apparently, choosing a couple of servers with mutually disjoint pool containments to contribute IMAGEs to a retrieval set enables the *additive* composition of sub-TARGETs, that is it alleviates META derivation considerably by avoiding the failure-prone data combination stage. In view of the complexity of subsequent transformations taking place during resolution plan derivation and the computational cost resulting thereof, a promising strategy favouring most parsimonious operations amounts to the following simple heuristic ordering of generated retrieval sets.

With respect to IMAGE eligibility in terms of target coverage (**Def. 2.5.1–15**), the formation of retrieval sets proceeds by three ranking criteria, viz. (r_1) the overall target coverage obtained by a retrieval set, (r_2) the relationship of servers contributing IMAGEs to this retrieval set with respect to disjointness of pool containments, and (r_3) the internal structure of *pool retrieval sets* which are the subsets of IMAGEs of a retrieval set contributed by each server in turn. These criteria are summarized in **Tab. 2.6**, giving a sum total of 27 discerned cases being identified uniquely by a code triplet '$r_1 r_2 r_3$'.

Table 2.6: Ranking Criteria for Retrieval Set Formation

	Target coverage	*Server disjointness*	*Pool retrieval set structure*
	r_1	r_2	r_3
1	full	one server	single IMAGE
2	good	mutually disjoint servers	mutually disjoint IMAGEs
3	poor	non-disjoint servers	other

Assuming a formal specification of "good" and "poor" target coverage, respectively, has been agreed upon the retrieval sets composed out of eligible IMAGEs can be classified with respect to these 27 types ranked roughly in lexicographical order of code triplets (for 'r_3' the least general – that is, smallest – code number with respect to all servers involved in a retrieval set must be chosen). Apparently, type **111** retrievals sets are most favourable ("full hits") since a REQUEST's TARGET T_* is covered entirely by a single IMAGE; closely following are type **112** ("local hits") as well as **121** and **122** ("additive hits") retrieval sets requiring a mounting of IMAGEs though still without a data combination (the latter being indispensable for "composite hits", that is type **1***x***3** or **13***x* retrieval sets). As to practical relevance, type *x***21** retrieval sets come next to type *x***11** ones since, particularly in non-local data integration REQUESTs, no single server may manage to cover T_*. Logically, types *x***23** and *x***31**/*x***32** are equivalent (all requiring data combinations) but type *x***23** retrieval sets, not disabling distributed computations, are still preferable. The worst, of course, are type *x***33** retrieval sets. Therefore, in many cases type **211** ... **223** retrieval sets – not insisting on full target coverage anymore – will in fact perform better than type **133** retrieval sets depending on specified acceptance regions and cost bounds for META derivations. Certainly less satisfactory will be type **3***xx* retrieval sets because of their low target coverage (implying that "poor" means a considerable deviation from T_* attaining, say, less than 90% coverage or so). Type **333** retrieval sets are a theoretical borderline case only, not deserving any attention in practice anyway. Though, as yet, it is not clear at all how to accomplish retrieval set formation efficiently according to such a ranking scheme, the computational cost for search stack creation could be reduced considerably by replacing the full-breadth search for eligible IMAGEs with a more selective strategy detecting better-ranking retrieval sets earlier than other ones.

Managing Distributed Request Processing

Strictly speaking, a META derivation consists of two consecutive passes, viz. a *resolution pass* to derive a feasible and acceptable resolution plan (cf. Subsection 2.5.3) and an *execution pass* to run the derived resolution plan at the data level in order to really obtain the requested META; both passes, of course, are controlled by the client. At the broker level, each pass demands a specific broker functionality: during the resolution pass, the broker is destined to compile search stacks (which is metadata level computing exclusively) as described above whereas during the execution pass a definite resolution plan is dismantled into its constituent "remote" database accesses and statistical computations delegated to the processing modules of local servers as far as possible (cf. [Sadreddini *et al.*, 1990]). In both passes, however, the broker is responsible for a proper scheduling of sub-tasks exploiting the potentials of (asynchronous) inter- as well as intra-operator parallelism: consisting of sub-processes executed at different network (server) sites, a pass can be finished only upon termination of all forked sub-processes by collecting and combining the sub-processes' results (data flow model). In contrast to the resolution pass where sub-process parallelism requires – some atypical and trivial cases excepted – synchronization and parameter exchange via shared global derivation environments, operations can be carried out independently in the execution pass. While in the resolution pass intra-operator parallelism dominates (applying the search operator to available servers concurrently), the execution pass is characterized by inter-operator parallelism causing possibly totally different computations (operator sequences) to take place at server sites. By the way, the distributed data-level computation contrasts the centralized (at the client processes) metadata-level derivation of resolution plans, thus highlighting the necessity for shared variable network communication allowing an unrestricted and transparent OBJECT access across processes.

REQUEST processing may assume either of two overall modes. The standard *dynamic* derivation mode comprises both resolution and execution passes implying that a resolution plan is derived anew each time a REQUEST is submitted. The second, *precompiled* derivation mode consists of selecting a possibly parameterized resolution plan template which is submitted to the execution pass of a META derivation immediately. Precompiled derivations save a lot of effort in case of standardized aggregates computed repeatedly (such as annual reports summarizing socio-economic indicators etc.) which, in general, are based on regular schedules of data production. Using a suitable precompiled derivation template, routine statistical aggregates can be obtained simply be re-running the execution pass on new or changed/updated data after instantiating the template's parameters (if any) appropriately. In case of unprecedented REQUESTs or the first-time creation of precompiled derivation templates, of course, only the dynamic derivation mode applies. In both of these processing modes, the METABROKER process acts as the client processes' gate to the federated METAMODEL database, branching data access spatially and gathering the partial responses returned by the enquired network servers.

By their very nature, METABROKERs are single-client/multi-server processes. Physically, these processes have to be run at both relay and server sites (cf. **Fig. 2.6–3**) because, from a functional point of view, the type of node to which a client is linked is irrelevant; in any case, there must be a mediating METABROKER interposed between clients and servers. At the METACOM level, the data interchange between clients, brokers, and servers is handled, according to OO terminology (cf. [Taylor, 1992] with respect to an easy-to-read overview) by object caches maintained locally at either side, viz. METAMANagers at client side and METASERvers at server side.

Just like clients, *servers* are linked to brokers; from a server's point of view, requests are submitted by brokers instead of clients. In contrast to METABROKERs which are responsible for just one client at a time, servers may interact with several brokers concurrently and, hence, are multi-processing nodes (very much like other multi-user time-sharing computer systems). Strictly speaking, the term 'server' has hitherto been used pars-pro-toto; a more detailed picture of METANET servers is drawn in **Fig. 2.6–4**. The core component of a server node, of course, is its statistical database (and database management) module (SDB) which, however, must be accompanied by a meta-information system module (SMIS) prepared to handle the metadata description of data comprised in the SDB module according, at least, to the METASTASYS requirements. In particular, source data must be described according to METAMODEL's component data model (CDM) standards (cf. Section 2.3). With respect to existing SDBs often grown over a long period of time the most part of which it was uncommon to document statistical data in terms of machine-readable metadata, this amounts to a laborious extension of existing data holdings by including the lacking metadata documentation at least for those data segments that should be exported and mapped to a METAMODEL. In future systems, SDBs probably will comprise an integrated meta-information component from the beginning; in such a way a consistent and complete capturing of relevant metadata can be realized which is also cost-effective since most metadata can be tapped directly from the data production or data generating processes.

Irrespective of the particular SDB/SMIS approach chosen or present, the actual interface of these *local* databases to METASTASYS is accomplished by a so-called *wrapper* layer [Taylor, 1992] converting local storage formats and standards into OBJECTs; more specifically, wrapping is used to remove *structural* (in the widest sense) data model differences by establishing standardized interface *behaviour* of local databases. Hence, in other words, this wrapper layer makes all data items *appear* uniform as objects conforming to METASTASYS standards and is, in turn, responsible for providing the translation functions between internal OBJECTs and proprietary SDB/SMIS formats in either direction. A pivotal role among those translation functions assume the source sample/property space mappings μ (cf. Section 2.3) or, more specifically, their inverses μ^{-1} to be used as data access

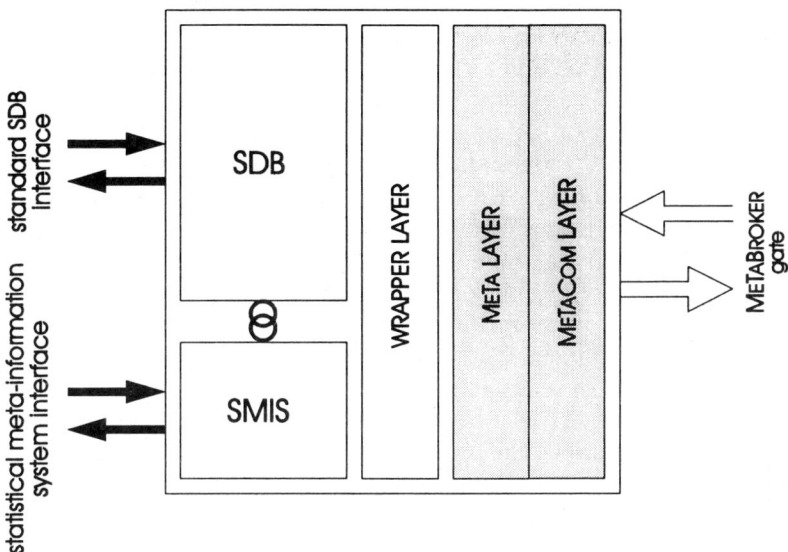

Figure 2.6–4: METANET Server Structure

functions. The functional META layer of a server (METASERver, to be precise) is a local shared workspace ("cache" memory) used particularly to mount partial results during the execution pass of a derivation before they are moved physically to the client sites; depending on the chosen distributed-memory management strategy, METASERvers can help to reduce actual SDB accesses by keeping retrieved IMAGES temporarily in local workspaces until a derivation is finished (this local management and pre-processing of output data will be of utmost importance particularly in case of bulky datasets kept on fragmented tertiary storage media; cf. [Chen *et al.*, 1995]). The OBJECT-based communication with METABROKERs is, of course, handled throughout and exclusively by the METACOM layer; rather naturally, depending on the respective processing pass, METASERvers provide metadata (resolution pass) or first order data (DATFILEs) predominantly; furthermore, during the resolution pass, pool containments – which, of course, is metadata either – may be shipped to brokers.

It is important to note that the inclusion of a local SDB into METANET does not affect its previous "local milieu"; all functions and interfaces prevail untouched. The only modification regards the addition of the wrapper layer which, viewed internally, is just one more front-end interface, or export schema, to yet another application software system fed by the database. As a matter of fact, the major distinction compared to other interfaces incurs from the necessity to handle SDB and SMIS subsystems simultaneously (unless the SDB is an integrated data/metadata information system already). Conversely, of course, after integration of a server into a METANET, local data and metadata may be accessed either by SDB's and

SMIS's standard (front-end) interfaces, respectively, or by the METASTASYS client interface.

The METACOM protocol makes use of the specific OBJECT structure of META-STASYS data objects which, as a design principle, package data and related metadata into logical units (according to the md/ds structure of Darius et al. [1993] termed "tandem" format). In particular, METAs – the carrier set of META algebra as well as the operators used for META derivation – are processed in tandem format throughout (although during a derivation's resolution pass only the metadata part of OBJECTs is considered). Apparently, encapsulating both data and metadata belonging together tightly, is an appropriate means to safeguard internal consistency which, as an accessory benefit, provides an appealing conceptual object model (extending the notion of *statistical entities* of Rafanelli and Ricci [1990]) as well. In view of this "object-oriented" representation of OBJECTs a preference for object-oriented software design is natural; although a mapping of all data and metadata components of OBJECTs to (flat) relations in a suitably extended relational design model is certainly possible, this would lead immediately to extensive and rather complicated nets of interdependent integrity rules clumsy to handle and maintain.

Moreover, the OO paradigm offers the elegant opportunity to attach methods to OBJECTs directly (in particular, METAGEN operators can be defined as genuine META methods), thus simplifying distributed processing enormously because OBJECTs and attached methods are distributed alike [Bukhres and Elmagarmid, 1996]. These benefits are further amplified by taking advantage of the semantic richness and plasticity of object-oriented data modelling in facilitating structure conversions required for arranging the federated system's component data models (cf. [Pitoura et al., 1995] and Section 2.3). By the way, regarding statistical datasets as objects (instead of, for instance, relations) enables the definition of methods incorporating the specific semantics of statistical database operations such as, say, case selection ("sub-sampling") in terms of a randomized retrieval of cases for the sake of sub-sample representativeness; it certainly is also somewhat easier to express statistical relationships (such as stochastic dependencies) in an object-oriented representation of statistical entities. Favouring an OO approach to METASTASYS development, statistical entities become visible globally in a METANET simply by creating an encompassing virtual OBJECT space managed by METACOM; by defining all logical entities (including "objects" such as derivation environments) consistently as OBJECTs, the tedious inter-process communication at low symbol-passing level – that is, character streams – can be avoided entirely; nevertheless, symbol-level data interchange (in particular, the EDIFACT/GESMES protocol for statistical data interchange) will still remain very useful for basic data import/export interfaces in various ways; cf. [Lebaube, 1992; Malmborg, 1992; Habermann and Waith, 1992].

In addition to data communication, METACOM is used for process synchronization. Basically, client, broker, and server processes can be conceived of as simple scripts carried out in a stepwise manner, linked by synchronization messages (in formal terms, each process behaves like a finite state machine where synchronizations mark state changes). An excerpt of a (simplified) client-server protocol is reproduced in Appx. C; a few explanatory remarks will be given in Section 2.7.

2.7 The DÖS'chen Prototype

An engineering discipline first and foremost, computer science is expected to give proof of its craftsmanship moulding the theoretical designs of computation structures into "real" program code every now and then. In research domains, software construction is undertaken mainly to gain tangible evidence and assure functionality in the ambitious information structure and algorithm designs otherwise hardly to grasp cognitively in view of their daunting complexity which, as a rule, inhibits the application of formal verification methods checking program consistency and termination in the large. Furthermore, information systems involving, by definition, some kind of man-machine interaction cannot be assessed confidently without more or less extensive hands-on experiences which, in turn, require at least an "emulated" run-time interface behaviour of the envisaged software pieces.

The METASTASYS project certainly ranges among the more ambitious software projects, and it calls for thoroughly considered man-machine interaction modes to enable an efficient, well-structured access to both statistical information and metainformation comprised in the system's distributed symbol store. In fact, METASTASYS not only needs an end-user interface but data and metadata import interfaces and system administration interfaces as well. Therefore, it has been decided to develop a small prototype version of a metadata-based statistical table derivation system focusing particularly on the formal management of taxonomic metadata in a restricted subject-matter domain and to create a few simple interface models necessary to handle metadata conveniently. Substantially, this program development emerged as a sub-task of a preceding research project, viz. the *Modelling Metadata* project (cf. Section 1.4) aiming at a formal metadata language helping to preserve official statistics data contexts by attaching suitably encoded data descriptions to statistical survey data. Since *Modelling Metadata* was one of several projects of the *Development of Statistical Expert Systems* research programme (acronym: DOSES) launched by EUROSTAT, the statistical office of the EU seated in Luxembourg, it was agreed upon to dub this little software prototype "DÖS'chen" (using the German diminutive to indicate roguishly its relation to DOSES). Starting in late 1989 and picking up the threads of earlier investigations in so-called statistical expert systems [Darius, 1986; Froeschl and Grossmann, 1988; Froeschl, 1989; 1995], the *Modelling Metadata* project laid the foundations of subsequent research leading up to the design of METASTASYS as expounded in the preceding sections of this chapter. DÖS'chen may be justly conceived as a first material yield of this long-term research effort, exerting no doubt a vigorous influence on the follow-up METASTASYS development.

With respect to the limited resources available, DÖS'chen never attempted to address more than a carefully selected function subset of a full-blown statistical aggregate derivation system. Moreover, many of the considered functions have been realized in a rather restricted and simplified way. Hence, before a scanty descrip-tion of DÖS'chen is given, the salient differences between this prototype and the 'grand design' of METASTASYS are summarized briefly.

A decision of major importance for DÖS'chen design was the deliberate omission of the system's data layer (that is, DÖS'chen operates at the metadata layer exclusively) emphasizing the *feasibility verification* step of REQUEST processing. Consequently, no (external) database – of statistical observation (micro-) data or aggregate (macro-) data derived thereof – was integrated into or interfaced with the prototype. Quite on the contrary, it was argued that, once the metadata layer calculus is accomplished satisfactorily, transferring the resolution plans derived at metadata calculus level to the data layer and translating them into corresponding (data layer) database operations in order to obtain the numerical output macrodata eventually would require nothing but rather straightforward implementation work. This, in fact, is the tenet of the METASTASYS proposal either.

Another crucial DÖS'chen design decision regards its restriction to *taxonomic* metadata as the pivotal functional metadata component in semantic data integration and, particularly, the harmonization of established data sources; essentially, this meant focusing on FRAME set-up (compiling nomenclatures of theoretical data languages) and the arrangement of data source mappings at the cost of an unbalanced and insufficient consideration of genuinely *statistical* metadata. Furthermore, only the simplest type of METAMODel structures have been accounted for, viz. there is a single – though fundamental – enumerative (ordinal) SCALETYPE DÖS'chen can deal with (corresponding to METAMODel's ENUMerative SCALE-TYPE; cf. **Def. 2.2.1–8** and **Tab. 2.2.1**). As a direct consequence, all DÖS'chen FRAMEs remain restricted to *categorical* sample spaces implying that the only SUMTYPE handled is COUNT (since DÖS'chen does not manage first order data, the distinction between SUMTYPEs 'COUNT' and 'MESO' becomes irrelevant). Correspondingly, all empirical data sources are assumed to be qualitative socio-economic surveys modelled after the *Labour Force* survey species which, by the way, has been used as a running DÖS'chen testing and demonstration example. Having ruled out quantitative SCALEs, all survey variables are treated as enumerative-categorical (nominal or ordinal) by default. In addition to this SCALETYPE uniformity enforced, DÖS'chen cannot discern different SCALE semantics; in particular, the universal QUalities *space* and *time* are in no way distinguished from other FRAMEQUalities. However, this does not cause any harm since only single-FRAME METAMODels are permitted; in other words, in DÖS'chen any METAMODel instance is restricted inevitably to a particular FRAMEUNIT as the prototype cannot cope with FRAME linkages explicitly. Likewise, DÖS'chen fails to manage SPECVALues properly, thus limiting the capabilities to specify FRAMEDits (in particular, DÖS'chen does not take care of FRAMEDITS consistency); if going to be used at all, the SPECVALs 'NULL' and 'MISSVAL' (cf. **Def. 2.2.1–12** and **–13**, respectively), have to be introduced just like ordinary MODalities (!).

DÖS'chen supports only rather simple types of survey variable mappings, viz. single or nested variables mapping directly to an associated FRAMEQUality; AXRECTification structures have not been implemented. As another (minor) simplification, VALMAPpings remain restricted to *1:1-* or *1:n-*MODality associations (the latter in DÖS'chen being called "map groups"). It must be remarked, however, that – by applying more or less elementary pre-processing operations to the original survey variables – still a great deal of mappings is accomplished quite easily. For the very same reason, DÖS'chen admits only the specification of domain concepts (that is, CONDEFs) which can be expressed directly in terms of "basic" FRAME-QUalities; it is impossible to arrange any kind of RECTDEFinitions as described in Subsection 2.2.3. Conversely, however, DÖS'chen supports the definition of *generic* CONDEFs (implying an internal FRAMEDit management) as well as the hierarchical nesting of super- and sub-CONDEFs.

DÖS'chen has been equipped also with rather limited networking capabilities. Although the prototype installation mimicked a client/multi-server environment (at message-passing level using TCP/IP socket communication) the built-in data distribution model, in fact, handles merely the case of servers with non-overlapping pool containments effectively. Similarly, DÖS'chen merges IMAGEs with disjoint TEXTures only because of its weak *merge* operator (cf. Subsection 2.5.2); in fact, these disjoint IMAGEs are *aligned* with respect to one or (at most) two dimensions such that, for instance, spatially distributed data are strung together along the ***space*** AXis, or repeated surveys are concatenated to (multivariate) ***time*** series, if necessary. Since IMAGE alignment ignores IMEDITs (cf. **Def. 2.3.3–1**) as well as TACONEDITs (cf. **Def. 2.4.1–7**), the *target coverage* achieved eventually may be rather poor. This simple-minded data combination strategy complies favourably with the crude *search* scheme selecting IMAGEs on account of (non-empty) *box overlap*s but ignoring the possibly devastating effect of *edits mismatch*es (cf. **Def. 2.5.2**). The set of DÖS'chen operators worked out in detail comprises the ACT-type operators (cf. Section 2.5) *select* (corresponding roughly to the METASTASYS *restrict* operator), *aggregate* and *re-group* (both of which encompassed in METASTASYS's *aggregate* operator); the latter operation is the more general one facilitating the collapsing of MODality groupings – in the sense of the subsumption relation (cf. **Def. 2.5.1–5**) – whereas *aggregate* refers to the borderline case of marginalizing over FRAME-QUalities. *Trim*ming is accomplished by concatenating DÖS'chen *select* and *re-group* operations. DÖS'chen does not possess a *disaggregate* operation.

Since a DÖS'chen METAMODel comprises a single FRAME only, each REQUEST submitted refers to this FRAME automatically. DÖS'chen REQUESTs are expressed in a simplified aggregate specification language consisting of three specification elements only which correspond roughly to METASPEL's **FOR**, **WITH**, and **BROKEN DOWN BY** clauses (cf. Subsection 2.4.3); moreover, it is neither possible to specify compound tables nor to make use of METASPEL's *folding* mechanisms. Although there has been a rudimentary parameterized *cost model* implemented to guide the selection of candidate IMAGEs, the selection of servers and IMAGEs cannot be controlled explicitly, that is

- DÖS'chen resolves REQUESTs always by issuing *non-informed request broadcasts* (cf. Section 2.6), and

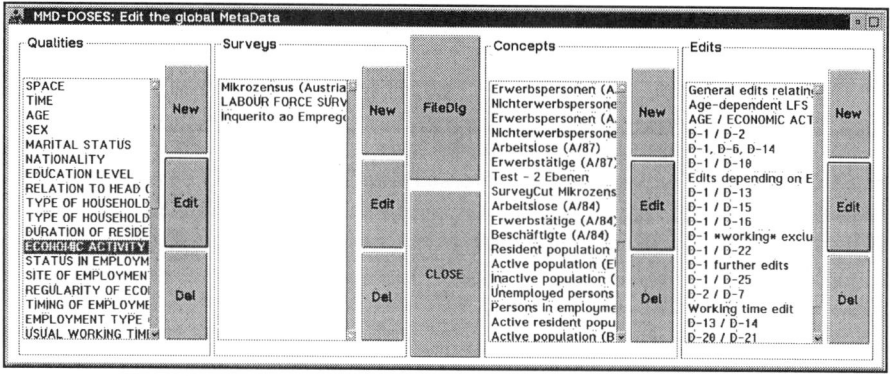

Figure 2.7–1: *GME* Main Screen

- no provision is made for delimiting the REQUESTs' *acceptance regions* (cf. Subsection 2.5.3).

Likewise, DÖS'chen does not maintain *semantic indices* helping to speed up IMAGE selections by reducing the search focus to a qualified IMAGE subset.

Functional Prototype Components

DÖS'chen consists of two major software components, viz. (i) the *Global Metadata Editor* (*GME*) and (ii) the *(statistical) Aggregate Generator* (*AGen*). Roughly, the *GME* provides the metadata administration interface of DÖS'chen used to define and edit the nomenclatures out of which FRAMEs, SOVARs and survey mappings are composed. At present, all kinds of metadata must be entered into the system manually via the *GME* user interface; in the longer run, of course, this method is rather impractical and ought to be substituted by metadata tapping procedures interfacing directly with data production systems. It must be noted, however, that particularly the arrangement of QUality domains, FRAMEdits, and (survey) mappings cannot, in general, be carried out automatically and will therefore need some kind of interactive administration interface anyhow. The *AGen*, in turn, is DÖS'chen's auxiliary end-user interface for composing and submitting REQUESTs referring to the METAMOdel maintained in the *GME*. In the DÖS'chen prototype program, *GME* and *AGen* share the very same internal (meta-) database (which is particularly convenient as long as both software components are running on the same CPU); however, *AGen* can read in *GME* metadata from an intermediary disk file as well. For the sake of simplicity, the METAMOdel terminology established in the *GME* is used directly as a "unified" global data language for REQUEST specification in the *AGen* interface (which, of course, could have been substituted easily by *mediating* interfaces adapted specifically to the demands of particular classes of end-users; for instance, it might be quite convenient in a multi-lingual context to have mediating *AGen* interfaces expressing global terminology in the end-users' respective mother

.7 THE DÖS'CHEN PROTOTYPE

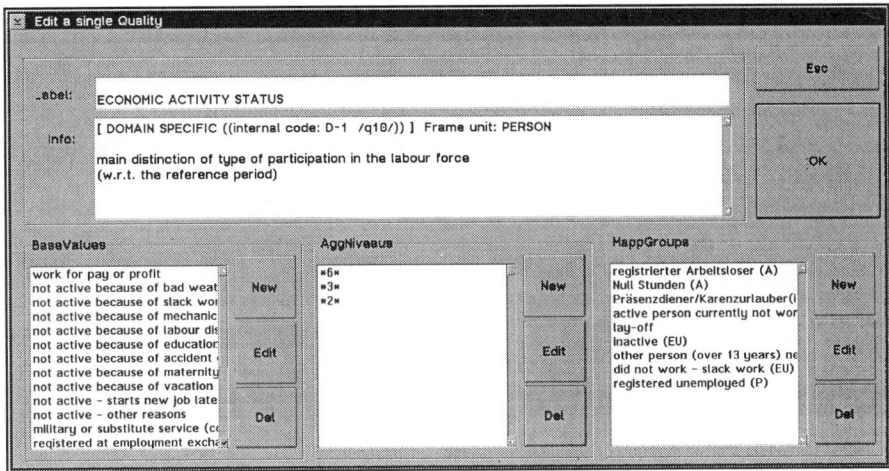

Figure 2.7–2: *GME* QUalities Definition Screen

tongues). There has never been created a particular user interface for the DÖS'chen (metadata) servers linked to "the" test client; the server configurations (including Internet addresses) and pool containments have rather been hard-coded into the programs emulating server behaviour.

Instead of giving a laborious description of each of the *GME* and *AGen* functions a couple of screenshots drawn from DÖS'chen's *graphical user interface* (GUI) will be used to illustrate the more salient operations and services. Following the well-established principles of direct manipulation man-machine interfaces (cf. [Wærn, 1989]), the DÖS'chen GUI relies on a few specific interaction patterns (praxemes) in addition to the usual *point-and-click* modes for window scrolling, menu selections ("markings"), etc., almost all computer users have become thoroughly accustomed to meanwhile. The screens reproduced in subsequent figures serve a dual purpose in that the panes visible on these screens are filled with actual *Labour Force* metadata discussed in Chapter 3; moreover, Appx. C gives a fairly detailed account of metadata structures underlying the DÖS'chen screens exhibited such that the interrelation between screens, stored metadata, and GUI functions can be reconstructed succinctly. Since DÖS'chen has always been regarded as an experimental device, screen design never received particular attention – the screenshots are faithful reproductions of dialogue snapshots without any kind of dressing.

The Global Metadata Editor

The *main screen* of the *GME* is shown in **Fig. 2.7–1**; basically, it comprises four larger panes (scroll windows) listing, in turn, the QUalities defined (for the sole FRAME DÖS'chen can deal with at a time), the surveys mapped to this FRAME presently, the predefined concepts (CONDEFs; cf. Subsection 2.4.1), and the global FRAMEDits arranged. Right to each pane three switch fields are attached which,

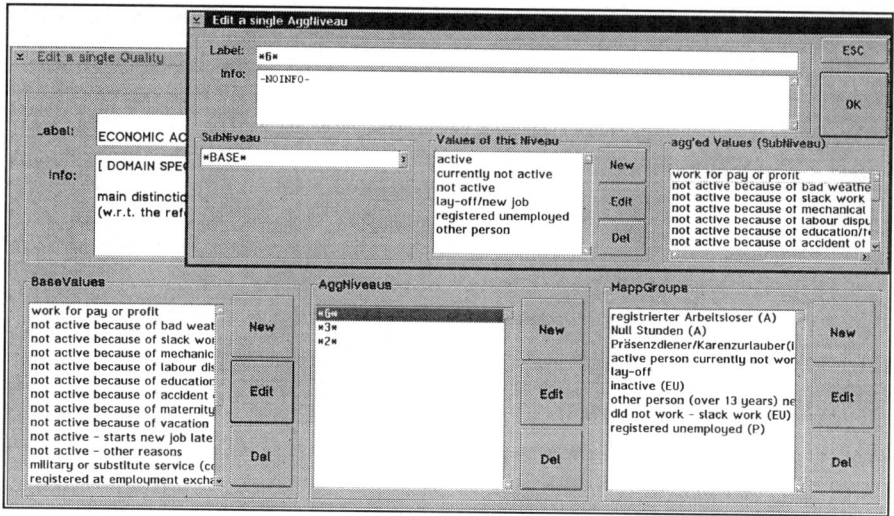

Figure 2.7–3: *GME* AGGLEvel Definition Screen

upon clicking, trigger sub-dialogues for either entering *New* items or *Edit*ing or *Del*eting the respective item marked in the pane (as indicated in the leftmost pane of the screen shown in **Fig. 2.7–1**); *Del*etions, of course, may bear disastrous consequences because of established internal cross-references becoming corrupted, and should therefore be avoided.

Clicking on *Edit* switch fields makes all preceding metadata entries of the marked item reappear for both inspection and update. For example, clicking on the *Edit* field attached to the leftmost pane of **Fig 2.7–1**'s screen displays the screen shown in **Fig. 2.7–2** (previous page) summarizing the metadata gathered for QUality 'ECONOMIC ACTIVITY STATUS'. In addition to an '`Info:`' pane indicating unformatted explanatory text this screen comprises three further scroll windows: the left pane lists the MOdalities (`BaseValues`) of the inspected QUality in definition order; the middle pane lists the AGGLEvels specified; the right pane lists the arranged "map groups" used to *map* individual SoVALs of survey variables to MOdality *sets* of the inspected QUality.

The valid definition of an AGGLEvel can be inspected (or edited) by clicking on the *Edit* field next to the '`AggNiveaus`' pane after marking a list entry as shown in **Fig. 2.7–3**. In the sub-window appearing thereupon, the AGGLEvel's AGGBASE is indicated in the lower left pane (where, when defining a *New* AGGLEvel, one of the previously defined AGGLEvels – among which the '*BASE*' AGGLEvel comprising the QUality's BASVALues is always present – must be selected as AGG-BASE); the middle pane lists the AGGLEvel MOdalities hitherto defined. The right pane indicates the arranged association between AGGLEvel MOdalities and AGGBASE MOdalities defining the grouping pattern of the AGGLEvel. For instance, marking the MOdality 'registered unemployed' of AGGLEvel '*6*' in **Fig. 2.7–3**

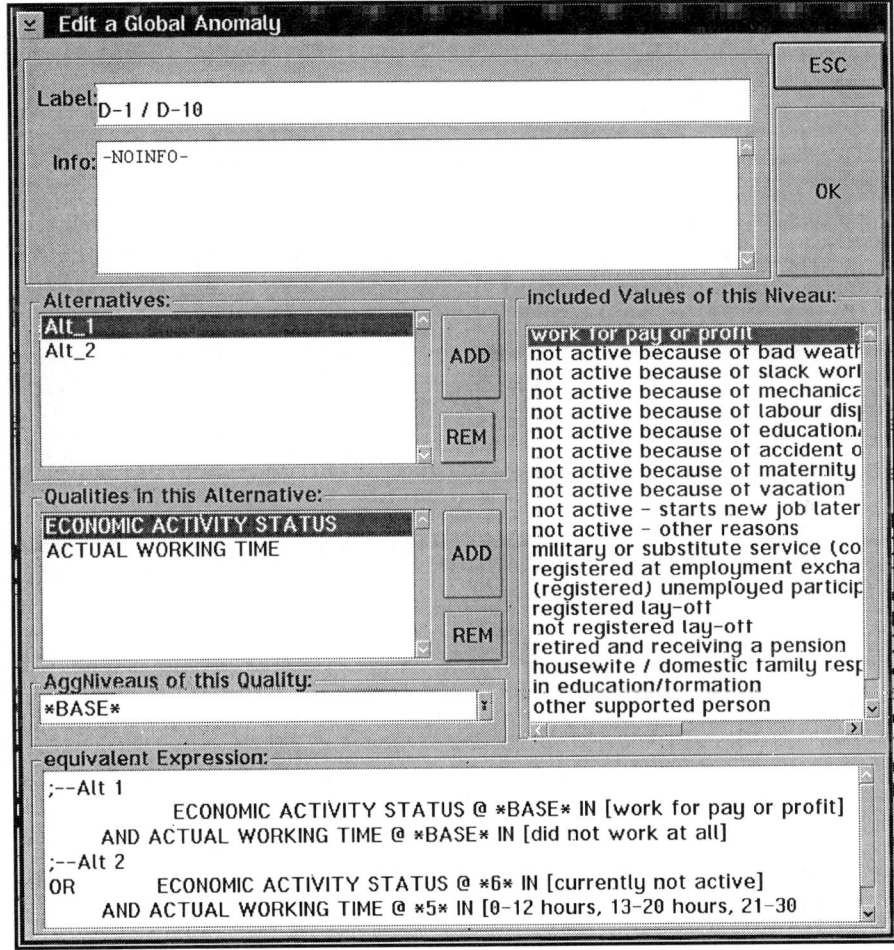

Figure 2.7–4: *GME* FRAMEdit Definition Screen

highlights a group of two BASVALues of QUality 'ECONOMIC ACTIVITY STATUS' in the right pane. Analogously, new or modified groupings are established simply by marking, directly in this pane, a subset of AGGBASE MODalities not already assigned to other AGGLEvel MODalities. Map groups are arranged accordingly.

It is always possible (and sensible) to attach Labels to the items exhibited in all *GME* panes although DÖS'chen discerns all items internally by hidden unique identifiers. Thus, the attached labels can be changed at will provided that these updates take place where the labels have been defined first. Likewise, the contents of 'Info:' fields can be modified or overwritten as convenient.

After a FRAME's QUalities have been arranged, the specification of FRAMEDITS follows next. Switching to the rightmost pane of the *GME* main screen, an existing

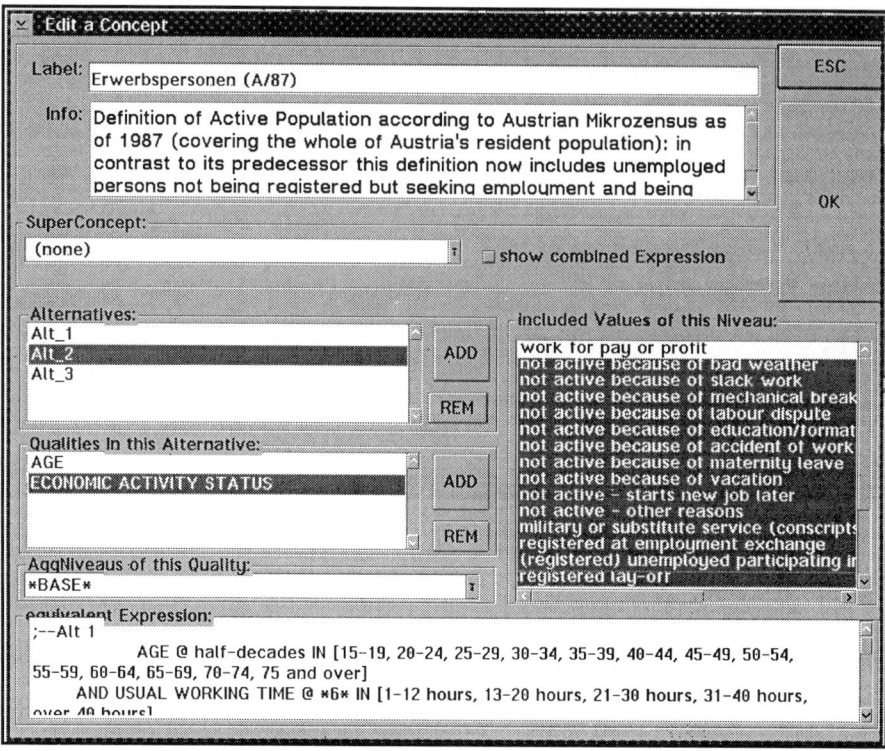

Figure 2.7–5: *GME* Domain Concept Definition Screen

FRAMEDit can be inspected (or edited) by marking it in the scroll window and clicking on the `Edit` field placed right to this pane. **Fig. 2.7–4** (previous page) shows a screen with FRAMEDit 'D-1 / D-10' marked; the sub-window laid over the main screen gives notice (in the '`Alternatives`' pane) of this FRAMEDit consisting of two REGCONs (cf. **Def. 2.4.1–2**; contrary to present METASTASYS definitions, DÖS'chen FRAMEDits formally are BLOCKCOLLections instead of single BLOCKs). The screen of **Fig. 2.7–4** tells furthermore that the first of this FRAMEDit's REGCONs is defined in terms of two QUalities, the first – and marked one – of which being restricted to MODality 'work for pay and profit'. FRAMEDits can be defined simply by repeatedly

- ADDing a PRIMCON (cf. **Def. 2.4.1–1**),
- ADDing QUalities the MODality domains of which become restricted, and
- choosing the MODalities of the AGGLEvel selected from the defined AGGLEvels listed (by AGGLAB) in the '`AggNiveaus of this Quality`' pane.

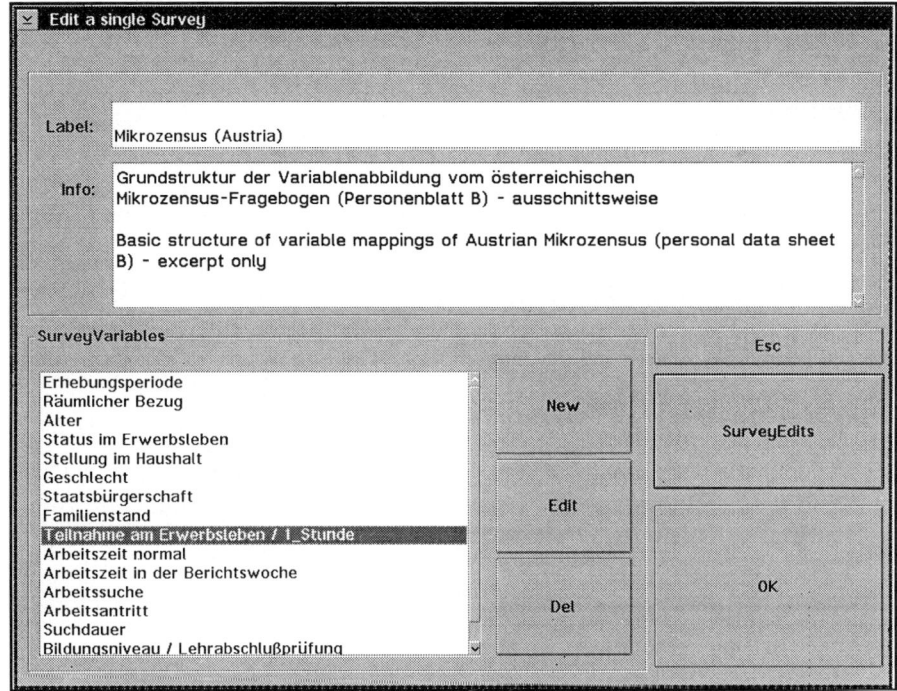

Figure 2.7–6: *GME* Survey Definition Screen

The compound expression resulting is indicated in the text pane placed at the bottom of the sub-window. Editing of CONDEFs is a bit more intricate; because of the hierarchical structure of selections updates always involve a two-stage *REMove–ADD* action; only the MODalities chosen for a restricting QUality may be replaced directly by modifying the markings in the 'Included Values of this Niveau' pane.

Completely analogous to FRAMEDits proceeds the specification of CONDEFs. For instance, as can be seen in **Fig. 2.7–5**, marking a CONDEF in the 'Concepts' pane of the *GME* main screen and clicking on the *Edit* field causes the 'Edit a Concept'-window to appear. In addition to an extra 'Info:' field, this screen is furnished with another pane indicating the 'SuperConcept', if any, the present CONDEF depends on. If a *New* CONDEF is created this pane lists all of the CONDEFs already defined. By ticking the small box labelled 'show combined expression' the bottom pane displays the plain CONDEF with the concept hierarchy flattened out instead of the incremental definition based on the indicated super-CONDEF.

Before survey mappings can be arranged formally, a general description of surveys and, in particular, survey variables must be provided. **Fig. 2.7–6** reproduces the screen displayed by *Edit*ing a marked survey in the *GME* 'Surveys' pane. In

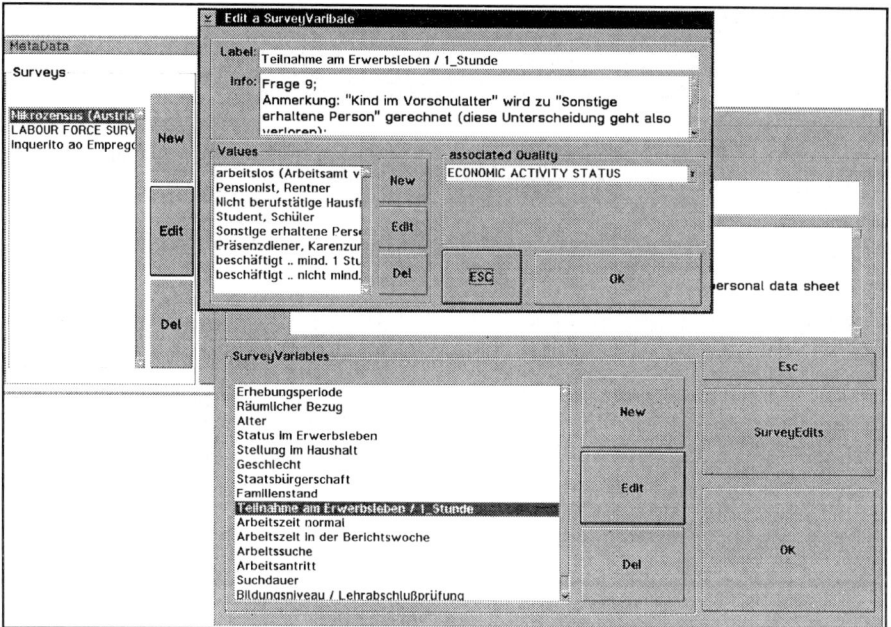

Figure 2.7–7: *GME* Survey Variable Definition Screen

addition to the standard 'Label:' and 'Info:' fields this screen contains a scroll window listing all variables included in the (possibly pre-processed) survey considered. As usual, *Edit*ing a marked variable (in this case, 'Teilnahme am Erwerbsleben / 1_Stunde') triggers a sub-window to appear which is reproduced in **Fig. 2.7–7**. In this window, the 'associated QUality' pane states the QUality the variable is associated with; the 'Values' pane lists all SOVALs of the considered variable. If a *New* survey variable is "created" the distinguished SOVALs have to be entered here while a QUality must be chosen by marking one of the not yet assigned FRAMEQUalities displayed in the 'associated Quality' pane.

The definite arrangement of mappings is now facilitated for each SOVAL in turn by clicking on the *New* or *Edit* switch fields, respectively, attached to the 'Values' pane; the association between a SOVAL and a BASVALue or a defined map group of the QUality the variable maps to is engendered (or updated) simply by choosing an item from the respective scroll windows. DÖS'chen will inform the user if a mapping definition dialogue is closed before all entered survey variables and the SOVALs of a survey variable, respectively, have been associated with FRAMEQUalities and MODalities or map groups.

.7 THE DÖS'CHEN PROTOTYPE

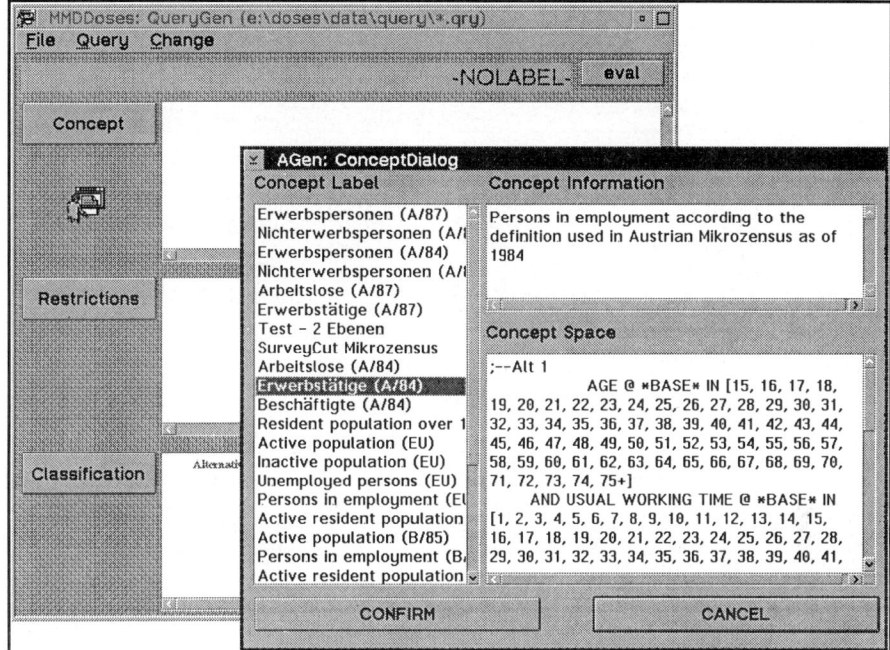

Figure 2.7–8: *AGen* Concept Selection Screen

The Aggregate Generator

Making use of the *GME* definitions, DÖS'chen's *AGen* is a simple end-user interface guiding the specification of REQUESTs within a rather restricted subset of METASPEL expressions. Basically, a DÖS'chen table request consists of three specification elements, viz.

- a CONDEF selected from the concept definitions arranged in *GME*'s 'Concepts' pane,
- an optional restriction applied to the selected CONDEF, and
- an optional cross-classification to become modulated upon the (restricted) CONDEF.

The first of these specification steps is illustrated in **Fig. 2.7–8** showing, in the background, the main *AGen* screen masked partly by the concept selection sub-window activated by clicking on the *Concept* switch field. In the sample screen, the CONDEF 'Erwerbstätige (A/87)' is marked in the left pane of this window; on the right-hand side panes both the 'Info:'-text attached to this CONDEF as well as the internal CONDEF representation are displayed. *CONFIRM*ing the choice returns control to *AGen*'s main screen again.

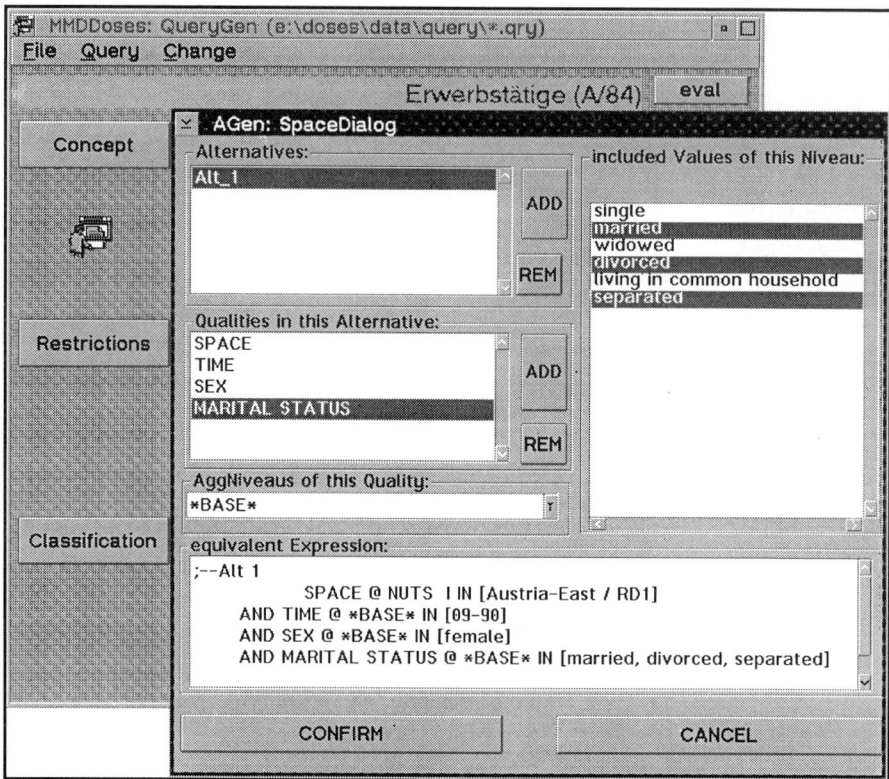

Figure 2.7–9: *AGen* Concept Restriction Screen

If desired, clicking on the `Restrictions` switch field activates the second specification step; the corresponding sub-window is shown in **Fig. 2.7–9**. Formally, restrictions are composed just like FRAMEDits or CONDEFs. Finally, by returning to the main screen and clicking on the `Classification` switch field QUalities may be ADDed one by one; **Fig. 2.7–10** exhibits a screen in the state of ADDing a cross-classification with respect to the FRAMEQuality 'STATUS IN EMPLOYMENT'. Once a QUality is chosen, the AGGLEVels defined for it in the *GME* are displayed in the upper right scroll window from which a suitable one must be selected by marking. The specified breakdowns are gathered and displayed in the lower right pane for verification. *CONFIRM*ing this choice and returning to the *AGen* main screen as shown in **Fig. 2.7–11** (overleaf), the table request is ready to be submitted by clicking on the `eval` switch field located in upper right corner of the *AGen* screen.

Although not shown explicitly, DÖS'chen is prepared to store table requests (note the `File` entry in the menu bar of *AGen*'s main screen) which can be retrieved and, if desired, modified before being submitted to evaluation.

.7 THE DÖS'CHEN PROTOTYPE

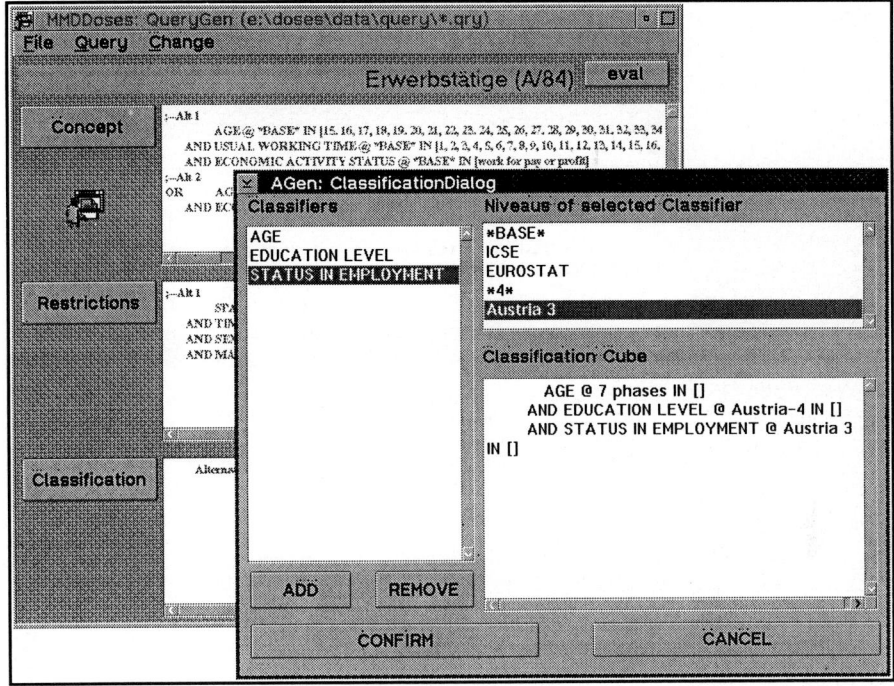

Figure 2.7–10: *AGen* Cross-Classification Screen

Implementation

DÖS'chen has been implemented in a SMALLTALK™ dialect close to the groundbreaking original language design [Goldberg and Robson, 1983] on a PS/2 platform. The application development kit of the SMALLTALK system used supported particularly a quick and effortless arrangement of DÖS'chen's GUI.

In contrast to the OBJECT types defined in METASTASYS (cf. Section 2.1) the software prototype uses considerably fewer constructs as basic data types (object classes); essentially, only five data structures are discerned in DÖS'chen, viz. VALues (BASVALues, in fact), AXes, XSETs (BLOCKs), BLOCKCOLLections, and BOXes (additionally, most of these data structures have been realized somewhat simplified and deviating slightly from METASTASYS definitions). A more detailed description of the final state of DÖS'chen development and an extended set of sample screenshots are comprised in a project report [Grossmann and Froeschl, 1994].

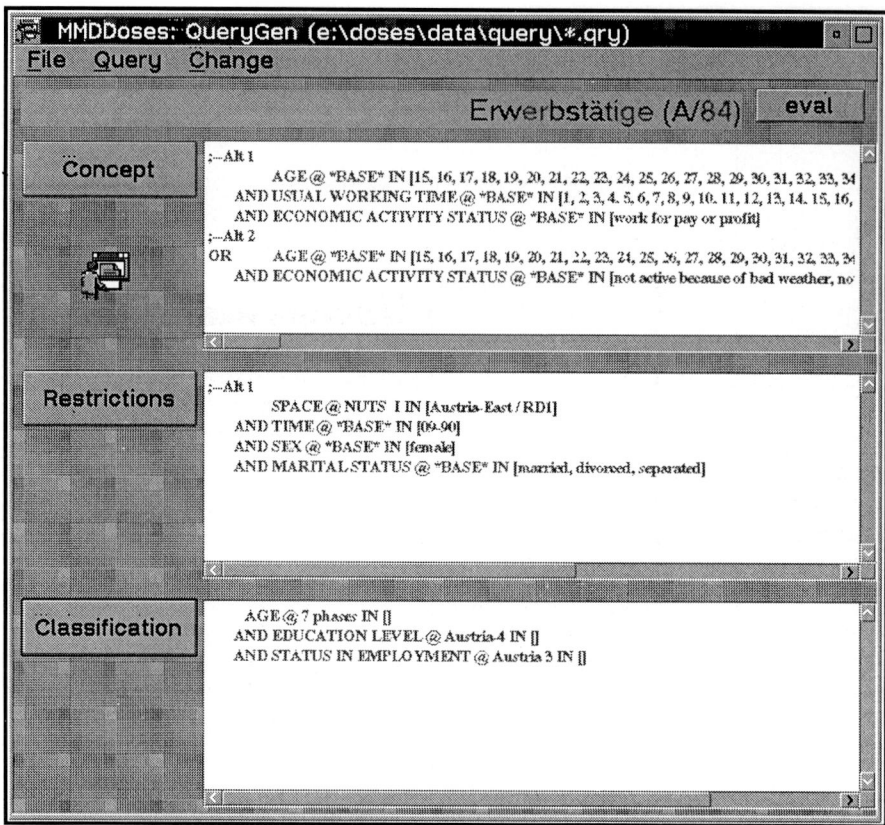

Figure 2.7–11: *AGen* REQUEST Screen

Chapter 3

Application Case I: Labour Force Statistics – Harmonizing Official Statistics Data

For apparent reasons, official statistics, with its huge data repositories compiled over long periods of time against changing backgrounds and comprising datasets interrelated in multifarious ways, provides an ideal testing site for the ideas and methodology developed in the METASTASYS framework for statistical data integration. As a particular application example, European *labour force statistics* is chosen in this chapter to illustrate in an exemplary way how data harmonization might be accomplished technically. There are several reasons suggesting the choice of this subject domain, not the least of which being the practical importance of employment statistics in political decision making against the background of a profound structural economic change taking place currently on a global scale transforming industrial societies into what is frequently termed the "global information society". Devising effective policies at European Union level obviously depends on integrated statistics to a large extent, calling into question the traditional means and approaches in fusing mainly national employment data into EUropean-wide data aggregates.

Semantic data integration, as it is attempted here, does not reduce to a mere technical affair, however. Quite on the contrary, utilizing this framework to the best of its capabilities depends crucially on the particular goals pursued; of course, these goals must be defined in subject-matter terms and agreed upon politically noting that the arrangement of harmonized terminological systems is subject to specific interests and reflects the differences in political power of the negotiating parties. This is all the more precarious as these interests, once inscribed into a technically "neutral" symbol processing system, tend to be lost out of sight in the general political discourse despite their subtle, though apparent, impact.

The chosen example of employment statistics also marks an inroad to the design space of advanced telematics applications within public administrations such as those of the European Union which in fact are in serious need of improving their informational infrastructures to keep up with the pace of social change and the pressing demand for far-sighted political strategies paving the way to an emerging economy, and society, supposedly quite different from the present ones.

3.1 Social Change and Economic Welfare

Within industrial societies, social orders are characterized to a large extent by the monetary exchange value of the work carried out by their citizens; essentially, the social status of the individual within the community is defined – in a society with strong money preference – in terms of the "productive" market-mediated contribution to the gross product. Plainly, this contribution equals the individuals' monetary incomes. Hence, being without employment means no less than lacking a productive income entailing, in turn, a substantial loss of social reputation (which, in particular, cannot be compensated by non-remunerated work or unemployment benefits granted more or less unconditionally). However, income from productive work is not sufficient to establish and maintain one's social status unless the gainful engagement is paid off on regular economic (labour) markets in cash or, to a lesser preferred degree, in kind.

In history, of course, this has not always been the case. From archaic subsistence economies up to those of the feudal societies of the Middle Ages and afterwards, there was no individual income or remuneration of labour; if those economies yielded any surplus to distribute at all, it was allocated "publicly" to the honour of the patriarch or court to represent power and grandeur (demonstrative extravagance), or – as a more mundane use of resources – to extend power, territory and hegemony by more or less militant activities. The sovereign's wealth, however, had little bearings on the material living conditions of plain people. In Western civilization, the economization of life – beginning long before the Ages of Renaissance and Enlightenment dawned – evolved from medieval urban marketplaces and particularly the increasing trade between towns (such as those of the Hanse), speeding up in a steady fashion right until today's hotly debated "structural change" determining economic thinking at the turn of the second millenium. It is the overall tendency of this development which the analysis of on-going socio-economic changes must focus upon in order to reveal its driving forces and determinants as a sound basis of understanding and thoughtful forecasting. Likewise, designing and introducing comprehensive information processing systems, such as supra-national socio-economic statistical database systems, requires an acute awareness of the general societal and cultural development either, lest implementation and utilization of these costly systems will prove inefficient and, perhaps, may affect the trace of a society's development adversely. Hence, this section tries to summarize very briefly the historical economic framework leading up to the currently on-going structural change in order to set the stage for any reasonable future large-scale socio-economic data integration and information processing system. More technically inclined readers, however, may skip this section safely, and resume reading with Section 3.2.

In Western societies, the key terms of this long-lasting continuous development are increase in *efficiency* and increase in *mobility* where the latter, for apparent reasons, has always been taken in tow by the former as its indispensable prerequisite. This increase in mobility can be traced along the evolution of traffic and telecommunication systems, the progressive division of labour, and – probably most important of all – the (re-)introduction of coined money and its early development to cash-less money transfers which emphasized the role of money as a *sign* of value rather independent of the material value of money itself [Foucault, 1966]: this certainly is the achievement of a literate society trusting in the meaning of abstract signs in a discourse mediated by script symbols. Indeed, the redefinition of wealth from the "inherent" value of physical property to the amount of goods it can be traded for is as much a leap in (intellectual) mobility as it is a fundamental stride towards modern market economy.

There is no doubt about the social effects exerted by this continuous structural change on the individual as well as on society as a whole. Sennet [1994], for instance, stresses the appearance of the university in the 13th century as the prototype of corporations privileged by the sovereign to give themselves statutes *including the right to change these statutes* autonomously as felt necessary by the corporations' representatives (at that time, universities were not necessarily restricted to or even involved at all in education). Today, it has become so much a matter of course that corporations exist independent of particular persons and may move from one place to another without compromising their integrity that it is hardly possible to think of it otherwise. Analogously, it have been the medieval trade fairs shaping the model of (regulated, by the way) markets giving rise to mercantilist model building and early starts in economic theory formation focusing predominantly on just pricing founded on ethical considerations. The development of markets as fields of economic competition coincides, no surprise, with the coming-up of institutions (such as the guilds) trying, with varying success though, to curb perilous competition by establishing prices and standardizing goods and services. In this milieu, *time* – getting ascribed the features of a commodity which can be offered and purchased – attains the characteristics of an economic category, too. Correspondingly, human labour enters a monetary exchange relation – workers more and more are paid off the time they are serving their masters, quite in contrast to the sole practice of former payments in terms of piece-wages [Sennet, 1994].

But, of course, it was the triumph of mechanics and its economic harnessing in mechanically empowered commodity production that led to the dramatic emergence of industrial society, shaping both technological progress and accompanying social upheavals of the late 18th and most of the 19th century. This era divided society, with historically unprecedented verve, into two rather unequal parts opposing each other fiercely – the capitalist and the underprivileged (proletarian) classes. While the former allocated private capital to implement and run "profitable" production plants for the sake of yielding maximal returns on investments, the unpropertied cads were thrown on wage work at deplorable conditions for bare survival. Factory production, by virtue of its competitive advantages originating from the drastic increases in production efficiency, quickly substituted much of traditional handicraft manufacturing and forced a large part of the resident population out of previously gainful work. Most factory work, though, required little skills or professional experience and, therefore, moulded the masses of dependent employees sharing essentially one major interest: cutting the economic power of the capitalists possessing the politically critical factor – production means – by establishing collective wage agreements and work time arrangements reducing both, the work hours per day (which, in early 19th century have risen up to 14 and more; cf. [Nowotny, 1993]) and the span of a person's working age (including the abolition of child work). By the end of the century, the basic pillars of social policy had been established, with the trade-unions negotiating wages and measures of social relevance (notably, emphasizing living standards and education) with industrial employers. Notwithstanding the persisting interest conflicts between employers and wage earners, the period up to the 1920ies brought forth relative prosperity for wider sections of the population; economic (and even more so: political) nationalism dominated "internal" labour disputes. This was the era of imperialism and particularly of national economies [Reich, 1993] sheltering their domestic markets from the vicissitudes of world trade by tariffs, thus opening good opportunities for high returns on investment in national key industries and particularly in high-volume manufacturing supplying active home markets (note that this has also been the time of the Schumpeterian entrepreneur introducing *innovative* products to consumer markets stimulating demand anew over and over again). As is well known, once those domestic markets became saturated the huge production capacities set up could no longer be utilized at profitable levels, and the rollback in sales and prices induced a finance crisis; fiercer competition spurred production efficiency, mainly by raising strongly the automation level of industrial production processes, unleashing a wave of (wage-work) unemployment, and the inevitable final shut down of many no longer profitable production plants left behind yet another flood of unemployment, historically unprecedented in size. Economic tensions, existential despair and patriotism created a politically explosive climate leading up, quite logically, to radical views splitting societies as well as strained foreign relations between nations.

Production Efficiency and Unemployment

Up until the Great Depression of the 1930ies economic analysis focused on the qualities and causes of wealth favouring static, ethical, or phenomenological approaches (cf. [Foucault, 1966] for a pertinent synopsis); the 19th century mainly added the notion of industrial production – "On the Economy of Machinery and Manufactures" was a then much respected book published by Charles Babbage in 1832 – and the concept of stepping up productivity by technical inventions and organization according to plan to the domain of (positivist) theory formation. Little consideration, however, was given to the *empirical* underpinnings of economic development (Babbage, better known as a pioneer in the development of mechanical computing engines long before the advent of electronic computing [Hyman, 1982; Eder *et al.*, 1994], probably being a notable exception in his vigorous pointing at the paramount value of succinct *data* in arguing) and the reasons of prosperity and proneness to crises; even less emphasis was laid on the social effects of economic and technical progress on the host of dependent and, hence, underprivileged wage workers which, somewhat cynically, figured as "production factor" in the well-worded statements of theorists. Against this intellectual and emotional background it comes as no surprise that contemporary governments and politicians were struck by the sudden impact of societal upheavals and, sometimes, turmoil of socially and economically deprived classes. Quite naturally, this political insensibility was paralleled by a lack of effective instruments to govern and control socio-economic development; contrarily, doubtful ideologies – like racial fascism – dominated more rational ways of political acting.

Despite several prolific theoretical developments in economic theory (for instance, the seminal work of Oskar Morgenstern) the widely established economic-political reasoning of the 1920ies/1930ies failed to catch up with the most pressing phenomena, mass unemployment in the first place, which called for vigorous governmental counteraction. It was John Maynard Keynes [1936] who argued conclusively that, under particular circumstances, a general economic equilibrium by no means contradicts the presence of substantial unemployment in a national economy and developed a pragmatic theory pointing out the public responsibilities to combat unemployment by creating publicly financed demands to stimulate consumption and, thus, help to re-integrate unemployed persons into the production process. Concurrently, the Roosevelt administration attempted to fight the Great Depression by its "New Deal" reshaping the landscape of the U.S. economy. Necessarily, the programmes and rational decisions of the New Deal administration had to be based on sound empirical models and reliable data what, consequently, fertilized immensely the development of *official statistics*; for instance, the compilation of GNPs (gross national products) and the arrangement of a system of social indicators are direct offspring of this development [Duncan and Shelton, 1992] (the emergence of labour force statistics is another offspring; cf. Section 3.2). Supposedly, the establishment of a well-organized statistical system describing the internal socio-economic structure of American society (contrasting sharply with the somewhat restricted focus of former economic statistics on external trade and payment balances) was of crucial advantage in wartime planning, and probably contributing significantly to the military and economic superiority of the USA in World War II; certainly, wartime logistics stressed even further the importance of statistical data for planning and decision making. It goes without saying, however, that statistics by itself could not create a significant contribution to economic recovery and growth; nevertheless, the diverging political and economic developments in the USA and Europe, respectively, are reflecting faithfully the rather opposing societal milieux and the different policy making approaches thriving. While the New Deal turned out moderately successful in reducing unemployment and attaining economic recovery, Europe slipped into disaster; in particular, the employment policy of Adolf Hitler's "Dritte Reich" was deliberately tailored for war – after a short period of successfully expanding demand by emitting governmentally certified, though economically fictitious, bills of exchange – as the German National Socialist government believed, nourished by an ideological conviction of the German race's natural superiority, that problems could only be solved ultimately by expansion to the East.

As a matter of fact, the devastation caused by war amount to a very effective, though tragic and agonizing, means of destroying economic surplus. Wars, even more than natural catastrophes, leave always behind an enormous demand for reconstruction and rehabilitation. Hence, economic development after World War II is characterized by a tremendous growth incited by large demands in capital and consumption goods. Furthermore, contrary to the period intervening both wars, the gained surplus was distributed, based on a broad political and social consensus, rather evenly across wide sections of society. The back-flow to Europe of a good deal of the capital having drifted away to the USA in the 1920ies in terms of lost rehabilitation loans, exerted an additionally stimulating upswing effect. As a consequence of distributing purchasing power in a well-considered way over all social strata, economic development could be maintained stable for quite a long time; real wages increased corresponding strongly with increases in labour productivity. Markets expanded smoothly, as new and "improved" products appeared (replacing their predecessors in *product life cycles*) and consumption attitudes changed continuously. The increasing labour productivity led to new patterns of time utilization; wage earners, as their real incomes rose, could substitute more and more work time by leisure time, the latter being used – to the mutual benefit of producers and employees – to consume the goods produced at work time [Nowotny, 1993]. Facing buoyant overall economic development and growth, public networks of social security – fostered by tax yields on the enormous profits gained by economics of scale in "national" industries and businesses [Reich, 1993] – were braided warranting a minimum of public welfare for everyone in trouble. Although women began to join the labour force to an extent worth mentioning, unemployment rates stayed exceptionally low, mostly because of the expansion of the tertiary sector and, particularly, the emergence of many new in-person services becoming provided on the market a good deal of which, in turn, was supplied by women (for instance, social services having long been the women's responsibility within families). Intellectually, this high-time of "Ford-ism" exhibited an uncurbed believe in a virtually boundless progress finally leading to a "victory of capitalism". Production efficiency and the power of market-oriented economy seemed to soar so dynamically that some observers began to feel uncomfortable with unbounded growth, emphasizing the ultimate limits of exploitable resources, calling into question the real worth of the abundance of consumable goods produced, and pointing at the ecological dimension, particularly the environmental pollution, incurred from raising production levels without restraint; all of which – in the eyes of quite a few – urged a thorough re-orientation of policies to prevent global disaster before long.

In the mid-1970ies, however, things irrevocably changed of themselves. National unemployment rates rose from below 2% to 7% and over; the share of income from dependent employment in GNPs declined again; although labour productivity still increased, real wages stagnated or even decreased. Economies did no longer go ahead as before, growth slackened down considerably. Having learned their Keynesian lesson, governments expanded public consumption, incurring formidable budgetary deficits, to step up employment again. In the longer run, this shifted profits to the finance capital and pushed the meanwhile traditional controversy between wage work and production capital somewhat into the political background.

Despite these massive endeavours and dedicated efforts, unfortunately, a high level of unemployment has prevailed ever since and, remarkably enough, unemployment rates uncoupled more and more from overall business cycles [EC, 1994]: as in phases of depression unemployment increases sharply, it remains at ever higher levels in periods of general economic recovery and growth. In 1993/94, summing up the persons out of employment in the Member States of the European Union gives the impressive figure of around about 18 million, more than a half of which without a job for one year at least [Blanpain and Sadowski, 1994], a tendency recognized for quite some time by now [EUROSTAT, 1988]. In some countries, like Italy and Ireland, even about two thirds of the unemployed persons are stricken with long-term unemployment [EUROSTAT, 1993b]. Another source of serious concern is the sharply rising level of youth unemployment affecting particularly persons leaving the education system and entering the job market with discouraging vocational outlooks. From an European point of view, the economic situation is even more precarious as the high rate of unemployment – about 10.4% according to [OECD, 1994] – is accompanied by a relatively low rate of *total* employment (measured in terms of the population of working age actually in work) – now below 60% [EC, 1994] – and a sneaking slowdown of business dynamics and growth compared to major competitors on world markets such as the USA and Japan (U.S. unemployment rate 1994: 6.4%, Japan 2.9%; employment rates above 70%). As there is no sign indicating a slackening or even inversion of this perilous development, measures of job creation and boosting global competitiveness – though certainly reasonable and necessary – will probably not lead back to full employment on regular labour markets; instead, much speaks in favour of a reorientation of policies, aiming at new societal models of (life-)time distribution – paid public work-time vs. unpaid private leisure time – by, for instance, reducing individual work-time or by extending "tertiary" non-profit labour markets [Rifkin, 1995] (say, in social or environmental domains), or by mixtures of like measures, in order to maintain social stability and establish a sustainable level of post-industrial societies' internal cohesion (see below).

Structural Change and Global Economy

Substantial unemployment and, in particular, long-term and youth unemployment endanger an economy's welfare and, moreover, threaten social stability by eroding the basic societal consensus being the political foundation of developed democracies. Since social security, to varying degrees, is financed by statutory charges on wages and salaries, high levels of unemployment heavily strain national budgets, narrowing down the states' political and financial rooms to move. Nevertheless, as has been argued compellingly by Reich [1993], the cherished picture of national economies comprising all of a nation's economic actors in a "common boat" strongly tying together each other's well-being is but a historical reminiscence, and certainly of no use anymore in responsible policy making. In spite of the pressing lack of compelling political solutions, economists and policy makers – facing the apparent inefficacy of national policies – must take cognizance of the bare fact that *structural* unemployment can no longer be banished by isolated national economic measures. Instead, the tremendous challenge of economic change – a further giant leap in efficiency and mobility, in fact – on a really global scale calls for *globally co-ordinated* and *integrated* approaches.

Prior to an effective employment policy and the implementation of well-targeted measures to improve economic dynamics and growth, something must be learned about the determinants of this on-going structural change. A relapse into traditional patterns of economic analysis (for which Reich [1993] coined the phrase of "vestigial thought"), however, will certainly not bear any promising insights and recipes; rather, fresh thinking, free of historical drag, is required once again.

What, now, are the characteristics of this structural change reshaping the industrial to the *information* society ?

First and foremost, national economies are superseded vigorously by the one and only *global economy*; national markets lose their importance as competition transgresses national borders. Many businesses, and particularly those creating added value to a significant extent, today operate on a global scale, linking organization and logistics by world-spanning tele-communication networks. Apparently, supranational operation opens a door to unprecedented flexibility in exploiting the economic, political, and social advantages of alternative locations, thus gaining the cutting edge in international competitiveness. Unlike the post-war economy where the economic success of a nation's largest businesses and this nation's welfare were closely tied together (Reich [1993] cites the example of General Motors which, in 1955, single-handedly created a share of about 3% of USA's GNP), the prosperity of multi-national businesses and nations no longer simply coincide. Domestic governmental policies have turned into plain decision parameters in the management of multi-national corporations (MNCs) which by no means determine their strategies as before. All of this, of course, spurs the dynamics of competition on both national and international markets even further.

Increasing competition, in turn, creates an economic pressure to streamline the internal operation and raise the efficiency of businesses. As competition in products becomes superseded more and more by competition in *capabilities*, businesses tend to restrict the scope of activity to a couple of indispensable operative core functions, flattening management hierarchies and "sourcing out" everything else. Reducing production depth lowers capital costs and raises flexibility, just as stock-free "just-in-time" production regimes and production on customer demand by fully automated flexible manufacturing systems and production lines do. Business process re-engineering cuts down management overheads, and staff re-sizing is facilitated by a massive introduction of electronic data processing and (digital) tele-communication in office automation as well as production planning and control.

These measures, as inevitable and rational they might be from a micro-economic point of view, nevertheless confront society with an aggravating problem, viz. the under-utilization of human resources or, in plainer words, structural unemployment. For one thing, as product markets globalize so does the labour market; in particular, low wage countries (such as the developing countries but also transition countries) suck off a large deal of unskilled work from industrialized zones. Although production then often happens to be far off final consumer markets, relatively low transportation costs hardly neutralize the remarkable reductions in production costs gained by transferring production to locations outside the industrialized hemisphere.

Hence, to regain a higher level of employment, new challenges must be met macro-economically. One pillar of an effective employment policy, particularly in Europe, are investments into a vigorously improved education system stressing vocational training and versatility particularly of young people. Moreover, more personal mobility is called for in view of both, new vocational topologies and chronologies. As to the former, there increasingly takes place a profound change in the "classical" distribution of roles between employers and dependent employees – the *lean management* model originating from the Japanese automobile industry [Rifkin, 1995] – blurring the once clear distinction between doing work as ordered and responsible, circumspect cooperation in the common interest of employee and employer. These new topologies also include new spatial work arrangements such as home (tele-) work or other decentralized modes of work (for instance, tele-cooperation) facilitated by tele-communication links. As to the new chronologies, an unprecedented variety of work time arrangements has already been introduced with the apparent background to better utilize invested capital and adapt work force to market situations (for instance, at the German car manufacturer *Volkswagen* the negotiations to safeguard jobs while reducing wage expenditures by about 15% led to some 140 different work time models [Blanpain and Sadowski, 1994]). In addition to a significant increase in part-time work, evening, night and weekend work, job-sharing models etc., ever more persons' work careers will be characterized by more frequent job changes, changes between jobs requiring different qualifications, more and longer breaks in employment, and life-long learning to cope with the pace of innovation and re-structuring. Expectedly, a growing number of professional occupations will become obsolete at all while others will change their profiles requiring a host of new qualifications such as foreign languages, computer literacy, social competence, analytical thinking, and – last but not the least – effective information handling.

At the macro-economic level, these changes become apparent by a re-weighting of shares of main economic sectors of activity. Referring to the USA, Reich [1993] summarizes this partitioning roughly as 25% *routine production work*, with a marked downward tendency, 30% *in-person services*, with a steeply increasing tendency, and about 20% *symbolic-analytic services* (compared to some estimated 8% in 1950); the final 25% comprising farmers, miners, extractors of natural resources and the sectors sheltered from market competition like government employees (including school teachers), employees in regulated industries, and government-financed workers. In his non-standard classification, Reich readily takes account of effects of structural change, having noticed that the traditional

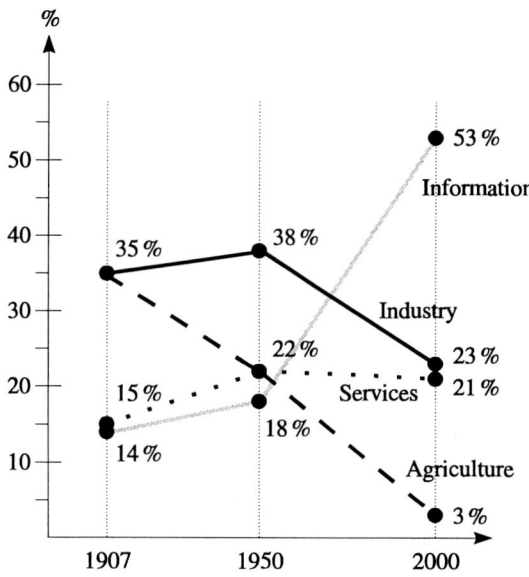

Figure 3.1: Development and Forecast of Economic Sector Shares in Germany

categories – subdividing economic activity into agriculture, industry, and services – no longer fit the emergent job profiles combining different characteristics of former job patterns cross-sectional. In Reich's terms, symbolic-analytical services – the job category including *"all the problem solving, problem-identifying, and strategic brokering activities"* – is the branch with highest growth prospects. As these services can be traded world-wide, they suffer from global competition (just like routine production work does) but *"they do not enter world commerce as standardized things. Traded instead are the manipulations of symbols – data, words, oral and visual representations."* Correspondingly, this quickly expanding information business sector will – cf. **Fig. 3.1** with respect to a German forecast [Polke, 1995] – be peopled with exceptionally well-trained scientists, engineers and researchers occupying a matrix of fanciful jobs as summarized in the synopsis of **Tab. 3.1** ([Reich, 1993]; combining the terms of each of the table's column gives a symbolic-analyst job title, each title may be prefixed additionally with "Senior", "Chief", or "Principal").

Empirically supported, Reich concludes compellingly that engaging heavily in symbolic-analytical services is in fact the sole option of industrialized countries to withstand global competition and maintain Western living standards, social security, and public welfare of citizens on condition of a successful surplus distribution among all sections of society. However, this option cannot be considered as a miracle cure to remedy unemployment (not to speak of regaining full employment) as this depends on huge investments in human capital and its effective translation into marketable goods and services.

Table 3.1: Synopsis of Symbolic-Analysis Job Titles

Communications	Management	Engineer
Systems	Planning	Director
Financial	Process	Designer
Creative	Development	Coordinator
Project	Strategy	Consultant
Business	Policy	Manager
Resource	Application	Adviser
Product	Research	Planner

Quite on the contrary, what can be observed in practice is a deepening separation between those being competitive, self-confident enough to accept the challenge, and apt to participate in information society on the one hand, and the less qualified and proficient suffering from social descent in the wake of under-qualification, redundancy, and marginalization on the other hand. There in fact is an enormous tendency to partition society into those "inside" and those "outside" employment; while the wealthy "inside" party faces quite rosy prospects, the unfortunate "outside" party is going to be left more and more behind. The big social issue therefore is, to which degree the economically strong part of the population, competing successfully on world markets, will still feel responsible for giving support to their less lucky compatriots [Reich, 1993]. In any case, contrary to previous historical experience national trade-unions will have little real impact on this development, as their influence decays hand in hand with the efficacy of any national policy [Blanpain and Sadowski, 1994].

In this particular situation, the social dimension of labour has to be placed in the foreground again [Rifkin, 1995]. Perhaps the historically grown concept of industrial wage work as the primary mode of gainful employment must be reconsidered thoroughly; a reassessment of the societal exchange relation between public (remunerated) and private (non-remunerated) *time* must define anew the distribution of "valuable" (and, hence, paid off) work [Nowotny, 1993] over the population in working age. This, of course, could include a relief of labour incomes from statutory charges financing social security. Since, apparently, remunerated work for everyone willing to work for pay or profit might never be attainable again, political creativity and social "re-engineering" is called for. For instance, Blanpain and Sadowski [1994] plead for a new social contract to safeguard the public welfare of all citizens for the sake of stability and social compromise; they particularly warn against ill-considered deregulation of labour markets in view of short-term advantages which, in the longer run, might deteriorate competitiveness rather than improve it and, as a consequence, neither salvage employment nor national welfare.

The European Union, having elected employment policy as the center-piece of its overall strategy, has set out to create jobs, stimulate growth and economic dynamics by establishing a single, homogeneous internal market supported by a variety of flanking measures (in particular, investments in traffic and tele-communication infrastructures), promote mobility, and increase human capital and innovative resources by launching educational and research programmes of considerable financial volumes [EC, 1994]. In the meantime, action must resort to more specific measures such as, for instance, encouraging the continued employment of persons prone to marginalization (because of advanced age or insufficient qualification) and promoting the mobility of young persons entering the labour market to speed up the adaptation to structural change [Wagner, 1993]. However, despite all political efforts of counteraction, automation and productivity increases may outpace in the longer run all employment promotion programmes, growth policies, and market deregulation [Rifkin, 1995]. After all, cognizance must be taken of the bare fact that unemployment is here to stay.

Policy Making and the Role of Data Integration

From these considerations it must have become clear that extensive socio-economic data is an indispensable prerequisite to cope rationally and responsible with the political and societal problems resulting from the current phase of structural change. On the one hand, data is badly needed to define targeted policies, on the other hand, data of similar type is used to monitor the efficacy of implemented policies. Moreover, established data production systems should, ideally, alert decision makers of emerging trends and developments causing concern and, possibly, calling for undelayed counteraction.

In this context, *employment statistics* certainly is of particular value and prominence; however, a closer look uncovers immediately that tracing the structure of both employment and unemployment is far more difficult than compiling plain demographic aggregates. In particular, traditional measuring systems tend to fall short of providing just the figures necessary to define the concepts of political focus and best operative leverage. After all, structural change and its social and political consequences depend, to a large extent, on relative points of view and, of course, measuring systems in social and economic statistics have always been reflecting such particular views (they for sure will continue to do so, anyway). As to past data produced by established statistical systems, this must be accepted but, in general, it is worthwhile to revise data production procedures in view of these emerging tasks and necessities.

It must also have become clear that – mirroring the general "globalization" of problems – the derivation of global answers is in need of a corresponding *global data context*. Apparently, national data left unrelated to each other will be of little use in devising supra-national policies. For instance, the increasing mobility and global operation of MNCs more and more renders meaningless the traditional ways of measuring economies (such as GNPs). As the value of statistical information is becoming essentially a function of combined data (national or otherwise), *data integration* – to be understood as a combination of data in subject-matter, or semantic, terms – attains a pivotal role in feeding the political decision making stage with *harmonized* socio-economic data. Furthermore, data will be appreciated the higher, the more timely and fine-grained it is delivered. This goal, of course, is achievable only if the most advanced techniques of information processing are applied to statistical data management as it has been suggested, by the way, as one of the measures mentioned in the White Paper of the European Commission [EC, 1994] to rationalize (sic!) official administrations by means of telematics.

3.2 Measuring Employment

In a wide sense, the measurement of employment belongs to the field of sociometry. Since, demographically, individual employment is the predominant means of earning a living in Western industrialized countries, employment statistics centers around *persons* as the main carriers of empirical characteristics. Of course, from a societal point of view emphasizing the macro-economic perspective, it is also of paramount interest to record parameters providing descriptive measures of overall employment structure and distribution. Taking its bearings of population statistics – termed "political arithmetics" in Halley's days ([Halley, 1693] published an account of mortality rates computed from the parish-registers of the town of Breslaw) – as well as the continental school of "state description" founded by Herman Conring [Sint, 1994] and promoted particularly by the renowned Gottfried Achenwall (1719-1792), employment statistics saw its infant appearance in surveys of *paid-off* work conducted in the second half of the 19th century. In fact, the U.S. Bureau of the Census made first inquiries about American jobs already in 1820, and devised a systematic way of job categorizing in 1870 [Reich, 1993]. Today, statistics of employment and labour has become a major pillar in making effective governmental policy; by now it is, in fact, hardly possible to imagine modern social economies without this firm foundation of democratic decision making on a well-developed system of social and economic statistics. Decisively stimulated by Roosevelt's New Deal (1933) fighting the Great Depression by implementing a multitude of novel programs for recovery and reform, each of which had to be based on reliable, timely, and accurate statistical figures for appropriate targeting, full-scale censuses became increasingly substituted by probability sampling methods, drawing particularly on Jerzy Neyman's seminal [1934] work on stratified finite population sampling (note that R.A. Fisher's "Design of Experiments" [1935] presupposes *infinite* populations throughout). Despite initial reservations about the use of sampling methods in official statistics (which, essentially, raised the fear of discrediting the profession by undue precision losses), it was particularly the ground-breaking work of Morris Hansen and Willam Hurwitz [1943] at the U.S. Bureau of the Census in proving both theoretically and practically the superiority of probability sampling in (un-) employment statistics that paved the way for subsequent world-wide establishing of Labour Force surveys [Hansen, 1987]. Using a rather simple and straightforward sampling procedure in the 1937 Enumerative Check Census of Unemployment not only revealed major flaws in the full Unemployment Census's outcomes – in particular, the Check Census sample estimated about 11 million persons unemployed in contrast to the 7.8 million persons who had registered [Duncan and Shelton, 1992] – but also gave rather detailed accounts of sampling errors highlighting both, cost-effectiveness of survey sampling and the role of sample design as an important factor of *quality control* in statistical data production. Based on this encouraging experience, preparatory work starting around 1940 and moulding Neyman's conception into a carefully designed multi-stage sampling scheme eventually led up to the Monthly Report on the Labor Force published by the U.S. Bureau of the Census from 1942 on using basically the methodology of Hansen and Hurwitz. Another important result of the methodological research carried on at the Census Bureau was the suggestion to choose *households* (instead of individual persons) as units of random sampling in employment surveys (likewise, later on Hansen and Hurwitz analyzed sources of bias caused by interviewer enumeration – the Enumerator Variance Study of the 1950s – and concluded compellingly that self-enumeration of respondents could in fact avoid bias introduced by interviewer enumeration which gave rise to a revision of the whole census procedure accordingly). Realizing the enormous gains in flexibility at comparatively low cost led to a quick adoption of sampling techniques in diverse areas of official statistics (at least in the USA [Duncan and Shelton, 1992]) such as family expenditure surveys or, somewhat later, the establishment of various social indicators in connection with the "Great Society" programs instituted in the 1960s.

Traditionally, employment statistics attempts to capture an economy's actual and potential work force as well as the persons of the work force not in employment; hence, three ratios are used mainly, viz.

- the *activity rate* representing the labour force as percentage of the population of working age;
- the *employment/population ratio* representing the persons in employment as a percentage of the population of working age; and
- the *unemployment rate* representing the unemployed persons as a percentage of the labour force.

Apparently, these ratios are not independent; the employment/population ratio (r) equals the product of activity rate (a) and the employment rate which is complementary to the unemployment rate (u): $r = a(1-u)$.

Looking deceptively simple, the difficulty really comes in defining the concepts underlying these ratios. Particularly the definition of unemployment rates has turned out to be a highly controversial issue because of its important role in political discourse and practical policy making. Broadly speaking, unemployment rates are used as *descriptive* indicators of the social and economic indigence of a (national) economy's persons in unemployment, of macro-economic cycles of growth and employment, of the overall degree of labour force utilization, and of trends in the imbalances on the labour market reflecting the adaptability of economies to forces of structural change [Biffl, 1994]; as *analytical* indicators they are used to support economic model building aiming at explanations of labour market dynamics and the dependency of employment and unemployment on market conditions and structural changes; finally, as *programmatic* indicators unemployment rates are used to monitor the effects of policies – such as job creation endeavours or vocational training programmes – on the (national) labour markets. In view of these rather diverging information demands, it hardly comes as a surprise that there cannot be "the" unemployment rate; nevertheless, for the sake of comparisons with respect to both, space and time, a convergent definition is necessary as to what should be actually measured as unemployment in which context. Hence, it has become established practice to use several unemployment rates concurrently. For instance, the U.S. Bureau of Labor Statistics maintains a range of seven indicators, *U1* through *U7*, since the mid-1970s. The apparent advantage of such a *system* of related unemployment rates – as opposed to unrelated indicators developed purposively for specific tasks or by different institutions – is its consistent use in a variety of contexts (this system approach has already proven worthwhile in measuring different quantities of money). With respect to international comparisons of employment it has been the International Labour Office (ILO) of the UN undertaking a long-lasting and still continuing effort to arrive at broadly accepted definitions of employment and labour force admitting at least rough comparisons of socio-economic development on a global scale. Despite its sound intentions, however, the ILO definitions are too vague for effective policy making in specific national or supra-national economic frameworks whence most governments and administrations have decided to use two (at least) types of unemployment measures, viz. one for international comparison and another one for internal labour market and social policy. In fact, even international comparisons are still compiled using different definitions such that, presently, there are unemployment rates according to OECD and EUROSTAT (the Statistical Office of the European Union) definitions, respectively, in addition to the ILO-based rates [Bartunek, 1994]; even though both, OECD and EUROSTAT, are starting from ILO definitions of employment and unemployment, they come up with slightly different outcomes (estimates) as the modifications to ILO concept definitions have been introduced to better reflect the specific socio-economic structure of the nations represented by the respective organizations.

In addition to the by now "classical" indicators of economic activity and employment and augmenting more traditional branches of economic statistics such as the statistics of labour cost and wages, employment statistics is acquiring a more and more broadening role of information supply encompassing, for example, indicators measuring the *duration* of unemployment (concerning particularly the persistently high level of long-term unemployment in many European countries [EUROSTAT, 1988]) or eliciting the *structure of employment* (accounting for the tremendous structural changes taking place in the labour market world-wide; cf. Section 3.1). After a phase of economic prosperity characterized by a shortage in labour supply in the post-war era, social welfare and democratic stability of Western political systems increasingly depends on a successful re-deployment of labour among society; for this reason the importance of employment statistics as an indispensable quantitative foundation of socio-economic decision making will rise definitely. Moreover, as national economies give way more and more to larger economic and, hence, political units – such as the European Union or the NAFTA – due to the on-going globalization of business and trade, a heavy emphasis is placed on the *integration* of socio-economic data across national borders. In case of employment and labour force statistics this certainly calls for vigorous attempts to achieve better comparability of pivotal economic indicators such as activity and unemployment rates both cross-sectional and longitudinal. Somewhat paradoxically, however, the improvements accomplished hitherto in cross-sectional comparability incidentally worsen longitudinal comparability rather than improve it because of discontinuities introduced either involuntarily or inevitably in the concepts measured (changes in basic definitions for the sake of harmonization).

3.2.1 International Development of Labour Force Statistics

Commencing in 1923, the International Labour Office (ILO) has arranged 15 international conferences up till now (the last taking place 1993) dedicated to various topics of employment statistics. During these seven decades of steady progress, the main concerns of the ILO have been the development and continuous update of a couple of fundamental concepts defining labour force, employment, and unemployment acceptable and applicable at a global level in order to ensure comparisons of national economies even if they are so divergent as those of industrialized and developing countries. In order to facilitate comparisons sensibly, the basic definitions must, of course, be accompanied by a set of internationally agreed upon classifications to be used for cross-sectional breakdowns. Despite several weaknesses, the ILO definitions of concepts and classifications have been accepted ubiquitously and provide since long the de facto standard of employment statistics in that even deviating approaches often take the ILO standard as their point of departure.

The first thorough attempt to find a workable approach to measuring unemployment and work-places, based on accounts of demographic population structures, has been undertaken at the Second International Conference of Labour Statisticians held in 1925. Essentially, this Conference adopted a definition of the "economically active population" and suggested to measure unemployment using data of unemployment insurance agencies, employment offices, population censuses or specific censuses or sample surveys; this proposal defining the "gainfully employed" was in use up to World War II. In 1938 the Committee of Statistical Experts of the League of Nations made a further effort to improve comparability of census data by giving the following definition of the *gainfully occupied* population [ILO, 1976]:

> *"For the purpose of international classification, any occupation for which the person engaged therein is remunerated, directly or indirectly, in cash or in kind – i.e. any principal remunerated occupation or any secondary occu-*

pation which is the sole remunerated occupation of the person concerned – is to be considered as a gainful occupation. Housework done by members of a family in their own homes is not included in that description, but work done by members of a family in helping the head of the family in his occupation is so included, even though only indirectly remunerated. The occupation of persons working in labour camps or other similar institutions or in unemployment relief projects is to be considered as a gainful occupation.

The particulars given should be based, generally speaking, on the occupation at the moment of the census. A person who has recently exercised a gainful occupation is to be considered as still engaged in that occupation even though, by reason of sickness, injury, vacation or inability to obtain work, he may, at the time of the census, be temporarily not working.

Young persons of working age and not at school, who have never actually exercised a gainful occupation, are not to be treated as part of the gainfully occupied population, even though they may be seeking work and consequently included in statistics of unemployment. It is, however, desirable that censuses should be taken that the number of young persons in this situation can be ascertained."

During and particularly after World War II a considerable expansion of employment statistics can be observed in most developed countries; especially the introduction of sample surveys helped to extend the comprehensiveness of topics covered. Stimulated first by wartime economics and its increased demand in socio-economic planning (coinciding with the proliferation of methods of operations research), it were the specific requirements and dynamic expansion of post-war economy driving the further development of employment statistics (and social statistics in general). The gap opening between labour supply and increasing labour demand – especially of "creative manpower" – called for refined statistical methods to forecast market behaviour and arrange policies suitably. It is the period of the 1950s and 1960s which brought the conception of social and economic welfare for everyone by redistributing growth and purchasing power more evenly among society than ever before, basing policies deliberately on rational planning. In this climate, official statistics attained a crucial role in supplying data for policy preparation as well as monitoring on a rather large scale, assuming the position of a mediator between decision making and societal reality. With respect to employment statistics, the concept of *labour force* had appeared by this time, and in 1947 the Sixth International Conference of Labour Statisticians reacted to this conceptual change by taking up anew the task of defining labour force, employment, and unemployment mainly on the basis of the economic activity of each individual during a specific period, thus departing from the previous concept of gainful work which did not tie employment as strictly to activity in a short time period only. Based on a proposal of the ILO the Eighth International Conference of Labour Statisticians eventually adopted an influencing resolution concerning statistics of the labour force, employment and unemployment in 1954 [ILO, 1976]:

"Definition of labour force

4. The civilian labour force consists of all civilians who fulfil the requirements for inclusion among the employed or the unemployed, as defined in paragraphs 6 and 7 below.

5. The total labour force is the sum of the civilian labour force and the armed forces.

Definition of employment

6. (1) Persons in employment consist of all persons above a specified age in the following categories:
   (a) at work; persons who performed some work for pay and profit during a specified brief period, either one week or one day;
   (b) with a job but not at work; persons who, having already worked in their present job, were temporarily absent during the specified period because of illness or injury, industrial dispute, vacation or leave of absence, absence without leave, or temporary disorganisation of work due to such reasons as bad weather or mechanical breakdown.

(2) Employers and workers on own account should be included among the employed and may be classified as "at work" or "not at work" on the same basis as other employed persons.

(3) Unpaid family workers currently assisting in the operation of a business or farm are considered as employed if they worked for at least one-third of the normal working time during the specified period.

(4) The following categories of persons are not considered as employed:
   (a) workers who during the specified period were on temporary or indefinite lay-off without pay;
   (b) persons without jobs or businesses or farms who had arranged to start a new job or business or farm at a date subsequent to the period of reference;
   (c) unpaid members of the family who worked for less than one-third of the normal working time during the specified period in a family business or farm.

Definition of unemployment

7. (1) Persons in unemployment consist of all persons above a specified age who, on the specified day or for a specified week, were in the following categories:
   (a) workers available for employment whose contract of employment had been terminated or temporarily suspended and who were without a job and seeking work for pay and profit;

(b) *persons who were available for work (except for minor illness) during the specified period and were seeking work for pay and profit, who were never previously employed or whose most recent status was other than employee (i.e. former employers, etc.) or who had been in retirement;*
(c) *persons without a job and currently available for work who had made arrangements to start a new job at a date subsequent to the specified period;*
(d) *persons on temporary or indefinite lay-off without pay.*

(2) *The following categories of persons are not considered to be unemployed:*
(a) *persons intending to establish their own business or farm, but who had not yet arranged to do so, who were not seeking work for pay or profit;*
(b) *former unpaid family workers not at work and not seeking work for pay or profit."*

In 1966, these definitions were amended by adding an age bound excluding all persons below 15 years from the labour force. Furthermore, the concept of reference week superseded the rather unspecific notion of "brief period" and it was also recommended to record information, for each surveyed person, about the representativeness of reference week data. These definitions have been reviewed thoroughly in 1982 (Thirteenth International Conference of Labour Statisticians), though leaving them virtually unchanged (cf. Subsection 3.2.2).

The 1954 session introduced also a standardized definition of *underemployment* discerning particularly visible and invisible underemployment. Contrary to the continued importance of these concepts in developing countries, its applicability in industrialized countries decreases steadily because of the emerging new types of (part-time) work contracts, home-work, collective reduction of regular work time, etc. which – in their tendency – give ground to a gradual change of the labour market blurring the classical distinction between employers, self-employers/own-account workers, and dependent employees. Although many persons no longer work the regular hours per week as regulated in collective arrangements, they cannot be classified as underemployed since, typically, labour productivity is rather high whereas the typical indicator of underemployment is a low productivity of labour. Quite on the contrary, it becomes more and more important to spot sources and extent of under-utilization of manpower since, apparently, the share of the population in industrialized countries being discouraged to seek work anymore or working actually less than desired is rapidly growing, thus giving serious cause of concern about the on-going segregation of society into $^2/_3$ of wealthy and $^1/_3$ of poor, less-educated, and underprivileged people drifting to or already living below the poverty line (Reich [1993], referring to the USA, goes even further speaking of the "fortunate fifth"). For obvious reasons, any assessment of under-utilization of manpower presupposes empirical data about the preferred type of work as well as the quantity of time persons would like to work under different (better) economic conditions compared to their current type of work and the quantity of time persons are working actually. Traditionally, only the work hours (hours worked either actually in the reference period or regularly in a period of same length as suggested by the ILO already as early as 1923) are registered in employment surveys; though being a natural starting point, this information is by no means sufficient to draw a satisfactory picture of time usage and under-utilization of manpower. Remarkably, to date the practice of labour force surveys has failed to adapt to these emerging information demands, thus certainly hampering effective political counteraction.

The Development of Nomenclatures

An essential precondition to achieve real comparability of supra-national statistics is the use of generally approved classification schemes. In particular, structural differences in labour force and unemployment can be assessed sensibly only on condition that all employment statistics are based on labour force surveys supporting the very same set of cross-sectional breakdowns. With respect to labour force statistics, three classification schemes have assumed a pivotal status, viz.

- the classification of economic activity;
- the classification of occupations, and
- the classification of status in employment (professional status).

As with concept definitions, the ILO has taken a leading role in arriving at standardized and internationally accepted classifications. In fact, it was the main topic of the First International Conference of Labour Statisticians in 1923 to develop and recommend for general application in economic statistics a comprehensive classification of economic activity the very first time. Drawing on a preparatory study of the ILO which summarized the principles used in contemporary classifications of industries, this Conference introduced the distinction between primary production (agriculture and mining), secondary production (manufacturing and construction), and services; it also recommended strongly to apply classifications by industry and by occupation to labour force statistics [ILO, 1976]. In the subsequent Conferences 1925 and 1926 the efforts in establishing a uniform list of industries had been intensified; in 1948 the Statistical Commission of the UN summarized the work done up till then and proposed the International Standard Classification of all Economic Activities (ISIC) receiving world-wide acceptance. The ISIC standard has proven very useful since, though having undergone several revisions meanwhile.

The topic of deriving a classification of occupations was taken up in 1949 at the Seventh International Conference of Labour Statisticians which proposed a division of occupations into 9 major groups. Subsequently, this proposal underwent several refinements until the International Standard Classification of Occupations (ISCO) was compiled in 1957 (Ninth International Conference of Labour Statisticians) [ILO, 1976]. This Standard consisted of major, minor, and unit groups identified by three-digit code numbers and was revised again in 1969; its present, and commonly used, version is ISCO-88.

In 1938, the Committee of Statistical Experts of the League of Nations suggested to subdivide the gainfully occupied population in 4 groups [ILO, 1976], viz.

"(1) employers (persons working on their own account with paid assistants in their occupation); (2) persons working on their own account either alone or with the assistance of members of their families; (3) members of families aiding the head of their families in his occupation; (4) persons in receipt of salaries or wages."

A similar proposal was drafted by the ILO in 1954 (Eighth International Conference of Labour Statisticians) comprising 4 groups either:

"(1) workers for public or private employers; (2) employers; (3) workers who work on their own account without employees; (4) unpaid family workers."

To this subdivision, the Statistical Commission of the UN added a further group, viz. *members of producers' co-operatives*, and in 1966 an international classification of employment according to work status was adopted by the same Commission [ILO, 1976]:

"Status (as employer, employee, etc.) refers to the status of an economically active individual with respect to his employment, that is, whether he is (or was, if unemployed) an employer, own-account worker, employee, unpaid family worker, or a member of a producers' co-operative, as defined below:
(a) **Employer:** *a person who operates his or her own economic enterprise or engages independently in a profession or trade, and hires one or more employees. Some countries may wish to distinguish among employers according to the number of persons they employ.*
(b) **Own-account worker:** *a person who operates his or her own economic enterprise or engages independently in a profession or trade, and hires no employees.*
(c) **Employee:** *a person who works for a public or private employer and receives remuneration in wages, salary, commission, tips, piece-rates or pay in kind.*
(d) **Unpaid family worker:** *a person who works a specified minimum amount of time (at least one-third of normal working hours) without pay, in an economic enterprise operated by a related person living in the same household. If there are a significant number of unpaid family workers in enterprises of which the operators are members of a producers' co-operative who are classified in category (e), these unpaid family workers should be classified in a separate sub-group.*
(e) **Member of producers' co-operative:** *a person who is an active member of a producers' co-operative, regardless of the industry in which it is established. Where this group is not numerically important, it may be excluded from the classification and members of producers' co-operatives should be classified to other headings, as appropriate.*

(f) **Persons not classifiable by status:** *experienced workers with the status unknown or inadequately described and unemployed persons not previously employed."*

In addition to these main classifications used in employment statistics, several further classifications play more or less important roles. First of all, regional subdivisions have to be mentioned here which, apparently, are fairly easy to handle. Rather contrarily, however, behaves the classification of education levels; although there is an International Standard Classification of Education (ISCED) it has turned out to indicate vocational qualification levels rather insufficiently and, hence, is not suited well for breakdowns in labour force statistics [Thomas, 1994]. A really useful classification of education levels covering vocational qualifications satisfactorily still has to be brought about.

Without calling into question the importance of established nomenclatures, it is all too apparent that, facing the impacts of structural change and globalizing commerce, the traditional classifications particularly of occupations and economic sectors are not adequate anymore to describe socio-economic reality. If the primary aim of labour force statistics is indeed the early spotting and tracing of change, this will be hardly achievable sticking to yesterday's categories. Considering the tremendous reshaping and redefinition of job profiles as well as management styles taking place globally in the transition from standardized high-volume production of post-war economy, with its static hierarchies and well-defined job categories, to the emerging post-industrial economy, with its "lateral", non-hierarchical management organization and rather unprecedented activity patterns (cf. Section 3.1), the misfit of established classifications really could not be a big surprise. For instance, the U.S. Bureau of the Census introduced its Major Occupational Groups nomenclature – distinguishing working class, business class, and service workers – in 1950 using it ever since [Reich, 1993] although meanwhile each of these Major Groups has lost its original economic meaning by and large due to the horizontal re-deployment of job profiles.

3.2.2 European Labour Force Statistics

Against the background of a globalizing economy, national employment statistics lose importance compared to consolidated statistics at supra-national levels; in particular, from a European point of view, the development of European Union labour force statistics receives predominant interest [Fürst, 1993]. Although the Member States run their own traditional systems of employment statistics, many of which are based on administrative registers (for instance, unemployment registers maintained by labour offices) or social security registers, the labour force data produced depends on national law and context making it rather difficult to obtain comparable data for deriving meaningful supra-national statistics [Thomas, 1994]. For this reason, EUROSTAT – the central statistical office of the European Union – has been appointed to implement a co-ordinated EU-wide Community Labour

.2 Measuring Employment

Force survey (CLFS, for short) on behalf of the European Commission such that [EUROSTAT, 1991]

> "... national statistical institutes are responsible for selecting the sample, preparing the questionnaires, conducting the direct interviews among households, and forwarding the results to Eurostat in accordance with the standard coding scheme.
> Eurostat devises the programme for analysing the results and is responsible for processing and disseminating the information forwarded by national institutes."

Thus, the CLFS in fact is a *concerted set* of national sample surveys of households including persons in employment, unemployed persons, and other persons living in the sampled households. More specifically [EUROSTAT, 1991], the

> "... survey is intended to cover the whole of the resident population, i.e. all persons whose usual place of residence is the territory of the Member States of the Community.
> For technical and methodological reasons, however, it is not possible in all countries to include the population living in collective households, i.e. persons living in homes, boarding schools, hospitals, religious institutions, worker's hostels, etc.
> Consequently, for the purposes of harmonizing the field of survey, the Community results are compiled on the basis of the population of private households only. This comprises all persons living in the households surveyed during the reference week. This definition also includes persons absent from the household for short periods due to studies, holidays, illness, business trips, etc."

Moreover, the

> "... labour force characteristics of each person interviewed refer to his situation in a particular reference week."

EUROSTAT suggests to choose a normal week in spring excluding (bank) holidays; since some Member States refer to a moving week this condition cannot be assured in every case, however.

Formerly, the CLFS was split into Member State samples by appointing a predetermined share of the households sampled (somewhat above 600.000 altogether, including around about 2 million persons) to each Member State; absolute (rounded off to hundreds) and proportional sample sizes are comprised in **Tab. 3.2.2–1** which also indicates the position of reference weeks in the Member States, respectively [EUROSTAT, 1993a].

Table 3.2.2–1: Reference Weeks and Sample Sizes of the CLFS

Member state	Reference week 1989	Sample size 1987	%	1989	%
Belgium	April	29 800	4,97	31 800	5,02
Denmark	March-April	16 000	2,67	16 300	2,57
FR of Germany	April	95 400	15,91	95 300	15,04
Greece	April-June	48 400	8,07	48 600	7,67
Spain	March-June	52 300	8,72	58 200	9,19
France	March	65 000	10,84	65 400	10,32
Ireland	April-May	45 100	7,52	45 400	7,16
Italy	April	128 400	21,42	140 700	22,21
Luxembourg	May	9 300	1,55	9 300	1,47
Netherlands	January-May	19 700	3,29	30 500	4,81
Portugal	January, April	27 600	4,60	26 700	4,21
U.K.	March-May	63 400	10,58	65 300	10,31
Σ		600 400		633 500	

As can be seen from **Tab. 3.2.2–1**, sample sizes have changed slightly from 1987 to 1989; major changes regard the increase in households sampled in Italy (+10%) and the Netherlands (+55%). Sample design is the responsibility of the Member States' national statistical offices and, since the prescription of sample sizes alone does by no means guarantee reliable estimates, meanwhile the policy has been changed to prescribe maximal bounds for sample errors instead [Thomas, 1994].

Analogous to sample designs, the estimation of population figures from sample data is delivered up to the national statistical institutes which, accordingly, use different weighting procedures; EUROSTAT simply claims that each individual data record gets attached a multiplicative weighting factor necessary to obtain the respective population totals.

Contrary to other developed countries like the USA, Canada, or Japan, where labour force surveys are conducted monthly for quite some time now, European countries have lagged behind considerably. In 1960 the first community labour force survey took place including the then six Member States of the EEC. Regular surveys were carried out annually from 1968 through 1971, followed by a period of biennial surveys from 1973 through 1981. Beginning with 1983, the CLFS was again conducted once a year. At present, the EU concerns to switch to a quarterly repetition of the CLFS in order to obtain more timely data; this initiative, however, has been postponed regardful of the strained budgetary situation of the FR of Germany after the German reunion hardly admitting a costly re-organization of the current data production system (in particular, it would be necessary to establish a professional salaried enumerator staff [Jäger, 1993]).

.2 MEASURING EMPLOYMENT

CLFS Concept Definitions and Nomenclatures

In order to be above any doubt and uncertainties about the validity of the statistical figures delivered by the CLFS, data production must conform to transparent and internationally approved rules [Fürst, 1993]. In particular, policies and decisions – which rather often involve the disposition over large public budgets – based on disputable criteria could hardly be sustained and justified. Thus, for the very same reason national employment statistics data is rejected as a valid source of consolidated EU labour force statistics, a general framework of definitions and classifications has to be used throughout. However, because of the natural interdependence between the questionnaires on the one hand and the concepts and classifications to be fed by questionnaires on the other hand the adaptation of national survey programs to concepts and classifications standardized at EU level is a fundamental requirement. As a direct consequence, each Member State – though being free in designing the national CLFS questionnaires and allowed to include further questions of national or regional interest if desired – is obliged to support the standardized concepts and classifications to the best of its abilities. Basically, the European Commission decided to adopt the ILO standards as approved at the Thirteenth International Conference of Labour Statisticians (1982) as methodological framework for the CLFS (the Conferences held in 1987 and 1993 essentially confirmed the 1982 standards); however, these ILO standards provide a rather coarse framework only. For instance, it is doubtful to consider a person employed if it is working as little as one hour per week since this definitely will be insufficient to earn a living in any European country; likewise, to be unemployed according to ILO definition, a person must be without work, available for work, and seek work actively. But what it precisely means to seek work *actively* may very much depend on specific criteria possibly even differing from country to country. Another source of discordance is the classification of persons not working but having what is called a "formal job attachment", that is a work-place to which a person may return definitely at some later time: in ILO terms those persons are always classified unemployed whereas both the OECD and EUROSTAT have decided to classify persons on (temporary) lay-off as unemployed only if they are both seeking work and available for work; otherwise they are considered inactive [Thomas, 1994; Bartunek, 1994]. Hence, it has been agreed to amend the ILO standards wherever felt useful and necessary such that the refined standards are still fully compatible with ILO standards. In particular, taking the ILO standards as a point of departure it becomes a practical possibility to arrange systems of related indicators (cf. Subsection 3.2.1) such that different concepts can be used as appropriate though still admitting consistent interpretation. At present, the main definitions used in the CLFS are [EUROSTAT, 1992]:

"**Employment**

9.(1) *The employed comprise all persons above a specified age who during a specified brief period, either one week or one day, were in the following categories:*

(a) "paid employment":

> *(a1) "at work": persons who during the reference period performed some work for wage or salary, in cash or in kind;*
> *(a2) "with a job but not at work": persons who, having already worked in their present job, were temporarily not at work during the reference period and had a formal attachment to their job. This formal job attachment should be determined in the light of national circumstances, according to one or more of the following criteria:*
>> *(i) the continued receipt of wage or salary;*
>> *(ii) an assurance of return to work following the end of the contingency, or an agreement as to the date of return;*
>> *(iii) the elapsed duration of absence from the job which, wherever relevant, may be that duration for which workers can receive compensation benefits without obligations to accept other jobs.*

(b) "self-employment":

> *(b1) "at work": persons who during the reference period performed some work for profit or family gain, in cash or in kind;*
> *(b2) "with an enterprise but not at work": persons with an enterprise, which may be a business enterprise, a farm or a service undertaking, who are temporarily not at work during the reference period for any specific reason.*

9.(2) *For operational purposes, the notion of "some work" may be interpreted as work for at least one hour.*

Unemployment

10.(1) *The unemployed comprise all persons above a specified age who, during the reference period, were:*
(a) "without work", i.e. were not in paid employment or self-employment, as defined in paragraph 9;
(b) "currently available for work", i.e. were available for paid employment or self-employment during the reference period;
(c) "seeking work", i.e. had taken specific steps in a specified recent period to seek paid employment or self-employment.

.2 Measuring Employment

In applying these definitions to the Community Labour Force survey, Eurostat and the Working Party on the survey have agreed on some minor departures from their precise meaning:

(i) Persons on lay-off, who, according to ILO definitions, should be classified as employed, are included in the unemployed on the grounds that their willingness to supply labour services is apparent in their expectation of returning to work. This very small group amounts to only 0.2% of total Community unemployment. The same argument is applied to those persons who have already found a job to start at a later date.

(ii) For persons intending to set up their own business or professional practice neither active job-seeking nor immediate availability is required, as both conditions are difficult to measure; job-seeking activities are of a particular nature for this group, while testing on immediate availability would be completely hypothetical.

(iii) It has been decided that in paragraph 10.(1) lit. (b) "currently available" should mean available to start work within two weeks of the reference period. In paragraph 10.(1) lit. (c) "specified recent period" is the four weeks preceding the survey interview, the reason being that delays inherent in job search (for example, periods spent awaiting the receipt of results of earlier job applications) require that the active element of looking for work may be measured over a period greater than one week, if a comprehensive measure of job-seeking is to be obtained.

...

Labour Force

The labour force comprises persons in employment and unemployed persons.

Inactive persons

All persons who are not classified as employed or unemployed are defined as inactive. Apart from showing pupils and students separately, no further breakdown is provided for this group.

Conscripts on compulsory military or community service are excluded from the compilation of the survey results."

Despite the profound efforts dedicated to the standardization of the subject-matter concepts measured in the CLFS it must be kept in mind while interpreting aggregate figures that perfect comparability of national data is not attainable in all but the most trivial instances.

Setting the "specified age" bound to 15 years the main concepts discerned in European labour force statistics can be represented diagrammatically as shown in **Fig. 3.2.2–1** (adapted from [EUROSTAT, 1992]; overleaf); the discriminating features attached to the circle and diamonds, respectively, and identified by capital letters are comprised in **Tab. 3.2.2–2** [EUROSTAT, 1992].

Table 3.2.2–2: Discrimination Criteria Used in CLFS Concept Formation

Node	Discriminating criterion
A	Person of 15 years or more living in a private household
B	Person did any work for pay or profit during the reference week
C	Person was not working but had a job or business from which absent in the reference week for reasons other than a new job to start in the future or lay-off
D	Unpaid family worker
E	Person was seeking employment
F	Person was looking to set up his own business or professional practice
G	Person had during last 4 weeks taken specific steps to find a job
H	Person could have started to work immediately (within 2 weeks)
I	Person was not seeking employment because a job which would start later had already been found
J	Person was on lay-off

For obvious reasons, these concepts must be expressible in CLFS terminology whence changes in concepts induce subsequent adaptations of CLFS variables. In its evolution, the CLFS has undergone several revisions, the last major revision having taken place in 1992. These changes to the CLFS reflect the general propensity to collect more and more detailed observation data captured more frequently than before aiming at a "richer" representation of reality – assuming this to be an inevitable consequence of societal development in general and of the establishment of the Single Market with its programmatically free circulation of persons, goods, commodities, capital and services in particular. Compared to its predecessor, the CLFS-83 (slightly revised in 1988), comprising 45 variables regarding demographic and employment-related data, the revised CLFS-92 contains some 70 variables not counting technical positions included formally in the questionnaire (note that, unlike the real questionnaires prepared by the national statistical institutes actually conducting the survey, the CLFS does not contain "questions" but variables only to which the questions asked contribute information eventually). This expansion in variables – causing various practical difficulties in conducting the now rather lengthy and tedious survey interviews – is put up with in the desire to

- achieve a better delineation of cardinal subject-matter concepts (for instance, to distinguish those seeking work actively from those being "merely" registered unemployed);
- support the formation of new concepts and concept variants (possibly not anticipated at survey definition time) to improve comparisons and analyses by increasing flexibility and comprehensiveness [Fürst, 1993];
- capture migration tendencies and mobility;

.2 MEASURING EMPLOYMENT

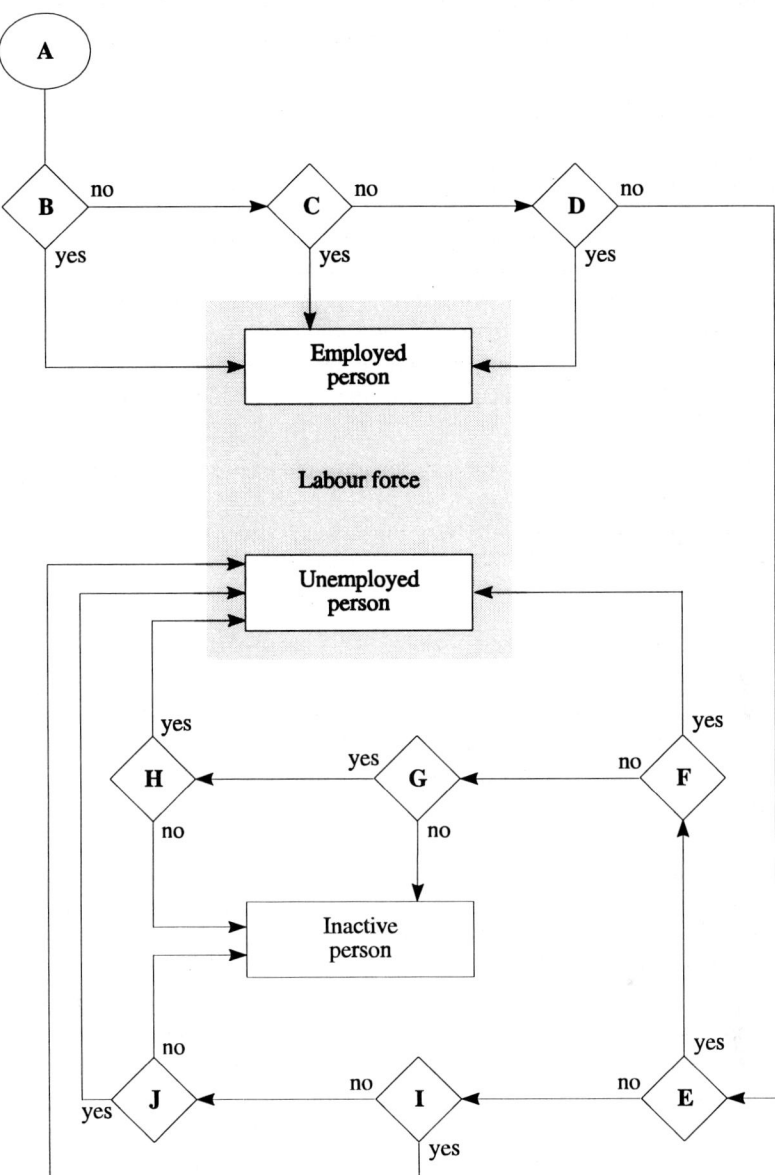

Figure 3.2.2–1: Main Concepts Discerned in European Labour Force Statistics

- more aptly record the structural change of employment (second jobs, increasing flexibility in work hours, home and tele-working, new types of work contracts, etc.);
- monitor education levels and vocational training measures (in view of qualifications attained and the relationships between education and employment);
- ascertain the methods of job search applied;
- better keep track of long-term unemployment.

The current CLFS-92 comprises a total of 77 variables – listed in **Tab. 3.2.2–3** [EUROSTAT, 1992] – including several technical positions used to identify the individual households and allowing to trace households across successive surveys with regard to the rotation structure of the samples according to which one quarter of households must be exchanged each time the survey is repeated [Jäger, 1993]. In **Tab. 3.2.2–3**, the symbols placed next to the variable numbers have the following meaning:

- a '*' indicates that coding conventions have changed compared to the 1983 survey series, whereas
- a '+' signals that this variable was introduced newly with the 1992 revision of the CLFS.

Table 3.2.2–3: Listing of the CLFS-92 Variables

No.		Variable name used in the Community Labour Force Survey
		Demographic background
1		Relationship to reference person in the household
2		Sex
3		Year of birth
4		Date of birth within year
5		Marital status
6		Nationality
7	+	Years of residence in this Member State
8	+	Country of birth
		Work status
9		Work status during the reference week
10		Reason for not having worked at all though having a job
		Employment characteristics of the first job
11		Professional status
12	*	Economic activity of the local unit of the establishment
13	*	Occupation
14	+	Number of persons working at the local unit of establishment
15	+	Country of place of work
16	+	Region of place of work
17	+	Year in which person started working in current employment
18	+	Month in which person started working in current employment
19		Full-time/Part-time distinction
20		Permanency of the job
21	+	Total duration of temporary job or work contract of limited duration
22		Number of hours per week usually worked

Table 3.2.2–3: Listing of the CLFS-92 Variables (cont'd)

No.		Variable name used in the Community Labour Force Survey
23		Number of hours actually worked
24		Main reason for hours actually worked being different from person's usual hours
25	+	Shift work
26	+	Evening work
27	+	Night work
28	+	Saturday work
29	+	Sunday work
30	+	Working at home
31		Looking for another job and reasons for doing so
		Information about second job
32		Existence of more than one job or business
33	+	Professional status
34	+	Economic activity of the local unit of the establishment
35	+	Occupation
36	+	Number of hours actually worked
37	+	Regularity
		Previous work experience of person not in employment
38	*	Experience of employment
39	+	Year in which person last worked
40	+	Month in which person last worked
41	*	Main reason for leaving last job
42		Professional status in last job
43	*	Economic activity of the local establishment in which person last worked
44	*	Occupation of last job
		Search of employment
45	*	Seeking employment for person without employment during the reference week
46		Type of employment sought
47	*	Duration of search for work
48	*	Main method used during previous four weeks to find a job
49	+	Date when person last contacted public employment office to find work
50	+	Willingness to work for person not seeking employment
51		Availability to start working within two weeks
52		Situation immediately before person started to seek employment
53		Registration at a public employment office
		Situation of inactive person
54		Situation of person who neither has a job nor is looking for one
		Education and training
55	*	Education and training received during previous four weeks
56		Purpose of the training received during previous four weeks
57	+	Total length of training
58	+	Usual number of hours training per week
59	*	Highest completed level of general education
60	+	Highest completed level of further education or vocational training
		Situation one year before survey
61		Situation with regard to activity
62		Professional status
63	*	Economic activity of local unit of establishment
64	*	Country of residence

Table 3.2.2–3: Listing of the CLFS-92 Variables (cont'd)

No.		Variable name used in the Community Labour Force Survey
65		Region of residence
		Technical items relating to the interview
66		Year of survey
67		Reference week
68	*	Member State
69		Region of household
70	+	Degree of urbanisation
71		Serial number of household
72		Type of household
73		Type of institution
74		Nature of participation in the survey
75		Weighting factor
76		Sub-sample in relation to the preceding survey
77		Sub-sample in relation to the following survey

With respect to concept formation, the pivotal variables are *Work status during the reference week* (no. 9), *Reason for not having worked at all though having a job* (no. 10), and *Professional status* (no. 11); the categories of these variables comprised in the CLFS coding scheme (but omitting the codes for 'missing/no response' and 'not applicable') are reproduced in **Tab. 3.2.2–4** [EUROSTAT, 1992]. Apparently, the categories of variable no. 11 conform to the 1954 ILO proposal (cf. Subsection 3.2.1; for clarification and disambiguating interpretations, the reference booklet of the CLFS issued by EUROSTAT [1992] includes extended *explanatory notes* to most of the (occasionally controversial) categories; according to the terminology introduced in Chapter 1 these notes – serving, in a purely descriptive way, as interpretative aid – could be justly regarded as *metatexts*).

Now, in terms of these categories the relationship between the subject-matter concepts used in the CLFS and the discriminating criteria stated in **Fig. 3.2.2–1/Tab. 3.2.2–2** can be established explicitly as shown in **Tab. 3.2.2–5**. The age condition (persons with age of 15 years or above) is derived easily from variables no. 3 and 4.

Table 3.2.2–5: Relationship between CLFS Concepts and Variables

Node	A	B	C	D	E	F	G	H	I	J
Variable(s)	3,4	9	10	11	45	46	48	51	9,10	9

Table 3.2.2–4: Categories and Codes of Cardinal CLFS Variables

No.	Code	Variable
9		*Work status during reference week*
	1	Did any work for pay or profit during the reference week – one hour or more (including family workers but excluding conscripts on compulsory military or community service)
	2	Was not working but had a job or business from which he/she was absent during the reference week (including family workers but excluding conscripts on compulsory military or community service)
	3	Was not working because on lay-off
	4	Was a conscript on compulsory military or community service
	5	Other (15 years or more) who neither worked nor had a job or business during the reference week
10		*Reason for not having worked at all though having a job*
	0	Bad weather
	1	Slack work for technical or economic reasons
	2	Labour dispute
	3	School education or training
	4	Own illness, injury or temporary disability
	5	Maternity leave
	6	Holidays
	7	New job to start in the future
	8	Other reasons (e.g. personal or family responsibilities)
11		*Professional status*
	1	Self-employed with employees
	2	Self-employed without employees
	3	Employee
	4	Family worker

Omitting the technical items (variables no. 66 through 77 in **Tab. 3.2.2–3**), the CLFS variables can be associated with the main population groups discerned in European labour force statistics as tabulated in **Tab. 3.2.2–6** [EUROSTAT, 1992].

Table 3.2.2–6: CLFS Variables and Main Population Groups

Main population group	*CLFS variables*
Every person	1–8,64,65
Every person 15 years or more	9,53,55–63
Persons in employment	10–37
Persons without employment	38–41 *if last worked less than 8 years ago:* 42–45
Persons in employment seeking another job	46–48,51
Unemployed persons	46–49,52
Inactive persons	50,51,54

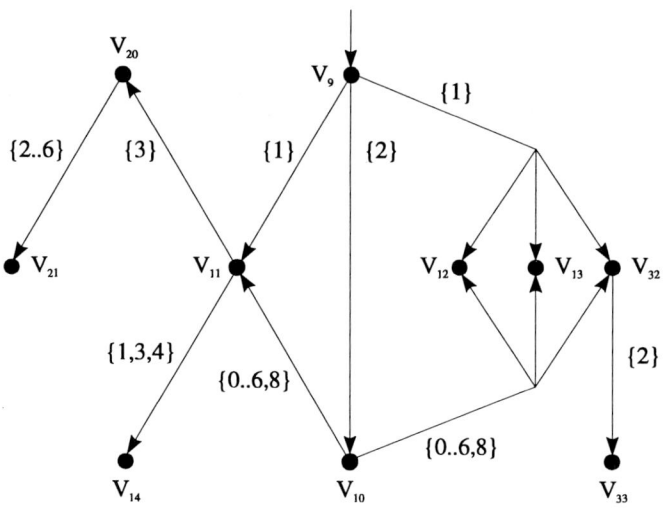

Figure 3.2.2–2: Dependency Structure of CLFS Variables (Excerpt)

An excerpt of the resulting dependency structure between the discerned subpopulations is explicated graphically in **Fig. 3.2.2–2**. The nodes of this dependency graph (cf. Appx. B) are labelled with CLFS variable numbers; the directed arcs are labelled with code sets stating the codes the conditioning variable must assume in order to let the conditional variable become applicable (for the sake of brevity, the notation '$x..y$' is used as shorthand for the complete range enumeration '$x,x+1,...,y$'). Note the following remarks to **Fig. 3.2.2–2**:

- variable no. 20, »*Permanency of the job*«: the codes ›2‹ through ›6‹ comprise various reasons for a work contract of limited duration including the case of a temporary job;
- variable no. 32, »*Existence of more than one job or business*«: code ›2‹ represents the category ›*Person had more than one job or business during the reference week (not due to change of job or business)*‹ ;
- arc V_{11}–V_{14}: according to the authoritative source of information about the CLFS-92 [EUROSTAT, 1992], variable no. 14 applies even if variable no. 11 does *not* apply – this, however, would mean that, for the sake of consistency, the category ›*No answer*‹ of variable no. 11 in fact is equivalent to a category ›*other*‹ (as included in the ILO classification proposal; cf. Subsection 3.2.1) decidedly different from category ›*No answer*‹ .

CLFS Classifications

The classifications used in the CLFS are either rather similar to ILO standards (NACE Rev. 1 of 1990, for instance, closely resembles ISIC) or conform to ILO standards at all (such as in case of the ISCO-88 being used to classify occupations). With respect to education and training the CLFS has set up its own provisional standard (the ISCED standard is used partly to encode the highest level of general education attained; variable no. 59). The main classifications of employment statistics are used several times within the CLFS, viz.

- NACE Rev.1 (operative since 1993): variables no. 12, 34, 43, and 63;
- ISCO-88: variables no. 13, 35, and 44 (for some reason the occupation is not asked with respect to the situation one year before the current survey);
- Professional status: variables no. 11, 33, 42, and 62.

The spatial classification used in the CLFS is specific to the European Union [EUROSTAT, 1992]; this Community Nomenclature of Territorial Units (NUTS) provides a 4-level classification of administrative units (level 0 = Member State, level 3 = region unit roughly corresponding to major political districts, in general) encoding the variables no. 16, 65, and 69.

CLFS Revisions and Longitudinal Comparability

A question of particular interest is the comparability of statistical data produced by the CLFS [Fürst, 1993] attained eventually with respect to both space and time. Spatial comparability will be remarked on in Subsection 3.2.3; as to temporal comparability it is reported [EUROSTAT, 1992; 1993a] that comparability is fairly satisfactory mainly because of the stability of the survey program (a minor revision in 1988 basically added a few variables) and the frequent (that is: annual) repetition of the survey. Problems recognized to affect adversely the temporal comparability of data are:

- the weighting factors for grossing-up sample figures to population figures are updated at rather large (10 year) intervals only and, hence, suffer from low accuracy;
- reference weeks are moving even within Member States;
- questionnaire forms as well as sample designs may be changed from time to time within Member States for various local reasons with unknown effects to the national data reported to EUROSTAT;
- there may be specific circumstances influencing the responses to the questions at interview time affecting part or all of the households sampled in a Member State;
- sampling errors may vary considerably over time.

The comparability of survey data is challenged most when the survey program itself is revised as this has happened last in 1992; for apparent reasons it is difficult to strike a balance between continuity and flexible adaptation to emerging information demands. Though giving priority to the latter, however, the following comparison points out that, excepting discordances of minor relevance, the changes introduced with the CLFS-92 revision left it fairly well compatible with the former CLFS-88 [EUROSTAT, 1990].

A fundamental change regards the shift of the age bound from 14 to 15 years, that is the CLFS-88 series included all persons of at least 14 years in the labour force (provided, of course, that all other conditions were applying). This adaptation, however, is easy to cope with since the persons' birth dates have always been recorded by years at least such that, for the sake of longitudinal comparisons, those of age 14 can be excluded simply from consideration.

With respect to individual variables, several cases need to be distinguished. First, the revision may introduce new variables not covered previously in any way; fortunately, this case is trivial – from the revised survey's point of view, the former surveys just generate marginal distributions of the extended sample space. In terms of the CLFS-92 series, quite a few variables have been introduced such as, for instance, the variables regarding the second job (variables no. 33 through 37) or the »*Degree of urbanization*« (variable no. 70).

Less trivial is the case of variables becoming split into several new variables replacing their forerunners. For instance, the variables no. 38, 39, and 40 have been a single one in the CLFS-88 series which encoded both that a person has never been in employment before (now code ›0‹ of variable no. 38) and, otherwise, the time that has elapsed since the person's last employment measured in six-month intervals up to three years or above; now, the CLFS-92 variables no. 39 and 40 record the point of time the person left/lost her last employment if there was a previous employment (code ›1‹ of variable no. 38). Evidently, except the refined measurement of the time elapsed since a person's last employment little has changed and, using the former temporal granularity, longitudinal comparisons are still supported without restriction.

The least convenient case regards surgical changes in the coding schemes of existing variables although, as far as pivotal classification (such as the NACE system) schemes are involved, change keys are arranged formally. Again, if the change consists of adding some new elements not interfering with previous categories, this is accomplished straightforwardly (for instance, adding the NUTS codes of new Member States entering the Union). Conversely, if categories are simply dropped no further adaptations are required; however, since this modification may invalidate the definitions of subject-matter concepts depending on dropped categories difficulties could arise (therefore, the dropping of categories should be avoided unless there is a statistical argument – for instance, the »*Work status*« category ›temporary lay-off‹ has been abandoned 1993 because it had turned out that there are a few persons only falling in this category). Otherwise, the relation between old and new coding schemes may vary from fairly simple to actually unresolvable. A simple case is variable no. 45, »*Seeking employment*«: the change is restricted essentially to a refinement of the discerned categories why a person is not looking for a job; in the corresponding CLFS-83 variable those categories had been coded as ›Other reasons‹ which, in terms of the CLFS-92 coding scheme becomes a grouping of discerned categories. Similarly, the time period ›24 months or more‹ becomes now divided into the categories ›24-47 months‹ and ›4 years or more‹ .

Somewhat more intricate is the formal relationship of the categories of variable no. 48 to its predecessor's ones; in this case, some categories become subdivided whereas others are merged:

- the category ›Other methods‹ of job search is split into several categories;
- the former codes ›2‹ and ›3‹ are turned jointly into code ›05‹, ›Inserted or answered advertisements in newspapers and journals‹.

For this reason, unless applicable disaggregations are found, longitudinal comparisons are feasible at the coarser levels of discerned categories only.

Analogously, variable no. 55 refines the former code ›2‹ into the categories coded ›2‹, ›5‹, ›6‹, and ›7‹ while the former codes ›4‹ and ›5‹ are now encoded jointly as code ›4‹ (the former code ›7‹, valid for U.K. only, seems to have vanished entirely).

Quite a delicate case represents the change of variable no. 41's coding scheme. For one thing, this variable now applies to *all* persons having been employed last less than 8 years ago (the former bound had been 3 years); secondly, the variable's categories comprised the full list for employees only whereas for persons with a different work status only the codes ›4‹ through ›7‹ applied. Both former and current categories are juxtaposed in **Tab. 3.2.2–7**.

Table 3.2.2–7: CLFS-83 and -92 Coding Schemes of Variable *"Main reason for leaving last job or business"*

CLFS-83: Main reason ...		CLFS-92: Main reason ...	
0	Dismissed	0	Dismissed or made redundant
1	Job of limited duration has ended	1	A job of limited duration has ended
2	Resigned	2	Personal or family responsibilities
3	Early retirement for economic reasons	3	Own illness or disability
		4	Education or training
4	Own illness or disability	5	Early retirement
5	Retired for other than economic or health reasons	6	Normal retirement
		7	Compulsory military or community service
6	Military or community service		
7	Other reason	8	Other reasons

A direct nominal assignment of old to new categories is somewhat problematic since, on the one hand, the distinctions between former categories and new ones are taxonomically blurred and, on the other hand, the former codes ›2‹ and ›7‹ go to the new codes ›2‹, ›4‹, and ›8‹ only for persons with professional status ›*employee*‹, as compiled in **Tab. 3.2.2–8** (for persons of other professional status, former code ›7‹ matches the new code ›8‹).

Table 3.2.2–8: The Employee Case of Category Association for Variable
"Main reason for leaving last job or business"

CLFS-92	0	1	2	3	4	5	6	7	8
CLFS-83	0	1	2	4	2	3	5	6	2,7

One possible solution to restore longitudinal comparability consists of combining bluntly the former codes ›2‹ and ›7‹ and associating them with the new code ›8‹; a more sophisticated approach could first try to disaggregate the former code ›2‹ into the shares going to new codes ›2‹, ›4‹, and ›8‹.

A special case are the variables no. 59 and 60: these variables have been introduced with the 1988 revision of CLFS-83. One of these variables, regarding the *Highest completed level of education or vocational training*, reappears in the CLFS-92 though completely modified because general education (variable no. 59) and vocational training (variable no. 60) have become separated now. The recoding of this former variable into the pair of new ones is rather intricate and indispensably involves disaggregations. The other variable introduced 1988 is dropped at all (but might be kept formally) whence no longitudinal comparisons are supported for apparent reasons.

Supplementary Employment and Unemployment Statistics

The Community Labour Force survey is the Member States' co-ordinated effort towards a reliable and comprehensive source of information about the labour market of the European Union. Of course, in addition to this dedicated survey there continue to exist the national employment statistics as supplementary sources of labour market data. Though being separated from the CLFS, these supplementary sources of employment data – comprising particularly the majority of dependent employed – are also used to prepare aggregates comparable at EU level; these "harmonized" employment statistics are published (by EUROSTAT) in a series of yearbooks called *Employment and Unemployment* (for instance, [EUROSTAT, 1989b]). Including a variety of national data sources mostly arranged for specific purposes (such as unemployment or social insurance registers), the national aggregates submitted are derived by "post-harmonization" for which reason the resulting figures quite often are less suited for direct cross-sectional comparisons in spite of the endeavours to convert national concepts into harmonized ones (some of these marked differences are discussed in Subsection 3.2.3). A particular weakness of this dual line approach to European employment statistics is the lack of data integration incurred – although there is plenty of data, it cannot be shared appropriately across institutional settings to improve the overall quality of statistics produced by exploiting the potential of variance reduction as well as the opportunity of "cross-validating" the consistency of estimates.

3.2.3 National Labour Force Statistics

In what follows, attention is directed towards specific aspects of national employment statistics systems of selected European countries only, as a more comprehensive overview would certainly burst the space available. Essentially, the discussion centers around Member States of the EU using Austria as a case in point to vividly illustrate the transition an established system of employment statistics has to undergo in order to attain the EU standards of labour force statistics.

Most European countries have established their national system of employment statistics soon after World War II; from the late 1950s on most countries also introduced either dedicated labour force surveys or less specifically targeted general socio-economic surveys (such as the German *Mikrozensus* [Jäger, 1993] or the welfare surveys conducted in the Scandinavian countries [Vogel, 1993]) including topics related to employment. Usually, however, the core parts of national employment statistics – such as unemployment rates or forecasts of labour market development – always referred to sources other than those more or less general-purpose surveys; many countries have arranged specific labour force surveys conducted four times a year (Italy, Spain, Portugal; lately also France, Denmark, and the U.K.) or even monthly (Netherlands). In particular, in view of the close-to-complete coverage and timeliness of employment data maintained in administrative registers, notably the national registers of unemployed receiving benefit payments as well as social security registers (comprising, in general, both dependent and self-employed persons), employment and labour force statistics are often based on official register data. Another statistical reason for preferring register data over sample data in national employment statistics can be found in the avoidance of both, (enumerator and panel) bias and affection by deteriorating responsiveness of inquired persons [Haslinger, 1989; Jäger, 1993]. Conversely, as these registers have been called into existence to serve the execution of national legislative regulations it is hardly surprising that national employment figures and specifically unemployment rates are often diverging markedly from international figures even in spite of being computed from the very same data.

For the sake of illustration, picking up the Austrian example, **Tab. 3.2.3–1** summarizes the data bodies relevant in Austrian employment statistics [Lackner, 1993]. Excepting the Austrian *Mikrozensus* (AMZ) which is a sample survey conducted every three months (comprising around about 29.000 households), all of the remaining data bodies originate from full censuses, registers, or periodical updates thereof. Data are supplied either by Austria's Central Statistical Office (ÖSTAT), the Federation of Social Insurance Authorities (Hauptverband der Sozialversicherungsträger), or the Bureau of the Labour Market Administration (Arbeitsmarktverwaltung – AMV, for short – now called "Arbeitsmarktservice" after privatization) on behalf of the Federal Ministry of Labour and Social Affairs (Bundesministerium für Arbeit und Soziales); the latter hosts the headquarters of the official employment exchange offices authorized to register the unemployed and administer the payment of unemployment benefits. For the computation of employment statistics, particularly ÖSTAT's AMZ and the Federal Ministry of Labour and Social Affairs (AMV)'s data is used (the latter for national employment figures only). As highlighted in **Fig. 3.2.3** (picking data from [Bartunek, 1994]), the derived unemployment rates for 1992 are rather divergent in size (but less so with respect to trend) although three of the four rates exhibited use AMZ survey data, viz. the AMZ rate published by ÖSTAT, the ILO/OECD rate published by the OECD, and the EUROSTAT rate. The AMV rate (published by the Federal Ministry of Labour and Social Affairs) is obtained [Flaschberger, 1994] by dividing the number of officially registered unemployed persons – including those seeking an apprenticeship – by the sum of both, insurance accounts (as supplied by the Hauptverband) and registered unemployed (note that, according to the Austrian Social Security Act, each work contract of a dependently employed person entails a distinct insurance account; in 1989, there were some additional 370.000 insurance accounts – roughly 10% – in excess of persons registered as dependent employed [Lackner, 1993]). While the differences between the AMV rate and the other ones can be explained simply by the dissimilarity in goals and methods of the different data bodies used the differences within the remaining rates result from applying somewhat divergent concepts (in fact, the Federal Ministry of Labour and Social Affairs publishes a monthly series of unemployment rates with different estimates yet obtained by applying a seasonal adjustment to raw register data [Flaschberger, 1994]). In particular, the AMZ rate is higher than those of OECD and EUROSTAT mainly because, contrary to ILO standards, ÖSTAT does not consider employed all persons working less than 12 hours a week (which are termed "insignificantly employed"), although the unemployed comprise only persons indicating to be registered as such (irrespective of their availability or evident search for work [Flaschberger, 1994]). The numerical differences between ILO/OECD and EUROSTAT rates arise from slightly divergent modifications of ILO definitions as expounded in previous subsections (due to a recent agreement, however, both OECD and EUROSTAT will use identical definitions and, thus, both rates will coincide henceforth [Bartunek, 1995]).

It certainly comes as no surprise that national data sources make use of more or less peculiar classifications hardly compatible with their international counterparts. For instance, most Austrian employment data refers to a national classification of economic activity called *Betriebssystematik 1968* (BS68 in **Tab. 3.2.3–1**) rather different from both ISIC and NACE (although from 1994 onwards the AMZ is using an Austrified version of NACE Rev. 1 called *ÖNACE* [Rainer, 1994]; unfortunately the opportunity has been missed to record, for some time, data with respect to both old and new classification systems side by side to bridge the otherwise rather insurmountable classification discontinuity [Hudec, 1995]).

This discussion made it apparent that an integration of employment data by no means has been accomplished at the national level, not to speak of supra-national data integration at all. Conversely, however, the production of employment data has been decided to remain a national responsibility even for the Member States of the EU for which reason most of them had to prepare either a specific labour force survey contributing CLFS data or to adapt an existing data production system to fulfil CLFS requirements (the latter approach was chosen in Germany by adapting the Mikrozensus accordingly).

Table 3.2.3–1: Austrian Sources of Employment Data

	Supplier	Periodicity	Coverage	Classifications used		
				econ. activity	spatial	other
General census	Hauptverband	biannual (January, July)	dependent employed (insurance accounts)	BS68	political districts	age, sex, wage class, work status, etc.
Periodical Update	Hauptverband	monthly	dependent employed (insurance accounts), self-employed with mandatory social insurance account, related family workers	BS68	Federal states	age, sex, wage class, work status, etc.
Domain census	ÖSTAT	5 yrs.	all employed except agriculture, independent (free-lance) self-employed including non-gainful private services, public administration	BS68	Federal state (partly political districts)	sex, work status, legal status of employer, etc.
Population census	ÖSTAT	10 yrs.	all persons employed	BS68	local communities	age, sex, work status, etc.
Work place census	ÖSTAT	10 yrs.	all persons employed except agriculture, housekeeping, caretaking	BS68	local communities, counting districts, etc.	work status (extended list)
Mikrozensus	ÖSTAT	every 3 months	all employed and unemployed persons	BS68	federal states	age, sex, nationality, work status, education level, etc.
Unemployment statistics	AMV	monthly	all persons registered unemployed	–	local communities	age, sex, duration of unemployment, etc.

National CLFS Contexts

As already pointed out, the *subsidiary principle* delivering data production up to national statistical institutes gives rise to various minor divergencies in data characteristics despite the rather succinct CLFS definitions. For one thing, the translation of CLFS terminology into national terminology turns out to be quite a challenging task if semantic precision should not be sacrificed. In particular, literal translation will simply fail very often because of different legal contexts and customs calling for "subject-matter" translations which sometimes may even depart significantly from CLFS concepts. As an example, for historical reasons national education systems are rather unique making it difficult to devise a single common classification; the "mappings" of national terminology to CLFS-92 variables no. 59 and 60, respectively (columns 86/87 technically) can be looked up in Annex V of the CLFS-92 reference booklet [EUROSTAT, 1992] (note that these mappings bear the heritage of their CLFS-88 predecessors as to code *4* of column 86 = variable no. 59). Quite surprisingly, even with respect to such simple features as »*Relationship to reference person in the household*« (CLFS-92 variable no. 1) national terminology may deviate as shown in **Tab. 3.2.2–2** for the case of the Portuguese labour force survey (*Inquérito ao Emprego* [Gomes and Miranda, 1991]): the corresponding question no. 13 of this survey, »*Affinity with the representative of the family*« comprises a further category not discerned in CLFS terminology.

Table 3.2.3–2: Categories of Question 13 of the Portuguese Labour Force Survey

Portuguese LFS – Question 13		*CLFS – Variable No. 1*	
code	value	value	code
1	Head of household	Head of household/Reference person	*1*
2	Spouse of head of household	Spouse/Spouse (or cohabiting partner) of reference person	*2*
3	Child of head of household (or his/her spouse)	Child of head of household or his/her spouse / of reference person (or of his/her spouse or cohabiting partner)	*3*
4	Ascendant of head of household (or his/her spouse)	Ascendant relative of head of household or his/her spouse / of reference person (or his/her spouse or cohabiting partner)	*4*
5	Daughter-in-law or son-in-law from the family's representative or from his consort	Other relative	*5*
6	Other relative		
7	Non-relative	Other	*6*

With respect to work status the Portuguese labour force survey deviates markedly from the CLFS-92 variable scheme; the national questionnaire splits the discerned categories of the CLFS-92 variable no. 9 (»*Work status*«) into a sequence of dichotomous questions as tabulated in **Tab. 3.2.3–3**: the *yes*-branch of questions no. 22 through 25, respectively, maps to variable no. 9 as shown in the rightmost column (questions 26 and 27, among others not shown, map to variable no. 10).

Table 3.2.3–3: Mapping of Portuguese Labour Force Survey Questions to CLFS-92

Question	Query text	Condition	Mapping
Q22	Are you making the compulsory military service ?	male, between 16 and 35 years, not handicapped	yes: code *4*
Q23	Did you work at least one hour last week, with the objective of a salary or a benefit ?		yes: code *1*
Q24	Though not having worked, is there any job you have been out temporarily (holidays, for example, etc.) ?	*Q23=no*	yes: code *2*
Q25	Didn't you work because your contract was temporarily interrupted ?	*Q24=no*	yes: code *3*
Q26	Were you studying ?	*Q25=no*	...
Q27	Did you do domestic tasks at your own home ?	*Q26=no*	...
...	...	...	...

A similar case arises for questions no. 33 through 36a (the latter has been denied an individual number in the source description of [Gomes and Miranda, 1991] used) where the Portuguese labour force survey discerns 6 categories of professional status (following the 1966 definition of the Statistical Commission of the UN; cf. Subsection 3.2.1) in contrast to the CLFS-92 discerning only 4 categories.

In addition to such taxonomic deviances, the national data production contexts may require the adoption of specific procedures for various reasons (for instance, in case of overseas territories etc.). The defined data production methodologies (including sampling and estimation schemes), in turn, interfere with established national data production procedures as a whole (including questionnaire designs, enumerator staff training, measures taken against non-response, etc.). Being aware of these nuisance factors, however, the level of comparability of CLFS data submitted by the Member States is considered fairly high because [EUROSTAT, 1992]

- each Member State records, in principle, the same set of characteristics in that there is a close correspondence between the list of CLFS variables and the national questionnaires;
- the same subject-matter concept definitions are used in all Member States;
- the national CLFS surveys take place "synchronized" in spring using a (normal) week as reference period;
- data are being processed uniformly by EUROSTAT.

Every now and then, national statistical offices revise their CLFS data submitted to EUROSTAT later (such as, for instance, Greece in 1983 and 1984 [EUROSTAT,

1993a]) calling for the explicit arrangement of versions of datasets (cf. Subsection 2.3.1) for the sake of unambiguous reference and distinction.

Somewhat annoying is the long delay in data submission by the national statistical institutes [Fürst, 1993]; even in view of the added work incurred by revising the CLFS programme 1992 it is hard to understand why it took 2 full years until all Member States had delivered the 1992 survey data to EUROSTAT [Thomas, 1994]. Referring to the period 1983–1991, a summary of national peculiarities in CLFS data production is given, for instance, in [EUROSTAT, 1993a].

Consolidated National Employment Statistics

Less optimistic as to comparability of national results is the situation with consolidated employment statistics – published annually (by EUROSTAT) in its *Employment and Unemployment* book series – based on "post-harmonization" of domestic employment data. Although aiming seriously at a deep integration of available labour force data, the data must be taken as heterogeneous as is whence several peculiarities are hard to bridge methodologically (if at all). To account for the major discordances in the source data used to compile the harmonized employment statistics of the EU, EUROSTAT [1989a] has issued a booklet quoting the main definitions as well as summarizing the tracked national departures from these definitions. In particular, there are two sets of definitions and, hence, departure summaries as two types of consolidated aggregates are prepared, viz. general statistics of labour force and employment as well as specific harmonized statistics of dependent employed (except for Greece which does not contribute national data to the harmonized statistics of dependent employed).

A first marked difference observable in national employment statistics regards the general population concept used. Basically, two types of concepts may be distinguished: the *domestic* and the *national* concept. In the former, the whole of a nation's resident population is included whereas in the latter the population considered comprises all persons being in employment on the territory of the nation (irrespective of their place of residence or nationality); hence, the main distinction between both concepts consists of the sub-population of frontier workers. With respect to measuring the labour force most European countries use the domestic concept; only Germany and the United Kingdom refer to the national concept (mainly because it better supports GNP computations). Belgium and Luxembourg have chosen a mixed approach in that with respect to the economically active population the domestic concept is applied whereas the national concept is used for the working population. The harmonized statistics of dependent employed are based predominantly on the national concept excepting Spain and France where the domestic concept is preferred; in Germany around about 90% of the dependent employed are surveyed according to the national concept [EUROSTAT, 1989a].

A further difference pertains to labour force coverage; in particular, soldiers (whether temporary or professional), conscripts, persons in education, casual and seasonal workers, home-workers, part-time workers, persons with specific work contracts, persons with multiple work contracts or employment relations, persons living in institutional households, etc. are dealt with differently in different countries (a comprehensive summary is contained in [EUROSTAT, 1989a]). Another source of divergence is the age bound for persons to be included in the labour force. Belgium, Spain, Ireland, and the U.K. set the bound to 16 years; Greece and Italy include persons of 14 years and above; other EU Member States use the ILO bound of 15 years except Portugal (12 years) and Denmark which does not have any such age bound at all.

Concerning all the other differences present, it comes as no surprise noting that most countries use different reference dates for the employment data produced. A summary (based on [EUROSTAT, 1989a] again) is compiled in **Tab. 3.2.3–4** distinguishing between both types of employment statistics involved; apparently, without seasonal adjustment national data are hardly comparable.

Table 3.2.3–4: Reference Dates Summary of National Employment Statistics

Nation	Reference period/Key date	
	Labour Force and Employment	*Harmonized Statistics of Dependent Employed*
Belgium, U.K.	*June 30*	*June 30*
Denmark	*January 1* (resident population) *last week of November* (professional status)	*January 1* (resident population) *last week of November* (professional status)
FR of Germany, France, Luxembourg	annual average	*March 31*
Spain	annual average	2nd *quarter*
Ireland	*mid-April*	*mid-April*
Italy	annual average over 4 quarters of a year	*March 31*
Netherlands	*January 1*	*March 31*
Portugal	2nd *quarter*	*March*
Greece	*April-June*	–

National statistics make use of classifications similar to but often not complying with the international ones. Sometimes, the differences can be bridged quite satisfactorily by suitable change keys; otherwise, classifications become rather coarse and, hence, less useful in international comparisons. With respect to the classification of economic activity, most countries either use (a version of) NACE or provide a NACE-compatible change key.

The Case of Austrian Mikrozensus

Much like other European countries Austria has developed its own system of employment statistics (as part of social and economic statistics). Since 1967 there also is a periodical sample survey called *Mikrozensus* (emulating its German forerunner which commenced a decade earlier) conducted every three months beginning with March 1; cf. **Tab. 3.2.3–1**. The Austrian Mikrozensus – AMZ, for short – is a one-stage rotation sample exchanging $1/8$ of the sampled addresses (dwellings taken from the official decennial housing census) each time the survey is repeated (note that a dwelling included in the sample may comprise more than one household [ÖSTAT, 1986]). Addresses (dwellings) are selected such that the numbers of persons interviewed in the dwellings within each of Austria's 9 Federal Countries are roughly of equal size; in a dwelling included in the sample all persons are interviewed by specifically trained enumerators [Lackner, 1993]. Like to its German counterpart, the AMZ is a multi-purpose survey comprising a mandatory standard part ("Pflichtprogramm", consisting of a fixed set of questions included in every survey repetition) as well as a voluntary special part ("Sonderprogramm", varying in response to short-term information demands; covering only part of the sampled households or even cancelled occasionally). All questions of relevance for a continued reporting on labour force and employment issues belong to the standard part of the AMZ as listed in **Tab. 3.2.3–5** ("Personenblatt B" of the 1992 questionnaire, with some technical positions omitted). Question Nr. 20 was added to the questionnaire from September 1988 onwards to accommodate for the ILO definition of employment (cf. Subsection 3.2.1). For obvious reasons, in view of Austria's join with the EU in 1994, the AMZ – analogous to the German Mikrozensus – had to undergo a major revision. Since March 1994, AMZ's standard part has been modified (comprising 24 questions now) by adapting both, questions and taxonomies, to the CLFS-92 standard. Furthermore, in March of every year the special part of the AMZ-94 is extended to cover all the remaining questions of the CLFS-92 not included in the standard part of the AMZ-94 (61 questions).

Table 3.2.3–5: AMZ Questions and Their Relation to the CLFS-92

Nr.	*Austrian Mikrozensus*	CLFS-No.
4	Geburtsdatum	3,4
5	Stellung im Haushalt	1
6	Geschlecht	2
7	Staatsbürgerschaft	6
8	Familienstand	5
9	Teilnahme am Erwerbsleben	9,45,48,54
10	Stellung im Beruf	11
11	Berufliche Tätigkeit	13
12	Betriebszweig	12
13	Normale wöchentliche Arbeitszeit	22
14	Tatsächliche Arbeitszeit in der letzten Woche	23
15	In den letzten 4 Wochen aktiv einen Arbeitsplatz gesucht ?	45,47
16	Kann Arbeit antreten	51
17	Wie lange Arbeitsplatz gesucht ?	47
18	Höchste abgeschlossene Schulbildung	59
19	Lehrabschlußprüfung	60
20	In der letzten Woche mind. 1 Stunde gegen Bezahlung gearbeitet oder im Familienbetrieb mitgearbeitet?	9

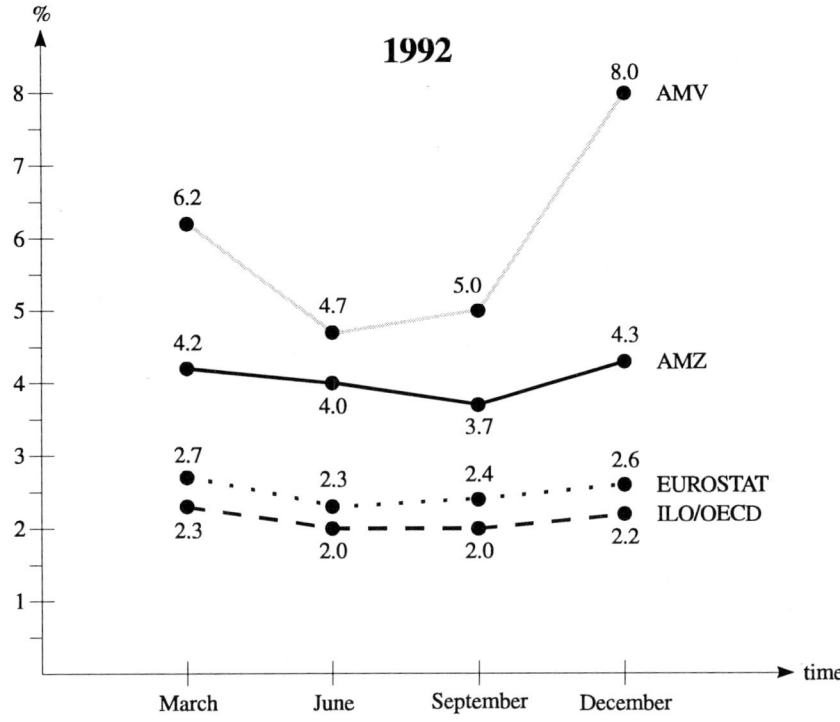

Figure 3.2.3: Four Different Austrian Unemployment Rates

Essentially, there is now a *1:1* correspondence between AMZ-94 questions and CLFS-92 variables except that occupations still use a national classification differing considerably from the ISCO taxonomy used in the CLFS [Bartunek, 1995].

Although the AMZ is based on a couple of definitions deviating considerably from CLFS and ILO standards, virtually all international employment data regarding Austria are derived from the AMZ (cf. **Fig. 3.2.3** and pertinent discussion) Hence, it is of particular interest to contrast the AMZ with European employment statistics, especially the CLFS-92. As can be concluded from **Tab. 3.2.3–5**, the AMZ variables cover only a small fraction of CLFS-92 variables (apparently, CLFS-83 coverage had been much better). It must be remarked either that, in general, AMZ questions discern fewer categories than the corresponding CLFS-92 variables do. Unfortunately, even in rather simple cases, no straight mapping of categories is achievable. For instance, the AMZ discerns 4 categories of »*Familienstand*« (marital status), viz. ›ledig‹ (single), ›verheiratet‹ (married), ›verwitwet‹ (widowed), and ›geschieden‹ (divorced); as Austria is a catholic country appraising traditional values the AMZ does not get hold of persons living separated: these are simply counted as married though, economically, they share a de facto status with divorced persons. A similar case in point is »*Stellung im Haushalt*« which can be mapped simply to the CLFS-92 variable »*Relationship to reference person in the household*« at least nominally as indicated in **Tab. 3.2.3–6**; this association, however, is not quite correct because, contrary to CLFS conventions, children of the head of the household's spouse are counted as ›*Sonstige Person*‹ (other person). All of this, of course, may have yet little impact in deriving labour force and employment aggregates.

Table 3.2.3–6: Comparison of AMZ and CLFS-92 Categories for Variable "*Relationship to reference person in the household*"

\multicolumn{2}{c	}{AMZ – Variable Nr. 5}	\multicolumn{2}{c}{CLFS – Variable No. 1}
code	value	value	code
1	Haushaltsvorstand	Head of household/Reference person	1
2	Ehegatte des Haushaltsvorstands	Spouse/Spouse (or cohabiting partner) of reference person	2
3	Kind des Haushaltsvorstands	Child of head of household or his/her spouse / of reference person (or of his/her spouse or cohabiting partner)	3
4	Mutter, Vater des Haushaltsvorstands	Ascendant relative of head of household or his/her spouse / of reference person (or his/her spouse or cohabiting partner)	4
5	Sonstige Person	Other relative	5
		Other	6

More tangible are differences regarding work status. In the AMZ (as of 1992), the main question, »*Teilnahme am Erwerbsleben*«, which every person is asked irrespective of her age, subdivides the population into persons who worked in the reference week and those who did not. According to AMZ concepts, a person is employed (›*Beschäftigt*‹) only if she is working 12 hours per week at least assuming that it is rather pointless to speak of employment unless it earns a person's living ("subsistence concept" [Bartunek, 1994]); this, of course, sharply contrasts the ILO labour force concept. Persons not employed in AMZ terms are subdivided further into 7 groups (cf. **Tab. 3.3.3–2**), viz.

- conscript on compulsory military or community service, woman on maternity leave (›Präsenzdiener, Karenzurlauberin‹);
- registered unemployed (›arbeitslos (Arbeitsamt vorgemerkt)‹);
- pensioner (›Pensionist, Rentner‹);
- housewife without occupation (›nicht berufstätige Hausfrau‹);
- student, pupil (›Student, Schüler‹);
- child of pre-school age (›Kind im Vorschulalter‹);
- other supported person (›sonstige erhaltene Person‹).

In particular, there is no distinction of reasons why a person, despite having a formal job attachment, has not been working in the reference week (CLFS-92 variable no. 10; cf. **Tab. 3.2.2–3**). This is particularly remarkable in view of the high share of seasonal workers in the tourism industry having temporary work contracts and receiving unemployment benefits off-season (in CLFS-92 terms, these persons would have to be counted as ›laid-off‹). From autumn 1988 onwards when questions Nr. 20 was added to the questionnaire it is possible to measure the persons working 1 to 12 hours a week, thus admitting to estimate the Austrian labour force according to ILO definitions although the separation of unemployed and inactive persons must be accomplished using several questions simultaneously; for instance, a person over 14 who is a housewife (question Nr. 1) not working in the reference week (questions Nr. 13, 14, and 20) but seeking work (question Nr. 15) and being available within a month (question Nr. 16) can be classified appropriately as unemployed except that with respect to availability the CLFS-92 bound of two weeks is not supported by the AMZ.

Compared to the CLFS-92 scheme, the AMZ (in its version prior to the EU revision) provides a rather coarse picture of the labour market which, additionally, is biased heavily towards dependent employment. Furthermore, the main classifications (such as the rather outdated classification of occupations, question Nr. 11, or *BS68* of economic activity classification, question Nr. 12; cf. **Tab. 3.2.3–1**) are difficult to adapt to the standard classifications used by the CLFS-92. An outstanding peculiarity of the AMZ is its »*Stellung im Beruf*« classification (question Nr. 10) which combines elements of classifications with respect to professional status, economic activity, education level, and the number of persons in the enterprise all at once, thus withstanding almost any direct interfacing with established classification standards.

In concluding, without a profound and encompassing overhaul of AMZ's standard part this survey would have fallen short of contributing CLFS-92 compatible labour force data in many ways. These divergencies have now been remedied by establishing the AMZ-94 as a CLFS-92 conforming annual survey. Conversely, there had been conducted several special part sub-samples dedicated to specific issues of employment statistics aperiodically, providing employment data roughly equivalent to or even more detailed than standardized in CLFS-92 (for instance, the survey on "employment issues" of September 1988 addressing especially the subjects of vocational activity, employment seeking, and job change [ÖSTAT, 1991]). As will be demonstrated in the following Section 3.3, it is still possible and worthwhile to include the "old" AMZ in a concerted European labour force data integration effort. Particularly from the Austrian point of view such an inclusion would be reasonable and desirable in order to safeguard the valuable stock of AMZ data for continued usage in both national and EU contexts; as otherwise all historical AMZ data would be lost inevitably for comparisons with earlier as well as present and future EU labour force data.

3.3 Labour Force Survey Integration

Statistical data integration – whether within or outside a system like METASTASYS – presupposes the existence of an agreed-upon data harmonization standard. In METASTASYS's terms (cf. Section 2.3), this standard amounts to the compilation of a *theoretical data language*. Eventually, such a standard must encompass both, the taxonomic layer capturing the semantics of subject-matter terminology and the production structure capturing the pragmatics of data production (how it is brought about *and* for what reason it is done at all) in the subject-matter field, although the former only has been considered thoroughly in the discussion and development of METASTASYS. In particular, with respect to labour force statistics, in the long run the theoretical data language must be prepared to relate semantically, at the metadata level, the *statistical* features – sample design, sample errors, non-response, major sources of bias, etc. – of labour force surveys and other data sources that should so become integrated just as it relates their taxonomies; at present, however, only the latter will be focused upon. Roughly speaking, the practical task of semantic data integration ramifies into two broad areas or stages:

- the *analysis* stage where it must be decided which data sources to include actually, what the final aim of the harmonization effort is and which goals should be particularly supported, and in which respects the considered data sources are differing to a relevant degree;
- the *synthesis* stage where a definite data model (more specifically, a META-MODel) as well as the corresponding *mappings* from data sources to this data model must be arranged.

Of course, the sub-tasks of the analysis stage are highly interdependent and must, therefore, be decided jointly. In the subsequent synthesis stage the crucial question arises how flexible the arranged data model should eventually be; in this respect, two points have to be taken into account primarily:

- Since societal reality is in constant change an evolving labour force data model is called for rather naturally. A specific case in point is the revision of the CLFS programme taking place in 1992. As prior experience teaches, and in view of the increasing pace of structural change in the socio-economic system and the labour market in particular, a major revision of a harmonized data model for employment statistics should be expected at least once a decade anyway. In coping with changes of the data model, two principal strategies are conceivable; either the data model is *updated* whenever necessary (this, in turn, implies a corresponding change in all affected mappings established before an update takes place) or, more elegantly but decisively harder to accomplish, the changes to come are *anticipated* properly (then, apparently, formerly arranged mappings

will not have to be adapted, only new ones reflecting the recent changes in data sources need to be set up).
- Furthermore, there are lots of data sources contributing subordinate information to a harmonized employment statistics data model which certainly should not be dismissed light-heartedly (for one thing, many short-termed research questions of special concern may be hard to answer using CLFS data alone). Specifically, it may be a tactical argument as well to support the inclusion of a fairly broad range of past, present, and future data sources since this measure might vigorously stimulate, for instance, the willingness of data suppliers to actually undertake the efforts necessary to arrange the mappings of local data languages to the global (theoretical) data language of the harmonized data model. With respect to the EU Member States, this could make available EU-wide a large repository of national or regional socio-economic and demographic data at little additional cost – if one regards the high degree of overlap and consistency in terminologies used in national statistics within individual Member States – once the national statistical offices have been convinced to participate for the sake of mutual benefit (cf. Section 2.6 arguing in favour of this hypothesis). Admittedly, it may not at all be straightforward to combine those national and, as is well known, rather heterogeneous data sources but the uniform access achieved is a gain in itself not to be underestimated.

In practice, the clairvoyance needed to anticipate the true future course of development even in a narrow domain such as employment statistics is hardly attainable – at best, short-term trends can be accounted for adequately. Conversely, however, it may be just as wrong to adopt a pure update strategy preserving present standards by encoding current terminology only and, thus, leaving little room for adaptation and evolution (although, as the transition from CLFS-83 to CLFS-92 has highlighted, those revisions often involve simple relationships easily captured in formal terms whence, for instance, the update of mappings may be facilitated mechanically or, at least, supported considerably).

Without conducting an in-depth subject domain analysis (which, of course, had to follow the lines sketched in Section 3.2 in a real attempt to set up a truly useful labour force data model), subsequent considerations remain restricted to the synthesis of a provisional METAMODel for employment statistics particularly taking its bearings of the former DOSES project *Modelling Metadata* (cf. Section 1.4). In that project already labour force statistics had been chosen as running example and, accordingly, labour force metadata have also been used to demonstrate the DÖS'chen prototype described in Section 2.7; cf. the excerpt of labour force metadata in Appx. C. Drawing on this metadata repository (referring to the CLFS-83 standard mainly), the labour force data model is recapitulated coarsely and, at the same time, adapted to changes having taken place since. Moreover, METAMODel has evolved as well giving rise to adaptations of yet a different sort. The presentation of the labour force data model, however, will restrict attention to (real) data sources to the degree just necessary to keep track of structures and terminologies relevant for arranging data source mappings. Though based on an empirical analysis of the present practice of labour force statistics and particularly the CLFS-92 standard, the actual METAMODel devised abstracts decisively from this standard which would in fact provide a conceivable choice of FRAME definition but also would, for obvious reasons, exclude the majority of data sources of employment statistics not arranged specifically to feed the CLFS from the outset. Instead, the METAMODel proposed avoids deliberately explicit references to any particular labour force data model or survey structure in current use, mainly for the sake of demonstrating how a less "biased" data model could be set up supporting data integration favourably as a neutral as possible platform for the mappings to be arranged. Specifically, this METAMODel will not assume any temporal or spatial constraint to hold, thus pretending universal and ubiquitous validity (in spite of being well aware of its late-20th century Western industrialized hemisphere contingency).

Technically, the sample labour force METAMODel developed acts as a spinal structure for an aggregate labour force information network (referred to as ALFIN henceforth) rendering superfluous the physical allocation of statistical labour force aggregates at a particular statistical agency such as EUROSTAT. Quite on the contrary, local and especially national data (as provided by national statistical offices in case of the CLFS) remain hosted decentralized at ALFIN *server* nodes (META-NET servers; cf. Section 2.6); EUROSTAT, just as any other data consumer, would be connected to ALFIN as a METANET client (this, of course, by no means precludes EUROSTAT from taking a specific METANET server role as regards specially compiled and processed labour force aggregates not readily provided by "first order" ALFIN servers or macrodata too expensive to derive anew each time from local server mesodata). Conversely, EUROSTAT might appropriately take a new responsibility as treasurer of the ALFIN METAMODel; in any case, EUROSTAT would continue to be a major disseminator, or rather: *broker*, of statistical EU-figures providing added-value information services.

3.3.1 An Employment Statistics Data Model

As reviewed in Subsection 3.2.2, the basic instruments to measure labour force and unemployment are sample surveys organized and conducted by the Member States' national statistical offices. But even in case of data sources other than sample surveys, employment data is collected in survey-like formats mainly as person-based microdata (cf. Subsection 1.3.1). As a consequence of this, the production structure of employment statistics distinguishes itself by quite a considerable degree of uniformity although, of course, particularly historical data may only be available as highly aggregated macrodata (for instance, time series), often without any breakdowns at all. By convention, CLFS survey data is available as microdata and, hence, can be converted lossless into METASTASYS *meso*data format (cf. Section 2.3).

The first decision in setting up a METAMODel regards the number of populations (types of observation units) discerned as well as the relationships between these populations. In employment statistics, a natural observation unit or, in META-

.3 Labour Force Survey Integration 331

MODel's terms, FRAMEUNIT (cf. Section 2.2) is *person*; however, in general (and with respect to sample surveys even exclusively), the statistical sampling units are not persons but (private) households or dwellings (cf. Section 3.2). Therefore, a two-FRAME data model consisting of a *person* FRAME as well as a *household* FRAME appears reasonable such that *household* and *person* FRAMEs are linked *1:n*. In reality, of course, things might turn out as somewhat more intricate because of persons residing in more than one household (for instance, secondary places of residence such as weekend lodges) or surveys referring to work-places instead of private domiciles; moreover, there may be several households in a single dwelling, etc. Irrespective of such objections, the devised labour force METAMODel takes a simplified view and resorts to a single *person* FRAME approach assigning household features directly to persons ("population instances") thus in fact losing the information about which persons together belong to a particular household (another "normalization" flaw of this simplification is the duplication of household data for a household's persons).

Having decided upon the single-FRAME model, the QUalities determining this FRAME come next. As apparent from the discussion in Section 3.2, practically all QUalities are of CATEGorical SCALETYPE (cf. Subsection 2.2.1), with a few exceptions regarding time-related dimensions such as age, work time, duration of an employment contract, duration of job search, etc. which, naturally, may be assigned INTeger SCALEs. If it is not intended to support arithmetical operations (such as averaging) at all, ENUMerative or even CATEGorical SCALETYPEs might be chosen generally. Here a compromise is suggested using CATEGorical SCALEs for almost all non-numerical dimensions (sometimes collapsing to DICHOtomous or BINary SCALEs) and INTeger SCALEs otherwise, assuming that all temporal quantities can be expressed in terms of non-negative integers, that is setting LoBound=0 and HiBound=HINF (cf. Subsection 2.2.1). Occasionally, however, CARDinal and ENUMerative SCALEs will be used as well. To the universal QUalities, ***space*** and ***time***, the SCALEs of type 'SPATial' and 'TEMPoral', respectively, must be assigned anyway; SPATial SCALEs are used additionally for all dimensions with a geographical reference (such as work-places or places of residence). As far as non-numerical SCALEs are involved the SCALEDOMains must, of course, be enumerated explicitly which might turn out as a somewhat tedious task in view of the extensive nomenclatures occurring in labour force statistics; notably the NUTS classification consists of a long list of basic administrative districts all of which must be entered as individual MODalities. Likewise, the spatial relationship between these basic units has to be established explicitly as well (defining a neighbourhood graph appropriately). Furthermore, most nomenclatures and classifications come with hierarchical structures calling for the definition of suitable AGGLEVels (cf. **Def. 2.2.1–9**) on top of the SCALEs arranged. More importantly, apart from a natural thematic subdivision of dimensions (demographic background, job characteristics, employment seeking, etc.; cf. **Tab. 3.2.2–3**) fortunately there is no substantial interdependency between them; hence, basically each dimension corresponds to an individual SOLQUality (cf. **Def. 2.2.1–17**) spanning its own SOLDIMension (cf. **Def. 2.2.1–19**) of the *person* FRAME. With respect to *additivity*, there are virtually no constraints other than those holding by default, that is the temporal additivity constraint '*time*.FIX' (cf. **Def. 2.2.1–16**) applies generally (since, technically, the interviewed persons are *persistent* sampling units observed *at* a point in time *over* the value ranges of all other dimensions). Otherwise, additivity is not constrained anyway except for some numerical dimensions (for instance, it is obviously pointless to sum up a person's usual and actual work hours per week). Omitting minor details, **Tab. 3.3.1–1** summarizes the QUalities included in the sample *person* FRAME for the labour force domain data model. For the sake of easy reference, its leftmost column assigns an identifying number to each QUality; the rightmost column indicates the respective QUality's SCALETYPE. The column second from right indicates the CLFS-92 variables (as listed in **Tab. 3.2.2–3**) contributing information to the respective *LFF* QUalities; a '+' is used to signal whenever variables jointly provide information whereas variable numbers in parentheses flag partial contributions only. Let, for convenience, this FRAME be assigned the FRAMELABel (cf. **Def. 2.2.2–1**) '*Labour Force Frame*'; *LFF*, for short.

This compilation comprises quite an impressive set of QUalities potentially relevant in present and future labour force statistics. The subsumption of QUalities under intermediary headings (which are not part of the FRAME definition) roughly follows the CLFS-92 structure reproduced in **Tab. 3.2.2–3**; however, the number of headings has been slightly increased as a couple of QUalities is introduced which do not find CLFS counterparts. The reason for including more and somewhat different QUalities in the *LFF* (compared to the CLFS variables) is twofold: first, as argued in Subsection 3.2.3, there exist many non-harmonized data sources contributing employment statistics data the semantics of which is rather difficult to capture in CLFS terms; secondly, it is by no means presumptuous to assert that before long the present CLFS nomenclature will fall short of meeting the sharply rising information demands of policy makers and socio-economic researchers. All of this can be accounted for only by devising an extended set of QUalities and associated DIMDOMains (cf. **Def. 2.2.1–11**) not coinciding particularly with any of the predecessing survey structures. After all, the proposed *LFF* may be extended even further as it is easily possible to find yet more descriptive dimensions to record (for instance, the suggested set-up does not comprise wage-related QUalities – although such questions are included in many national labour force survey questionnaires), not to speak of the plenty of descriptive dimensions conceivable in a universal (cf. Subsection 2.2.2) *person* FRAME having little or no bearing on labour force and employment statistics whatsoever.

Although **Tab. 3.3.1–1** lists somewhat above 100 QUalities nominally, the resulting *LFF* in fact may have considerably more dimensions since three groups of QUalities, namely those under the headings of employment, job, and work characteristics, respectively, have to be repeated for each of a number of jobs accounted for. As more and more persons earn their livings by having several jobs or gainful engagements concurrently, it might soon be desirable to extend the characteristics recorded to second and even third jobs (note that the revision of the CLFS programme of 1992 gave much attention to second job characteristics; cf. Subsection 3.2.2). Presumably, $n = 3$ might be a reasonable choice for the next decade or so,

.3 LABOUR FORCE SURVEY INTEGRATION

thus giving a total *LFF* FRAMESIZE (cf. **Def. 2.2.2–2**) of 161. Practically, of course, the QULABs of those QUalities would have to be augmented to indicate explicitly the job to which they actually refer.

Table 3.3.1–1: Qualities of *Labour Force Frame*

No.	MOD*ality*	CLFS	Type
	universal Qualities		
Q_1	space	68+69	S
Q_2	time	66+67	T
	demographic characteristics		
Q_3	sex	2	B
Q_4	age	3+4	I
Q_5	marital status	5	C
Q_6	place of birth	8	S
Q_7	nationality	6	C
Q_8	usual place of residence (household)		S
Q_9	present household different from usual one		B
Q_{10}	duration of residence in usual place/household		I
Q_{11}	duration of residence at present place (household)	7	$I_\diamond$
	household characteristics		
Q_{12}	relation to head of household	1	$C_\diamond$
Q_{13}	type of household in usual place of residence		$C_\diamond$
Q_{14}	type of household in present place of residence (reference week)	72+73	C
Q_{15}	number of persons living in (present) household		N
Q_{16}	number of children under 15 living in (present) household		N
	employment status		
Q_{17}	work status (reference week)	9+10	$C_\diamond$
Q_{18}	regularity of employment	(37)	$C_\diamond$
Q_{19}	number of (current) jobs	(32)	$E_\diamond$
Q_{20}	main reason for having more than one job		$C_\diamond$
Q_{21}	total time elapsed since person first started to work		$I_\diamond$
	employment characteristics (*n*-th job)		
$Q_{22\text{-}n}$	professional status	11,33	$C_\diamond$
$Q_{23\text{-}n}$	occupation	13,35	$C_\diamond$
$Q_{24\text{-}n}$	leased worker		$B_\diamond$
$Q_{25\text{-}n}$	type of establishment		$C_\diamond$
$Q_{26\text{-}n}$	economic activity of local unit of establishment	12,34	$C_\diamond$
$Q_{27\text{-}n}$	number of persons working at local unit of establishment	14	$N_\diamond$
	job characteristics (*n*-th job)		
$Q_{28\text{-}n}$	permanency of the job	20	$C_\diamond$
$Q_{29\text{-}n}$	duration of the job since commencement	17+18	$I_\diamond$
$Q_{30\text{-}n}$	type of employment contract	(30)	$C_\diamond$
$Q_{31\text{-}n}$	total duration of employment contract with limited duration	(21)	$I_\diamond$
$Q_{32\text{-}n}$	full-time/part-time distinction	19	$C_\diamond$
$Q_{33\text{-}n}$	circumstances opening the way to job		$C_\diamond$

Table 3.3.1–1: Qualities of *Labour Force Frame* (cont'd)

No.	Modality	CLFS	Type
work characteristics (n-th job)			
$Q_{34\text{-}n}$	location of work-place	15+16	$S_\diamond$
$Q_{35\text{-}n}$	work time arrangement	25+26	$C_\diamond$
$Q_{36\text{-}n}$	job profile		$C_\diamond$
$Q_{37\text{-}n}$	number of usual work days per week		$I_\diamond$
$Q_{38\text{-}n}$	usual work hours per week	22	$I_\diamond$
$Q_{39\text{-}n}$	actual work hours (reference week)	23,36	$I_\diamond$
$Q_{40\text{-}n}$	main reason for difference between usual and actual hours work	24	$C_\diamond$
$Q_{41\text{-}n}$	hours usually worked at night		$I_\diamond$
$Q_{42\text{-}n}$	hours usually worked on Saturdays		$I_\diamond$
$Q_{43\text{-}n}$	hours usually worked on Sundays/holidays		$I_\diamond$
$Q_{44\text{-}n}$	occasional night work	27	$C_\diamond$
$Q_{45\text{-}n}$	occasional Saturday work	28	$C_\diamond$
$Q_{46\text{-}n}$	occasional Sunday/holiday work	29	$C_\diamond$
$Q_{47\text{-}n}$	main reason for difference between regular work hours in profession and usual work hours		$C_\diamond$
work experience of previously employed person			
Q_{48}	previous employment experience	38	$B_\diamond$
Q_{49}	time elapsed since last employment	39+40	$I_\diamond$
Q_{50}	main reason for terminating last job	41	$C_\diamond$
Q_{51}	professional status of last job	42	$C_\diamond$
Q_{52}	occupation of last job	44	$C_\diamond$
Q_{53}	economic activity of local unit of establishment of last job	43	$C_\diamond$
Q_{54}	number of persons in the local unit of establishment of last job		$N_\diamond$
Q_{55}	total duration of last employment		$I_\diamond$
Q_{56}	type of employment contract of last job		$C_\diamond$
Q_{57}	job profile of last job		$C_\diamond$
Q_{58}	status of inactive person	54	$C_\diamond$
employment seeking			
Q_{59}	looking for (an)other job	45 (+31) (+50)	$C_\diamond$
Q_{60}	main reason for seeking employment/another job	31	$C_\diamond$
Q_{61}	type of employment sought	46	$C_\diamond$
Q_{62}	desired occupation of employment sought		$C_\diamond$
Q_{63}	type of employment contract sought		$C_\diamond$
Q_{64}	desired job profile		$C_\diamond$
Q_{65}	duration of search	47	$I_\diamond$
Q_{66}	main method used to find work	48+53	$C_\diamond$
Q_{67}	time elapsed since last active step undertaken to find work	49	$I_\diamond$
Q_{68}	availability for work in ... weeks	51 (+50)	$I_\diamond$

Table 3.3.1–1: Qualities of *Labour Force Frame* (cont'd)

No.	MOD*ality*	CLFS	Type
Q_{69}	reasons for not being available earlier		C_*
Q_{70}	situation immediately before person started to seek work	52	C_*
Q_{71}	reason for not being registered unemployed at official employment exchange or office		C_*
	education and training		
Q_{72}	financial assistance/unemployment benefit payments	(53)	C_*
Q_{73}	highest level of general education attained	59	C_*
Q_{74}	highest completed level of further education or vocational training	60	C_*
Q_{75}	level of computer/data processing skills attained		C_*
Q_{76}	relevance of education and vocational skills in getting main (first) job		C_*
Q_{77}	education/training received during last month	55	C_*
Q_{78}	purpose of training received during last month	56	C_*
Q_{79}	qualification type mainly aimed at in the training received during last month		C_*
Q_{80}	obligatory/voluntary training		C_*
Q_{81}	total length of training	57	I_*
Q_{82}	usual number of training hours per week	58	I_*
Q_{83}	time elapsed since last training in main job		I_*
Q_{84}	time elapsed since last education/vocational training generally		I_*
	employment situation one year before		
Q_{85}	place of residence one year before	64+65	S
Q_{86}	work status one year before	61	C_*
Q_{87}	number of jobs one year before		E_*
Q_{88}	professional status (main job) one year before	62	C_*
Q_{89}	occupation (main job) one year before		C_*
Q_{90}	economic activity of local unit of establishment one year before	63	C_*
Q_{91}	type of establishment one year before		C_*
Q_{92}	same establishment (or economic successor) as one year before		C_*
Q_{93}	number of persons working in local unit of establishment one year before		N_*
Q_{94}	type of employment contract one year before		C_*
Q_{95}	job profile one year before		C_*
Q_{96}	full-time/part-time distinction one year before		C_*
	work characteristics (main job) one year before		
Q_{97}	location of work-place one year before		S_*
Q_{98}	work time arrangement one year before		C_*
Q_{99}	number of usual work days per week one year before		I_*
Q_{100}	usual work hours per week one year before		I_*
Q_{101}	actual work hours (reference week) one year before		I_*
Q_{102}	main reason for difference between usual and actual hours work one year before		C_*

Table 3.3.1–1: QUalities of *Labour Force Frame* (cont'd)

No.	MODality	CLFS	Type
Q_{103}	hours usually worked at night one year before		I_*
Q_{104}	hours usually worked on Saturdays one year before		I_*
Q_{105}	hours usually worked on Sundays/holidays one year before		I_*
Q_{106}	occasional night work one year before		C_*
Q_{107}	occasional Saturday work one year before		C_*
Q_{108}	occasional Sunday/holiday work one year before		C_*
Q_{109}	main reason for difference between regular work hours in profession and usual work hours one year before		C_*

Whether being equal to 109 or 161, the FRAMESIZE of the *LFF* is far too large to admit here a succinct description of each of the QUalities involved. Therefore, only a selected set of QUalities will be discussed shortly in what follows, mainly focussing on the reasons for including the QUalities and the purpose they are serving; to a lesser extent, MODalities and AGGLEVels are dealt with.

As can be seen from **Tab. 3.3.1–1**'s rightmost column, most of the DIMDOMains comprise the SPECVALue 'NULL' (cf. Subsection 2.2.1) for the simple reason that the QUalities they are attached to – some general QUalities under the headings of demographic and household characteristics excepted – apply only to *persons* over some specified age bound (which, in the EUropean context, may be set to 15 years) unless there are even more specific conditions QUalities depend on.

Spatial References and Nationality Enumeration

Several *LFF* QUalities refer to regional classifications whence it is advisable to set up a uniform SPATial SCALE valid for all of them. A natural choice for this SCALE's BASVALues is a subdivision of the earth's surface according to grown administrative regional units or juridical circuits since these subdivisions generally bear a long tradition outlasting, as has been proven, many of the coming and going political federations as well as social upheavals (for instance, Lechner [1992] refers to the political – and topographically well-founded – partition of Lower Austria by now being stable since its inception for almost a millenium). In a European context, the regional division must at least conform to the NUTS III level (comprising 828 elements in EUR-12 each of which, in general, is a union of still smaller administrative units) but may have a more fine-grained structure as well. For instance, Austria is subdivided into 2 337 political communities condensed to 35 NUTS-III units [Fahrn-gruber, 1993]. From a sampling point of view, it might be tempting to use blocks or addresses as smallest SCALE units but this would lead to both huge and unstable MODality sets hard to arrange and maintain. Conversely, regional classifications always call for an elaborated hierarchy of aggregation levels most of which are predetermined by political federations and national borders. Again, within EU boundaries, the NUTS hierarchy, comprising the levels NUTS-II to NUTS-0, successively condense NUTS-III units to increasingly larger units (NUTS-0 corresponds to Member States); cf. [EUROSTAT, 1992], Annexes I and IV. For thematic mappings the bottom-level SCALE units must be arranged in a neighbourhood graph to preserve topological relationships (which may also comprise coordinate information for interfacing with geographic information systems; cf. Subsection 2.2.1) inherited implicitly to all AGGLEvels defined. Established this way, the SPATial SCALE can be assigned not only to the universal *space* QUality (Q_1) but to other QUalities such as *place of birth* (Q_6), *usual place of residence* (Q_8), *place of residence one year before* (Q_{85}), *location of work-place* (Q_{34-n}), and *location of workplace one year before* (Q_{97}) as well.

A specific case is the CLFS-92 variable (no. 70), »*Degree of urbanization*«, excluded deliberately from the *LFF* definition: this refers to a feature of regions rather than of households (or, even stranger, persons). That it is among the CLFS-92 variables may be explained by the coarse regional classification of households used there since, otherwise, the three-way distinction into densely-populated, intermediate, and thinly-populated areas is a direct function of population density (cf. the Explanatory Notes section in [EUROSTAT, 1992]) and, thus, *must* be the same for all households – and *person*s – sampled in a particular geographical unit lest the dataset will become inconsistent formally (in database terminology, »*Degree of urbanization*« is functionally dependent on the *space* attribute). In the "clean" *LFF* approach, population density of regions would either be calculated explicitly from demographic data and spatial information attached to the *space* QUality or retrieved directly from attribute data accompanying the neighbourhood graph.

Another remark is in place regarding the *nationality* QUality (Q_7). In contrast to spatial nomenclatures, nationality refers to a logical classification (incidentally conforming to geographical units, of course; as a pertinent example cf. [EUROSTAT, 1992], Annex IV) the elements of which may arise or cease to exist rather independently from persistent *space* (as it would be the case, for instance, if a European citizenship would replace the hitherto national citizenships of the Member States of the EU; another case in point is the disintegration of former Yugoslavia into a couple of successor nations). Although there might be clear translation rules relating former and new nationality MODalities – for instance, in terms of MODality groupings – any changes in the set of *nationality*'s BASVALues by no means affect automatically the territorial subdivisions of SPATial SCALEs. As a consequence, the *LFF* definition assigns a separate CATEGorical SCALE to Q_7.

Unlike SPATial SCALEs encoding a given topography, *nationality*'s SCALE comprises an unstructured list of items ("nation labels"). However, MODalities are grouped frequently such that particular MODalities (nations) are distinguished from groups of other MODalities; in most cases this distinction is twofold – *natives* and *foreigners*. This is somewhat delicate semantically, since – in *LFF* terminology – despite the logical soundness of the principle adopted, for a *person* to belong to either of the cases her place of residence is completely irrelevant; *LFF* users are forced to distinguish crisply between resident, native, and resident native population and make a careful choice in preparing cross-national comparisons. By the way, native–foreigner groupings can be arranged as AGGLEvels in METAMOD only with respect to one *'native'* MOdality (or MOdality group) at a time whence for each nationality a special (set of) AGGLEvel(s) must be provided on demand – even if the AGGDomains of these AGGLEvels are identical nominally. Additionally, of course, other hierarchies – by geographic neighbourhood, ethnic classification, etc. – could be compiled as well in the usual way.

Temporal Reference

A principal decision of *LFF* design regards the temporal resolution of the *time* SCALE. Since, practically, the reference period used in data production is a week, it makes little sense to introduce a temporal unit smaller than a week to define the BASVALues of the TEMPoral SCALE assigned to the universal *time* QUality (Q_2). In view of the temporal granularity used practically in statistical aggregates, a monthly resolution of time is suggested implying that data pertaining to reference weeks or key-dates can always be assigned straightforwardly to a certain month unambiguously chosen from *time*'s SCALEDomain (sometimes, of course, weeks may cross monthly boundaries, thus either calling for some adjustment of observation data in a pre-processing stage or the association of data with both of the intersected months). Proposing a monthly temporal unit apparently restricts seasonal adjustments applied to observed data to months as well.

The SCALEDomain of *time* can be arranged favourably using the mechanism of COMVALues (cf. **Def. 2.2.1–2**); the resulting template then is *month.year* where *month*=⟨Jan, Feb, Mar, Apr, ..., Nov, Dec⟩ and *year* denotes the respective four-digit number of the year as usual (the BASET of *year* may be chosen suitably; note that it is always possible to extend the range of *year* by adding further BASVALues at either end of the sequence). Defining months as basic temporal unit, the corresponding periodicity graph (cf. Subsection 2.2.1) simply connects respective *month*s between successive *year*s, thus supporting particularly annual comparisons between reference months.

On top of this TEMP scale typical AGGLEvels such as 'season' (that is, quarters of a year), 'semesters', and (full) 'years' can be arranged easily; of course, other patterns of MOdality groupings are feasible as well (though, in general, calling for somewhat different aggregation procedures afterwards; note that, by default, aggregation with respect to all *'time*.Fix'-QUalities over subsequent *time* MOdalities is facilitated by simple averaging; cf. Section 2.5).

Other Time-Related Scales

In addition to the universal *time* dimension, the *LFF* comprises several QUalities referring to time though with different resolutions. In contrast to *time*, however, all of these SCALEs do not refer to *absolute* time but measure temporal distance with respect to a reference period or key-date. Conversely, since summing over durations (for instance, to compute averages, etc.) is a well-defined operation, these scales should be equipped with metric semantics. Instead of devising a single (relative) metric time scale with a temporal resolution fine-grained enough to express all measurement granularities occurring, the choice of a couple of formally unrelated scales probably is more handsome for practical purposes. Thus, essentially, five resolutions must be accounted for, viz. *hours* (for QUalities $Q_{38\text{-n}}$, $Q_{39\text{-n}}$, $Q_{41\text{-n}}$, $Q_{42\text{-n}}$, $Q_{43\text{-n}}$, Q_{82}, Q_{100}, Q_{101}, Q_{103}, Q_{104}, and Q_{105}), *days* ($Q_{37\text{-n}}$, Q_{81}, and Q_{99}), *weeks* (Q_{65}, Q_{67}, and Q_{68}), *months* (Q_{10}, Q_{11}, Q_{21}, $Q_{29\text{-n}}$, $Q_{31\text{-n}}$, Q_{55}, Q_{83}, and Q_{84}), and *years* (Q_4, *age*). QUality Q_{49}, *time elapsed since last employment*, might be either measured in weeks or months (the latter will do in most cases). In spite of the different scale units, the resulting SCALEs are identical nominally being INTeger SCALEs with a SCALEDOMain of natural numbers throughout as stated in the introduction to this subsection – with one exception: for apparent reasons, for $Q_{38\text{-n}}$ and Q_{100} (usual work hours) a different INTeger SCALE with LoBound=1 must be used. Contrary to the generic indication of SCALETYPEs in **Tab. 3.3.1–1**, however, the actual assignment of SCALEs to QUalities must distinguish between the SCALEs on account of the units meant (that is, for instance, the I_ϕ-SCALE assigned to Q_4, *age* in *years*, is indeed formally different from the I_ϕ-SCALE assigned to – say – Q_{65}, *duration of search*, in *weeks*).

As convenient, for each of these SCALEs a set of AGGLEvels may be arranged. In particular, several *age* classifications may be very handsome providing MODality groupings according to various criteria – half-decades, decades, subdivisions according to education career (such as, for instance, pre-school age, school age, working age, and retirement age), personal employment life cycle, etc.

Counting Scales

In a few cases, the *LFF* assigns CARDinal SCALEs to QUalities such as $Q_{27\text{-n}}$ (*number of persons working at local unit of establishment*) and its replications, Q_{54} and Q_{93}. In contrast to numbers of *person*s where summations are perfectly reasonable, the QUalities Q_{19} and Q_{87} are counting a *person*'s jobs in the range of up to three or so; for this reason, ENUMerative SCALEs with a SCALEDOMain $\langle 1, ..., n \rangle$ are assigned to these QUalities in the *LFF*.

A rather specific status enjoy the QUalities Q_{15} and Q_{16} because they are artefacts incurred from adopting a single-FRAME data model – if there was a *household* FRAME in addition to the *person* FRAME, both QUalities, *number of persons living in (present) household* and *number of children under 15 living in (present) household*, would disappear entirely (and reappear dynamically, if necessary, as adjoined N-SCALE DIMensions from the *1:n*-linkage of *household* and *person* FRAMEs; cf. Subsection 2.4.1). In the *LFF* set-up, Q_{16} is particularly unpleasant to deal with because of its explicit reference to the age bound of 15 (years) which – though being a reasonable choice presently for EUropean countries – makes it quite rigid and resistant to adaptation over time; furthermore, it may be difficult to arrange mappings to this QUality appropriately.

Work Status

The central criterion in labour force statistics is *work status* (cf. Section 3.2) subdividing the population of *person*s into the major groups of economically active and inactive persons and, hence, playing a pivotal role in formal concept definitions (cf. Subsection 3.4.1). Reflecting the well-developed methodology of the CLFS-92 (cf. **Tab. 3.2.2–4**), the *LFF* proposal devises a list of MOdalities for the work status QUalities (Q_{17} and Q_{86}) as comprised in **Tab. 3.3.1–2**.

Table 3.3.1–2: MOdalities of *LFF* QUality *work status*

	LFF-QUality Q_{17} (*work status*)	CLFS-92	
pos.	MOdality	*variable*	*code*
1	work for pay or profit	9	*1*
2	not active because of bad weather	10	*0*
3	not active because of slack work for technical reasons (technical breakdown etc.)	10	*1*
4	not active because of slack work for economic reasons		
5	not active because of labour dispute	10	*2*
6	not active because of education or vocational training	10	*3*
7	not active because of own illness, accident of work, or other temporary disability	10	*4*
8	not active because of maternity leave	10	*5*
9	not active because of vacation	10	*6*
10	not active because of new job starting in (near) future	10	*7*
11	not active because of personal or family responsibilities	10	*8*
12	not active because of other reasons		
13	lay-off (seasonal worker)	9	*3*
14	lay-off (otherwise)		
15	compulsory military or community service	9	*4*
16	participating in salaried (in cash or in kind) public employment promotion scheme	9	*5*
17	participating in public vocational training scheme		
18	domestic family responsibilities (housekeeping etc.)		
19	other		

The 19 MOdalities of *work status* in *LFF* provide a superset of CLFS-92 values for the combined »*Work status*« (no. 9) and »*Reason for not having worked ...*« (no. 10) variables; in a few cases, the CLFS-92 values are refined while other MOdalities do not have a CLFS-92 counterpart. Note that the CLFS-92 variables are nested because no. 10 applies only on condition that no. 9 attains value (code) ›2‹ and, hence, map to a single QUality in a natural way (cf. also to Appx. C.1 for the DÖS'chen definition of QUality ECONOMIC ACTIVITY STATUS). With respect to variable no. 10, the value ›Slack work for technical or economic reasons‹ gets split mainly since 'slack work because of economic reasons' (pos. *4*) ensues quite different implications than 'slack work because of incident technical reasons' (pos. *3*)

which, in general, will not be very important as to orders of scale. Furthermore, the BasValue 'personal or family responsibilities' (pos. *11*) is singled out from ›Other reasons (e.g. personal or family responsibilities)‹. The ›Was not working because on lay-off‹ is subdivided, now distinguishing if the lay-off regards seasonal work (pos. *13*) or not; in the latter case (pos. *14*), the reasons for being laid-off deserve further scrutiny since it may have one of several causes. Noting that unpaid domestic family responsibilities such as housekeeping are becoming more and more substituted by remunerated in-person services it appears reasonable to include a pertinent MODality in the *work status* QUality – performing domestic family services certainly is a kind of economic activity and not just a status of an economically inactive person. However, it is assumed that only those *person*s belong to class 'domestic family responsibilities' (pos. *18*) who deliberately consider this activity as a kind of job and not as involuntary pastime activity while looking for other gainful work.

Two categories are introduced newly, viz. positions *16* and *17* of **Tab. 3.3.1–2**. Referring to the discussion in Section 3.1 regarding the increasing pressure to find ways and means of employing persons dropped out from the labour market because of their advanced age or insufficient qualifications particularly in domains of useful social or other community work not being paid for on the regular (labour) marketplace, it might be very useful to classify *person*s participating in public employment promotion schemes (the "second labour market", cf. [Blanpain and Sadowski, 1994]) separately. Likewise, it is of specific interest to measure the unemployed finding temporary shelter in public vocational training programs (pos. *17*) established to improve their participants' chances to regain employment. The participation in a public employment promotion or vocational training scheme, in general, presupposes the registration at an official employment exchange service (cf. Q_{66}; see below); for all of the four categories encoded in the MODalities of positions *16* to *19*, however, further criteria (QUalities) are necessary to discern unemployed from inactive persons (cf. Subsection 3.4.1).

For easy usage, several AGGLEVels may be arranged to reduce the number of discerned categories. A natural candidate is the compression of MODalities in positions *2* through *12* (corresponding to the nesting of variable no. 10 in the CLFS-92) named, conforming to variable no. 9, code ›2‹, ›Was not working but had a job or business …‹ (cf. **Tab. 3.2.2–4**). Likewise, the lay-off distinction as well as the MODalities in positions *16* to *19* may be collapsed into single AGGLEVel values (cf. Appx. C.1 for analogous examples).

Employment Status Description

In addition to *work status*, the general employment status of an employed person involves some more descriptive dimensions. First, being economically active does not particularly elucidate how many gainful engagements a *person* actually has (Q_{19}) and what the reasons for having more than one job actually are (Q_{20}). Since the work status refers to a short reference period (week) only, it may not be representative for a person's employment status at all; specifically, a person with several concurrent formal job attachments may be employed regularly even if these jobs are temporary or seasonal ones; likewise, despite having several jobs, currently or otherwise, employment may not be continuous after all. For this reason, the *LFF* comprises the QUality *regularity of employment* (Q_{18}) which, contrary to the CLFS-92, is not assigned particularly to any of a *person*'s jobs (except there is a single job). Q_{19}, the *number of (current) jobs* QUality also controls the repetition factor n of QUalities allotted to employment, job, and work characteristics (QUalities $Q_{22\text{-n}}$ through $Q_{47\text{-n}}$; cf. Subsection 3.3.2).

Professional Status

Unlike *work status*, the QUality *professional status* ($Q_{22\text{-n}}$; Q_{51} and Q_{88}) is arranged with 9 positions, six of which are conforming to the 1966 definition of the Statistical Commission of the UN (cf. Subsection 3.2.1) comprising the MODalities 'employer', 'own-account worker', 'employee', 'unpaid family worker', 'member of a producers' co-operative', and 'other', augmented with three further MODalities, viz. 'self-employed with paid employees (small independent)', 'unpaid assistant worker', and 'unpaid social or community worker' (cf. also to Appx. C.1; QUality STATUS IN EMPLOYMENT) refining the basic 6-fold classification. The term "unpaid" is to be understood as not being remunerated in cash.

Collapsing 'employer' and 'self-employed with paid employees (small independent)' as well as 'unpaid assistant worker', 'unpaid social or community worker' and 'other' gives the original UN classification as a (first) AGGLEvel. The CLFS-92 categories (cf. **Tab. 3.2.2–4**) are obtained by collapsing MODalities as shown in **Tab. 3.3.1–3** (cf. Explanatory Notes of [EUROSTAT, 1992]).

Table **3.3.1–3**: AGGLEvel 'CLFS-92' for *LFF* QUality *professional status*

	LFF – professional status		CLFS-92 (variable no. 11)	
pos.	MODality	*code*	*value*	
1	employer,	1	Self-employed with employees	
2	self-employed with paid employees (small independent)			
3	own-account worker,	2	Self-employed without employees	
8	member of a producers' co-operative			
4	employee,	3	Employee	
6	unpaid assistant worker,			
7	unpaid social or community worker			
5	unpaid family worker	4	Family worker	

As another example, Austrian employment statistics discerns between ›Selbständige‹, ›Unselbständige‹, and ›Mithelfende Familienangehörige‹, giving rise to an

AGGLEVel collapsing positions *1, 2, 3,* and *8* (›Selbständige‹), as well as positions *4, 6,* and *7* (›Unselbständige‹); pos. *5* directly corresponds to ›Mithelfende Familienangehörige‹. In either example, MODality 'other' is not included in the AGGLEVel arranged.

An aspect of increasing importance, particularly in fighting the continuing unemployment of persons seeking work as employees in EUropean countries, is the instrument of employee *leasing* [Blanpain and Sadowski, 1994]. Thus, it appears reasonable to introduce a pertinent QUality ($Q_{24\text{-}n}$) in the *LFF* indicating if an employee is leased (currently) to an establishment legally different from the employer's one the employee is appointed to formally. It may also be reasonable to introduce a further distinction between occasionally and permanently leased workers (the latter referring to employees hired for the sole purpose of being leased to other establishments).

Occupations, Job Profiles, and Economic Activity Classification

Indispensably, the *LFF* accounts for a variety of standard classifications such as those of occupations or economic activity (QUalities $Q_{23\text{-}n}$, Q_{52}, Q_{62}, Q_{89}; $Q_{26\text{-}n}$, Q_{53}, Q_{90}). These will follow basically the suggestions of ILO, OECD, or EUROSTAT as discussed in Subsection 3.2.1. However, as there is a variety of rather "incommensurable" taxonomies in use in different data sources, it might turn out soon that the existing nomenclatures must be extended with an intermediary bottom layer of categories to facilitate mappings (change keys) appropriately.

Furthermore, the *LFF* definition envisages yet another type of job classification complementary to the established ones: since the traditional subdivisions of occupations according to standardized schemata of industrial bureaucracy are failing more and more to capture the economically relevant content of jobs it is advisable to draw a different distinction in terms of, say, the three-fold classification into routine work, in-person service, and symbolic analysis [Reich, 1993] which could, of course, be refined as to the main character of the work (manual, clerical, managerial) or to specific qualifications connected to it (for instance, information technology skills), to state just a few examples. Thus, with an eye to the analysis of structural change and its major effects on jobs and the labour market in general, each occurrence of the occupation QUality is paralleled with a job profile QUality ($Q_{36\text{-}n}$, Q_{57}, Q_{64}, Q_{95}) in the *LFF*.

With respect to leased workers ($Q_{24\text{-}n}$) it will be particularly interesting to record the economic activity ($Q_{26\text{-}n}$) of the establishments to which workers are leased (and not the obvious economic activity of worker leasing establishments).

The newly proposed *job profile* QUalities excepted, standard classifications of occupations and economic activity come with elaborated and codified hierarchies simple to arrange in AGGLEVel terms. The MODalities of *job profile* QUalities may be grouped in various ways according to different views at job profiles. Among the higher AGGLEVels will certainly be the above-mentioned three-fold classification in routine work, in-person service, and symbolic analysis.

Employment, Job, and Work Characteristics

In order to gain evidence about actual work conditions and the structure of employment a couple of further QUalities needs to be arranged. Three major groups of QUalities may be discerned in this respect, viz. QUalities referring to features of economic establishments (Q_{25-n}, Q_{27-n}; Q_{54}, Q_{91}, Q_{93}), QUalities describing the day-to-day work conditions (Q_{37-n} through Q_{47-n}; Q_{99} through Q_{109}), and QUalities reporting on the nature of employment contracts (Q_{28-n}, Q_{30-n}, Q_{31-n}, Q_{32-n}, Q_{35-n}; Q_{55}, Q_{56}, Q_{94}, Q_{96}, Q_{98}). Several of these QUalities have been mentioned previously whence only the remaining ones will be dealt with here.

As to the first group, the QUality *type of establishment* (Q_{25-n} and Q_{91}, respectively) may be used to capture the legal as well as the organizational forms of enterprises. This may be increasingly important to know in view of the blurring distinction between employers and dependent employees; as more and more establishments adopt lean management principles, flatten internal hierarchies, and outsource "peripheral" activities, sub-processes and tasks it is no longer the big companies (national corporations; cf. [Reich, 1993]) determining the statutory – and rather static – structure of business organization but the smaller firms and partnerships with their much more flexible internal structures and legal as well as financial arrangements. Additionally, this QUality could also be used to record whether the enterprise is local one, a national (either private or public) one, or a branch site of a multi-national corporation or enterprise.

The inevitable decline of accustomed distinctions between production and management workers (the *working class* vs. *business class* concept of Lynd and Lynd [1929]) is also reflected in novel types of employment contracts and work arrangements. Although the CLFS-92 has already reacted to emanating changes (by introducing the variables no. 25 through 30) it has done insufficiently so; a more systematic approach is envisaged by arranging two well-defined QUalities *type of employment contract* (Q_{30-n}; Q_{56}, Q_{94}) and *work time arrangement* (Q_{35-n}; Q_{98}). While the former addresses the distinction of contract types (such as collective contract, individual contract, home working, mixed dependent/independent employment contracts, etc.), the latter focuses on the particulars of temporal appointments (such as definite work time, floating work time, shift work, capacity oriented work time, overtime remuneration or time compensation, 4-day weeks, rotation schedules, etc.). The remaining QUalities of this group by and large could correspond to their CLFS-92 counterparts (cf. **Tab. 3.3.1–1**). The SCALE of the *type of employment contract* QUality may also be used for QUality *type of employment contract sought* (Q_{63}).

Many of the QUalities describing work characteristics have been commented on previously (since they are time-related metric QUalities); these QUalities are augmented with a few categorical ones regarding night work and weekend work (Q_{44-n} through Q_{46-n} and Q_{106} through Q_{108}, respectively) copied simply from the CLFS-92 variables no. 27 to 29 except that code ›1‹ of these variables (›Person usually works at night‹ etc.) does no longer apply; after all, the QUalities' SCALEs could be adapted to provide more case distinctions as to how frequent occasional night or weekend work is done actually (then, perhaps, these QUalities' SCALETYPEs should be changed from C to E).

Excepting the time-related QUalities, most of the discussed QUalities have CATEGorical SCALEs comprising quite small MODality sets; therefore, AGGLEVels play a minor role only. For metric QUalities, AGGLEVels may be defined ahead of time (such as, for instance, classifying *usual work hours per week*, $Q_{38\text{-n}}$, into the groups of 'insignificantly employed', 'part-time employed', and 'full-time employed') but will be outranked in general by dynamic classifications driven by empirical data distributions.

Employment Seeking

The counterpart to QUalities referring to persons in employment are QUalities describing the situation of the unemployed. Moreover, employed persons may seek yet further employment for various reasons. Hence, it is essential in the *LFF* to account for both, the type of employment sought and the specific characteristics of job search.

QUality *looking for (an)other job* (Q_{59}) functions as a switch, stating also the reasons why a person, employed or not, does *not* seek a job (this will be instructive not only for unemployed persons). Conversely, QUality *main reason for seeking employment/another job* (Q_{60}), covers a conceivable set of categories eliciting the primary motive for job search; of course, in case of unemployed persons the reason most likely will be 'to earn a living', a MODality being added to the list inherited, for instance, from CLFS-92 variable no. 31 (cf. **Tab. 3.2.2–3**). The *type of employment sought* (Q_{61}) is a mixture of *professional status* ($Q_{22\text{-n}}$) as well as *full-time/part-time distinction* ($Q_{32\text{-n}}$) MODalities, at least in the homonymous CLFS-92 variable (no. 46); perhaps, it should in fact have been separated accordingly in the *LFF* definition. Many QUalities describing the job sought in more detail refer to occupation (Q_{62}), type of employment contract (Q_{63}), and job profile (Q_{64}), with MODalities as discussed above. Specific criteria of job search are the *duration of search* (Q_{65}) and the *main method used to find work* (Q_{66}) which, in particular, comprises a MODALITY 'standing registration at public employment office' in addition to the MODalities corresponding to the values of CLFS-92 variable no. 48 (cf. **Tab. 3.2.2–3**). To be classified as unemployed according to ILO definitions (cf. Subsection 3.2.1), persons not only must seek work but be available as well; the temporal horizon of availability (in weeks) is captured in Q_{68}. This QUality, together with Q_{59}, accounts also for the attribute of a person's willingness to work despite not seeking work actively: an unemployed person not willing to work simply will not be available in finite time. Active search of work implies some steps undertaken by the person seeking employment which may be determined jointly by the method of search applied (Q_{66}) and the *time elapsed since last active step undertaken to find work* (Q_{67}). Q_{72} registers whether unemployed persons receive public (for instance, social security) or other financial assistance other than that obtained from participating in a salaried employment promotion scheme (cf. *work* *status*, Q_{17}, pos. *16* in **Tab. 3.3.1–2**). Q_{50}, *main reason for terminating employment*, provides additional information about previously employed persons helpful in tracing the economic reasons for becoming unemployed for both reasons, voluntary job termination and dismissals. Since either cause of unemployment sheds light on the dynamics of job mobility against the background of structural economic change particularly affecting the labour marketplace, this type of information – in conjunction with other QUalities such as Q_{55} – will become more and more crucial in targeting and assessing the efficacy of, for instance, specific governmental employment programmes [Wagner, 1993].

Education and Vocational Training

There is little controversy about the impact of education and vocational training on the international competitiveness of national or supra-national economies [EC, 1994]; in fact, there no doubt is a high correlation between a person's level of – general and vocational – education and professional skills on the one hand and the probability to stay in employment on the other hand. Therefore, QUalities related to education, training, and professional skills receive considerable attention in *LFF* design. In addition to the familiar descriptive dimensions established, for instance, in the CLFS-92 (corresponding to Q_{73}, Q_{74}, Q_{77}, Q_{78}, Q_{81}, and Q_{82} in *LFF*), the *LFF* comprises some more QUalities specifically addressing data processing skills (Q_{75}), the new qualifications – again including data processing skills – recent training has aimed at (Q_{79}), and the frequency of such training taking place (Q_{83}, Q_{84}). Acknowledging the growing importance of professional qualifications on the increasingly competitive labour marketplace, Q_{76} refers to the relevance of a person's education and/or vocational skills in getting her present main job. The time-related QUalities excepted, the arrangement of MOdality sets for all of the newly introduced QUalities certainly requires careful consideration (whence they are omitted from presentation here).

As most of the QUalities in the education and training group (will) possess quite specific MOdality sets of categorical type, AGGLEvels again are of minor importance.

Miscellaneous

The *LFF* comprises a couple of further QUalities not yet considered. Some of these QUalities need no particular explanation (like *sex*, Q_3), may be arranged according to more general principles (such as *marital status*, Q_5, as discussed in Subsection 3.2.3), or can be adapted or copied from their respective CLFS-92 counterparts (like *situation immediately before person started to seek work*, Q_{70} vs. variable no. 52, or *previous employment experience*, Q_{48} vs. variable no. 38). A few comments, however, are in place with respect to QUalities related to household characteristics. For one thing, as remarked in Subsection 3.2.2 an extended list of MOdalities (cf. **Tab. 3.3.3–1**) is required to enable appropriate mappings to Q_{12}, *relation to head of household* (cf. particularly to **Tab. 3.2.3–2** and **Tab. 3.2.3–6**). Furthermore, extended scales for Q_{13} and Q_{14} comprising various types of private (primary/secondary, domestic/foreign, stationary/mobile, etc.) and institutional (welfare, educational, religious, military establishment, hospital, asylum, worker hostel, etc.) households are mandatory for an easy handling of divergent survey coverage of data sources considered. For convenience, this multitude of MODalities may be compressed to several AGGLEVels for subsequent usage in table specifications; for instance, one such AGGLEVel may simply discern 'private' and 'collective' households. Although not provided for in the *LFF* proposal, the set of QUalities concerning household characteristics might be enlarged by, for instance, a QUality recording whether the sampled household is one of several households in the same dwelling or (one-family) house, or another QUality regarding the household's *equipment* useful or in fact already in use for gainful work; in particular, telecommunication and electronic data processing facilities (such as telefax, PCs, etc.) as these have become highly characteristic for many kinds of own-account employment and even indispensable for most type of home work (for both employees and persons gainfully employed otherwise).

3.3.2 Global *LFF* Edits

In the previous subsection, the DIMensions of a theoretical labour force statistics sample space have been arranged giving proof of METAMODel meeting the emanating modelling requirements. However, sample space – or FRAME – definition encompasses the definition of structural zeroes as well. As typical for FRAMEs consisting of mostly categorical QUalities, the *LFF* comprises an extensive collection of FRAMEDits (cf. **Def. 2.2.2–4**) because a few QUalities only apply generally to all *persons* (sampling units). This may be seen readily in **Tab. 3.3.1–1**'s rightmost column where all QUalities restricted to specific subsets of *persons* are flagged by subscripting the SCALETYPE symbol with 'ϕ' (the 'NULL' flag).

In devising *LFF*'s FRAMEDITS, only those TRAITs (cf. **Def. 2.2.2–2**) ought to be included which denote infeasible MODality combinations for sure, that is where it is definite that no data source whatsoever may use a combination of measurable values mapping to a TRAIT defined as FRAMEDit. Thus, for instance, defining *LFF* TRAITs denoting persons registered unemployed (Q_{66}) and, at the same time, working one or more hours a week usually ($\sum_n Q_{38-n}$) as FRAMEdit is not sound noting that such combinations might be perfectly legal in some countries according to, say, national Social Acts [Fürst, 1993].

In order to simplify the description of *LFF* FRAMEDits, the following shorthand notation will be used. First, QUalities will be denoted by numbers most of the time; if the range of a QUality is restricted to a subset of MODalities defined in the QUality's DIMDOMain, these MODalities – or the position numbers attached (if any) – are enumerated, separated by commas, in a parenthesized term attached to the QUality number. Hence, the general expression looks like

$$quality\_ref(\text{MODality\_set})$$

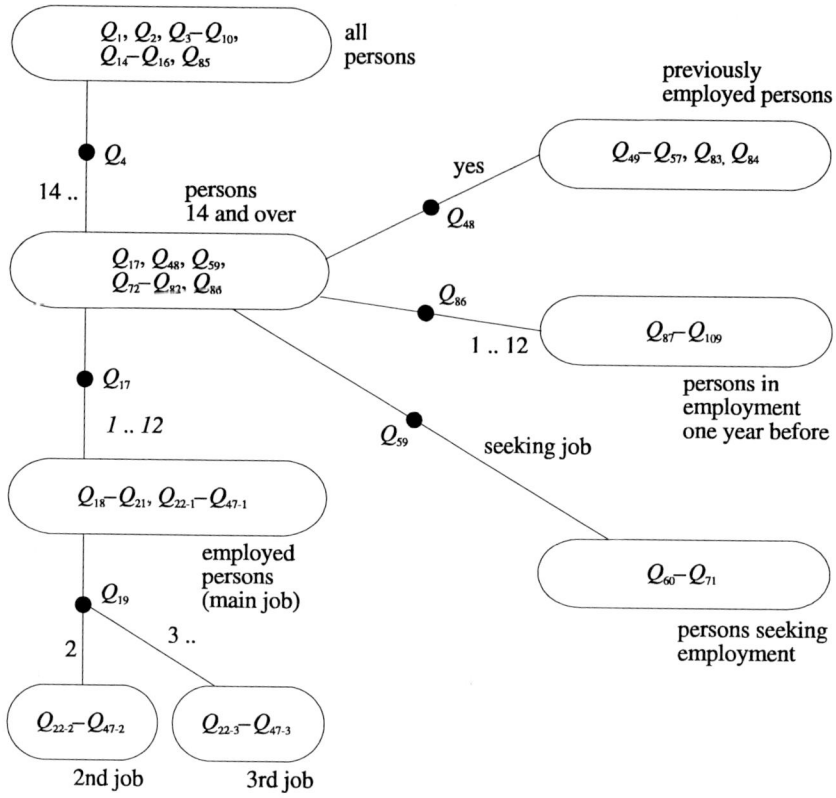

Figure 3.3.2: Main *LFF* Quality Dependencies

where the parenthesized term may be dropped if MODality_set coincides with the Quality's full DimDoMain. Furthermore, MODality_sets may be stated using the interval notation introduced in Subsection 2.4.3, that is '*lob* .. *hib*', '*lob* ..', or '.. *hib*' letting *lob* and *hib* denote the range's lower and higher bounds, respectively. As usual, 'ϕ' denotes the NULL value, and '*' is used to represent the set of all DimDom values except 'NULL'. A horizontally crossed MODality denotes the MODality set complement (excluding SPECVALues 'NULL' and 'MISSVAL'). According to the notation introduced in **Def 2.4.1–2**, the (primitive) concept product is denoted by operator '$\otimes$'.

Basically, *LFF* FRAMEDITS can be subdivided into hierarchical and non-hierarchical edits. By and large, this formal distinction coincides with a layering of *LFF* Qualities according to the main population groups similar to those summarized in **Tab. 3.2.2–6** with respect to the CLFS scheme (as to the hierarchical FRAMEDITs) on the one hand and a couple of further, more or less unrelated subject-matter edits on the other hand.

Dependency Edits

First turning to the hierarchical FRAMEDits, **Fig. 3.3.2** gives an overview of most of the *LFF* QUality dependencies in terms of a "generalized" dependency graph (cf. Appx. B); in fact, this graph omits very few of the hierarchical dependencies present in the *LFF*. Two of these refer to QUalities Q_{11} and Q_{13} depending on Q_9 (cf. **Tab. 3.3.1–1**): for apparent reasons, neither of Q_{11} and Q_{13} applies unless Q_9 takes on the MOdality 'yes', giving rise to the FRAMEDits

$$Q_9(\text{no}) \otimes Q_{11}(*) \qquad Q_9(\text{no}) \otimes Q_{13}(*)$$

as well as

$$Q_9(\text{yes}) \otimes Q_{11}(\phi) \qquad Q_9(\text{yes}) \otimes Q_{13}(\phi)$$

(note that Q_9 is not conditional whence the sequencing rule, cf. Appx B, does not apply). Another dependency omitted is that of Q_{12} which applies only in case of Q_{14} assuming one of the private household MOdalities.

The QUalities Q_{11} to Q_{13}, and Q_{58} excepted, **Fig. 3.3.2** attaches each of the remaining *LFF* QUalities to one of seven groups of hierarchically dependent QUalities unless they are unconditional ones going into group "all persons". Now, according to the algorithm expounded in Appx. B, the dependencies shown in **Fig. 3.3.2** can be converted into FRAMEDits mechanically as illustrated in the following examples.

Evidently, all dependent QUalities contained in the dependency graph of **Fig. 3.3.2** depend on Q_4, *age*. Setting the age bound to 14 (which, by the way, may be an unwise choice in view of several countries using or having used lower age bounds in several data sources), each of the QUalities contained in the group of "persons 14 and over" will induce two FRAMEDits analogous to the ones for the case of Q_{17}, *work status*, viz.

$$Q_4(0 .. 13) \otimes Q_{17}(*) \qquad Q_4(20 ..) \otimes Q_{17}(\phi)$$

as instances of a simple dependencies with unconditional entries (cf. Appx. B). Note that the lower age bound of 20 years for restricting the range of the dual edit is chosen somewhat arbitrary; however, there must be some room left for data sources' SOEDits to adapt to actual age bounds between 14 and 20 years.

Although being instances of simple dependencies either the QUalities of the group "persons seeking employment" are *conditionally* dependent because Q_{59} is already a conditional entry; hence, the FRAMEDits induced, for instance, for Q_{65}, *duration of search*, are as follows:

$$Q_{59}(\text{seeking job}) \otimes Q_{65}(*)$$
$$Q_{59}(\text{seeking job}) \otimes Q_{65}(\phi)$$

and, by virtue of the sequencing rule,

$$Q_{59}(\phi) \otimes Q_{65}(*)$$

This pattern is repeated over and over again for all conditional simple dependencies. For instance, picking $Q_{22\text{-}2}$ (from group "2nd job"), the FRAMEDits induced are

$$Q_{19}(\phi, 1, 3 \,..) \otimes Q_{22\text{-}2}(*)$$
$$Q_{19}(2) \otimes Q_{22\text{-}2}(\phi)$$

if the sequencing rule is collapsed with the first of the two edits.

The one and only different case regards Q_{58}, *status of inactive person*, depending on both Q_{17} and Q_{59} as follows:

- if Q_{17} assumes a MODality out of positions *1 .. 18*, the person certainly cannot be inactive;
- otherwise, if Q_{17} assumes MODality pos. *19*, the person to be inactive must not seek a job or have already found a job starting later.

More specifically, Q_{58} applies on condition that Q_{17} assumes MODality pos. *19* and Q_{59} assumes a MODality different from 'seeking job' or 'starts job already found later' (this range being denoted here briefly as *'not seeking'*). Since, in turn, both conditioning QUalities are depending on age (Q_4), a *multiple* dependency with two conditional entries results inducing three FRAMEDits:

$$Q_{17}(1 .. 18) \otimes Q_{59}(\text{not seeking}) \otimes Q_{58}(*)$$
$$Q_{17}(19) \otimes Q_{59}(\sout{\text{not seeking}}) \otimes Q_{58}(*)$$
$$Q_{17}(1 .. 18) \otimes Q_{59}(\sout{\text{not seeking}}) \otimes Q_{58}(*)$$
$$Q_{17}(19) \otimes Q_{59}(\text{not seeking}) \otimes Q_{58}(\phi)$$

as well as (FRAMEDits originating from applying the sequencing rule)

$$Q_{17}(\phi) \otimes Q_{58}(*) \qquad Q_{59}(\phi) \otimes Q_{58}(*)$$

viz. one FRAMEDit for each conditional entry of the multiple dependency.

Among the QUalities contained in **Fig. 3.3.2** there are some further dependencies within these main groups of QUalities. For instance, in the group "employed persons (main job)" it is rather pointless to consider Q_{20}, *main reason for having more than one job*, unless Q_{19} indicates that a *person* indeed has more than one job. Similarly, in group "person seeking employment", Q_{67}, *time elapsed since last active step undertaken to find work*, is senseless if a person does not undertake any steps to find work at all (Q_{66}). Both cases, of course, are instances of simple (conditional) dependencies inducing FRAMEDits analogous to those described above (that is including sequencing rule edits).

A somewhat more intricate case is the multiple dependency of $Q_{40\text{-}n}$, *main reason of difference between usual and actual work hours*, presupposing that there in

.3 LABOUR FORCE SURVEY INTEGRATION

fact is a difference between *usual work hours per week*, $Q_{38\text{-}n}$ and *actual work hours (reference week)*, $Q_{39\text{-}n}$. Since both $Q_{38\text{-}n}$ and $Q_{39\text{-}n}$ have INTeger SCALEs not enumerable explicitly, FRAMEDits can be stated in terms of an *edit schema* only; letting m denote an INTeger MODality, this schema comprises FRAMEDits of following shape:

$$Q_{38\text{-}n}(m) \otimes Q_{39\text{-}n}(m) \otimes Q_{40\text{-}n}(*)$$
$$Q_{38\text{-}n}(\cancel{m}) \otimes Q_{39\text{-}n}(m) \otimes Q_{40\text{-}n}(\phi)$$
$$Q_{38\text{-}n}(m) \otimes Q_{39\text{-}n}(\cancel{m}) \otimes Q_{40\text{-}n}(\phi)$$

as well as the sequencing rule edits

$$Q_{38\text{-}n}(\phi) \otimes Q_{40\text{-}n}(*) \qquad Q_{39\text{-}n}(\phi) \otimes Q_{40\text{-}n}(*)$$

According to the *lazy induction* principle (cf. Section 2.1), this edit schema is extended as far as necessary to fit the demands of actual computations.

Non-hierarchical Domain Edits

The larger part of FRAMEDITS consists of hierarchical QUality dependencies; there are, however, further edits holding for pure subject-matter reasons which cannot be dealt with in such a mechanical way. The arrangement of these FRAMEDits is nevertheless important to assure consistency of subsequent concept definitions typically supposing that FRAMEDITS have been established appropriately and with great care (cf. Subsection 2.4.1).

Non-hierarchical FRAMEDits may concern either QUalities between or within the major groups depicted in **Fig. 3.3.2**. An example of the former case is the edit regarding QUalities Q_{17} (*work status*), Q_{48} (*previous employment experience*), and Q_{49} (*time elapsed since last employment*): a currently employed person (Q_{17}) who also has had some previous engagement (Q_{48}) cannot be out of employment for any time period whatsoever (Q_{49}); thus,

$$Q_{17}(1 .. 14) \otimes Q_{48}(\text{yes}) \otimes Q_{49}(*)$$
$$Q_{17}(15 .. 19) \otimes Q_{48}(\text{yes}) \otimes Q_{49}(\phi)$$

(note that, unlike dependency edits, non-hierarchical edits do not come in pairs or larger sets of related FRAMEDits). The range of Q_{17}-MODalities has been extended to positions *13* and *14* to account for persons on lay-off (cf. **Tab. 3.3.1–2**) who, albeit not employed according to adopted definitions, still have a formal job attachment.

A similar inconsistency arises if persons are classified to be either at work or absent from work only temporarily (Q_{17}, MODalities in positions *1* through *12*) but not working usually at least one hour per week; thus

$$Q_{17}(1 .. 12) \otimes Q_{38\text{-}1}(0)$$

(note that $Q_{38\text{-}1}$ is irrelevant for other MODalities of Q_{17} which is taken account of by another – dependency – edit; see above).

An example of a FRAMEDit defined involving two QUalities of the same group is the one between Q_{18} (*regularity of employment*) and $Q_{28\text{-n}}$ (*permanency of the job*): a person having a permanent job must, by definition, work regularly (except for temporary interruptions irrelevant in this context). This gives rise to the FRAMEDit

$$Q_{18}(\text{regular}) \otimes Q_{28\text{-n}}(\text{permanent job or work contract of unlimited duration})$$

A similar case is the incompatibility of conscripts receiving unemployment benefits which can be expressed simply as FRAMEDit

$$Q_{17}(15) \otimes Q_{72}(\text{does not receive financial assistance or unemployment benefit})$$

recalling that pos. 15 of Q_{17} refers to persons of *work status* 'compulsory military or community service'.

Apparently, there is a couple of further non-hierarchical FRAMEDits not going to be discussed here such as, for instance, the FRAMEDits between *age* (Q_4) and *marital status* (Q_5) or *highest level of general education attained* (Q_{73}) because persons under, say, 15 years neither may get married nor obtain a university degree.

Compared to the transcript excerpts of Appx. C.2 presenting several FRAMEDits of the FRAME definition used in demonstrating the DÖS'chen prototype (cf. Section 2.7), many of the examples analyzed diverge considerably; this is mainly due to the profound revision of FRAMEQUalities (note that *LFF*'s predecessor comprised some 30 QUalities only) having taken place; furthermore, all FRAMEDits induced by sequencing rules have been omitted from the transcripts in Appx. C.2.

3.3.3 Survey Mappings

The real viability test of the *LFF* consists in the arrangement of data source mappings which, in view of the specific data sources discussed in Section 3.2, will be confined to *survey* data sources. However, since this restriction doesn't make a serious difference at the taxonomic level considered here, the mappings of labour force statistics data sources could justly be referred to as *survey* mappings.

In order to illustrate how labour force surveys in fact may be mapped to the *LFF*, a couple of examples is taken from the CLFS at Community as well as the national level (cf. Subsection 3.2.2); with respect to the latter the Portuguese Labour Force survey (based on an English transcript [Gomes and Miranda, 1991]) – P-LFS, for short – is chosen as a typical representative of a national survey destined to provide employment statistics data conforming to the CLFS-standard. On the other hand, the Austrian Mikrozensus (AMZ; cf. Subsection 3.2.3 specifically) will be used to highlight the difficulties in arranging a reasonable mapping for a non-harmonized national survey contributing labour force data; these intricacies notwithstanding, even such an odd data source can be mapped quite satisfactorily to the *LFF*.

At first glance it is apparent that, in general, both the CLFS itself as well as the national surveys prepared to contribute CLFS data can be mapped straightforwardly to the *LFF*; in particular, excepting several technical positions of the CLFS, all CLFS-92 variables are associated to at least one *LFF* QUality – typically, the relationship between CLFS variables and *LFF* QUalities is *1:1* at all (cf. **Tab. 3.3.1–1**) – unless there are explicit objections raised against the inclusion of CLFS variables into survey mappings (a case in point has been the *Degree of urbanization* variable; cf. Subsection 3.3.1). Although national data sources are forced, by EU law, to conform to the CLFS schema and it would, therefore, be sufficient to consider the mappings of CLFS schemata to the *LFF*, some examples will make it evident that direct mappings from national data sources to the *LFF* produce, in general, results preferable to mappings obtained by formally composing the two mapping stages. Nevertheless, for datasets (SOSECTions of data sources; cf. **Def. 2.3.1–5**) already mapped comprehensively to one of the CLFS schemata a "short-cut" mapping of CLFS-versions is likely to be economically more efficient than the establishment of mappings from original source data structures.

Anyway, all of the following examples given are rather sketchy in view of the limited space available. Thus, in particular, virtually every data source would require some kind of pre-processing of "raw" survey variables (as corresponding to survey questionnaires) almost completely omitted from the subsequent discussion. Furthermore, other preparatory steps – such as, for instance, the arrangement of dummy mappings for *lurking* variables (cf. Subsection 2.3.1) – will be disregarded except for a selected few cases. Likewise, the transformation of SOEDITS into IMEDITS (cf. **Def. 2.3.3–1**) will not be dealt with in general since a large share of EDits of, for instance, the CLFS-92 survey scheme (cf. **Fig. 3.2.2–2**) is covered readily by global *LFF* FRAMEDits introduced in Subsection 3.3.2 above. As there is no specific FRAMEARRangement (cf. **Def. 2.2.2–2**) stipulated, QUality ordering of the *LFF* will be assumed as given implicitly in **Tab. 3.3.1–1** (note that this ordering reflects any dependencies between QUalities by incidence only).

Facet Mappings

The first step in mapping a data source – expressed in terms of a SOFRAME – to a FRAME is the formal association of the labour force surveys' SOFRAMEs to the *LFF*. In METAMOD, this is stated in terms of a FACMAP (cf. **Def. 2.3.3–3**) on account of the same FRAMEUNIT shared by both the considered SOFRAMEs and the *LFF*. Assigning (provisionally) the SOFRAMELABels 'CLFS83', 'CLFS92', 'PLFS', and 'AMZ' to CLFS-83, CLFS-92, P-LFS and AMZ surveys, respectively, all of these FACMAPs can be denoted symbolically as

$$\mu_V(CLFS83) = LFF \qquad \mu_V(PLFS) = LFF$$
$$\mu_V(CLFS92) = LFF \qquad \mu_V(AMZ) = LFF$$

More specifically, such a mapping must be defined formally for each *facet* of a data source (for instance, each dataset obtained from one of the annually or quarterly conducted surveys) noting that changes in the (taxonomic) survey structure – like those introduced in revising the CLFS schema in 1992 – require the arrangement of distinct data source facets.

Universal QUalities

Once the FACMAPs have been settled, VARMAPs and VALMAPs (cf. **Def. 2.3.3–4** and **2.3.3–5**, respectively) come next. Somewhat specific roles play the universal QUalities, the mappings to which, in general, presuppose more or less special pre-processing steps.

Considering *time* first, the CLFS-92, for instance, records the reference periods of surveys in variables no. 66 (year) and 67 (week within year). Formally, variable no. 67 is a refinement of the scale used in variable no. 66 and, hence, represents a nested variable. In view of *time*'s temporal granularity of months (cf. Subsection 3.3.1), the pre-processing bringing about the variable nesting could, in one and the same step, recode the reference weeks into respective months, thus facilitating a subsequent *1:1* mapping of SOVALues to MODalities at the VALMAP level. This mapping, of course, preserves the information about the Member States' differing survey reference periods (cf. **Tab. 3.2.2–1**) though at the coarser monthly resolution only. Formally, letting *RefMonth* denote the pre-processed SOVAR obtained from variables no. 66 and 67, and letting further t denote a specific SOVALue, VARMAP and VALMAPs can be stated as

$$\mu_V^{(CLFS92)}(RefMonth) = \langle time \rangle$$

and

$$\mu_{|RefMonth}^{(CLFS92)}(t) = \langle t \rangle$$

(CLFS-83 mappings are completely analogous). In the P-LFS, Questions no. 3 (year) and no. 11 (date of interview) provide the reference period information; this is somewhat redundant, of course, and the respective mappings may be expressed in terms of Question no. 11 directly. The AMZ is conducted quarterly at fixed months (March, June, September, and December) using these months' first day, respectively, as key-date. Since this key-date is not part of the questionnaire data, this temporal reference represents an instance of an *implicit* variable (cf. Subsection 2.3.1) to be created in the pre-processing stage; the resolution of this implicit SOVARiable is months readily.

Similar to the temporal reference, both CLFS-83 and CLFS-92 encode the spatial reference by two separate variables, viz. no. 68 (Member State) and 69 (regions at NUTS-II level). Before mappings to *space* can be established formally, again an intermediary nesting step is required, after which SOVALs can be mapped directly to MODalities or suitably arranged MODality groups (if a NUTS-II SOVALue represents a set of *space*'s BASVALues). In contrast to temporal mappings, however, the spatial contributions of data sources are subdivided naturally into state territories, that is – with respect to EU countries – each Member State's mapping to *space* covers only those MODalities referring to the NUTS-II regions of its own territory. For instance, in the P-LFS Question no. 1 provides the region code according to NUTS-II taxonomy whence the VALMAP is *1:1* to MODality groups representing NUTS-II region units; conversely, the AMZ records district codes beneath NUTS-III level which possibly could be mapped directly to *space* MODalities. Occasionally, facet *space* mappings deviate from the "general" format or receive even major changes. For instance, the Portuguese labour force surveys of 1986 and 1987 did not discern between the NUTS-II regions 'Alentejo' and 'Algarve' [EUROSTAT, 1990], and the German re-union in 1990 extended the territory covered by the 5 "Neue Bundesländer"; in both cases, this gives rise to specific data source facets redefining the VALMAPs of spatial references (this is not to be confused with the case of *regular* omissions of territorial units; for instance, Spain does not report labour force data for the NUTS-II region 'Ceuta y Melilla' [EUROSTAT, 1990]).

General Domain QUalities

Many of the survey variables belonging to rather general person attributes are not particularly intricate to map to their *LFF* QUality counterparts; an example of utmost simplicity is the mapping of »*Sex*« to Q_3 (however, there are counterexamples such as the AMZ variable »*Familienstand*«, Nr. 8, not so easy to map to Q_5 as discussed in Subsection 3.2.3).

How data sources may differ with respect to a general domain QUality and how these differences are captured formally in VALMAPs is highlighted for the case of Q_{12}, *relation to head of household*. **Tab. 3.3.3–1** (overleaf) summarizes in tabular format the VALMAPs of respective CLFS-92 (which is identical to the CLFS-83 one), P-LFS, and AMZ SOVARiables.

As can be seen in connection with **Tab. 3.2.3–6** for the AMZ variable (Nr. 5), »*Stellung im Haushalt*«, the direct mapping gives a different result compared to the concatenated mapping via CLFS-92 variable no. 1; likewise, the direct mapping differs for the P-LFS as well (Question Q13; cf. **Tab. 3.2.3–2**). Appx. C.4 gives a brief account of how these mappings have been realized in the DÖS'chen prototype demonstration.

Work Status

A case of particular interest is the arrangement of mappings to *LFF* QUality Q_{17}, *work status*, because most domain concepts used for statistical tabulation and analysis are expressed in terms of this QUality (cf. Subsection 3.4.1). As different data sources have been set up with rather diverging views of relevant concepts to be measured it comes as no surprise that the resulting mappings are varying to a considerable degree.

With respect to the CLFS schema, the mapping is accomplished very easily once variable no. 10 (»*Reason for not having worked ...*«) has been nested into variable

Table 3.3.3–1: SoVar Mappings to *LFF* Quality Q_{12}

	relation to head of household (Q_{12})	VALMAP *Definitions*		
pos.	Modality	$\mu_{IV1}^{(CLFS92)}$	$\mu_{IQ13}^{(PLFS)}$	$\mu_{\text{Stellung im Haushalt}}^{(AMZ)}$
1	reference person (head of household)	1	1	1
2	spouse (or cohabiting partner)	2	2	2
3	child of reference person	3	3	3
4	child of spouse (or cohabiting partner)			5
5	mother/father of reference person	4	4	4
6	mother/father of spouse (or cohabiting partner)			—
7	daughter or son in law of reference person or spouse (or cohabiting partner)	5	5	5
8	other relative of reference person or spouse (or cohabiting partner)		6	
9	other	6	7	

no. 9 (»*Work status*«), code ›2‹, in a suitable pre-processing step. Letting this intermediary SoVar be denoted by »*WorkStat*«, the corresponding VarMap is

$$\mu_{IV}^{(CLFS92)}(WorkStat) = \langle \text{work status} \rangle$$

and the resulting ValMap encodes the association of SoVals (cf. **Tab. 3.2.2–4**) to Modalities as listed in **Tab. 3.3.1–2**.

This mapping holds for both CLFS-83 and CLFS-92 notwithstanding the general change in the age bound defining a person's working age (which has been raised from 14 to 15 years in 1992). Although this change does not affect the mappings, IMEdits, of course, must be adapted accordingly; specifically, the former IMEdit

$$Q_4(14 \mathinner{..} 19) \otimes Q_{17}(\phi)$$

of the CLFS-83 mapping must be replaced with

$$Q_4(14) \otimes Q_{17}(*) \qquad Q_4(15 \mathinner{..} 19) \otimes Q_{17}(\phi)$$

in order to fill the "gap" left in the FrameEdits to accommodate for specific data source mappings (cf. Subsection 3.3.2).

As the P-LFS splits up the *work status* QUality in yet more variables in the Portuguese questionnaire (cf. **Tab. 3.2.3–3**) a pertinent pre-processing step is mandatory before MetaMod mappings can be established formally.

Table 3.3.3–2: AMZ Variables Needed for *work status* Mapping

Nr.	Code	Variable
9		*Teilnahme am Erwerbsleben*
	1	Beschäftigt
	2	Arbeitslos (Arbeitsamt vorgemerkt)
	3	Pensionist, Rentner
	4	Nicht berufstätige Hausfrau
	5	Student, Schüler
	6	Kind im Vorschulalter
	7	Sonstige
	8	Präsenzdiener, Karenzurlauberin
15		*In den letzten 4 Wochen aktiv einen Arbeitsplatz gesucht ?*
	0	Nein
	1	Arbeitsplatz
	2	Lehrstelle
20		*In der letzten Woche mind. 1 Stunde gegen Bezahlung gearbeitet od. im Familienbetrieb mitgearbeitet ?*
	0	Nein
	1	Bezahlt
	2	Mitgearbeitet

Less trivial is the formal arrangement of a *work status* mapping in case of the AMZ (which, in this respect, might be quite representative of other non-harmonized labour force data sources). First, excepting the auxiliary variable »*Alter*« (computed from the interviewed person's birth date and the survey date), three AMZ-variables must be combined in a pre-processing step to achieve a SOVAR ready for mapping. These variables and their code lists are comprised in **Tab. 3.3.3–2**. Variable Nr. 9 applies generally to all persons while variables Nr. 15 and 20 are asked persons over 14 years only (inducing a corresponding SOEDit).

Now, taking care of the interdependencies between these three variables, a mapping can be stated – provided that an AXIS (cf. **Def. 2.2.3–3**) is composed out of QUalities Q_{17}, Q_{58}, Q_{59}, and Q_{66} – as summarized in **Tab. 3.3.3–3** (overleaf). The left-hand part of the table comprises the supercodes induced jointly from variables Nr. 9, 15, and 20; the right-hand part lists the component values of the 4-dimensional IMAXis using the notation for value ranges introduced in Subsection 3.3.2. Fields shaded dark-grey in the left-hand part of the table denote values determined by functional dependencies (SOEDits). Note, furthermore, that in order to build supercodes from three variables, variable Nr. 9 must be pre-processed individually collapsing its codes ›*1*‹, ›*4*‹, ›*5*‹, and ›*7*‹ in a couple of cases. The rightmost column of **Tab. 3.3.3–3** enumerates the IMVALs of the resulting *LFF* IMAXis.

Table 3.3.3–3: AMZ *work status* Mapping

AMZ *Variable*			LFF ImAxis				
Nr. 9	Nr. 15	Nr. 20	Q_{17}	Q_{58}	Q_{59}	Q_{66}	*pos.*
1	0	0	2..7,9..12	ϕ	~~seeking job~~	ϕ	1
1,4 5,7	0	1,2	1	ϕ	~~seeking job~~	ϕ	2
1	1,2	0	2..7,9..12	ϕ	seeking job	*	3
1,4 5,7	1,2	1,2	1	ϕ	seeking job	*	4
2	0	0	13,14,16 17,19	ϕ	~~seeking job~~	ϕ	5
2	1,2	0	13,14,16 17,19	ϕ	seeking job	standing registration at public employment office	6
3	ϕ	ϕ	19	retired	ϕ	ϕ	7
4	0	0	18	ϕ	~~seeking job~~	ϕ	8
4	1,2	0	18	ϕ	seeking job	*	9
5	0	0	19	in education	~~seeking job~~	ϕ	10
5	1,2	0	19	in education	seeking job	*	11
6	ϕ	ϕ	ϕ	ϕ	ϕ	ϕ	12
7	0	0	19	permanently disabled, other	ϕ	ϕ	13
7	1,2	0	19	ϕ	seeking job	~~standing registration at public employment office~~	14
8	ϕ	ϕ	8,15	ϕ	ϕ	ϕ	15

Although quite intricate, this mapping is not yet correct unless persons under 15 years have been recoded generally to category ›6‹ of variable Nr. 9; this way, pupils of regular school age (6 to 14 years) are separated appropriately from pupils ("Schüler") of age 15 and over. The age bound of 15 years applying to the AMZ induces, of course, the same type of IMEDits as the CLFS-92 does; see above. Note also that Q_{59}(~~seeking job~~) must always be accompanied by $Q_{66}(\phi)$ because of the Q_{59}/Q_{66}-FRAMEDit (cf. Subsection 3.3.2). A specific deficiency of the AMZ compared to the CLFS schema is its shortcoming to distinguish between people at work and those not at work temporarily though having a job (CLFS-92 variable no. 10); as a consequence, code ›1‹ of variable Nr. 9 becomes associated with the group of Q_{17}-MODalities comprising the positions *1 .. 7* as well as *9 .. 12*. Only code ›8‹ of this variable –maternity leave – corresponds explicitly to $Q_{17}(8)$. It is just another peculiarity of the AMZ that it subsumes persons doing their compulsory military or community service under the same code as persons on maternity leave: in former days, when women only had the right to suspend work for taking a maternity leave it had been easy to separate them perfectly from the males doing compulsory military or community service by resorting to the sex variable (»*Geschlecht*«, Nr. 6). In this case, if variable Nr. 6 was included in the pre-processing step of the SOVAR then attached to the augmented IMAXis comprising Q_3, *sex*, as well it would thus be possible to split IMVALue *15* accordingly; in practice, this no longer works because in the meantime Austrian law permits also *males* a maternity leave.

Professional Status

A case rather similar to that of *work status* described above regards the mappings to Q_{22-1} (and Q_{22-2}, respectively), *professional status*. Again, the mapping of the respective CLFS-83 and CLFS-92 variables is achieved easily by encoding the VALMAPs according to the associations listed in **Tab. 3.3.1–3**. Somewhat more involved is the mapping with respect to the P-LFS because of this survey's preference to use a cascade of dichotomous variables to enumerate the discerned categories. However, after a simple pre-processing of the P-LFS Questions 33 through 36a (cf. Subsection 3.2.3) and 80 through 83a, respectively, the resulting intermediary SOVARs can be matched with Q_{22-1} (Q_{22-2}) as outlined in **Tab. 3.3.3–4**.

Table 3.3.3–4: P-LFS Mappings to *LFF*-QUality *professional status*

P-LFS *Question*	Value	Q_{22-1}/Q_{22-2} MODality
Q33/Q80	self-employed with employees	1,2
Q34/Q81	self-employed without employees	3
Q35/Q82	employee	4
Q36/Q83	family worker without remuneration	5
Q36a/Q83a	active member of production co-operative	8

Q_{22-n} MODality positions 6 ('unpaid assistant worker') and 7 ('unpaid social or community worker') do not get assigned a SOVALue according to this definition (which, hence, is in demand of further clarification).

Particularly unpleasant is the arrangement of an AMZ mapping of variable »*Stellung im Beruf*« (Nr. 10) to Q_{22-1} (there is no variable related to Q_{22-2} in the AMZ) as has been indicated in Subsection 3.2.3. Because of the high degree of semantic overlap with AMZ variables Nr. 12 (»*Betriebszweig*«), Nr. 18 (»*Höchste abgeschlossene Schulbildung*«), and Nr. 19 (»*Lehrabschlußprüfung*«) – which, by the way, gives rise to a multitude of SOEDits hard to overlook – an IMAxis comprising no less than six QUalities – Q_{22-1} accompanied by Q_{24-1}, Q_{25-1}, Q_{27-1}, Q_{48}, and Q_{73} – must be moulded before the VALMAP can be set up. Without delving into details of this awkward case, the inevitable introduction of a 6-dimensional IMAxis considerably reduces the utility of this data source in subsequent table derivations since particularly combinations with other IMAGEs will be difficult to achieve (cf. Subsection 2.5.3).

Other Domain Specific QUalities

In general, the extension of the variable catalogue in 1992 excepted, both CLFS-83 and CLFS-92 schemata are highly conforming. However, in particular cases the transition from the 1983 version to its 1992 successor is everything but smooth; such a case in point is Q_{50}, *main reason for terminating last job* (cf. Subsection 3.2.2 and specifically **Tab. 3.2.2–7**). Although the situation is improved slightly by suggesting an extended MODality set as comprised in **Tab. 3.3.3–5**, the mapping of the respective CLFS-83 variable still cannot be accomplished straightway (cf. the arguments put forth in Subsection 3.2.2 and, in particular, **Tab. 3.2.2–8**).

Table 3.3.3–5: Q_{50} MODalities and CLFS/P-LFS Code Mappings

LFF-Q_{50}		CLFS-92 code	CLFS-83 code	P-LFS code
pos.	MODality			
1	dismissed individually	0	0	2
2	dismissed collectively			1
3	abdication because of personal or family responsibilities	2	2	5 12
4	cancellation on reciprocal agreement			4
5	terminated because job had limited duration	1	1	3
6	education/training	4	2	10
7	early retirement for economic reasons	5	3	6
8	early retirement for other reasons		5	8
9	normal retirement	6		9
10	compulsory military or community service	7	6	11
11	illness, disability	3	4	7
12	other	8	2,7	13

Note that the CLFS-83 mapping involves also a special IMEDit connected to Q_{22-1} and Q_{50} since the codes ›0‹ to ›3‹ of the CLFS-83 »*Main reason ...*« variable applied only for employees (code ›3‹ of CLFS-83's »*Professional status*« variable).

The rightmost column of **Tab. 3.3.3.–5** includes the mapping of P-LFS's variable (Question) no. 110, too, which apparently is accomplished simply requiring a preparatory collapsing of categories coded ›5‹ and ›12‹, respectively (Q110 of the P-LFS makes a – questionable – distinction between abdication and job termination because of personal or family responsibilities). The AMZ does not contain a pertinent variable contributing to Q_{50} at all; this, incidentally, provides an example of a *lurking* variable. Likewise, $Q_{33\text{-}n}$ (*circumstances opening the way to job*), is not fed by the CLFS schemata (neither is it by the AMZ); only the P-LFS comprises a variable (Question no. 50) mapping to $Q_{33\text{-}1}$.

In a few cases, no direct association of a CLFS variable to a *LFF* QUality is possible. For instance, considering the CLFS-92 variable no. 50, »*Willingness to work for person not seeking employment*«, the mapping presupposes an IMAxis comprising both Q_{59} (*looking for (an)other job*) and Q_{68} (*availability for work in ... weeks*) because the willingness to work implies short-term availability of the person. To Q_{59} and Q_{68}, in turn, map the variables no. 45 (»*Seeking employment for person without employment during the reference week*«) and 51 (»*Availability to start working within two weeks*«), respectively; therefore, a pre-processing step is necessary generating supercodes for feasible triplets of codes from variables no. 45, 50, and 51 becoming mapped to this IMAxis as shown in **Tab. 3.3.3–6**. Again, fields shaded dark-grey indicate functionally dependent values (SoEDits). Code ›1‹ of variable no. 50 represents the value ›would like to have work‹ while code ›2‹ represents the value ›does not want to have work‹; the '+' is used to denote availability within two weeks for variable no. 51, otherwise a '–' symbol is inserted. The '*otherwise*' entries represent the remaining categories of variable no. 45 and MOdalities of Q_{59}, respectively, assuming that they are in strict *1:1* correspondence. Not being available for work at all can be expressed by setting Q_{68} to 'HINF' (meaning "not within finite time").

Table 3.3.3–6: Mapping CLFS-92 Variable No. 50 to the *LFF*

CLFS-92			LFF	
No. 45	No. 50	No. 51	Q_{59}	Q_{68}
Person is seeking employment	φ	+	seeking job	0 .. 2
Person is seeking employment	φ	–	seeking job	3 ..
Person has already found a job which will start later	φ	φ	starts job already found later	φ
otherwise	1	+	otherwise	0 .. 2
otherwise	1	–	otherwise	3 ..
otherwise	2	φ	otherwise	HINF

A similar case is the CLFS-92 variable no. 53 (»*Registration at a public employment office*«) mixing up the facts of being officially registered unemployed and of receiving an unemployment benefit. While the former corresponds to a specific MODality of Q_{66}, *main method used to find work*, the reception of unemployment benefits is recorded separately (and independently) by QUality Q_{72}. Hence, also the mapping of variable no. 53 involves an IMAXis comprising both of Q_{66} and Q_{72}.

Classification Mappings

The availability of classifications is of apparent importance in deriving statistical labour force aggregates. Therefore, nomenclatures of breakdown dimensions used in data sources must be mapped carefully to the *LFF*. Although, in general, there is a *1:1* correspondence between classification variables (SOVARs) and classification QUalities at the VARMAP level, the arrangement of VALMAPs may be hampered by various subtle semantic divergencies which are usually settled by specifically convening supra-national nomenclature harmonization boards. It is no doubt advisable to incorporate the agreed-upon nomenclatures and suggested change keys in the set-up of *LFF* mappings; this, however, still may not do away completely with further adaptations (for instance, if mappings involve IMAXis comprising classification and other QUalities at the same time). Where no recommendations for harmonizing classifications can be taken recourse to, *ad hoc* mappings must be devised anyway. In case of the *LFF*, a case in point is the arrangement of classifications with respect to education and vocational training (cf. Subsection 3.2.2); here, for instance, even the mapping of pertinent CLFS-83 variables is a rather challenging task as there may, most probably, be no way to prevent a "loss" of data because of unsurmountable taxonomic incompatibilities.

3.4 Harmonized Data

Once the data source mappings have been accomplished successfully, the specification of statistical labour force aggregates can commence referring to the *LFF* exclusively. More specifically, since statistical labour force aggregates imply the preceding formation of domain concepts, the first step in *LFF* data view specification consists of arranging a couple of CONCEPTs (cf. Section 2.2.3) such that, for instance, the main indicators of labour force and employment statistics – viz. *activity rates, employment/population ratios*, and *unemployment rates* – can be expressed in terms of these CONCEPTs (cf. Section 3.2). Further CONCEPTs of interest certainly are *underemployment* and *long-term unemployment*, to state just a few. As a matter of fact, there may well be different opinions, reflecting diverging interests, as to what a domain concept really means; therefore, CONCEPTs will most likely come in several variants making semantic distinctions explicit at the level of taxonomic metadata, that is by using different expressions built of terms taken from the *LFF* data language consisting of QUalities, AXes, AGGLEvels and MODalities.

It is the tenet of the METASTASYS approach to data integration that its major advantage is the freedom given to the data consumer to specify the domain concepts of his interest rather independently from considerations underlying particular labour force survey designs (or the sample spaces of other relevant data sources). Of course, this freedom to customize domain concepts as individual demands call for does by no means preclude the possibility to establish agreed-upon concept definitions such as, for instance, those of the ILO or the couple of CLFS concepts set up for internal use in the EU (cf. Section 3.2). Evidently, in order to obtain figures comparable over both space and time the very same domain concepts must be applied. For this reason mainly, the following paragraphs will refer predominantly to standardized concept definitions used in international labour force statistics; after all, *LFF* design itself has been based largely on these definitions (cf. Section 3.3). Additionally, however, some concept definitions deviating from international standards will be discussed to highlight *LFF*'s flexibility in supporting a wide range of domain concepts such that, for instance, international comparisons can be facilitated using specific – in particular, national – definitions of, say, unemployment rates (cf. [Biffl, 1994]) since this, every now and then, has turned out to be a fiercely debated topic of nations' internal affairs, especially if national and international figures diverge to a considerable degree (as a case in point with respect to Austrian unemployment in 1992, cf. **Fig. 3.2.3**). Apparently, METASTASYS allows a "disinterested" comparison of statistical indicators based on related though distinct CONCEPT definitions as well, thus giving valuable support in the assessment and development of well-targeted indicators.

Most statistical labour force aggregates are either *tables* or *time series*; unless these aggregates provide absolute figures (counts of *person*s in the *LFF* case), usually percentages (ratios) are tabulated or indices compiled to time series. Noting that METASTASYS neither makes a formal distinction between tables and time series nor treats percentages and indices differently in internal computations, all standard labour force statistics are safely comprised in METASTASYS's scope of derivable macrodata structures – an extensive survey of statistical tables produced in labour force and employment statistics [Fahrngruber, 1993] indeed confirmed METASPEL's over-sufficient coverage of the table structures provided (also cf. [Grossmann and Froeschl, 1992] with respect to a coarse classification of statistical aggregates used in the labour force domain). With respect to classifications, tables usually are broken down by general demographic criteria (such as sex, age, or general education level), by economic taxonomies (professional status, occupation, sector of economic activity), by spatial subdivisions (using particularly the NUTS nomenclature), and, of course, by time in case of time series. As all of these classification criteria are readily supplied by the *LFF*, table structures and time series are easily specified on condition that the subject-domain concepts underlying the statistical indicators of concern have been composed properly.

3.4.1 Labour Force Concept Definitions

The essential domain concepts of labour force statistics build a logical hierarchy of four main layers, three of which are shown diagrammatically – overleaf – in **Fig.**

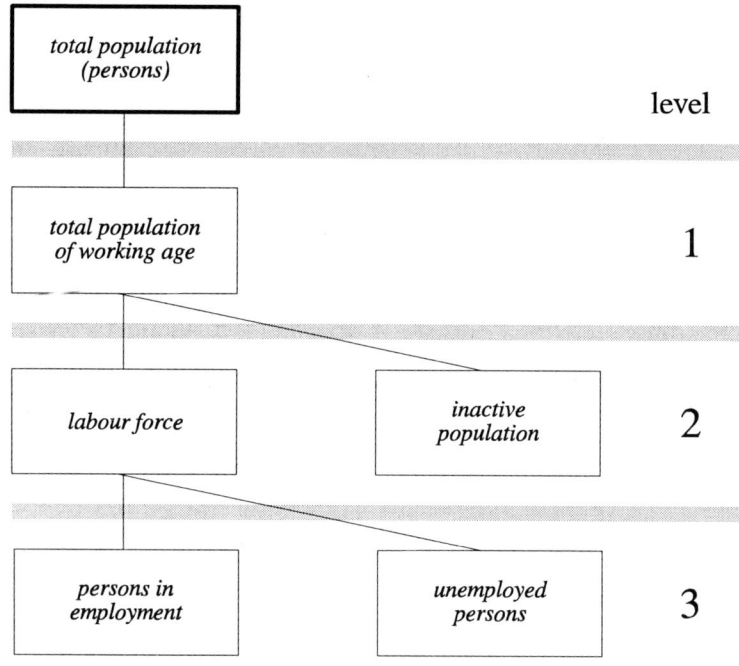

Figure 3.4.1: Logical Hierarchy of Labour Force Domain Concepts

3.4.1 (for the sake of comparison, also cf. **Fig. 3.2.2–1** as well as [Lackner, 1993; EUROSTAT, 1992]).

At the top of the hierarchy resides the *ur*-concept (cf. Subsection 2.4.1) of the *LFF*, viz. the population comprising the whole collective of instances of FRAME-UNIT *person*. This "concept" per se is not very interesting since persons below working age do not belong to the labour force by definition. Hence, the first (level 1) domain concept of subject-matter relevance is the *population of working age*. Obviously, there must be some age bound(s) defining what the working age of persons is; as a consequence, this concept already provides a first branching point for divergent definitions. Furthermore, the total population of working age may be delimited in terms of the place of persons' residence – sometimes, the resident population is meant while at other times it may be the domestic population, etc.

Splitting the population of working age into the *labour force* on the one hand and the sub-population of *inactive persons* on the other hand establishes the second level of the concept hierarchy. Again, there is much room for arbitrariness in delineating active from inactive persons; in particular, the views upon distinguishing inactive from unemployed persons have been a constant source of controversy because the available discerning criteria – such as a person's willingness to work or her active steps undertaken to find work – are of rather vague nature and, besides,

.4 Harmonized Data

difficult to measure reliably. Anyway, the labour force is further decomposed into the disjoint sub-populations of *persons in employment* and *unemployed persons*, thus constituting the hierarchy's third level of domain concepts. This distinction is accomplished quite easily as there are obvious criteria at hand to identify the persons at work for the large share of dependent employed; somewhat more intricate is the delineation of self-employed persons, unpaid workers, or persons working for a few hours a week or even occasionally only. Somewhere half-way between employed and unemployed persons are those persons seeking employment in spite of having some job; under qualified circumstances, these *underemployed persons* could be considered as yet another sub-population of the labour force but, usually, no such distinction is introduced formally (or shifted the hierarchy downwards to level 4 by splitting the sub-population of persons in employment into underemployed persons and a complementary sub-population).

Beneath level 3 of this concept hierarchy, both domain concepts of persons in employment and unemployed persons are subdivided further in various ways, thus bringing forth a couple of "structural" concept levels elucidating specific socio-economic features present within level 3 concepts. For instance, persons in employment may be subdivided according to professional status (discerning mainly dependent employed and self-employed persons/employers) or according to the (usual) work time (full-time/part-time distinction); conversely, unemployed persons could be subdivided especially with respect to the duration of job search, thus splitting off the socially crucial sub-population of persons in long-term unemployment.

Yet a further step lower down in the concept hierarchy there will be subdivisions according to the major economic classifications such as occupation and sector of economic activity; these subdivisions, however, more and more overlap with standard breakdowns of labour force macrodata whence it is no longer reasonable to refine the hierarchy explicitly in terms of domain concept definitions. In particular, it is not advisable to include spatial and temporal specifications into these concept definitions unless individual concepts apply to certain geographical regions only and/or have a limited temporal validity for some peculiar reason.

Despite the apparent national differences in the concepts defined and used in labour force and employment statistics (cf. Subsection 3.2.3) and the multitude of existing concept definitions resulting thereof, virtually all of them are reducible to the sketched concept hierarchy without any major difficulty (in favour of evidence for this proposition cf. [Lackner, 1993] and the program transcripts of the DÖS'chen prototype reproduced in Appx. C).

Now, returning to the domain concepts of levels 1 to 3 as shown in **Fig. 3.4.1**, the cardinal types of statistical indicators are obtained as quotients from pairs of these concepts as follows:

- the *activity rate* is defined as quotient of the level 2 concept of *labour force* and the level 1 concept of the *total population of working age*;

- the *employment/population rate* is defined as quotient of the level 3 concept of *persons in employment* and, again, the level 1 concept of the *total population of working age*;
- the *unemployment rate* is defined as quotient of the level 3 concept of *unemployed persons* and the level 2 concept of *labour force*.

Thus, to complete the definition of any of these quotients, only some formal definition of each of the domain concepts involved must be inserted appropriately. To this end, several variants of the four concepts mentioned will be reviewed below.

Recalling that, by virtue of METASPEL conventions, a formal CONCEPT definition includes all TRAITs generated by *LFF* DIMensions not constrained explicitly, some care must be exercised to delineate domain concepts correctly by the compiled CONDEFs (cf. **Def. 2.2.3–1**); conversely, however, TRAITs coinciding with or being contained in FRAMEDits are excluded automatically from CONDEFs (during conversion to CONBOX representations) as explained in Subsection 2.4.1.

The following sample of formal CONDEFs will make use of a stylized language combining description conventions (for FRAMEDits) introduced in Subsection 3.3.2 with a layout emphasizing the logical structure of concept definitions in a rather self-explanatory way. Bracketed numbers enumerate the REGCONs (cf. **Def. 2.4.1–2**) within CONDEFs where it is understood that the enumeration sequence is immaterial functionally. The character string next to the 'CONDEF'-token denotes the defined concept's CONLABel; the character string next to the 'superconcept:'-token refers to the particular CONDEF of the concept hierarchy the present CONDEF is going to refine incrementally (cf. Subsection 2.4.1) by imposing further constraints to the 'superconcept:'-CONDEF so referred to. In order to flag the semantic differences between concept variants, CONLABs are arranged to include a referential suffix in parentheses. Throughout, the *LFF*'s ur-concept will be denoted as '#'. A '⌐' symbol terminates a CONDEF paragraph.

CLFS Domain Concepts

A translation of EU labour force domain concepts – cf. Subsection 3.2.2 for a review – is brought about quite easily by copying the logical hierarchy of domain concepts in terms of CONDEFs. First, the level 1 concept of the total population of working age must be captured. With respect to the CLFS-92 standard, this involves an age bound set to 15 years; furthermore, only persons living in private households of EU Member States are included; this amounts to define

    CONDEF   population in working age (EU-92)
      superconcept: #
        [1] $Q_4(15\ ..) \otimes Q_8(\text{EU-Member State}) \otimes Q_{14}(\text{private})$
⌐

where the MODalities 'EU-Member State' and 'private', respectively, are assumed to belong to suitably defined AGGDOMains. Similarly, the respective definition corresponding to the CLFS-83 standard could be phrased as

CONDEF   population in working age (EU-83)
   superconcept: #
   [1] $Q_4(14\ ..) \otimes Q_8$(EU-Member State) $\otimes Q_{14}$(private)
⌐

with the only apparent change of the age bound being lowered to 14 years.

Mainly because of the intricate delineation of unemployed and inactive persons the definition of level 2 concepts is better postponed until the level 3 concepts of persons in employment and unemployed persons have been established as CONDEFs; the labour force-concept then merely consists of the "addition" of both level 3 concepts. The former, 'persons in employment (EU-92)', is exceptionally simple to define (and is, CONLABel and 'superconcept:'-reference adapted accordingly, exactly identical for the CLFS-83 variant):

CONDEF   persons in employment (EU-92)
   superconcept: labour force (EU-92)
   [1] $Q_{17}(1\ ..\ 12)$
⌐

The concept of unemployed persons is more difficult to describe as this requires a three-way distinction:

CONDEF   unemployed persons (EU-92)
   superconcept: labour force (EU-92)
   [1] $Q_{17}(13,14,16,17,19) \otimes$
           Q_{59}(seeking job) $\otimes Q_{61}$(~~self-employment~~) $\otimes$
           Q_{66}(~~no method used~~) $\otimes Q_{67}(0\ ..\ 3) \otimes Q_{68}(0\ ..\ 2)$
   [2] $Q_{17}(16,17,19) \otimes$
           Q_{59}(seeking job) $\otimes Q_{61}$(self-employment)
   [3] $Q_{17}(16,17,19) \otimes$
           Q_{59}(starts job already found later)
⌐

All of the BLOCKs of this CONDEF comprise persons participating in public employment promotion schemes (pos. *16*) or in vocational training programmes (pos. *17*); cf. **Tab. 3.3.1–2**. Case [1] adds persons on lay-off (positions *13* and *14* of Q_{17}) or entirely out of employment actively seeking a job as dependent employee and being available within two weeks (Q_{68}). Case [2], in turn, addresses unemployed persons seeking self-employment. The final case [3] deals with unemployed persons having found a job already which will start in the (near) future and who, in general, will not seek another job accordingly. Note that, in case [1], depending on the modes of job seeking considered as *active* ones, different formulations could result.

Having arranged these CONDEFs, now the 'labour force (EU-92)' CONDEF can be stated simply as

CONDEF   labour force (EU-92)
   superconcept: population in working age (EU-92)
     [1] $Q_{17}(1 .. 12)$
     [2] $Q_{17}(13,14,16,17,19) \otimes$
                Q_{59}(seeking job) $\otimes$ Q_{61}(~~self-employment~~) $\otimes$
                Q_{66}(~~no method used~~) $\otimes$ $Q_{67}(0 .. 3) \otimes Q_{68}(0 .. 2)$
     [3] $Q_{17}(16,17,19) \otimes$
                Q_{59}(seeking job, starts job already found later) $\otimes$
                Q_{61}(self-employment)

Note that combining of the Q_{59}-MODalities 'seeking job' and 'starts job already found later' in a single REGCON is admissible because of the Q_{59}/Q_{61} dependency edit (cf. Subsection 3.3.2) rendering Q_{61} applicable only in case of 'Q_{59}(seeking job)'.

Less trivial to state is the CONDEF of 'inactive persons (EU-92)' complementary to 'labour force (EU-92)' with respect to their common superconcept 'population in working age (EU-92)'; favourably, the derivation of "concept complements" should be supported by predefined user operations (which amounts to implement no more than *DeMorgan*'s rules of set complement adapted to BLOCKCOLLections). In whatever way obtained, the 'inactive persons (EU-92)' concept would comprise all persons with *work status* (Q_{17}) MODalities of positions *15* (conscripts) and *18* (persons doing housekeeping as primary activity) as well as persons being unemployed (pos. *19*) but not seeking work or seeking work as dependent employee without being available within two weeks.

There is virtually no difference between the CONDEFs presented and those resulting from the former CLFS-83 standard; essentially, the changes affect the modified mapping for Q_{66} (the CLFS-83 predecessor of which comprised fewer SOVALues) not becoming tangible at CONDEF level. Hence, the CONDEFs established for the CLFS-92 merely have to be repeated once again replacing the CONLABels' suffixes suitably.

ILO Domain Concepts

Since the CLFS domain concepts are based, by and large, on concepts recommended for use by the ILO, it comes as no surprise that deviations between respective definitions are not very numerous. Basically, the ILO concept of the total population of working age conforms to the 'population in working age (EU-92)' CONDEF using an age bound of 15 years. In contrast to its CLFS counterpart, however, the ILO definition explicitly excludes persons of the armed forces from the labour force which, of course, might be captured simply by *LFF* QUality $Q_{23\text{-}1}$, *occupation* (of first job); after all, since only private households are included, this distinction is likely to have little quantitative impact assuming that members of the armed forces usually reside in collective households. Whatsoever, according to ILO definitions, there is a domain concept 'total labour force (ILO)' comprising both civilian labour force and armed forces whence, formally, 'civilian labour force (ILO)' is a sub-concept of the 'total labour force (ILO)'.

Breaking down the civilian labour force into persons in employment and unemployed persons, the ILO-conforming CONDEFs read as follows:

CONDEF persons in employment (ILO)
superconcept: civilian labour force (ILO)
[1] $Q_{17}(1 .. 12) \otimes$
$Q_{23\text{-}1}$(employer,self-employed with paid employees
(small independent),own-account worker,
employee)
[2] $Q_{17}(1 .. 12) \otimes$
$Q_{23\text{-}1}$(unpaid family worker,unpaid assistant worker) $\otimes$
$Q_{38\text{-}1}(13 ..)$
⏎

This definition assumes (somewhat arbitrarily) that the normal working time referred to in ILO proposals (cf. Subsection 3.2.1) is somewhere between 37 to 40 hours a week whence $1/3$ of this, rounded-off, yields an effective bound of 13.

CONDEF unemployed persons (ILO)
superconcept: civilian labour force (ILO)
[1] $Q_{17}(13,14)$
[2] $Q_{17}(16,17,19) \otimes$
Q_{59}(seeking job) $\otimes Q_{67}(0 .. 1) \otimes Q_{68}(0 .. 2)$
[3] $Q_{17}(16,17,19) \otimes$
Q_{59}(starts job already found later)
⏎

Apparently, compared to the CLFS-based CONDEFs the major departures regard (i) the treatment of persons on lay-off which are classified as unemployed irrespective of their availability for work or attempts to find work, and (ii) the loosened requirements as to the criteria of active job search; basically, a person is simply considered unemployed provided she is seeking a job and is available for work. Again, by "additive" combination of both concepts of 'persons in employment (ILO)' and 'unemployed persons (ILO)' the superconcept of 'civilian labour force (ILO)' is obtained.

Domain Concepts Used in Austrian Employment Statistics

The Austrian employment statistics, based on the *Mikrozensus* (AMZ) as described in Subsection 3.2.3, uses a slightly different hierarchical schema of domain concepts. The major departure from the hierarchy as depicted in **Fig. 3.4.1** regards the inactive population ("Nichterwerbspersonen") which includes both persons of working age and persons of pre-working age; the age bound used for discerning both sub-populations is 15 years again. As usual, the Austrian labour force (in German: "Erwerbspersonen") is composed additively of two level 3 concepts, viz. "Erwerbstätige" (persons in employment) and "Arbeitslose" (unemployed persons). According to [Lackner, 1993], these domain concepts are arranged as follows:

>   CONDEF   Erwerbspersonen (A-93)
>   superconcept: Wohnbevölkerung (A)
>   [1] $Q_4(15\ ..) \otimes Q_{17}(1\ ..\ 7, 9\ ..\ 12) \otimes$
> $Q_{23\text{-}1}$(employer,self-employed with paid employees
>   (small independent),own-account worker,
>   employee,unpaid family worker) $\otimes$
> $Q_{38\text{-}1}(13\ ..)$
>   [2] $Q_4(15\ ..) \otimes Q_{17}(14,16,17,19) \otimes Q_{59}$(seeking job) $\otimes Q_{68}(0\ ..\ 4)$
>   [3] $Q_4(15\ ..) \otimes Q_{17}(14,16,17,19) \otimes Q_{59}$(seeking job) $\otimes$
> Q_{66}(standing registration at employment office)
>   [4] $Q_4(15\ ..) \otimes Q_{17}(13)$

Persons on maternity leave, '$Q_{17}(8)$', as well as conscripts, '$Q_{17}(15)$', are excluded from the Austrian labour force.

>   CONDEF   Erwerbstätige (A-93)
>   superconcept: Erwerbspersonen (A-93)
>   [1] $Q_{17}(1\ ..\ 7, 9\ ..\ 12) \otimes$
> $Q_{23\text{-}1}$(employer,self-employed with paid employees
>   (small independent),own-account worker,
>   employee,unpaid family worker)

Strictly speaking, the second restriction (regarding $Q_{23\text{-}1}$) could be dropped safely since it is already determined unambiguously by the first one.

.4 HARMONIZED DATA

CONDEF   Arbeitslose (A-93)
  superconcept: Erwerbspersonen (A-93)
    [1] $Q_{17}(13)$
    [2] $Q_{17}(14,16,17,19) \otimes Q_{59}$(seeking job) $\otimes Q_{68}(0 .. 4)$
    [3] $Q_{17}(14,16,17,19) \otimes Q_{59}$(seeking job) $\otimes$
            Q_{66}(standing registration at public employment office)
⌐

Again, formally it would have been sufficient to replace the three cases of this CONDEF by a single one, viz.

    [1] $Q_{17}(13,14,16,17,19)$

letting the further restrictions "sink down" from the CONDEF's super-concept, thus making the distinction of BLOCKs [2] to [4] reappear (note also that the second and third BLOCKs of this CONDEF are not disjoint). In this concept definition, seasonal workers on lay-off, '$Q_{17}(13)$', are always considered unemployed whereas other persons not at work are classified as unemployed only on condition that they (i) are seeking a job and (ii) are available within one month at most. Persons registered unemployed at a public employment office are included automatically.

The domain concept of 'Erwerbspersonen (A-93)', in turn, is a sub-population of the resident Austrian population ("Wohnbevölkerung") of 15 years and over, viz.

CONDEF   Wohnbevölkerung (A)
  superconcept: #
    [1] Q_8(Austria)
⌐

where 'Austria' is assumed to be an AGGDOMain MODality representing the national territory of Austria. Persons living in collective households are included.

For the computation of national unemployment rates instead of 'Erwerbstätige (A-93)' a more restricted concept 'Unselbständig Erwerbstätige (A)' is actually used [Bartunek, 1994] as (level 3) concept for persons in employment:

CONDEF   Unselbständig Erwerbstätige (A)
  superconcept: Wohnbevölkerung (A)
    [1] $Q_4(15 ..) \otimes Q_{17}(1 .. 7, 9 .. 12) \otimes$
            $Q_{23\text{-}1}$(employee) $\otimes Q_{38\text{-}1}(12 ..)$
⌐

Furthermore, unemployed persons now include – in a sense, paradoxically – persons in employment though usually working less than 12 hours a week:

CONDEF   Arbeitslose einschließlich geringfügig Erwerbstätiger (A)
superconcept: Wohnbevölkerung (A-93)
[1] $Q_4(15\ ..) \otimes Q_{17}(1\ ..\ 7,9\ ..\ 12) \otimes Q_{38\text{-}1}(1\ ..\ 11) \otimes$
        $Q_{59}(\text{seeking job}) \otimes Q_{67}(0\ ..\ 3) \otimes Q_{68}(0\ ..\ 4)$
[2] $Q_4(15\ ..) \otimes Q_{17}(13,14,16,17,19) \otimes Q_{59}(\text{seeking job}) \otimes$
        $Q_{67}(0\ ..\ 3) \otimes Q_{68}(0\ ..\ 4)$
[3] $Q_4(15\ ..) \otimes Q_{17}(1\ ..\ 7,9\ ..\ 12) \otimes Q_{38\text{-}1}(1\ ..\ 11) \otimes$
        $Q_{59}(\text{seeking job}) \otimes$
        $Q_{66}(\text{standing registration at public employment office}) \otimes$
        $Q_{67}(4\ ..)$
[4] $Q_4(15\ ..) \otimes Q_{17}(13,14,16,17,19) \otimes Q_{59}(\text{seeking job}) \otimes$
        $Q_{66}(\text{standing registration at public employment office}) \otimes$
        $Q_{67}(4\ ..)$
[5] $Q_4(15\ ..) \otimes Q_{17}(1\ ..\ 7,9\ ..\ 12) \otimes Q_{38\text{-}1}(1\ ..\ 11) \otimes$
        $Q_{59}(\text{seeking job}) \otimes$
        $Q_{66}(\text{standing registration at public employment office}) \otimes$
        $Q_{67}(0\ ..\ 3) \otimes Q_{68}(5\ ..)$
[6] $Q_4(15\ ..) \otimes Q_{17}(13,14,16,17,19) \otimes Q_{59}(\text{seeking job}) \otimes$
        $Q_{66}(\text{standing registration at public employment office}) \otimes$
        $Q_{67}(0\ ..\ 3) \otimes Q_{68}(5\ ..)$

Self-employed persons or unpaid family workers working more than 11 hours a week are excluded from the concept of 'Unselbständig Erwerbstätige (A)'; if they are working less than 12 hours and are seeking work, however, they are considered unemployed. Joining both CONDEFs yields the level 2 concept of 'Arbeitskräftepotential (A)' comprising dependent employed as well as unemployed (including insignificantly employed) persons seeking work which, of course, differs markedly from the ILO labour force definition.

Albeit referring to a rather provisional and crude FRAME design, Appx. C.3 provides a further couple of variants of Austrian labour force domain concepts.

Further Examples of Labour Force Domain Concepts

As in practically each country of the world there has been established some system of employment statistics, numerous examples of level 1 to level 3 domain concepts could be reproduced here. Moreover, national concept definitions are revised every now and then, thus evoking a multitude of definition variants rather difficult to oversee. Hence, instead of giving further examples of the major labour force domain concepts, a couple of less familiar concepts is presented.

A topic of growing importance certainly is *long-term unemployment* which, in fact, can be captured simply by adding a further constraint to the level 3 concepts of unemployed persons imposing a lower bound on the *duration of job search* Quality (Q_{65}) of the *LFF*. For instance, setting this lower bound to one year [EUROSTAT, 1988] (or 52 weeks since Q_{65} has been proposed to measure time in weeks; cf. Subsection 3.3.1), the concept of 'unemployed persons (EU-92)' could be refined as follows:

> CONDEF   long-term unemployment / 1
>   superconcept: unemployed persons (EU-92)
>   [1] $Q_{65}(53\ ..)$

This definition of long-term unemployment, however, would include persons only who are either seeking work or are preparing themselves for returning to work by participating in vocational training programmes or other types of employment promotion schemes; in other words, the measurement of long-term unemployment based on labour force concepts tends to disregard "marginalized" manpower which, unfortunately, has risen to a sizable share of total unemployment [EUROSTAT, 1988]. Conceivably, the notion of long-term unemployment might therefore be extended to particularly include discouraged persons with previous work experience having ceased to search for work actively; in that case, the former definition should be rephrased, as one possibility, like this:

> CONDEF   long-term unemployment / 2
>   superconcept: #
>   [1] $Q_4(16\ ..) \otimes Q_{17}(13,14,16,17,19) \otimes Q_{59}(\text{seeking job}) \otimes$
> $Q_{65}(53\ ..) \otimes Q_{66}(\text{no method used}) \otimes Q_{68}(0\ ..\ 2)$
>   [2] $Q_4(16\ ..) \otimes Q_{17}(13,14,16,17,19) \otimes$
> $Q_{59}(\text{starts job already found later}) \otimes Q_{48}(\text{yes}) \otimes Q_{49}(53\ ..)$
>   [3] $Q_4(16\ ..) \otimes Q_{17}(13,14,16,17,19) \otimes Q_{48}(\text{yes}) \otimes$
> $Q_{49}(53\ ..) \otimes Q_{59}(\text{not seeking because discouraged}) \otimes Q_{68}(0\ ..\ 2)$
>   [4] $Q_4(16\ ..) \otimes Q_{17}(19) \otimes Q_{48}(\text{no}) \otimes Q_{59}(\text{seeking job}) \otimes$
> $Q_{65}(53\ ..) \otimes Q_{70}(\text{full-time education})$

Since persons under 16 years cannot be unemployed for more than one year (setting the age bound for the working population to 15), all Q_4-expressions of this CONDEF are adapted to '16 ..' accordingly. The Q_{59}-MODality 'not seeking because discouraged' represents the group of BASVALues enumerating different reasons for not seeking work such as, say, 'believing that no work is available', 'had looked for work, but could not find any', 'believes to lack necessary skills, training, or experience', 'presumably too old or too young to get a job', and the like.

Similarly, domain concepts for *youth unemployment* could be set up which, in the simplest case, is accomplished by introducing an additional upper age bound to the concepts of unemployed persons. Analogous to definitions of long-term unemployment, however, it might be desirable to discern unemployed young persons without any previous work experience from those having lost or terminated a job. Apparently, youth unemployment and long-term unemployment concepts could be merged in order to measure the young persons suffering from long-term unemployment.

    CONDEF   long-term youth unemployment / 2
      superconcept: long-term unemployment / 2
        [1] $Q_4(.. 24)$

This definition proposes an upper age bound of 24 years in accordance with the familiar convention to regard those aged 25–44 as "prime-age" workers [EUROSTAT, 1988].

Prolonged joblessness is an extreme case of *underemployment* if viewed from the perspective of persons seeking additional gainful work. Unfortunately, underemployment is a concept even harder to grasp formally than the concept of unemployment. This difficulty arises mainly from the relativity of underemployment – if, for instance, the number of regular work days per week is reduced to, say, 4 in a particular branch of economic activity (due to increases in productivity or saturated consumer markets) then persons employed in this branch would hardly be considered underemployed. Consequently, the measuring of underemployment cannot rest on absolute criteria such as the usual work hours per week but rather on relative criteria such as part-time work or an unusual difference between usual and actual work hours per week, if combined with persons' active search for additional work. In general, underemployment concepts exclude unemployed persons by definition. A provisional CONDEF of underemployment could look like:

    CONDEF   underemployed persons (EU)
      superconcept: persons in employment (EU-92)
        [1] $Q_{19}(1) \otimes Q_{22\text{-}1}(\text{employee}) \otimes Q_{32\text{-}1}(\text{part-time}) \otimes$
                $Q_{59}(\text{seeking job}) \otimes Q_{61}(\text{self-employment}) \otimes$
                $Q_{66}(\text{no method used}) \otimes Q_{67}(0 .. 3) \otimes Q_{68}(0 .. 2)$
        [2] $Q_{19}(1) \otimes Q_{22\text{-}1}(\text{employee}) \otimes Q_{32\text{-}1}(\text{part-time}) \otimes$
                $Q_{59}(\text{seeking job, starts job already found later}) \otimes$
                $Q_{61}(\text{self-employment})$
        [3] $Q_{19}(1) \otimes Q_{22\text{-}1}(\text{employee}) \otimes$
                $Q_{40\text{-}1}(\text{less hours due to slack work,}$
                          $\text{less hours due to variable hours}) \otimes$
                $Q_{59}(\text{seeking job}) \otimes Q_{61}(\text{self-employment}) \otimes$
                $Q_{66}(\text{no method used}) \otimes Q_{67}(0 .. 3) \otimes Q_{68}(0 .. 2)$

[4] $Q_{19}(1) \otimes Q_{22-1}$(employee) $\otimes$
        Q_{40-1}(less hours due to slack work,
                less hours due to variable hours) $\otimes$
        Q_{59}(seeking job,starts job already found later) $\otimes$
        Q_{61}(self-employment)

[5] Q_{22-1}(~~employee~~) $\otimes$ ($\Sigma_{n=1,2,3}\ Q_{38-n}< 40$) $\otimes$
        Q_{59}(seeking job) $\otimes$ Q_{61}(~~self-employment~~) $\otimes$
        Q_{66}(~~no method used~~) $\otimes$ $Q_{67}(0\ ..\ 3) \otimes Q_{68}(0\ ..\ 2)$

[6] Q_{22-1}(~~employee~~) $\otimes$ ($\Sigma_{n=1,2,3}\ Q_{38-n}< 40$) $\otimes$
        Q_{59}(seeking job,starts job already found later) $\otimes$
        Q_{61}(self-employment)

⌐

Of course, this definition relies on several questionable assumptions; for instance, persons working as full-time employees are never considered underemployed (the MODality 'part-time' may in fact represent a group of BASVALues discerning different modes or causes of part-time work). Perhaps, if felt necessary, this deficiency could be resolved by including Q_{47-1} explaining any peculiar circumstances why a full-time employed person may in fact be underemployed. Likewise, this definition deems all employed persons not being employees underemployed provided that they (i) usually do not work 40 hours or more a week and (ii) are seeking yet another job as employees or otherwise.

Eventually, there will be defined a more or less extensive set of *ad hoc* domain concepts for various specific questions to solve against the *LFF*. For instance, it could be of interest to investigate how many persons there are in the tertiary sector of economic activity being hired as leased workers with individual working contracts. Apparently, such kind of concepts is easily established using *LFF* terminology:

CONDEF   leased workers with individual work contract
                                            in tertiary economic sector
    superconcept: persons in employment (EU-92)
        [1] Q_{22-1}(employee) $\otimes$ Q_{24-1}(yes) $\otimes$ Q_{26-1}(services) $\otimes$
                Q_{30-1}(individual work contract)

⌐

In general, in contrast to standardized domain concepts it is not very reasonable to provide ad hoc concept definitions in a store of predefined CONDEFs unless the compilation of CONDEFs becomes very intricate and there is some chance that such a contrived CONDEF might be re-used at all. Quite on the contrary, since many of these ad hoc concepts will consist of a single BLOCK only, they could be specified elegantly "on the fly" within METASPEL sentences using the **WITH** clause as discussed in Subsection 2.4.3.

3.4.2 Labour Force Aggregates

The ultimate purpose of the *LFF* consists in providing a terminological "baseline" for the specification of statistical labour force aggregates. Hence, this chapter is concluded by a compilation of sample tables as they are typically encountered in employment statistics. For the sake of a coherent presentation, the following examples of labour force aggregates refer to CONDEFs introduced in Subsection 3.4.1. Although representing a small fraction of conceivable table structures only, these sample tables should illustrate to some degree how the sequence of processing steps starting from source data to (raw) tables finds its completion in the table specification stage. Using METASPEL (cf. Subsection 2.4.3), the set of sample tables and time series is stated in terms of METASPEL sentences; to illustrate the point, in a few cases these formal expressions are accompanied by reproductions (excerpts) of tables as printed in publications of statistical agencies or published by statistical offices. Obviously, there are table layouts different from those shown (but still consistent with respective METASPEL sentences); however, the layout of reproduced tables resembles closely the publication formats preferred in source documents such as those of EUROSTAT [1993ab].

With respect to ALFIN (cf. Section 3.3), since the *LFF* is a *virtual* data structure, a client connected to the *LFF* by virtue of METACOM (that is, the client is running METASTASYS's genuine communication protocol; cf. Section 2.6) can compose REQUESTs locally irrespective of where the source data needed to generate the response METAs actually reside. In a practical setting, of course, the submission of METASPEL sentences as shown below will not be quite sufficient; in general, either further layout specification clauses had to be added or some post-processing of obtained results (editing) would be in place to accomplish the final aggregates as desired.

The domain concept definitions exemplified in the previous subsection have been stated in a *generic* format, that is without explicitly restricting the universal QUalities *space* and *time*. Therefore, if these CONDEFs are used in aggregate specifications it might happen by incidence that there cannot be spotted data sources covering the submitted REQUESTs' CONBOXes (cf. Subsection 2.4.1); unless ways are found to impute data (by applying data propagations; cf. Subsection 2.5.3) in such cases, TARGETs will have to be "relaxed" – that is: truncated – accordingly as described in Section 2.5. Any such relaxation of a REQUEST is, of course, reported minutely by METAGEN in an utterly transparent way.

For the sake of clarity, the following sample tables are grouped into paragraphs covering a range of topics such as overall demographic aggregates, main population groups with respect to labour force, activity rates and related ratios, tables of "summable" features, and tables reporting on special concepts; finally, a compound table with nested concepts will round off this presentation. In view of the limited space available, each paragraph comprises only a couple of table instances each of which is highlighting several aspects of aggregate specification at once. As to summary types, tables of absolute and relative figures (counts and percentages) dominate in employment statistics (just as they do, in fact, in almost any other branch of official statistics); rather naturally, numerical aggregates occur for temporal attributes (duration of job search, work hours per week, etc.) only. Apparently, averages or higher-moment aggregates are not published in "standard" tables like those encountered in official statistics [Fahrngruber, 1993].

Demographic Population Totals

A demographic aggregate of elementary type is a table of population totals. Following a preferred presentation format of EUROSTAT, **Tab. 3.4.2–1** (overleaf) exhibits an excerpt of a table of "Total population in private households" broken down three-fold, viz. with respect to sex, Member State, and time (in fact; one could speak of a couple of juxtaposed time series broken down with respect to sex and country). Structurally, this table is very simple, of course, and can be stated in METASPEL terms as follows:

T3.1 ←
    COMPUTE table; **OF** n; **ON** LFF;
    FOR #;
    WITH usual place of residence (household): =EUR-12 &
            type of household in present place of residence
                      (reference week)@private/collective: =private;
    BROKEN DOWN BY Σ | sex;
    OVER EU/non-EU: EUR-12 _2;
    IN year: 1983 .. 1991 ⌡

This specification assumes that several AGGLEvels have been arranged for LFF's FRAMEQUalities, viz.
- the AGGLEvel 'private/collective' for Q_{14}, *type of household in present place of residence (reference week)*, with two MODalities 'private' and 'collective';
- the AGGLEvel 'year' for Q_2, *time*;
- a hierarchy of AGGLEvels comprising, inter alia, the AGGLEvel 'EU/non-EU', providing a MODality 'EUR-12' representing the Union of the twelve Member States, which is based on AGGLEvel, say, 'EUR-10+2' supplying the MODalities 'EUR-10', 'España', and 'Portugal'; this AGGLEvel, in turn, could be based directly on an AGGLEvel termed 'NUTS-0' representing the regional classification of *space* by countries for the remaining 10 Member States of 'EUR-10'.

On condition that the ALFIN consists of the *LFF* only, the '**ON** *LFF*'-clause could be dropped, of course. Using the depth indicator (cf. Subsection 2.4.3) '_2' in the **OVER** clause of T3.1 generates the spatial breakdown of **Tab. 3.4.1–1** although the totals for EUR-10 and EUR-12, respectively, may appear to the right of the table depending on how the ordering of sub-tables is established internally. An explicit arrangement of columns would be obtainable by composing the table explicitly replacing the present **OVER** clause by an **OVER** clause like

OVER [EU/non-EU: EUR-12, EUR-10+2: EUR-10,
NUTS-0: {Belgique-Belgie .. United Kingdom}]

inducing three tables by implicit unfolding of this clause.

The '-' entries in **Tab. 3.4.2–1** would be reproduced faithfully by METASTASYS unless further data sources could be found allowing to fill in these BLANKs (cf. Subsection 2.5.2). Of course, the BLANKs in the EUR-12 column are a direct consequence of the figures missing for Spain and Portugal.

If, for some reason, a table of the total EUropean population of 65 years and over were of interest, this could be a achieved by adapting the **WITH** clause of T3.1 yielding

T3.2 ←
    COMPUTE table; **OF** n; **ON** *LFF*;
    FOR #;
    WITH *age*: 65 .. &
        usual place of residence (household): =EUR-12 &
        type of household in present place of residence
            (reference week)@*private/collective*: =*private*;
    BROKEN DOWN BY Σ | *sex*;
    OVER EU/non-EU: EUR-12 _2;
    IN year: 1983 .. 1991 ⏎

The percentage of persons 65 years and over based on the total population is now obtained easily by META algebra, viz.

T3.3 ← T3.2 ÷ T3.1 ⏎

If, conversely, the population distribution within EUR-12 is of interest, a table of percentages must be specified:

T3.4 ←
    COMPUTE table; **OF** %(n); **ON** *LFF*;
    FOR #;
    WITH *usual place of residence (household)*: =EUR-12 &
        type of household in present place of residence
            (reference week)@*private/collective*: =*private*;
    BROKEN DOWN BY Σ | *sex*;
    OVER EU/non-EU: EUR-12 _2;
    IN year: 1986 .. 1991;
    DROPPING *space* ⏎

Table 3.4.2–1: Sample CLFS Table [EUROSTAT, 1993a]

COUNT	Eur 12	Eur 10	Belgique Belgie	Danmark	...	Luxembourg	Nederland	Portugal	United Kingdom
					Males and females				
1983	–	265 856	9 587	5 049	...	357	14 068	–	55 528
1984	–	266 705	9 807	5 044	...	358	14 123	–	55 549
1985	–	267 280	9 806	5 064	...	356	14 103	–	55 769
1986	315 695	267 980	9 797	5 076	...	358	14 224	10 167	55 914
1987	317 383	268 870	9 789	5 093	...	363	14 297	10 214	56 099
1988	318 393	269 511	9 772	5 097	...	365	14 355	10 241	56 216
1989	319 739	270 631	9 879	5 103	...	368	14 483	10 269	56 372
1990	321 653	272 841	9 886	5 112	...	379	14 585	10 301	56 559
1991	323 822	274 984	9 925	5 127	...	384	14 716	10 311	56 705
					Males				
1983	–	128 703	4 705	2 487	...	175	6 978	–	27 032
1984	–	129 017	4 781	2 477	...	175	6 999	–	27 046
...	...	...	...	...	...	...	...	...	...
1990	156 067	132 509	4 814	2 514	...	185	7 215	4 924	27 614
1991	157 132	133 528	4 836	2 523	...	188	7 287	4 941	27 700
					Females				
1983	–	137 153	4 882	2 562	...	182	7 090	–	28 495
1984	–	137 688	5 026	2 567	...	183	7 124	–	28 502
...	...	...	...	...	...	...	...	...	...
1990	165 586	140 332	5 072	2 598	...	194	7 370	5 378	28 945
1991	166 690	141 456	5 089	2 605	...	196	7 429	5 370	29 005

In table T3.4 the temporal range is restricted deliberately to '1986 .. 1991' because the computation of percentages is not possible for earlier years noting the absence of figures for Spain and Portugal in **Tab. 3.4.2–1**.

Other population totals, possibly using a variety of breakdowns, can be arranged analogously to T3.1 through T3.4 in an apparent way.

Main Population Groups in Employment Statistics

According to the table structure of T3.1 as specified above, by simply exchanging the argument of the **FOR** clause, aggregates for the main population groups discerned in labour force statistics are obtained. For instance, a summary of the EUR-12 labour force as expressed in terms of CONDEF 'labour force (EU-83)', broken down by sex, is requested by

    T3.5 ←
        COMPUTE table; **OF** n; **ON** LFF;
        FOR labour force (EU-83);
        BROKEN DOWN BY $\Sigma \mid sex$;
        OVER EU/non-EU: EUR-12 _2;
        IN year: 1983 .. 1991 ↵

Note that the **WITH** clause of T3.1 is no longer necessary as these restrictions are already enclosed in the CONDEF used. Now, similarly, inserting a

    FOR persons in employment (EU-83);

clause would induce a table having a structure identical to T3.5 though for this restricted population only. Likewise, instead of *sex* other breakdowns could be specified such as, first of all, main economic classifications like *professional status*, *occupation*, and sector of *economic activity*. Although two or more of these classifications may be used jointly in one **BROKEN DOWN BY** clause, this leads to difficulties in achieving legible table layouts and, therefore, should be avoided generally.

A somewhat more complex table as depicted schematically in **Tab. 3.4.2–2** (overleaf) would call for a METASPEL sentence as follows:

    T3.6 ←
        COMPUTE table; **OF** n; **ON** LFF;
        FOR persons in employment (ILO);
        WITH nationality: {Norway,Sweden,Finland,Iceland,Denmark};
        BROKEN DOWN BY $\Sigma \mid sex * \Sigma \mid age$@broad age groups
              * *full-time/part-time distinction (main job)*@full-part $\mid \Sigma$;
        OVER nation: ={Norway,Sweden,Finland,Iceland,Denmark};
        IN month: =Mar.1990 ↵

Table 3.4.2–2: Sample Three-Way Breakdown Table of Rates

Norway, Sweden, Finland, Iceland, Denmark   **March 1990**

%		persons in employment (ILO)	
sex	age	full-time/part-time disdinction (main job)	
		full-time	part-time
Σ	Σ		
	14 - 24		
	25 - 49		
	50 - 64		
	65 and over		
male	Σ		
	14 - 24		
	25 - 49		
	50 - 64		
	65 and over		
female	Σ		
	14 - 24		
	25 - 49		
	50 - 64		
	65 and over		

In this three-way table, all domestic Scandinavian persons (15 years and over) in employment are cross-classified by sex, age group and the extent of work time in first job for March 1990. The AGGLEvel 'broad age groups' is assumed to comprise the classes '14–24', '25–49', '50–64', and '65 and over' which, of course, gets modified internally by cutting off MODality '14' from the first group due to the age bound of 15 enforced by the CONDEF used.

T3.6 could be converted to a table of percentages based on *full-time/part-time distinction* simply by changing the **OF** clause and adding a **BASED ON** clause:

T3.7 ←
    COMPUTE table; **OF** %(n); **ON** *LFF*;
    FOR persons in employment (EU-83);
    BROKEN DOWN BY Σ | *sex* $*$ Σ | *age*@broad age groups
                        $*$ *full-time/part-time distinction (main job)* | Σ;
    OVER nation: ={Norway,Sweden,Finland,Iceland,Denmark};
    IN month: =Mar.1990;
    BASED ON *full-time/part-time distinction (main job)* | Σ ⏎

In T3.7, for each MOdality of the *full-time/part-time distinction (main job)* QUality the total of 100% is broken down by the remaining cross-classification dimensions of sex and age groups as indicated in **Tab. 3.4.2–2**.

Using the CONDEF of 'long-term unemployment / 2', T3.8 is an instance of a time series broken down by age groups and education level for regional units at NUTS-II level (including batch sums for NUTS-I level regions) in Belgium:

T3.8 ←
    COMPUTE table; **OF** n; **ON** *LFF*;
    FOR long-term unemployment / 2;
    BROKEN DOWN BY Σ | *age*@broad age groups
                           * *highest level of general education attained* | Σ;
    OVER NUTS-0: ={Belgique-Belgie} _2;
    IN year: 1985 .. 1995 ↵

As no data for 1995 may be available at request time, the internal TARGET could be relaxed to something like '1985 .. 1994'.

Tables of Rates

The tables investigated hitherto have all been either of summary type COUNT or PERCENT. As far as the statistical indicators of activity rates, employment/population ratios, and unemployment rates are concerned, however, the natural summary type to be used is RATE. For instance, repeating again the table structure of T3.1, the *activity rates* of persons between 14 and 24 years are requested as follows:

T3.9 ←
    COMPUTE table; **OF** RATE; **ON** *LFF*;
    FOR labour force (EU-92) ÷ population in working age (EU-92);
    WITH age: .. 24;
    BROKEN DOWN BY Σ | *sex*
    OVER EU/non-EU: EUR-12 _2;
    IN year: 1989 ↵

Likewise, using the concept definitions of the ILO, the same source data may be used to compute *unemployment rates*:

T3.10 ←
    COMPUTE table; **OF** RATE; **ON** *LFF*;
    FOR unemployed persons (ILO) ÷ civilian labour force (ILO);
    WITH age: .. 24;
    BROKEN DOWN BY Σ | *sex*;
    OVER EU/non-EU: EUR-12 _2;
    IN year: 1983 .. 1991 ↵

In fact, both RATEs may be juxtaposed for the sake of comparison in a single (compound) table by collecting both FOR clauses' parameter phrases in a single bracketed clause like

FOR [labour force (EU-92) ÷ population in working age (EU-92),
unemployed persons (ILO) ÷ civilian labour force (ILO)]

In order to better assess the differences between both rate definitions, it might be wise to include the resulting table into a further one with identical breakdown structure providing the respective population totals.

A temporal comparison on unemployment rates calls for a specific parameter phrase in the sentence's **OF** clause; considering an Austrian example comparing unemployment in spring (March to June) of years 1989 and 1990 with a spatial breakdown at the level of Federal Countries (Bundesländer), a specification like the following one might result:

T3.11 ←
    COMPUTE table; **OF** Δt RATE; **ON** *LFF*;
    FOR Arbeitslose einschließlich geringfügig Erwerbstätiger (A) ÷
                                  Unselbständig Erwerbstätige (A);
    WITH age: .. 65;
    OVER nation: =Austria _1;
    BETWEEN season: spring.1989 **and** season: spring.1990 ↵

This aggregate will comprise the (signed) numerical differences for Austrian unemployment rates obtained by subtracting the rates computed for spring 1990 from those computed for spring 1989.

Numerical Aggregates

Since categorical features dominate in labour force statistics, there is only a small share of numerical aggregates to be considered. In most cases, these aggregates either refer to averages of hours worked usually or actually in a given time period (a week, generally) or they measure duration such as average time unemployed persons stay in the state of job seeking. A typical representative of tables of the former type (cf., for instance, [EUROSTAT, 1991]), conforming to the table structure of T3.1 except that the temporal breakdown is replaced by a nested breakdown in terms of main groups of economic activity and professional status, is obtained as follows:

T3.12 ←
　　COMPUTE table;
　　OF [Σ *usual work hours per week (main job)* $\div n$];　**ON** *LFF*;
　　FOR persons in employment (EU-83);
　　BROKEN DOWN BY
　　　　Σ | *sex* $*$ Σ | *professional status (main job)* $*$ *economic*
　　　　　　activity of local unit of establishment (main job) @ main groups
　　OVER EU/non-EU: EUR-12 _2;
　　IN year: 1989 ⏎

This would induce a META isomorphic to **Tab. 3.4.2–3** except that the restriction to "Employees" had to be brought about by table editing, that is: eliminating all entries belonging to MODalities of QUality $Q_{22\text{-}1}$, *professional status (main job)*, other than 'employee' by hand.

If an analogous table for actual hours worked per week is asked for, the QUality summed over must be exchanged in the **OF** clause; furthermore, to make the real scope of the relevant domain concept apparent, an additional **WITH** clause is inserted:

T3.13 ←
　　COMPUTE table;
　　OF [Σ *actual work hours (reference week) (main job)* $\div n$];　**ON** *LFF*;
　　FOR persons in employment (EU-83);
　　WITH work status (reference week): =work for pay and profit;
　　BROKEN DOWN BY
　　　　Σ | *sex* $*$ Σ | *professional status (main job)* $*$ *economic*
　　　　　　activity of local unit of establishment (main job) @ main groups
　　OVER EU/non-EU: EUR-12 _2;
　　IN year: 1989 ⏎

Omitting the **WITH** clause wouldn't change the generated META anyway except that a (foot-)note would have been issued highlighting that all TRAITs of the CONDEF were cancelled for MODalities other than 'work for pay or profit' because of a FRAMEDit.

If the usual work time of persons is scrutinized, the results may be somewhat different considering all jobs a person may have. In this case, apparently, the usual work hours in each of the up to three jobs recorded in the *LFF* must be summed up before averaging can take place. In METASPEL, the corresponding specification is arranged easily by introducing a CONAXis (cf. **Def. 2.2.3–4**) named, say, *total work hours in all jobs* of SCALETYPE 'INTeger' defined in terms of the RECTFORMula

$$\Sigma_{n=1,2,3} Q_{38-n} = m$$

for all INTeger $m > 0$.

Table 3.4.2–3: Sample CLFS Table with Numerical Aggregate [EUROSTAT, 1989c]

HOURS	Eur 12	Eur 10	Belgique Belgie	Danmark		Luxembourg	Nederland	Portugal	United Kingdom
					Males and females				
Total	39.3	39.0	38.5	35.7	...	40.1	33.2	43.8	38.9
Agriculture	47.1	47.1	54.2	45.0	...	52.0	43.5	50.2	50.9
Industry	40.4	40.3	39.3	37.2	...	40.3	35.7	43.8	42.7
Services	37.9	37.5	37.5	34.4	...	39.3	31.5	41.8	36.5
Employees	37.6	37.3	36.0	34.3	...	38.7	32.1	41.3	37.7
Agriculture	41.1	39.9	38.3	36.4	...	–	32.9	46.9	43.3
Industry	39.8	39.6	38.4	36.3	...	40.0	35.4	43.4	42.3
Services	36.2	35.9	34.6	33.3	...	38.1	30.7	39.1	35.3
					Males				
Total	42.4	42.3	41.0	38.9	...	41.8	37.7	45.3	45.1
Agriculture	50.0	50.8	58.6	47.9	...	56.5	49.9	51.1	55.3
...	...	...	...	...	...	...	...	...	...
Employees	40.6	40.5	38.3	36.9	...	40.3	35.8	43.2	44.0
Agriculture	43.0	42.1	39.6	38.3	...	–	35.0	48.6	48.0
...	...	...	...	...	...	...	...	...	...
					Females				
Total	34.5	33.8	34.3	31.9	...	36.9	25.5	41.6	30.6
...	...	...	...	...	...	...	...	...	...
Employees	33.3	32.8	32.2	31.4	...	35.7	25.7	38.5	30.3
...	...	...	...	...	...	...	...	...	...

T3.14 ←
 COMPUTE table;
 OF [Σ *total work hours in all jobs* ÷ *n*]; **ON** *LFF*;
 FOR persons in employment (EU-83);
 BROKEN DOWN BY
   Σ | *sex * economic activity of local unit*
        of establishment (main job) @ main groups
 OVER EU/non-EU: EUR-12 _2;
 IN year: 1989 ⏎

A simple table summarizing the average duration of job search for persons in employment (according to ILO definition) with previous work experience is determined by a MetaSpeL sentence like the following one:

T3.15 ←
 COMPUTE table;
 OF [Σ *duration of search* ÷ *n*]; **ON** *LFF*;
 FOR unemployed persons (ILO);
 WITH *previous employment experience*: =yes;
 BROKEN DOWN BY
   Σ | *sex * occupation of last job* @ main groups | Σ;
 OVER NUTS-II: # **with** NUTS-0: **except** {Ireland,United Kingdom};
 IN year: 1989 ⏎

This example also makes extensive use of special range specifications in its **OVER** clause; the resulting META's spatial breakdown will enclose all NUTS-II region units (within the EU territory as currently defined) except those of Ireland and the U.K. Otherwise, the breakdown of the table is twofold, viz. sex and main groups of occupation of the unemployed persons' last jobs.

Some More Specific Concepts

In addition to the class of tables being based on the standard domain concepts of labour force statistics there is, of course, a myriad of custom-tailored tables emanating from specific inquiries. Naturally, only a small and arbitrary selection of illustrating cases can be presented here.

 As a first one, assume that for some reason a table of women in paid dependent part-time employment working at most 15 hours per week usually in their one and only job in the services sector of economic activity are of interest; furthermore, the scope of scrutiny should be restricted to the city of Berlin and comprise a quarterly comparison for years 1989 through 1991 in order to spot any significant changes during that time period (perhaps in a subsequent inferential-statistical analysis):

T3.16 ←
    COMPUTE table; **OF** *n*; **ON** *LFF*;
    FOR persons in employment (EU-92);
    WITH *sex*: =female & *number of (current) jobs*: =1 &
          usual work hours per week (main job): .. 15 &
          full-time/part-time distinction (main job)@full-part:
                                                        =part-time &
          economic activity of local unit of
                establishment (main job)@main groups: =services;
    BROKEN DOWN BY *age*@broad age groups | Σ;
    OVER NUTS-II: {Ehemaliges Berlin Ost,Ehemaliges Berlin West};
    IN season: {spring,autumn}.1989 .. 1991 ⌐

The resulting table would have a three-fold breakdown, viz. age (4 groups), space (2 groups), and time (6 groups). The determination of an appropriate table layout facilitating comparisons visually is supported by METAMOD's periodicity graph by means of which METAGEN is "aware" of the temporal relation between value combinations 'spring.1989', 'spring.1990', 'spring.1991', and, analogously, for the 'autumn'-combinations. Instead of the chosen table concept, 'persons in employment (EU-92)' its national counterpart could have been inserted as well, of course.

As a second example, the share of long-term unemployed persons among unemployed persons is considered. Formally, a table of RATEs is asked for relating, say, the domain concepts 'long-term unemployment / 2' and 'unemployed persons (EU-83)':

T3.17 ←
    COMPUTE table; **OF** RATE; **ON** *LFF*;
    FOR long-term unemployment / 2 ÷ unemployed persons (EU-83);
    BROKEN DOWN BY *age* | Σ;
    OVER EU/non-EU: EUR-12 _2;
    IN year: 1988 ⌐

A variant of T3.17 might replace the breakdown with respect to *age* in years by a double breakdown by age in age groups and sex, and, additionally, specify a time series covering the range 1983 .. 1991 for persons being unemployed for at least *two* years (cf. [EUROSTAT, 1993b],

T3.18 ←
    COMPUTE table; **OF** RATE; **ON** *LFF*;
    FOR LTU-2+years ÷ unemployed persons (EU-83);
    BROKEN DOWN BY *sex* | Σ * *age*@below/above 25;
    OVER [EU/non-EU: EUR-12,{Belgique-Belgie ,..., United Kingdom}];
    IN year: 1983 .. 1991 ⌐

where the domain concept is assumed to be defined as

> CONDEF   LTU-2+years
>   superconcept: long-term unemployment / 2
>   [1] $Q_{65}(104\ ..)$
> ↵

noting that *LFF* QUality Q_{65}, *duration of search*, is measured in weeks. AGGLEvel 'below/above 25' of QUality *age* would consist, of course, of the MODality groups '.. 24' and '25 ..'.

In a similar way, a RATE of underemployed persons based on persons in employment could be derived. To this end, only the **FOR** clause of T3.18 must be replaced with

> **FOR** underemployed persons (EU) ÷ persons in employment (EU-92)

to obtain a table structurally identical to T3.18 otherwise.

To illustrate the difference the choice of domain concepts can make, T3.19 specifies a numerical comparison of aggregates both measuring unemployment,

> T3.19 ← T3.19.1 − T3.19.2 ↵

where the operand tables T3.19.1 and T3.19.2 might be established as follows:

> T3.19.1 ←
>   **COMPUTE** table; **OF** *n*; **ON** *LFF*;
>   **FOR** Arbeitslose einschließlich geringfügig Erwerbstätiger (A);
>   **BROKEN DOWN BY** *sex* | Σ * Σ | *nationality*@domestic/foreign;
>   **OVER** nation: Austria;
>   **IN** year: 1984 .. 1994 ↵

> T3.19.2 ←
>   **COMPUTE** table; **OF** *n*; **ON** *LFF*;
>   **FOR** unemployed persons (ILO);
>   **BROKEN DOWN BY** *sex* | Σ * Σ | *nationality*@domestic/foreign;
>   **OVER** nation: Austria;
>   **IN** year: 1984 .. 1994 ↵

By computing absolute differences element-wise in a four-fold classification (by sex and domestic/foreign persons), T3.19 will elicit the numerical deviations of unemployed persons measured according to ILO definitions in relation to unemployed persons measured according to the Austrian definition used in T3.19.1. It must be noted, however, that in both aggregates the annual figures are estimates

Compound Tables

As a matter of fact, most of the table examples discussed in this subsection are instances of compound tables. In general, table composition follows implicitly by combining kernel and margin tables (cf. Subsection 2.4.3) in the **BROKEN DOWN BY** clause of METASPEL sentences. Occasionally, however, compound tables are requested in explicit terms as illustrated by expression T3.20. This METASPEL sentence denotes a table corresponding to a layout structure as sketched in **Tab. 3.4.2–4**: in this table, the main domain concept, 'Wohnbevölkerung (A)', restricted to persons 15–60 years living in private households, is cross-classified three-fold, viz. by sex, highest level of general education attained, and a distinction as to active and inactive persons.

> T3.20 ←
> **COMPUTE** table; **OF** *n*; **ON** *LFF*;
> **FOR** Wohnbevölkerung (A)
> >> Erwerbstätige (A-93) **wrt** *professional status (main job)*
> >> Arbeitslose (A-93)
> >> Nichterwerbspersonen (A) **wrt** *status of inactive person*;
> **WITH** *type of household in usual place of*
> *residence*@private/collective: =private &
> *age*: 15 .. 60;
> **BROKEN DOWN BY** *sex* | Σ *
> Σ | *highest level of general education attained*;
> **OVER** nation: Austria;
> **IN** year: 1990 ⏎

Compared to **Tab. 3.4.2–4** (overleaf), T3.20 only loses the combined batch sum of 'Erwerbstätige (A-93)' and 'Arbeitslose (A-93)' – conforming, in fact, to the domain concept of 'Erwerbspersonen (A)' – because, according to the definitions given in Subsection 2.4.3, the hierarchical nesting of sub-concepts is not supported by present METASPEL language structures. Conversely, however, each of the sub-concepts stated in the **FOR** clause absorbs the sentence's **WITH** clause automatically.

Although a bit clumsy to specify, table structures like the one presented last can be achieved by table editing as well (on condition that the constituent sub-tables have been derived already, of course). After all, it may not be a good idea to cram a single (layout) table with too many conceptually different (internal) sub-tables in view of cognitive comprehension and succinct presentation.

Table 3.4.2–4: Sample Compound Table

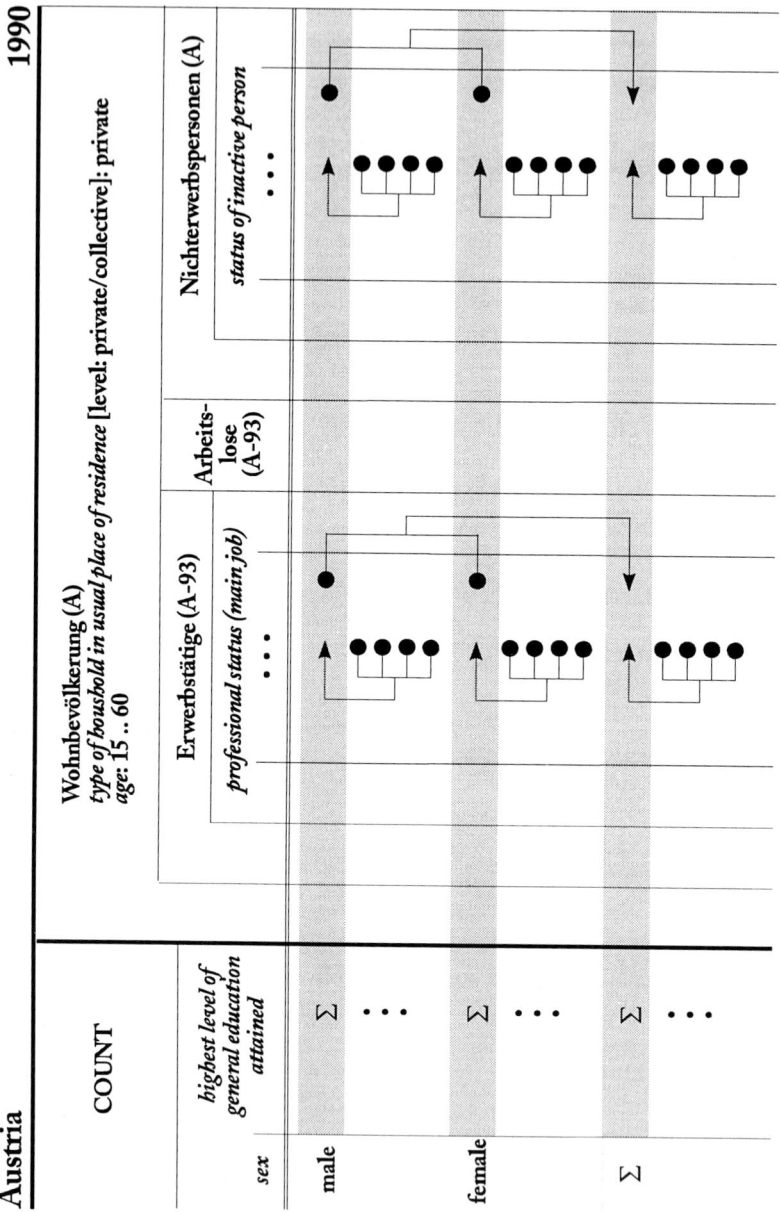

Chapter 4

Application Case II: Tourism Statistics Strategic Management Information

Viewed abstractly, there is little difference between official statistics macrodata and management information systems macrodata; formally, there is no difference at all. Hence, with respect to data integration both areas are facing equivalent and technically identical problems in terms of data modelling and data management, for which reason a transfer of technical approaches and experiences to the field of strategic management information systems seems rather natural.

As a specific application domain in the field of strategic management information systems, *tourism statistics* has been chosen because (i) tourism business is among the first (after stock and commodity exchanges, international bank clearinghouses, and flight reservation/booking systems) characterized by a globalizing market becoming established on world-wide connected electronic tele-communication media, a situation calling for unprecedented means and innovative ways of strategic management, and (ii) more and more tourism data is conveyed purely electronically, although to date improving advertising and distribution mainly by means of so-called tourism information systems.

After reviewing the tourism business in the wake of the electronic revolution, laying particular emphasis on the information aspects of strategic tourism *marketing*, this chapter takes up the rather close relation of tourism macrodata processing to official tourism statistics in order to develop a framework for tourism data integration according to METASTASYS principles. Starting from a recapitulation of traditional tourism data aggregate production schemes, above all the fairly typical Austrian example, a couple of data models is investigated and their effective use demonstrated by several sample aggregate specifications.

In contrast to official statistics aggregate data networks where data is broadcast mainly according to legal regulations and enforced semantic standards, tourism data networks will operate on a more or less commercial basis and, hence, will be governed by market demands. However, leaving aside the conceivable impacts of this crucial difference, the presentation of the proposed data integration approach concentrates predominantly on issues of its technical arrangement.

4.1 Electronic Markets and Decision Making

In a sense, the nomadic way of life may be seen as a prehistoric root of tourism, followed by settlement and emerging civilization where excursions along established paths and routes may have introduced the over and over repeating travelling episode of leaving home, staying away, and returning [Leed, 1991]. A necessity of subsistence, or economy, moving around surely has also been accompanied by a feeling of curiosity ever since – exploring unknown territory (physically, but psychological factors cannot be overlooked) amounts to extending one's experience and feeding or reassessing the mind's own self-affirmation of being a powerful creature having the world at command. Particularly in modern times, when the European culture of Enlightenment changed the self-image of man in his world, travelling attained a fundamentally new quality with the expeditions of great discoverers such as Marco Polo, Cristoforo Colombo, Fernão de Magalhães, or James Cook, to name just a few; certainly it is no exaggeration to state that those travel experiences and impressions led to the insight of the earth in fact being a globe rotating in the universe rather than a disc bounded by an infinite ocean. It goes without saying that Charles Darwin's expedition on board of the Beagle and the manned Apollo-11 mission to the moon fit well into *this* tradition of tourism with evidently far-reaching implications for our self-understanding. In fact, travelling and thereby becoming aware of the differences between different parts of the world has so profoundly changed our perception of space (and, as a further consequence, our entire personality as well) that, today, we could hardly imagine to spend a whole life at a single isolated spot – the proverbial "island" – on earth; quite on the contrary, visiting as many unknown places as possible in one's life seems to have become almost an obsession in (industrialized) societies – and be it for the sole (tacit) purpose of encountering *oneself* in other, unfamiliar locations.

In fact, developed societies crucially depend on the spatial mobility of their citizens for a couple of reasons; being once a privilege of upper classes, travelling has become a mass phenomenon both calling for and being stimulated by high-capacity traffic systems (ship, railway, aeroplane, and particularly automobile). Traffic and transportation services have grown into important business branches and, as already pointed out in Section 1.1, traffic of persons and goods is supplemented increasingly with the traffic of symbols in world-spanning communication networks. Basically, in the 19th century the evolution of both, transportation and tele-communication systems has been driven economically but, presumably as a side effect, the idea of travelling had permeated Western society and – due to capacity growth of traffic systems – the marginal travelling costs of individual journeys had sunk so much at this time that it gave birth to tourism as a pastime activity as well. A further prerequisite for the emerging mass tourism was the establishment of a – standardized – tourism infrastructure, particularly urban gastronomy and lodging facilities, within a generally evolving economic infrastructure; typically, early tourism centred around major events such as the series of world exhibitions starting in 1851 but extended soon – along the growing railway networks – to rural and seaside recreation resorts preferedly visited by urban upper class families. Nevertheless, 19th century tourism remained of limited economic importance in terms of national economies since travelling still was a pleasure of wealthier parts of the population only. Real mass tourism and tourism industry took shape in the wake of social change and increasing welfare particularly after World War II when the enormous productivity gains were more evenly distributed among citizens and, specifically, the broad class of dependent employees received an increasing share of national income. Conversely, industrial capital investments could be turned into profits by mass consumption only which, in turn, was enabled by a social redistribution of *time*: both production and consumption of goods takes time, so people must be given time to consume what is produced [Nowotny, 1993]. Today, a great portion of spare time (that is, time not used to make money or to earn a living) is devoted to tourism activities whence holiday tourism and the leisure travel industry has developed into a huge economic sector (for instance, Austria's 1993 excess of imported commodities could be financed to 90% by a surplus of foreign tourism exchange payments, at a total level of about 156 billion Austrian shillings – approx. 11 billion ECU – of foreign exchange payments earned in tourism; [ÖSTAT, 1994]). In other words, tourism has become a market of national importance at least for some countries and receives due attention as a major national resource of economic wealth. However, as society is in constant change, so are markets, products, and consumer behaviour – tourism markets being no exception anyway. For instance, recent forecasts in world tourism predict around about 1 billion tourists annually by the year 2004 with a strong preference of destinations in the Far East, Central America, and the Carribean islands [Vialle, 1995]. In this situation, effective means to recognize changes in the tourism market, to anticipate tourism trends and to respond in time are mandatory preconditions – at each organization level: national, regional, and local – to stay competitive. More profoundly, a thorough understanding of tourism market structure and dynamics marks the starting point for a discussion of contributions and future role of tele-communication and computing technologies in the tourism business which themselves, as has become apparent already, are among the driving forces reshaping this business vigorously.

Tourism markets share rather specific characteristics; for one thing, product structure is quite complex, thus resulting in high information cost at consumer side – typically, a holiday booking consists of three components (transfer, accommodation, and entertainment programme on site) which need to be combined. Furthermore, journeys are so-called confidence goods, that is product quality cannot be assessed actually before consumption is taking place. At supplier side, tourism products typically are non-storable goods (such as a bednight) forcing to sell capacity at almost any price lest it be written off entirely. Moreover, in many places profits have been re-invested rather short-sightedly in the blunt expansion of lodging capacities far beyond average demands, thus weakening further the market position of accommodation entrepreneurs. This has led to the establishing of specific downstream information activities to propagate advertising and distribution of tourism products: non-profit tourist boards representing particular geographic areas and entrepreneurial information brokers (such as travel agents) actually selling tourism products who, basically, get paid for taking over the consumer's information burden. From a local tourism service supplier's view (for instance, a hotel), **Fig. 4.1–1** depicts the main communication flows in tourism business. Apart from direct marketing activities (arc o_1) destination marketing is carried out mainly by non-profit tourist boards (arc o_2) at destination level or higher-level representatives at regional or national levels. In most cases, these tourist boards have been restricted legally to marketing activities and, therefore, prohibited from reservation and selling (for instance, in Austria; cf. [Froeschl and Werthner, 1994]). In fact, tourist boards are responsible not only for advertising tourism destinations but for *branding* (that is, creating distinguishing features and suggestive presentations helping to identify a destination and thus enabling effective marketing) and regional socio-economic development as well [Froeschl and Werthner, 1994]. Actual distribution of tourism products is performed by information brokers (travel agencies, bureaux of airlines, etc., currently handling around about 80% of total turnover) which resort either to packaged products offered by tourism wholesalers (tour operators) or utilizing computerized reservation/distribution systems (CRS/GDS). Especially the latter initiated considerable changes of tourism market structure [Ernst and Walpuski, 1994] since they operate on-line – that is, real-time information and booking is provided – and admit almost world-wide access and, hence, product distribution. Despite apparent limitations with respect to marketing capabilities (basically, they were conceived as booking and reservation systems oriented towards business tourism) global distribution systems (GDS) provide a natural cornerstone of electronic tourism marketing [Schmid, 1994] focusing on holiday tourism as well. In the long run, electronic media by and large will substitute conventional tourism information media (such as brochures, travel catalogues, etc.; for instance, cf. tourism information systems like the Tyrol's TIS, [Werthner, 1993]) and enable electronic bargaining by comprising closely coupled product presentation and booking components (for instance, cf. [Maartmann-Moe *et al.*, 1994; Byerley *et al.*, 1995]). Rather naturally, already established destination information systems will evolve into electronic tourism marketplaces once the present (and apparently justified) reservations are giving way to increasing confidence in the legal reliability of transmitted data by appropriate data protecting, authentication, and other credibility assuring measures [Wildhaber, 1995].

From a pure marketing point of view, a shift to electronic marketing media gains overwhelming advantages over traditional means – information is delivered quickly and can be kept up-to-date easily, more as well as more detailed information can be conveyed, presentation is enhanced (for instance, by multi-media systems), information is accessible cheaper and easier (for instance, by integration of heterogeneous information subsystems as exemplified by [Austin *et al.*, 1994]), targeting of consumer strata is facilitated favourably, there is better leverage to influence consumer behaviour, market coverage can be improved [Marshall, 1994] – simply speaking, destination promotion and marketing become much more effective.

.1 Electronic Markets and Decision Making

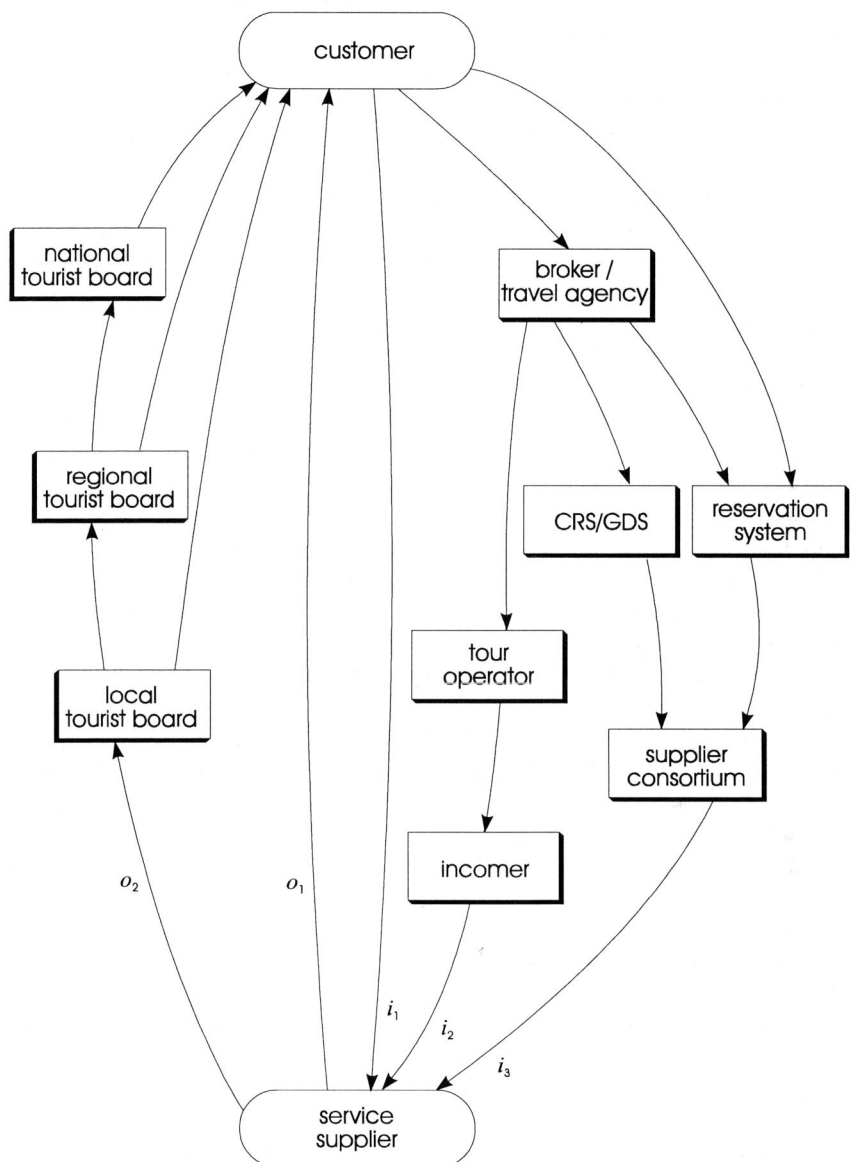

Figure 4.1–1: Main Tourism Communication Flows [Werthner, 1993] (adapted)

Hence, electronic marketing is seen as an appropriate means to cope with changing consumer attitudes and behaviour (for instance, later bookings, a greater number of shorter holidays, more selective choice, more detailed and more specialized information requests, more frequent changes in service demands, less regular and decreasingly loyal guests, etc.) in a phase of increasing competition in the globalizing tourism marketplace. This expansion of the market (defined as an economic space where competition is effective), which in turn is stimulated by electronic marketing, enforces a reconsideration of marketing *strategies*. In particular, electronic markets offer an opportunity to destinations – and the typically rather small tourism enterprises within those destinations – to increase their share of the value-adding chain by exploiting the direct communication paths (particularly arc i_1 in **Fig. 4.1–1**) between service suppliers and – prospective – guests. Technically, digital electronic end consumer networks will make possible next time a delivery of tourism information right into the consumers' living rooms by converting familiar TV equipment to interactive electronic shopping malls including holiday supermarkets comprising a really global supply of products [Froeschl and Werthner, 1994]. Hence, any service supplier can, in principle, be *tele*-present world-wide by means of electronic tele-communication media. Practically, however, traditional service suppliers – airlines and large professional supplier organizations excepted – will neither have sufficient means nor know-how to make good use of this opportunity; in fact, without purposive counteracting market power will concentrate further in the hands of a few large, often multi-national vendors optimizing their profits without taking any care of national or destination interests (to cite an example, supplied services such as flight seats and bednights are taking on the features of commodities as traded on international commodity exchanges) [Archdale, 1994].

On the eve of fully electronic tourism markets there are quite a few stressing the importance of tourism information systems as *strategic* destination marketing tools (for instance, cf. [Haines, 1994]). As product competition gives way more and more to distribution competition the strategic role of electronic marketing increases significantly especially in responding quickly to market behaviour based on, for instance, real-time sales monitoring. In addition to promoting destinations and their tourism offer to catch the interest of prospective tourists the traditional service-orientation of information suppliers, particularly non-profit tourist boards, ought to be substituted by consumer-led information systems to account for the changing social customs in the usage of leisure time which themselves are determined fundamentally by the nature of working processes and socio-economic production structures [Nowotny, 1993]: the economization of leisure time and the professionalization of leisure time activities. In short, electronic marketing bears the potential of flexible tourism offers custom-tailored to *individual* demands ("mass-customization") on a world-wide scale which will certainly become the hallmark of competitiveness in a globalizing market with increasing product transparency as a consequence of tremendously sinking information cost. Flexible tourism marketing (reminiscent of flexible manufacturing in commodity production), in turn, demands a source of comprehensive and timely market information. Evidently, traditional methods of tourism data collection will no longer be sufficient in this respect for both reasons, lagging behind and failing to meet actual information requirements [Wöber, 1994]. However, in view of destination marketing systems run, basically, by local tourist boards [Froeschl and Werthner, 1994] it is quite natural to augment those systems' functionality with a *management information* component comprising electronically the decision making information necessary for strategic marketing tasks.

Highlighting this added functional layer, **Fig. 4.1–2** (overleaf) refines the diagram of **Fig. 4.1–1** with respect to incoming and outgoing communication flows of tourism management/marketing information systems (MIS). Incoming flows represent, in general, case-oriented data (business transaction logs) whereas outgoing flows typically comprise data aggregates. An essential source of market information to be gathered is booking data (arc c_1); particularly, as soon as destination marketing systems are extended from "simple" one-way tourism information broadcasting media to bi-directional, interactive booking and reservation systems, either by inclusion in CRS/GDS or similar large-scale distribution media, they constitute a prime data tapping source of market information hardly available otherwise (not even by expensive market research). At management information system level, booking data can be supplemented with service supplier data (such as service, equipment, or category descriptions; arc c_2) as well as guest responses obtained, for instance, from tourist inquiries. Replacing traditional data reporting paths, local tourism data will be passed on to downstream information systems (management information systems at regional or national tourist boards, national tourism statistics systems) and will provide also a basis for lateral market information suppliers (market research institutions, information wholesalers, data subscription services, etc., among which traditional statistical offices will be subsumed either) distributing market information to a wider audience of commercial and non-commercial market participants. A rather subtle question regards the direct backflow of market information at detailed levels to commercial vendors of tourism services (arcs d_2 to d_4) since this will heavily influence the vertical distribution of market power: the greater the strategic information advantage of a destination's service suppliers (advocated by local tourist boards) the better their position to gain larger shares of the tourism value-adding chain.

Whether destination marketing systems operate at destination or a higher level, strategic marketing draws on aggregate tourism data combining a multitude of destinations. Hence, there emerges the data integration problem. Traditionally, tourism statistics is organized at national levels whence integration of tourism aggregates within the boundaries of nations is rather straightforward – at least as long as conventional tourism data is concerned only. On a global scale, in fact, transborder data flow integration turns into yet another data harmonization task making tourism macrodata obtainable from local information systems compatible with each other semantically such that consolidated global market information can be fed back to local or non-local management (marketing) information systems. In this respect, semantic data integration provides the enabling technology for effective strategic decision making on electronic tourism markets.

At present, admittedly, it is rather difficult to assess the impact of information technology on the overall structure of the tourism market and especially on the changes in the tourism value-adding chain. Even more difficult is predicting seriously how marketing policies will respond in the longer run. Less doubtful, however, appears the role of aggregate market information in future tourism marketing: before long, any tourism marketing policy will make use of tourism macrodata and even better so, if information is readily available *within* established electronic data processing media. Although the proposed data integration model tries to anticipate an answer to this emerging managerial demand, the outlined approach indicates a possible information-technological solution only; just as in other application areas, intricate economic, social and legal questions are left unanswered. For instance, the then still present market imperfections may call for specific legal regulations as to who will be admitted access to which type of available tourism market information in what time and at which price – apparently, the ubiquitous presence of global market information will definitely change market dynamics and offer hitherto unknown opportunities to those having means and know-how to exploit the new tourism information resources. Thus, the development of technical solutions certainly must be accompanied with serious economic as well as ethical considerations if this information technology should prove really beneficial.

4.2 Traditional Tourism Data Structures

In pre-electronic ages, tourism data aggregates were processed paper-based in a multi-stage procedure beginning with enterprise data passed on to district/community administrations before becoming merged into national tourism statistics by statistical offices. On its way, data was condensed at each stage primarily to cut down the volume of data reported to subsequent processing stages. Evidently, this – still predominant – procedure is hopelessly old-fashioned because (i) it tends to be rather slow and tedious and (ii), perhaps even more severely, much of the collected information is lost on its way due to intermediary aggregation steps. From an open systems view, neither should available data be destroyed voluntarily – since already incurred data production costs are wasted unduly – nor should data be given up simply because of differing technical or semantic standards: at national level, tourism data traditionally has, by law, to be reported periodically according to compulsory legal standards but despite the advantage of national uniformity it remains a fact that different data sources (most evidently at the international level) conform to differing terminologies, definitions, and feature patterns of interest; this, in turn, has led to different data reporting formats, technical standards, and logical data models being adhered to. Shifting to electronic data processing certainly does not do away with this multitude of terms and standards – quite on the contrary, diversity is even likely to increase in view of the broadening multi-cultural scope and the certainly continued attempts to split markets by pushing proprietary product standards. However, speed and transmission rates indeed can be raised enormously by electronic media, and computing technology supports data integration effectively by establishing *algorithmic* data interfaces linking the currently rather isolated "data islands".

.2 Traditional Tourism Data Structures

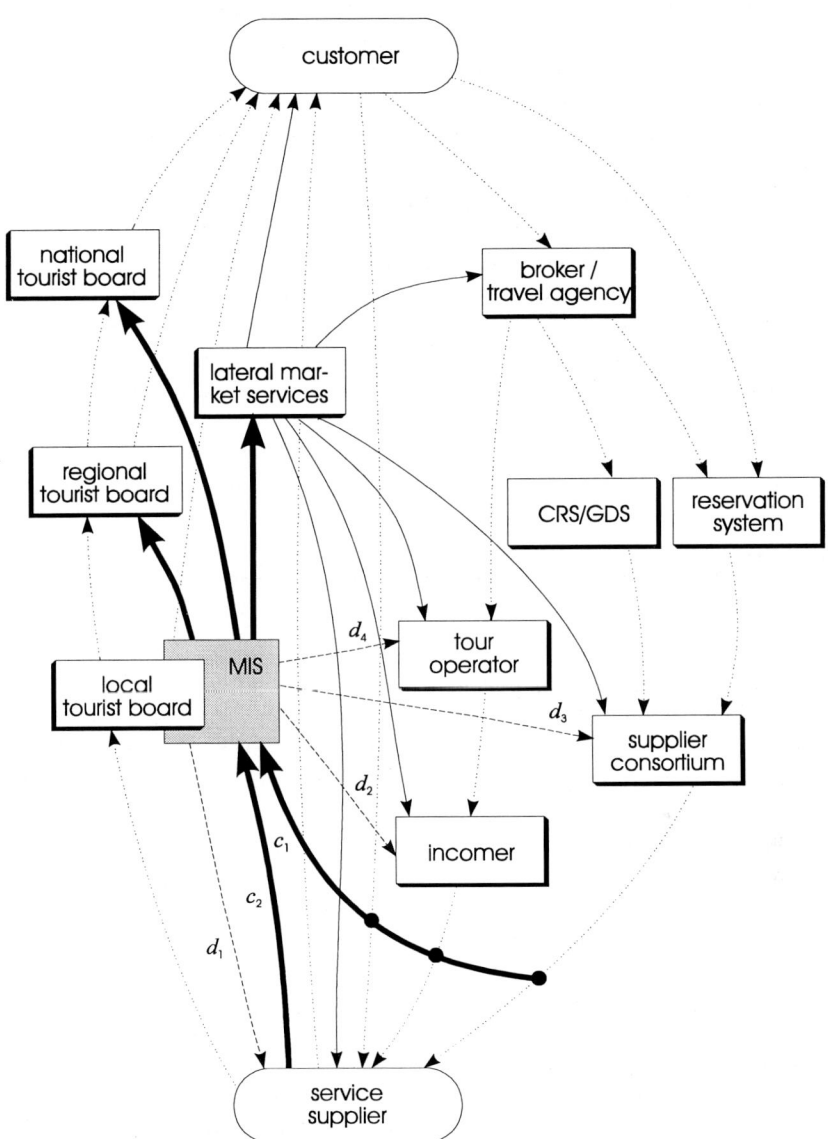

Figure 4.1–2: Tourism Management Information Capturing and Dissemination

Statistical tourism data comprise two different kinds of data to be distinguished carefully: *log* data and *transaction* data. The latter refers to data necessary to carry out present or future transactions (such as the booking of a hotel room) whereas log

– that is: time-tagged – data records a statistically relevant skeleton of past transactions (for instance, actual hotel bookings with "product data" etc. removed) mainly for the purpose of statistical aggregation over large collections of transaction data records. Hence, in what follows, only log-time data will be concerned. By the way, no distinction is drawn between pure holiday and business tourism.

Traditionally, tourism statistics concerns two aspects of the tourism business, viz. (i) a quantitative description of tourism offers in terms of lodging enterprises and their equipment and service features (lodging enterprise statistics), and (ii) a recording of tourist streams in terms of arrivals and bednights (bednight statistics). Typically, other aspects of the tourism business such as price levels, transportation facilities (for instance, funiculars) or gastronomic services are not covered hitherto by tourism statistics (but instead are included, for instance, in traffic or trade statistics). In the long run this certainly is rather unfavourable since, at management level, much can be gained by an integrated statistical view of *all* relevant aspects of the tourism business. Apparently, data integration has got to add something to conventional tourism statistics (except a mere electronic replacement of paper-based data reporting) before it can really support the strategic management information function outlined in Section 4.1 above. Particularly in view of already established destination marketing systems providing destination data (such as data describing destination infrastructure) it will become fairly easy to interrelate destination data with tourism statistics in order to improve the decision makers' data basis.

To provide a starting point in what follows, Austria's present system of tourism statistics is taken as a baseline of discussing the preconditions of a data integration approach. Despite apparent peculiarities of national tourism statistics systems, the Austrian system can be regarded quite representative. After reviewing this traditional data capturing and production structure, Section 4.3, based on this outline, introduces suitable candidate data models for tourism data actually enabling data integration.

4.2.1 Lodging Enterprise Statistics

In Austria's system of tourism statistics lodging enterprises are counted once each year (key-date May 31) distinguishing commercial enterprises, private accommodations and other lodging facilities (spa hotels run by social security institutions, other spa hotels, children and youth convalescent homes, sanatoria, youth hostels, shelter huts with service). With respect to commercial enterprises three accommodation quality categories are discerned, viz. 1/2-star (low standard, including enterprises not categorized officially at all), 3-star (medium standard), and 4/5-star (high/top standard), indicating equipment and service levels offered. For each commercial lodging enterprise or other lodging facility (i) the days of service in each month of the year of reporting and (ii) the number of available rooms – subdivided into room types (single room, double room, room with 3 or more bed-places, apartment) and broken down with respect to room standard (cold/hot water, shower/bath, WC, central/other heating system) – are recorded. Additionally, spare beds (in total) available for each room type as well as additional equipment (outdoor swimming pool/private lido, indoor swimming pool, sauna, tennis/squash court) are included in the inquiry. Altogether, a single lodging enterprise is characterized by a microdatum record comprising 53 features – accommodation type (1), service days per month (12), room type and standard (36), and additional equipment (4). It must be noted, however, that quality categories, room standards and additional equipment are not independent of each other (for instance, a hotel is assessed 4-star only if, at least, it is equipped with an indoor swimming pool and all rooms have automatic heating). For private accommodations only accommodation type (lodging on farmstead/not on farmstead, holiday apartment/house, other) and the number of available guest bedplaces and spare beds – separately for winter season (November to April) and summer season (May to October) – are recorded (that is, 5 features only). Thus, essentially, lodging enterprise statistics depicts the available quantity of tourist bedplaces broken down by accommodation type, room type, and room standard.

In a next step, individual enterprise or private accommodation data records are collected by communal administrations (villages, municipal districts) and aggregated to a tabular macrodata arrangement (inventory sheet) showing the number of enterprises ("Betriebe"), bedplaces ("Betten"), and spare beds ("Zusatzbetten") broken down with respect to accommodation type ("Unterkunftsart") – in the table stub – for key-date, winter and summer seasons as shown in **Fig. 4.2.1** (overleaf). Additionally, the number of camping sites is also included in this inventory form.

Finally, at national level, inventory forms are aggregated spatially to give a national enterprise and bedplace inventory. To this end, Austria's 1498 political communities, comprised in 9 federal countries, are subdivided into 1409 ordinary communities, 9 municipalities (federal capital Wien, 8 province capitals) and 80 health resorts (communities with mineral springs, stimulating climate, etc.), thus giving rise to a distinction of holiday, city, and health tourism.

Specimen aggregates (excerpts) of Austria's tourism enterprises and room standards are reproduced (from [ÖSTAT, 1994]) for period 1992/93 in **Tab. 4.2.1–1** and **Tab. 4.2.1–2** (cf. following pages) as published in the annual tourism statistics report of Austria's Central Statistical Office. In **Tab. 4.2.1–1**, the symbols' meanings are as follows: F = number of enterprises, B = number of bedplaces, N = number of spare beds.

4.2.2 Bednight Statistics

In contrast to enterprise statistics giving a picture of available lodging capacity, bednight statistics documents economic utilization of available capacity. Hence, the basic information unit of bednight statistics is the individual booking record. More specifically, in Austria a passenger form ("Meldezettel") must be filled in for each arrival and departure.

		AM STICHTAG 31. MAI 1993			IM WINTERHALBJAHR 1992/93				IM SOMMERHALBJAHR 1993		
	9 ②	Be-triebe	Betten	Zusatz-betten	4 ②	Be-triebe	Betten	Zusatz-betten	8 ②	Be-triebe	Betten	Zusatz-betten
Gewerbliche Beherbergungs-betriebe der Betriebsgruppe (Kategorie) 5 bzw. 4 Stern	01				01				01			
3 Stern	02				02				02			
2 bzw. 1 Stern	03				03				03			
Privatquartier nicht auf Bauernhof	60				60				60			
Privatquartier auf Bauernhof	61				61				61			
Kurheim eines Sozialversicherungsträgers	71				71				71			
Sonst. Kurheim u. Erholungsheim f. Erwachsene	72				72				72			
Sanatorium, Heil- u. Pflegeanstalt	75				75				75			
Kinder- u. Jugenderholungsheim	73				73				73			
Jugendherberge bzw. Gästehaus	81				81				81			
Bewirtschaftete Schutzhütte	83				83				83			
Ferienwohnung, Ferienhaus (Privat)	85				85				85			
Sonstige Fremdenunterkunft	84				84				84			
Campingplatz	70		XXX	XX	70		XXX	XX	70		XXX	XX

Figure 4.2.1: Inventory Form Used in Austria´s Tourism Statistics

Table 4.2.1–1: Tourism Enterprises and Beds in Austria 1992/93 [ÖSTAT, 1994]

UNTERKUNFTSARTEN		FREMDUNTERKÜNFTE UND BETTEN IM WINTERHALBJAHR 1992/93									
		BURGENLAND	KÄRNTEN	NIEDERÖSTERREICH	OBERÖSTERREICH	SALZBURG	STEIERMARK	TIROL	VORARLBERG	WIEN	ÖSTERREICH
1 BETRIEBSGRUPPE 5/4*	F	17	197	93	91	328	112	479	121	108	1546
	B	1797	18593	10132	7650	28556	9465	47345	9760	20557	153855
	N	69	1533	640	310	983	554	2775	329	764	7957
BETRIEBSGRUPPE 3/*	F	61	778	396	275	1200	692	1295	297	109	5103
	B	3004	28818	16308	12441	46668	24048	61764	12635	8281	213967
	N	110	1192	826	388	1886	1285	2401	300	421	8809
BETRIEBSGRUPPE 2/1*	F	188	990	798	945	1398	1097	4251	633	100	10400
	B	4312	18830	16828	20975	32189	20084	102738	13756	4883	234595
	N	186	948	787	1089	1577	1351	6117	689	319	13063
ZUSAMMEN	F	266	1965	1287	1311	2926	1901	6025	1051	317	17049
	B	9113	66241	43268	41066	107413	53597	211847	36151	33721	602417
	N	365	3673	2253	1787	4446	3190	11293	1318	1504	29829
2 KURHEIME DER SOZIAL VERSICHERUNGSTRÄGER	F	2		8	16	11	8	1			48
	B	297	532	968	1542	1084	737	45			5204
	N										38
3 SONSTIGE KURHEIME UND ERHOLUNGSHEIME F. ERW.	F	6	14	33	29	13	27	17	4		143
	B	650	1197	2566	1907	1028	1331	701	122		9502
	N	51	47	49	62	104	118	28			462
4 PRIVATQUARTIERE NICHT AUF BAUERNHOF	F	105	1442	984	993	4623	2489	9454	1899	6	21995
	B	832	9490	5724	5678	29340	13883	63028	12354	107	140436
	N	51	383	511	444	1443	930	2989	486	2	7239
...		...	...	...	...	...	...	...	...	...	...

This passenger form records name, sex, birth date, occupation, place (particularly, *nation*) of residence, nationality, names and birth dates of accompanying spouse and children, date of arrival, and date of departure. In case of travelling parties, with exception of the travelling manager only a subdivision of party members with respect to nations of residence is indicated.

Passenger forms must be filled in irrespective of accommodation type by Austrian federal law (to indicate a size of magnitude, 1993 about 25 million arrivals have been registered in Austria). However, further processing of passenger forms is somewhat different with respect to commercial and non-commercial lodging enterprises. Commercial enterprises are obliged to report guest arrivals and bednights day by day subdivided by the tourists' nations of residence; non-commercial enterprises, that is private accommodations and other lodging facilities, need to report tourist arrivals and bednights (subdivided by a reduced list of residence nations) on a monthly basis only. This way, each commercial lodging enterprise reports 47 (discerned nations of residence) by 2 (arrivals and bednights) by d (number of days in month of reporting) microdatum elements per month (that is, around about 2850 numbers). Since, presently, there are around about 17000 commercial lodging enterprises in Austria, bednight statistics is fed with at least 6 billion numbers each year. In order to cope with this myriad of data, microdatum records are aggregated monthly with respect to a three-way breakdown including nations of residence, arrivals/bednights, and accommodation types, at communal administration level. Other passenger form data is omitted from generated aggregates though it might be of apparent interest for tourism marketing: occupation (of the person filling in the form, at least), age of travellers, as well as information about the number of persons travelling together (single, pair, family with children, travelling parties) – all of this being strongly relevant to infer socio-economic class membership of tourists. At this stage, by the way, the connection of lodging enterprise microdata and bednight microdata is ripped up either since the identifying (microdata) enterprise labels are dropped.

As with lodging enterprise inventory sheets, communal administrations report bednight aggregates to Austria's Central Statistical Office (ÖSTAT) month by month where they become aligned spatially, keeping the three-way breakdown of communal data. Further on, the Central Statistical Office generates higher-level aggregates (macrodata) such as the one shown in **Tab. 4.2.2** (overleaf) which reproduces a sample table (excerpt) of total bednights for a couple of Austria's top ranking communities from 1986 through 1993 [ÖSTAT, 1994].

4.2.3 Statistical Tourism Aggregates

Based on lodging enterprise and bednight data as collected and reported by community administrations, tourism data aggregates – macrodata – can be derived to give information about state and changes of (Austria's) tourism business. Since tourism data capturing is a full registration (as opposed to sampling), aggregates are obtained simply by adding up counts reported by communities.

Table 4.2.1–2: Room Standards in Austrian Tourism at May 31, 1993 [ÖSTAT, 1994]

AUSSTATTUNG DER ZIMMER IN GEWERBLICHEN BEHERBERGUNGSBETRIEBEN AM 31. MAI 1993									
IN GEWERBLICHEN BEHERBERGUNSBETRIEBEN ZUSAMMEN	BUR-GEN-LAND	KÄRN-TEN	NIEDER-ÖSTER-REICH	...	...	...	VOR-ARL-BERG	WIEN	ÖSTERR-REICH
EINBETTZIMMER INSGESAMT	1023	5165	4533	...	...	...	2742	3611	45633
davon ohne Fließ.W. od. m. Fließ. Kaltw.	2	25	7	...	...	...	20	15	147
mit Fließ, Kalt- und Warmwasser	174	1522	956	...	...	...	506	577	9964
mit Fließ, Kalt- und Warmwasser u. WC	26	74	117	...	...	...	46	4	801
mit Bad oder Dusche	60	253	257	...	...	...	95	407	2504
mit Bad oder Dusche u. WC	761	3291	3196	...	...	...	2075	2608	32217
davon mit Zentralheizung	992	4605	4454	...	...	...	2712	3610	44678
mit sonstiger Heizung	7	78	29	...	...	...	18	1	244
ohne Heizung	24	482	50	...	...	...	12	-	711
ZWEIBETTZIMMER INSGESAMT	3710	29882	15134	...	...	...	12201	14613	214003
davon ohne Fließ.W. od. m. Fließ. Kaltw.	1	156	26	...	...	...	28	38	573
mit Fließ, Kalt- und Warmwasser	266	4675	2088	...	...	...	1116	522	23524
mit Fließ, Kalt- und Warmwasser u. WC	42	169	110	...	...	...	79	15	1278
mit Bad oder Dusche	184	1519	913	...	...	...	444	707	10632
mit Bad oder Dusche u. WC	3217	23363	11997	...	...	...	10534	13331	177996
davon mit Zentralheizung	3500	26802	14885	...	...	...	12087	14494	208557
mit sonstiger Heizung	29	591	111	...	...	...	80	114	1716
ohne Heizung	181	2489	138	...	...	...	34	5	3730
MEHRBETTZIMMER INSGESAMT	866	5139	2465	...	...	...	1188	1397	32818
davon ohne Fließ.W. od. m. Fließ. Kaltw.	-	120	15	...	...	...	14	18	407
mit Fließ, Kalt- und Warmwasser	70	1005	446	...	...	...	309	112	5606
mit Fließ, Kalt- und Warmwasser u. WC	6	32	22	...	...	...	19	42	298
mit Bad oder Dusche	20	343	138	...	...	...	46	149	1923
mit Bad oder Dusche u. WC	770	3639	1844	...	...	...	800	1076	24584
...									

Of particular interest, of course, are tables highlighting developments in lodging capacity (changes in number of lodging enterprises and number of guest bedplaces) as well as market portfolios indicating guest mix and guest mix changes in terms of tourists' nations of residence subdivided by arrivals and bednights (from which average duration of stays can be estimated). Another significant indicator of market development is the so-called *bed-utilization* rate telling to which degree an available guest bedplace has been sold effectively on the average in a season or year. Obviously, computation of this rate requires a – spatial – combination of enterprise lodging macrodata and bednight macrodata.

Surveying Austria's annual tourism statistics report published by the Central Statistical Office (for instance, [ÖSTAT, 1994]) reveals that a large section is devoted to tables of arrivals and bednights broken down with respect to various criteria, viz. geographically (communities, federal countries, nation-wide), temporally (monthly, seasonally, annually), nations of residence (detailed, foreign/domestic), accommodation type (commercial, private, other), type of tourism (holiday, city and health tourism), and combinations thereof. For instance, **Tab. 4.2.3** (overleaf) reproduces an excerpt of foreign tourist share of bednights broken down by community and accommodation type. A considerably smaller portion of the report is dedicated to lodging enterprise aggregates (enterprises, bedplaces, room standards). However, the parts of the report looked up most frequently refer to tabulations of tourist arrivals/bednights by communities and (selectively collapsed) nations of residence and of bed-utilization rates by communities and winter/summer seasons, respectively.

As is well-known, printed tables necessarily represent a compromise between anticipated information demand and page layout. In general, to fit on a page, most details (that is, breakdown dimensions) must be omitted from final output, thus constraining the usability of printed tables considerably. This disadvantage can be fixed rather simply, of course, by replacing pre-computed printed tables by electronic database systems or CD-ROMs providing low-level aggregates (such as community administration-level macrodata): given some suitable tabulating software, the derivation of specific target tables is fairly straightforward as long as only a concatenation of basic data is required. In particular, as soon as access to low-level aggregates is admitted and supported, data analysis at destination level becomes a real possibility since, then, a destination actually can compare its own tourism data with other destinations' data in a very detailed way.

Despite these apparent improvements, however, it is still impossible to integrate tourism data on a larger scale. For one thing, as has been pointed out in previous subsections, even within a single nation data reporting conforms to slightly different standards of granularity (for instance, temporal resolution of bednight statistics switches from days for commercial lodging enterprises to months for non-commercial enterprises in Austria's official data capturing scheme); the plethora of terminologies has already stimulated dedicated standardization efforts in classification either within multi-national vendors or at national level (for instance, the German "tourism information standard" [Blencke, 1994]).

.2 Traditional Tourism Data Structures

Table 4.2.2: Austrian Bednights by Communities, 1986-1993 [ÖSTAT, 1994]

	GEMEINDE	ÜBERNACHTUNGEN IN ALLEN FREMDUNTERKÜNFTEN IN DEN WICHTIGSTEN BERICHTSGEMEINDEN IN DEN KALENDERJAHREN 1986 BIS 1993 — INSGESAMT							
		1986	1987	1988	1989	1990	1991	1992	1993
1	90001 WIEN	5,239,035	5,878,000	...	...	...	6,718,056	6,882,843	6,558,047
2	50618 SAALBACH-HINTER.	1,913,574	1,907,766	...	...	...	1,901,019	2,132,708	2,105,220
3	70220 SOELDEN	1,463,962	1,520,763	...	...	...	1,951,101	2,022,907	2,036,749
4	80228 MITTELBERG	1,643,204	1,608,075	...	...	...	1,820,465	1,852,444	1,822,381
5	50101 SALZBURG	1,467,399	1,527,915	...	...	...	1,875,859	1,725,224	1,647,094
6	50628 ZELL AM SEE	1,336,876	1,365,388	...	...	...	1,493,016	1,475,333	1,369,080
7	50402 BAD HOFGASTEIN	1,302,664	1,290,137	...	...	...	1,337,240	1,318,905	1,329,971
8	70920 MAYRHOFEN	1,074,682	1,081,383	...	...	...	1,324,044	1,295,305	1,307,847
9	70351 SEEFELD	1,141,346	1,105,702	...	...	...	1,288,211	1,376,717	1,262,975
10	70101 INNSBRUCK	1,253,721	1,305,014	...	...	...	1,311,082	1,347,789	1,211,371
11	20813 ST. KANZIAN	1,061,300	1,045,372	...	...	...	1,226,634	1,177,062	1,165,056
12	70334 NEUSTIFT	873,052	798,059	...	...	...	1,083,002	1,007,163	1,089,373
13	50403 BADGASTEIN	1,112,885	1,074,614	...	...	...	1,074,386	1,094,976	1,070,804
14	20201 VILLACH	1,020,361	1,069,428	...	...	...	1,173,386	1,147,109	1,051,328
15	70608 ISCHGL	715,647	724,708	...	...	...	928,101	1,019,851	1,043,497
16	80113 LECH	854,162	888,804	...	...	...	957,228	1,012,808	1,010,070
17	20305 HERMAGOR	876,559	881,164	...	...	...	1,101,095	1,041,290	997,930
18	70409 KIRCHBERG	930,037	925,923	...	...	...	1,020,132	1,036,430	978,203
19	70621 ST. ANTON	772,422	811,289	...	...	...	977,102	992,264	976,410
20	70907 EBEN	882,785	856,513	...	...	...	993,736	974,243	923,425
21	20601 BAD KLEINKIRCHHEIM	968,759	942,662	...	...	...	1,063,503	1,006,393	923,392
22	70530 WILDSCHOENAU	801,236	795,485	...	...	...	928,154	943,121	920,999
23	20711 FINKENSTEIN	941,900	923,755	...	...	...	924,055	899,650	871,253
	...								

Even if nomenclature is homogeneous by law or similar compulsory regulations, no direct comparisons may be possible yet between, for instance, different accommodation types because of heterogeneous resolution of taxonomic hierarchies (that is, the elementary accommodation types actually discerned). Furthermore, since associations between lodging enterprise statistics and bednight statistics is lost by intermediary (microdata) aggregation, equipment and service changes (such as infrastructure investments) cannot be correlated directly with changes in bed-utilization or tourist frequencies. A further obstacle to data integration are changes in concepts and classifications. For instance [ÖSTAT, 1994], when Austria's tourism statistics was set up for the first time in 1947, only four accommodation types were distinguished; later on, in the 1970ies, accommodation types extended to 12 and quality categories as well as room types and standards (sanitary equipment, heating) were included. From 1970 to 1979 commercial lodging enterprises were further subdivided into four sub-classes (hotels/motels/inns/pensions, boarding houses, spa hotels, others); from 1980 on private accommodations were dissected into those on farmsteads/not on farmsteads, holiday houses/apartments, and others. Expectedly, similar taxonomic discontinuities will occur over and over again.

Internationally, tourism data integration faces a multitude of nomenclatures, taxonomies, and concepts from the outset. Just to mention a few examples, different countries actually use different accommodation quality categories for commercial lodging enterprises, apply different criteria for room standard descriptions, and often make different subdivisions with respect to the travellers' nations of residence. In view of global tourism markets, however, it is of paramount interest to combine destination data also internationally in order to figure out trends in lodging standards as well as changes in tourist preferences. Of like importance might be a global monitoring of destination policies with respect to quality levels, product innovation, infrastructure development, and additional tourism services offered.

Apart from integration of tourism data already existing, future requirements may call for the inclusion of unprecedented types of tourism data. For instance, traditional tourism statistics does not provide any information about travelling motives although this may bear a considerable impact on a destination's tourism development (such as congress tourism, for example, may be very important for a city) and, hence, its marketing activities. More generally, at destination marketing level, tourism information systems integrating tourism statistics with client databases facilitate a very efficient and quick implementation of marketing strategies: just as tourism statistics helps to draw a larger picture of market trends, client databases might or might not mirror these trends locally, thus opening a unique opportunity to detect marketing potentials and react accordingly (for instance, by direct – electronic ? – mailing initiatives for well-targeted consumer populations such as frequent stayers, sports enthusiasts, senior tourists, etc., cf. [Haines, 1994]; but also by steering investment allocations in the longer run).

Once tourism data combination is possible in a flexible way, the cardinal prerequisite to create strategic tourism management information systems is fulfilled. Of course, even the bare electronic replacement of traditional tourism statistics data processing would improve decision making in the tourism business considerably.

Table 4.2.3: Foreign Share of Tourist Bednights in Austria 1993 [ÖSTAT, 1994]

ÜBERNACHTUNGEN IM KALENDERJAHR 1993
IN DEN BEDEUTENDSTEN FREMDENVERKEHRSGEMEINDEN NACH UNTERKUNFTSARTEN
Ausländeranteil in %

GEMEINDE	GEWERBLICHE BEHERBERGUNGSBETRIEBE				PRIVAT-QUAR-TIERE	FERIEN-WOHNUN-GEN	CAMPING-PLÄTZE	ÜBRIGE UNTER-KÜNFTE	INS-GESAMT
	5/4 *	3 *	2/1 *	ZU-SAMMEN					
90001 WIEN	88,9	86,6	78,5	86,9	92,8	-	97,0	72,3	86,5
50618 SAALBACH-HINTER.	93,6	92,8	87,7	92,2	82,0	87,1	-	61,5	84,2
70220 SOELDEN	96,4	96,6	97,1	96,8	97,1	98,3	98,4	22,7	95,9
80228 MITTELBERG	99,6	99,5	99,9	99,7	99,5	99,8	99,9	99,9	99,7
50101 SALZBURG	80,2	76,5	66,5	77,3	65,5	-	92,6	81,2	78,1
50628 ZELL AM SEE	92,5	91,9	87,8	91,6	86,1	74,6	93,5	71,1	88,9
50402 BAD HOFGASTEIN	84,3	63,0	53,4	70,8	47,7	52,7	74,3	1,1	55,3
70920 MAYRHOFEN	92,8	90,5	95,7	93,1	94,9	96,7	97,8	5,6	93,4
70351 SEEFELD	97,1	98,1	97,3	97,4	94,8	95,5	-	-	97,0
70101 INNSBRUCK	82,2	82,6	76,0	81,3	28,6	60,9	87,9	70,6	79,8
20813 ST. KANZIAN	67,9	54,8	55,2	56,9	35,1	98,0	54,4	1,4	53,1
70334 NEUSTIFT	97,5	97,8	97,1	97,4	97,6	74,7	99,1	82,4	96,9
50403 BADGASTEIN	82,4	83,8	74,3	81,7	68,7	75,2	84,3	9,3	59,8
20201 VILLACH	53,2	59,2	61,9	56,7	59,9	93,0	88,9	24,0	62,1
70608 ISCHGL	93,3	93,7	93,3	93,4	91,3	69,5	-	-	93,1
80113 LECH	87,0	84,6	80,1	84,6	71,6	67,4	-	15,6	83,0
20305 HERMAGOR	74,3	64,8	57,9	68,5	70,2	83,4	92,2	31,8	71,4
70409 KIRCHBERG	90,9	93,3	90,0	91,8	81,2	90,8	-	38,9	87,7
70621 ST. ANTON	92,0	93,3	91,2	92,1	87,7	98,3	-	27,2	88,5
70907 EBEN	97,4	97,9	95,8	97,1	95,4	72,1	96,0	93,8	97,0
20601 BAD KLEINKIRCHHEIM	83,1	72,2	69,2	79,5	57,3	97,5	-	8,9	76,3
70530 WILDSCHOENAU	98,9	97,7	97,8	97,8	97,7	75,5	-	95,8	97,6
20711 FINKENSTEIN	88,0	66,7	59,8	72,6	68,3	-	79,1	6,1	71,9
⋮	⋮	⋮	⋮	⋮	⋮	⋮	⋮	⋮	⋮

4.3 Tourism Data Frame Definition

The preceding discussion suggests a replacement of centralized (national) tourism data production by a network organization comprising local tourism information systems collecting tourism data already electronically as depicted in **Fig. 4.1–2**. Viewed statistically, each local tourism information system – more precisely, the management information system component of it – acts as a data source responsible for the delivery of tourism data in its (geographical) sphere of operation. As handled up till now, these decentralized data sources may transmit only aggregate data (that is, tourism *macro*data) to the network whereas internally, of course, local tourism microdata are still fully available for any kind of processing. According to the METASTASYS approach, hence, local data sources are METANET (cf. Section 2.6) *server* nodes of an *aggregate tourism information network* (ATIN, for short); together, network servers build the ATIN's distributed (statistical) tourism database. Although METASTASYS prescribes a particular data model to be obeyed strictly by participating ATIN servers, it remains each server's private decision – within the boundaries of legal commitments – at which aggregation level and to which extent it actually exports data to the ATIN. By its very nature, the ATIN is a "public" information service open to anyone having access to it and possessing the ATIN client software (that is, the tourism domain METAMODel and the METANET client REQUEST specification interface). In particular, management information systems located at higher tourist board levels (cf. **Fig. 4.1–1**) can be attached to the ATIN as qualified clients to supply these tourist boards' strategic task management appropriately with timely and comprehensive tourism statistics data. Additionally, it should be kept in mind that, though the ATIN is conceived as a *tourism* data network basically, it may very well include other data servers and thus, for instance, provide access to social, economic, trade and industry, etc., statistics as available by national statistical offices or other macrodata brokers.

Before the envisaged ATIN can be set up really, the structure of the underlying METASTASYS domain data model, in METAMOD terms, must be decided upon; this, in turn, presupposes a definite commitment as to the theoretical data language going to be implemented by this data model as well as an appropriate FRAME conception.

4.3.1 Preparatory Considerations

As tourism statistics systems hitherto have been established at national level, a rather natural starting point for the definition of a common tourism data language is provided by current national subject-matter terminologies. In order to achieve semantic integration of national tourism data, the arranged theoretical data language layer, first of all, must be capable to encompass the differences in tourism terminologies and concepts of (national) data suppliers included in the ATIN. Moreover, this common language should also make provision for conceivable future requirements as far-sighted as possible. However, since neither an analysis of national tourism concepts nor the investigation of future tourism scenarios is at the heart of present considerations, in what follows more or less traditional tourism data structures – as introduced in Section 4.2 – are modelled in METASTASYS terms. For the sake of generality, property space and FRAME definitions will be restricted to QUalities, leaving open SCALEDOMains (MODality sets) by and large, assuming that terminological differences remain constrained essentially to domain values distinguished *within* description dimensions. The resulting "skeleton" model, of course, does not preclude any principal extensions in scope.

The first step in setting up property spaces is an enumeration of descriptive dimensions used in the application domain. With respect to lodging enterprises, natural candidates for QUalities are: *accommodation type*, *quality category*, *number of rooms* available in enterprise, *sanitary room standards*, *heating system*, and *general facilities* provided by the lodging enterprise. Bednight statistics, on the other hand, typically counts *arrivals* and *bednights* by *nations of residence*. In addition to these, FRAME definition also requires the adjunction of the universal QUalities ***space*** and ***time***. Considering ***time*** first, bednight statistics at least needs a resolution in days; even in case of electronic data reporting there is little need to switch to time units less than a day as base values (at most, the time at which tourists actually arrive might be of – subordinate – interest). If, in turn, days are indeed used as time units it could be reasonable also for lodging enterprise statistics to report the *dates* of service days instead of numbers of service days per month.

Spatially, scale resolution is yet somewhat more intricate. Hitherto, administrative communities have been the base units of geography, but this might not be sufficient in the long run. Particularly larger communities (such as cities or rural communities spreading over a large area) could be subdivided further, thus supporting the composition of destinations out of units smaller than political communities. In that case, local data sources would already provide aggregates with a spatial decomposition according to community subdivisions. Whatsoever, it will be assumed that enterprises can be associated with one and only one spatial unit (thus losing, of course, the possibility to trace spatially separated branches of enterprises).

The *nations of residence* QUality, obviously, is yet another spatial classification subject to reconsideration. In view of more efficient, better targeted destination marketing a resolution at the level of nations is hardly adequate anymore; additionally, in many countries – such as Austria – passenger forms already register the tourists' full home addresses whence it is relatively easy to replace nations by smaller geographical units (for instance, federal states or provinces) as base values. Since such a kind of subdivision inevitably increases the set of discerned geographic units enormously, a subdivision basically depending on population sizes of nations is advisable. In this sense, in subsequent FRAME definitions the QUality (*place of*) *residence* will be used instead of *nation of residence*; the latter, of course, will represent an AGGLEvel of this QUality.

Choosing *enterprise* as an apparent FRAMEUNIT candidate, the definition of enterprise QUalities is rather straightforward. Since category labels, in general, are assigned to commercial lodging enterprises only, discerned category labels (such as '3*','4*','5*', ...) are simply enumerated in the SCALEDOMain of the *accommodation type* QUality (of course, those category labels may be grouped in a suitably defined AGGLEvel). However, difficulties may incur in finding common SCALEs for the base values to be distinguished actually; acceptable SCALEDOMains probably can be derived only by extensive semantic comparison of classifications currently in widespread use. Room descriptions will cover several QUalities since *number of bedplaces, sanitary standard, heating system*, and – possibly to be included – additional room equipment QUalities (such as *telephone, minibar, TV-set*, etc.) represent individual features which can be combined almost unrestrictedly. Anyway, room description SCALEDOMains either are determined rather naturally or compiled simply by gathering occurring values. For most tourism statistics systems, the *number of bedplaces* dimension in fact distinguishes only a couple of *room types* (single, double, room with three or more bedplaces, apartment); as a consequence, since the number of guest bedplaces available in an enterprise is of interest, *bedplace count* needs to be introduced as a further FRAMEQUality.

To give a more detailed picture, arrivals and bednights could be augmented with further QUalities referring to tourists' social class membership (*sex, age, occupation*) and *companionship* (for instance, single traveller, pair, family with 1, 2, ... children, small/medium/large travelling party). Particularly the QUalities *residence, occupation* and *age* would allow a statistical integration of tourism data with other socio-economic aggregates (such as Labour Force statistics data, cf. Chapter 5; or household budget survey data) sub-classified partly along the same description DIMensions.

Once these QUalities are agreed upon and FRAMEs are set up accordingly in a compulsory METAMODel, preparatory measures to actually integrate a local data source into the ATIN with respect to METASTASYS interface conventions comprise (i) the compilation of export data (source data) from local microdata or of already (pre-)aggregated data and (ii) the generation of local metadata describing export data formally. In general, source data preparation – that is: pre-processing entailing structure conversion – amounts to arrange local property spaces (that is SOFRAMEs using local terminology) in terms of ATIN component data models whereas formal mappings (cf. Section 2.3) associate local metadata/property spaces with "global" (ATIN) FRAMEs. Additionally, whenever a data source contributes in several, structurally different ways to one and the same global FRAME, separate export source data and source mappings must be prepared for each of these data source facets (cf. Section 2.3). These mappings, of course, have to take care of the spatio-temporal relationship of local source data with respect to the global domain METAMODel, too.

Even in view of the enlarged tourism data model discussed in Subsection 4.3.2, quite a few local data sources will surpass committed semantic data reporting requirements and possibilities (for internal information services, at least) by orders of magnitude. Obviously, this surplus information is sort of *private property* not transparent to other participants of the ATIN.

4.3.2 Frame Definitions: A Couple of Approaches

Once QUalities and associated SCALEDOMains (AGGDOMains) have been settled, there still are different ways of FRAME definition conceivable. A rather terse data model, for instance, could consist of a single FRAME encompassing *all* of the envisaged QUalities. More satisfactory from a semantic point of view, however, may be data models comprising several FRAMEs: in addition to lodging *enterprises, rooms, arrivals,* and *bednights* are further obvious FRAMEUNIT candidates. *Destination* description certainly calls for yet another FRAME being linked to the *enterprise* FRAME. In any case, lodging enterprises and destinations are the cornerstone FRAMEUNITs and, hence, pivotal FRAMEs of the tourism data model, though subsequently attention is focused on *enterprise* and *enterprise*-related FRAMEUNITs only.

Setting up a single enterprise FRAME model – with *individual* enterprises as instances – a lossless representation of instance data would have to include a separate DIMension for each independently recorded enterprise feature. In view of Subsection 4.3.1's discussion and remembering (cf. Section 2.2) that codes $C, B, N, S,$ and T denote scale types *categorical, binary, (finite) cardinal, spatial,* and *temporal,* respectively, clusters of SOLQUalities and COMQUalities are stated UPPER case, and factor composition of COMQUalities is expressed in dot notation, the resulting FRAME, taking its traditional property space bearings, would consist of DIMensions as listed in **Tab. 4.3.2**:

Table 4.3.2: Sample FRAME Definition

DIMensions	dimension type	quality type	scale type	dimension attributes
accommodation type	DSOLID	solid	C	
ROOM-DESCRIPTION	DCLUSTER	composite	$N_\diamond$	
BEDS	DCLUSTER	composite	$N_\diamond$	
ENTERPRISE-FACILITIES	DCLUSTER	solid	B	
TOURISTS	DCLUSTER	composite	N	*time*.FLUX
space	DSOLID	solid	S	
time	DSOLID	solid	T	

In this FRAME, ROOM-DESCRIPTION represents a CLUSTDIM of COMQUalities,

ROOM-TYPE.SANITARY.WC.HEATING.TEL.TV.BAR,

BEDS a CLUSTDIMension

ROOM-TYPE.⟨*bedplace count, spare bed count*⟩,

and, finally, TOURISTS a CLUSTDIMension

RESIDENCE.⟨*arrivals | bednights*⟩.

If the *enterprise* FRAME is defined this way, room descriptions are encoded in 7-fold COMQUalities (such as '*double.shower.wc.auto.phone.no-tv.no-bar*: N_ϕ', representing double rooms with shower, WC, and telephone, but neither a TV-set nor a minibar) each counting an *enterprise*'s number of rooms fitting the respective description. The CLUSTDIM 'ENTERPRISE-FACILITIES' comprehends a couple of binary QUalities indicating if a facility (such as *indoor-swimming-pool*) is present or not. 'RESIDENCE' is a QUFACTor listing the discerned places of residence and, thus, 'TOURISTS' results in QUalities such as '*UK.arrivals*: N' and '*USA.bednights*: N'. Furthermore, composite 'RESIDENCE.{*arrivals | bednights*}'-QUalities are defined '*time*.FLUX', that is counts can be added with respect to the ***time*** dimension; however, simultaneous summarization with respect to both, *arrivals* and *bednights*, is inhibited properly by DIMension definition. If camping sites are included (as MODality in *accommodation type*) then, for the sake of consistency, several DIMDOMains will have to contain the NULL value ('ϕ') because camping sites typically do neither contribute room description data nor bedplace counts (but, depending on circumstances, possibly arrival and bednights counts): hence, for *accommodation type* MODality 'camping site' the BLOCK

accommodation type(camping site) ⊗
ROOM-DESCRIPTION(*) ⊗
BEDS(*)

is defined as FRAMEDit; conversely,

accommodation type(~~camping site~~) ⊗
ROOM-DESCRIPTION(ϕ) ⊗
BEDS(ϕ)

marks the dual FRAMEDit (cf. Subsection 3.3.2 with respect to the notation used). Camping sites may also require rather specific values in the 'ENTERPRISE-FACILITIES' cluster leading possibly to further FRAMEDits in combination with *accommodation type* MODalities.

Assuming that ROOM-TYPE comprises t, SANITARY s, WC 2, HEATING 3, TEL 2, TV 2, BAR 2, ENTERPRISE-FACILITIES e, and RESIDENCE r dimensions, this *enterprise* FRAME bears a sum total of $3+e+48st+2(r+t)$ QUalities for each of which an individual instance may carry a value (although, for a particular enterprise, most of the time only the values of $2r$ dimensions will change for each ***time*** unit, viz.

.3 Tourism Data Frame Definition

day). Evidently, even for moderate-sized *r*, a rather high-dimensional FRAME will result.

Practically, due to differences in local data source property spaces only part of the FRAME's DIMensions might be covered. Concerning Austrian tourism data, for example, room descriptions generally do not include the attributes 'TEL', 'TV', and 'BAR'; in this case, of course, each 'ROOM-TYPE.SANITARY.WC.HEATING'-combination is mapped to a *set* of 'TEL.TV.BAR'-QUalities such that each source value is associated with a set of *enterprise* TRAITs expressed by a simplex constraint (cf. Subsection 2.2.3). For example, if a lodging enterprise reports a value *n* for '*double.bath.wc.auto*', then this SOVARiable will be mapped to simplex

$$\sum_{i_1, i_2, i_3} double.bath.wc.auto.i_1.i_2.i_3 = n$$

Similarly, simplex constraints will also occur whenever enterprises report arrivals and bednights broken down by fewer places of residence as defined in the *enterprise* FRAME. To take an Austrian example again, unlike commercial lodging enterprises non-commercial ones have to report arrival ("Ankünfte") data monthly ("Berichtsmonat"), not discerning Spain, Portugal and Ireland ("Übrige EG"). Assuming that these three nations are among the elements of QUCLUSTer 'RESIDENCE' and *time* unit is day, VARMAPpings as follows would result:

$$\mu_V^{(Tourism-A.2)}(Berichtsmonat) = \langle time \rangle$$

$$\mu_V^{(Tourism-A.2)}(Ankünfte\ Übrige\ EG) = \langle Spain, Portugal, Ireland \rangle.\langle arrivals \rangle$$

(meaning that, in fact, '⟨*Spain.arrivals,Portugal.arrivals,Ireland.arrivals*⟩' is the IMAXPROFile of an IMAxis the SOVAR »*Ankünfte - Übrige EG*« maps to); likewise, the respective VALMAPpings would look like

$$\mu_{|Berichtsmonat}^{(Tourism-A.2)}(July93) = (1July .. 31July).1993$$

$$\mu_{|Ankünfte-Übrige\ EG}^{(Tourism-A.2)}(n) = \left(\sum_{c \in \{Spain, Portugal, Ireland\}} c.arrivals = n\right) =$$
$$= \langle (n_1, n_2, n_3) : n_1 \in D_{Sp}, n_2 \in D_{Po}, n_3 \in D_{Ir}, n_1 + n_2 + n_3 = n \rangle, n \in N$$

where D_{Sp}, D_{Po}, and D_{Ir} stand for '*D(Spain.arrival)*', '*D(Portugal.arrival)*', and '*D(Ireland.arrival)*', respectively. The mappings of other dimensions already present – as SOVARiables – in local source data, in general, will be of *1:1*-type, possibly introducing groupings (that is, non-singletons) of MODalities (for instance, Austrian tourism statistics does not distinguish between enterprise categories '4*' and '5*', thus giving rise to a mapping group '{4*,5*}' with respect to the *accommodation type* QUality). Rather frequently, local data sources will not contribute information to several FRAMEQUalities at all; for instance, Austrian private lodging accommodations do not report room description data. In these cases of *lurking* dimensions, of course, a *default* mapping takes place (cf. Subsection 2.3.1). Since, in general, each ATIN server (as part of some local tourism information system, for instance) provides tourism data for a certain geographical area, data sources basically become strung together in a non-overlapping fashion along the ***space*** dimension.

In view of reasonable and extensible data representations, the single-frame model – despite its formal elegance – might not be adequate. Moreover, a data model discerning several "natural" entity types may be easier to handle actually although then, formally, defining target concepts occasionally requires a somewhat greater number of METASPEL specification clauses. A specific advantage of a multi-entity model, however, is its extensibility without increasing the number of FRAME DIMensions artificially. **Fig. 4.3.2–1** depicts a conceivable tourism data model consisting of five entity types (including *destination*) but still omitting, for example, information about the enterprises' days of service which might be modelled as yet another FRAME linked to *enterprise*. *Destination* and *enterprise* are modelled as strong entities whereas *room*, *arrival*, and *bednight* are weak entities (depending in their existence on *enterprise* instances). *Destination* and *enterprise* are in *1:n* relationship since, physically, an enterprise can belong to just one destination. *Room*, *arrival*, and *bednight* are linked to *enterprise* by *1:n*-linkages as well though, contrary to *destination* and *enterprise*, those entities' instances do not possess individual labels but carry *enterprise* instance labels. To these entities, FRAME DIMensions are associated as indicated in **Fig. 4.3.2–1**. In addition to terminology introduced previously, 'DESTINATION-FACILITIES' is used as – QUCLUSTer – placeholder for a couple of enumerative or binary QUalities (left unspecified here). Naturally, weak entities inherit *enterprise*'s ***space*** dimension. Since enterprises and rooms are likely to change over longer periods only, it would probably be sufficient to have years or seasons (half-years) as ***time*** units in *enterprise* and *room* FRAMES, in contrast to *arrival* and *bednight* FRAMES which necessarily have to have a ***time*** unit of day. Further DIMensions could be attached appropriately to the different FRAMES – **Fig. 4.3.2–1** illustrates an extension of the *arrival* FRAME where further guest attributes such as *age*, *sex*, *occupation*, and *companionship* are appended to standard *residence* and ***time*** QUalities. The irregularities (viz., edits) of the single-frame model, mainly caused by the 'camping site' MODality of QUality *accommodation type*, recur in the multi-frame model's *enterprise* FRAME. The other FRAMES, fortunately, are not affected since they will never be touched by *enterprise* instances of type 'camping site' if not applying to it; instead, this MODality is recorded in *enterprise*'s LINKEDITS components.

The number of FRAME DIMensions of the multi-frame model is kept surprisingly small: taking the definitions introduced for the single-frame model variant, *enterprise* ($3+e+2t$), *room* (8), *arrival* (2), and *bednight* (2) FRAMES together comprise $15+e+2t$ explicit and a sum total of $18+e+2t$ QUalities. This dimensionality reduction results from turning the single-frame model's composite QUFACTors (in QUCLUSTers) into ordinary QUalities. Now, an individual *room* instance is de-

.3 Tourism Data Frame Definition

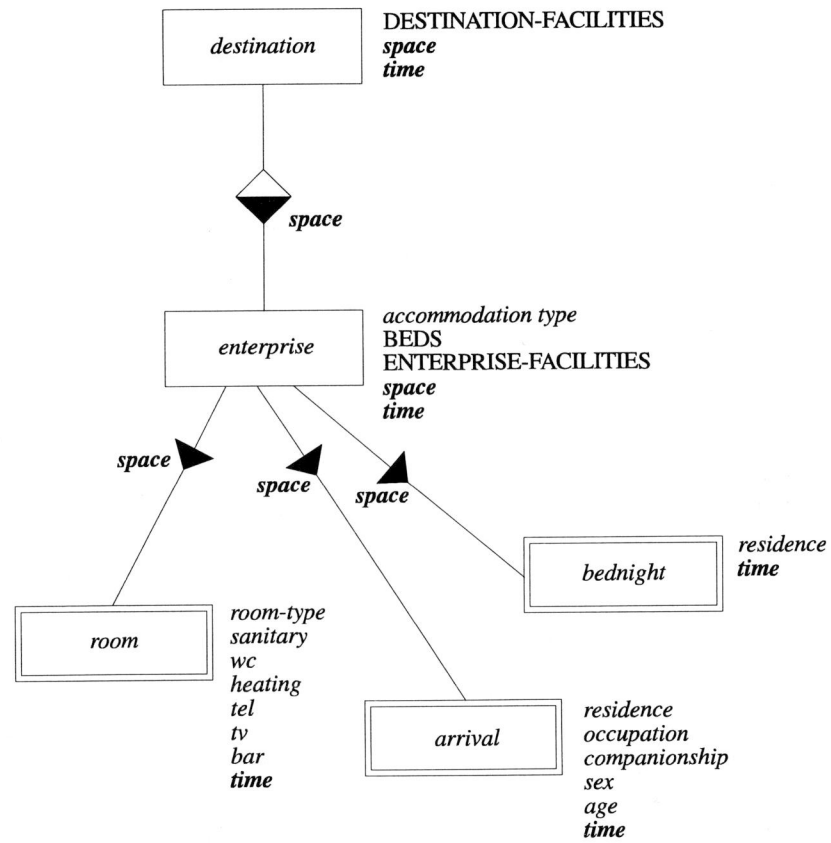

Figure 4.3.2–1: Tourism Data Multi-FRAME Model

scribed by eight DIMensions, a *bednight* instance by two DIMensions, and an *arrival* instance – in the extended FRAME – by six DIMensions (note, however, that implicitly the QUalities of linked FRAMEs become *adjoined* and thus, the FRAMEs' real dimensionality does not reduce at all).

In this data model, the *residence* DIMensions of *arrival* and *bednight* FRAMEs become '*time*.FLUX' in order to permit aggregations over *time*. Furthermore, *residence* could be assigned an *S* SCALE semantics (using favourably a suitable aggregation level of the universal *space* QUality), thus supporting spatially oriented querying. With respect to *bednight*, for instance, **Fig. 4.3.2–2** (overleaf) exhibits an excerpt of this 2-dimensional FRAME; in the example shown, enterprise (instance) e_1 counted 3 bednights from U.K. and 1 from Italy on July 14th, 1993, whereas enterprise e_2 had 1 bednight from U.K. on July 13th and 14th, respectively, as well

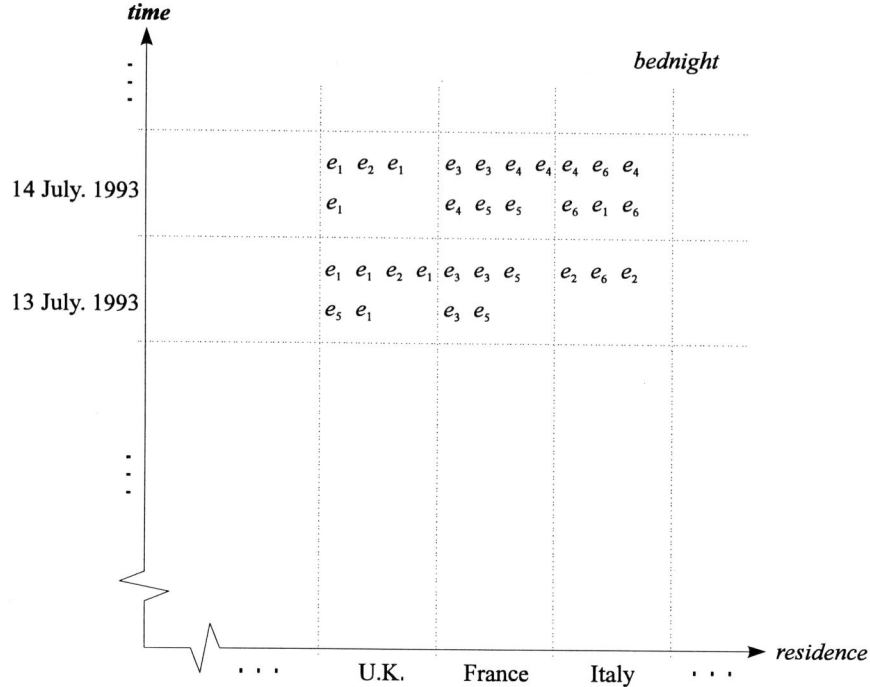

Figure 4.3.2–2: Sample *bednight* FRAME

as 2 Italian tourists staying overnight on July 13th. Although entries in this FRAME, in general, carry non-unique *enterprise* labels, each single entry represents an individual *bednight* instance (however, note that enterprise labels must be unique within *time* slices in the *enterprise* FRAME, since *enterprise* is defined as strong entity type). Separating *arrival* and *bednight* records, of course, does not remove the additivity constraint between this now different FRAMEUNITs; hence, enterprise's LINKADDTYPE component has got to be initialized with 'Q.FIX' entries for each combination pair of *arrival-bednight* QUalities (except the universal ones).

In contrast to the parsimony and efficiency of the multi-frame data model, data source embeddings require a little more effort. In general, preparation of local data source IMAGEs contributing to the data model's FRAMEs amounts to a splitting and rearrangement of (micro-) data records before actual mappings can be determined. Regarding Austrian tourism statistics, for instance, according to established roles and facets separate export datasets need to be arranged for mapping inventory forms (cf. Subsection 4.2.1) to the *enterprise* and *room* FRAMEs, namely for private accommodations as well as for commercial/non-commercial (non-private accommodation) lodging enterprises to each of both FRAMEs; another couple of source data is mapped to *arrival* and *bednight* FRAMEs where mappings differ between

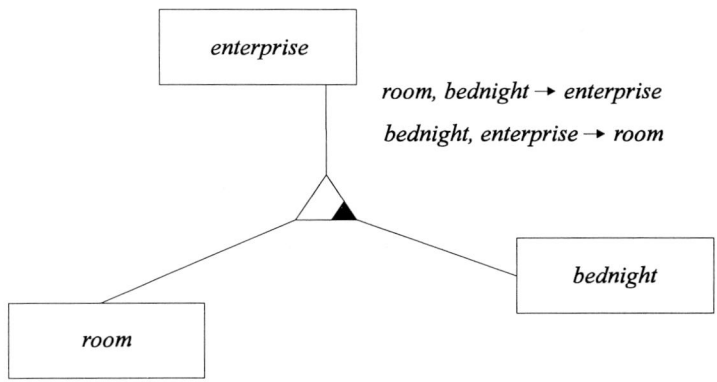

Figure 4.3.2–3: Ternary FRAME Linkage

commercial and non-commercial accommodations (giving rise to different *facets*). Moreover, import (source) data belonging to weak entity types must contain appropriate *enterprise* labels to ensure correct FRAME linkages; since this linkage information presumably is no longer available for historical tourism data, source data must of course be mapped to the *enterprise* FRAME augmented by adjoining the respective QUalities from linked FRAMEs (thus, basically, re-establishing the single-FRAME situation). The FRAME linkage between *destination* and *enterprise* FRAMEs has to be established explicitly in a similar way.

The multi-frame data model as shown in **Fig. 4.3.2–1** links all of the weak entities to the *enterprise* FRAME. As a consequence, no relation *between* weak entities is representable in this model. If, for instance, the data model should maintain a relation between *bednight* and *room*, both entities have to be turned into strong ones in order to establish an explicit FRAME linkage. This way, then, aggregates such as bed-utilization rates broken down by *room-type* and *enterprise category* are derivable since each *bednight* instance is linked to both, an *enterprise* instance and a *room* instance. Unfortunately, as this example highlights, the inclusion of further linkages may introduce functional dependencies as well: a *bednight* instance b associated with a particular enterprise e and a particular room r implies that r is also associated with e. Apparently, *enterprise*, *bednight*, and *room*, FRAMEs now are in a *ternary* relationship as depicted in **Fig. 4.3.2–3**. Likewise, establishing a linkage between *arrival* and *bednight* would cause a further couple of non-binary functional dependencies. Since the METASTASYS data model does not deal with other than binary FRAME linkages, ternary (or *n*-ary, $n > 2$) relationships must be taken care of already before source data mapping.

4.4 Information Delivery at the Management Level

In the ATIN, tourism (meso-) data is dispersed over the whole network and, typically, each participating server node is responsible for a – geographically – limited amount of data only. Conversely, each ATIN client perceives a single uniform tourism data model abstracting from physical network structure, data distribution, and server responsibilities: a client may specify any "tourism concept" of statistical interest to him which can be expressed in the global language of METASTASYS's tourism domain data model and see if – or, at least, to which degree – the ATIN's (active) servers indeed provide (meso-) data to feed the inquired concept. Obviously, the better a concept fits the data sources' actual contributions to the global data model or matches carefully established *domain views*, the higher gets the chance to obtain exactly what is requested. Despite this elementary and inevitable limitation, however, any ATIN client with sufficient ATIN data access rights can exploit all tourism data available within the network (to own a right of data access, of course, does not imply *cost-free* access; retrieval, in fact, may be charged by server sites depending on legal regulations or contracts). In particular, since meso-data is processed and provided on a local scale electronically, produced data is accessible everywhere in the network and, in general, with fairly short processing delays. For instance, a tourist board running its local tourism information system can retrieve rather detailed and most recent ATIN tourism aggregates to combine them statistically with *internal* tourism databases, or tourism executive managers can derive tourism aggregates of competing destinations to perform strategic market analyses, sift out short-term trends in tourism service demand patterns, or check simply if implemented marketing strategies are in line at all with observable market behaviour [Marshall, 1994]. As a matter of fact, this type of statistical tourism information delivered at the tourism management level really starts to pay back the investments in data capturing and tourism statistics system maintenance; in addition to the retrospective documentation function of traditional tourism statistics, the very same (basically) data now attains an operational role in competitive decision making in the tourism branch.

The following paragraphs aim at illustrating by selected examples, how fairly conventional statistical tourism aggregates can be specified with respect to rather "standard" tourism domain views. This presentation, of course, is not meant to preclude useful future extensions; however, in favour of highlighting the differences in concept specifications resulting from the alternate choices of the underlying data model as discussed in the Section 4.3, more or less common tables of tourism data are preferred over more ambitious ones which, occasionally, are only pointed at. Moreover, subsequent examples are restricted to domain concepts of utmost simplicity (generally, the FRAMEs' *ur*-concepts are considered); certainly, for practical purposes, a couple of useful domain concepts, arranged in hierarchical relationships, could be defined ahead of time to ease daily work and provide standard domain views once the details of data model set-up are settled fully. In fact, parameterized or even completely specified METASPEL table templates (cf. Subsection 2.4.4) may be prepared and submitted to routinely evaluation according to a defined schedule.

Before the series of examples is actually stated, a few conventions and definitions are introduced. First, for the sake of easy reference, the symbols S and M, respectively, will be used to denote the single-frame data model and the multi-frame data model (as shown in **Fig. 4.3.2–1**) introduced in Subsection 4.3.2. Secondly, since the categorical *accommodation type* is a frequently applied classification criterion in tourism statistics, **Tab. 4.4** (overleaf) summarizes a couple of useful aggregation levels of this QUality which will be encountered in several example tables later on. The leftmost column of **Tab. 4.4** lists a set of assumed MODalities arranged such that columns 1 to 4 permit a simple representation of AGGLEvels (thus, this MODality ordering does not necessarily reflect a real enumeration of MODalities). **Tab. 4.4** exhibits Austrian labels of aggregation values as well as the labels of AGGLEvels as they may be used in an Austrian interface version for concept specification and table output though columns 3 and 4 are assumed not to be specific to Austrian usage. Incidentally, column 1 of **Tab. 4.4** defines appropriately a mapping of the Austrian tourism SOVARiable »*Art der Fremdenunterkunft*« to QUality *accommodation type*.

Likewise, it is assumed that for *residence* and RESIDENCE, respectively, AGGLEvels fitting current Austrian BASETs are arranged, namely

- level 'Austria-1' comprising 47 VALues ('Wien', 'Übriges Österreich', 44 nations and nation groups other than Austria, 'Übriges Ausland'),
- level 'Austria-2' comprising 29 VALues ('Wien', 'Übriges Österreich', 26 nations and nation groups, 'Übriges Ausland') based on AGGLEvel 'Austria-1', and
- level 'Austria-3' comprising 2 VALues ('domestic-Austrian', 'non-Austrian') only.

Analogous to Chapter 3, each table is assigned to a symbolic METASTASYS table variable numbered increasingly in order to facilitate easy referencing of table specifications in text descriptions. Although, in many cases, table specifications could have been shortened in view of defined default settings for various clauses, full specifications are given throughout for the sake of completeness.

Lodging Enterprise Aggregates

Beginning with a simple example, the first METASPEL sentence describes a table of Austrian tourism enterprises subdivided by *accommodation type* and federal country for the winter season 1992/93 paraphrasing published official statistics tables such as the one shown in **Tab. 4.2.1–1**. Incidentally, the specification of T4.1 is identical irrespective of the chosen data model (with respect to model S, of course, the **ON** clause may be omitted). This specification assumes that federal countries immediately sum up to nations and, hence, the spatial breakdown of nation 'Aus-

Table 4.4: Sample AGGLEvels for QUality *accommodation type*

0	1	2	3	4
based on	column 0	column 1	column 0	column 3
MODalities	base (A)	reduced (A)	7types	5types
5*	5/4*	gewerbliche Beherbergungsbetriebe	5/4*	gewerbliche Berherbergungsbetriebe
4*				
3*	3*		3*	
2*	2/1*		2/1*	
1*				
private accommodation on farmstead	Privatquartier am Bauernhof	Privatquartiere	Privatquartiere	Privatquartiere
private accommodation not on farmstead	Privatquartier nicht am Bauernhof			
private holiday apartment or house	Ferienwohnung, Ferienhaus (nur privat)	andere Privatunterkünfte	Ferien- wohnungen	Ferien- wohnungen
other private accommodation	sonstige Fremdenunterkunft		Übrige Unterkünfte	Übrige Unterkünfte
spa hotel of social security institution	Kurheime der Sozialversicherungsträger	Kurheime der Sozialversicherungsträger		
other spa hotels	Sonstige Kurheime und Erholungsheime	Sonstige Kurheime und Erholungsheime		
sanatoria/adult convalescent homes	Sanatorien, Pflege- und Heilanstalten	Sanatorien, Pflege- und Heilanstalten		
children/youth convalescent homes	Kinder- und Jugenderholungsheime	Kinder- und Jugenderholungsheime		
youth hostels	Jugendgästehäuser, Jugendherbergen	Jugendgästehäuser, Jugendherbergen		
mountain inns	bewirtschaftete Schutzhütten	bewirtschaftete Schutzhütten		
camping site	Campingplatz	Campingplatz	Campingplätze	Campingplätze

tria' by the depth indicator '_1' directly yields Austria's federal countries; likewise, *time* is assumed to have an AGGLEvel 'season' defined appropriately such that season 'winter' comprises the months November to April, and season 'summer' the months May to October.

.4 INFORMATION DELIVERY AT THE MANAGEMENT LEVEL   423

T4.1 ←
    COMPUTE table; **OF** *n*; **ON** *enterprise*; **FOR** #;
    BROKEN DOWN BY *accommodation type*@reduced(A) _1 | Σ;
    OVER nation: =Austria _1 | Σ; **IN** season: =winter.92-93 ↵

The ancestor AGGLEvel of '*accommodation type*@reduced(A)' – indicated by the depth indicator '_1'– is '*accommodation type*@base(A)' as defined in **Tab. 4.4** above; thus, the **BROKEN DOWN BY** clause induces enterprise counts at three levels: 'base(A)', 'reduced(A)', and a sum total.

The next sample table, T4.2, specifies another standard tourism table, viz. the number of tourist bedplaces available in Austrian tourism enterprises and private accommodations during the winter season of 1992/93, with little else changed compared to T4.1:

T4.2 ←
    COMPUTE table; **OF** Σ #.*bedplace count*; **ON** *enterprise*;
    FOR #;
    BROKEN DOWN BY *accommodation type*@reduced(A) _1 | Σ;
    OVER nation: =Austria _1 | Σ; **IN** season: =winter.92-93 ↵

In this 2-dimensional external table (***space*** by *accommodation type*) summation for each breakdown compartment is achieved by weighting of enterprises *e* by $\Sigma_{Q_e \text{ROOM-TYPE}}\, e[Q.bedplace\ count]$ before summing up actually commences (note that $e[Q.bedplace\ count] \in N$ according to introduced QUality definitions). Since '#.*bedplace count*' uniquely determines the QUalities of 'ROOM-TYPE.{*bedplace count*}', it can be used as shorthand notation in this sentence's **OF** clause.

Table T4.2 will not include the MODality 'camping site' since, implicitly,

    accommodation type(camping site) ⊗ ROOM-TYPE.{*bedplace count*}(*)

coincides with a FRAMEDit; thus, this table concept definition leads to a footnote accordingly which, however, can be prevented by excluding camping sites from the TACONCEPT explicitly as shown in T4.3:

T4.3 ←
    COMPUTE table; **OF** Σ #.*bedplace count*; **ON** *enterprise*;
    FOR #;
    WITH *accommodation type*: **except** =camping site;
    BROKEN DOWN BY *accommodation type*@reduced(A) _1 | Σ;
    OVER nation: =Austria _1 | Σ; **IN** season: =winter.92-93 ↵

If a compound table of bedplace counts and spare bed counts is desired (cf. **Tab. 4.2.1–1**), in T4.3 only the **OF** clause needs to be rephrased as

OF [n, Σ #.*bedplace count*, Σ #.*spare bed count*];

to obtain the number of enterprises coupled with number of bedplaces and spare beds.

Supposing that the underlying tourism FRAME is established EU-wide and supposing further that with respect to this tourism FRAME an EU *domain view* is arranged such that an AGGLEvel '*accommodation type@EU*' is supported by all contributing data sources, T4.3 could be restated to compare enterprise and bedplace counts for a couple of competing EU tourism countries:

T4.4 ←
    COMPUTE table; **OF** [n, Σ #.*bedplace count*]; **ON** *enterprise*;
    FOR #;
    BROKEN DOWN BY *accommodation type@EU* | Σ;
    OVER nation: {FRG,France,Austria,Italy,Denmark} | Σ;
    IN month: =April.1994 ⌐

However, this TACONCEPT might deliver misleading figures, since Denmark reports data only for (commercial) enterprises with more than 40 bedplaces; to prevent pertinent footnotes and achieve a fair basis of national comparison, a sub-concept of *enterprise* called, say, '*enterprise-41+*' could be arranged suitably (assuming a BASVALue ordering of *accommodation type* as given in **Tab. 4.4**) as

    accommodation type(5* .. 1*) ⊗
    BEDS($\Sigma_{Q \in \text{ROOM-TYPE}, B \in \{bedplace\ count,\ spare\ bed\ count\}}$ $Q.B > 40$)

defining, for the sake of brevity, the simplex-constrained BEDS-Axis in situ. Now, T4.4 can be restated in terms of CONCEPT '*enterprise-41+*':

T4.5 ←
    COMPUTE table; **OF** [n, Σ #.*bedplace count*]; **ON** *enterprise*;
    FOR *enterprise-41+*;
    WITH *accommodation type@EU*: =hotel;
    BROKEN DOWN BY *accommodation type@7types* | Σ;
    OVER nation: {FRG,France,Austria,Italy,Denmark} | Σ;
    IN month: March .. April.1994 ⌐

The resulting table T4.5 now comprises the number of hotels with at least 41 guest bedplaces as well as total number of bedplaces available, both broken down by category levels and nations (but note that hotels may have fewer than 41 guest bedplaces actually because of spare beds pushing the hotel bedplace counts beyond 40). A further comment to T4.5 is in place here: since neither *bedplace count* nor *spare bed count* is defined '*time*.FLUX' generated output depends on the temporal

.4 INFORMATION DELIVERY AT THE MANAGEMENT LEVEL

basis of data reporting – if, for instance, data is reported monthly *two* structurally identical sub-tables are generated (one for each month); if, conversely, data is reported less frequently (such as annually) only a single value covering 'March.1994' and 'April.1994' *at once* for each table compartment will result. Which numbers are generated, after all, depends on whether data sources contribute to *time* slices covering or, at least, intersecting the indicated *time* range, or not; in the latter case, a data propagation method applies whereas in the former case data can be simply retrieved prior to aggregation.

A modified version of table T4.5 can be used to compute the average number of bedplaces in a lodging enterprise: employing the functional folding principle, this is achieved simply by replacing the **OF** clause by

$$\mathbf{OF}\ [\ \Sigma\ \#.bedplace\ count \div n\ ];$$

Turning to aggregates regarding entities different from enterprises, of course, the choice of the data models matters. For instance, considering tourism aggregates of room equipment in lodging enterprises, table specification differs slightly with respect to models S and M as sketched in Subsection 4.3.2. Particularly in model M now at least two FRAMEs, viz. *room* and *enterprise*, get involved. Taking **Tab. 4.2.1–2** as an initial template, T4.6s specifies a table of rooms broken down by room type and sanitary standard/heating as well as federal countries (of Austria) for commercial lodging enterprises, referring to model S:

T4.6s ←
    COMPUTE table; **OF** Σ ROOM-DESCRIPTION; **FOR** #;
    WITH *accommodation type*@reduced(A):
                                    =gewerbliche Beherbergungsbetriebe;
    BROKEN DOWN BY ROOM-TYPE | Σ *
                                Σ | (SANITARY * WC + HEATING);
    OVER nation: =Austria_1 | Σ; **IN** =31May.1993 ⌐

T4.6s does not reproduce **Tab. 4.2.1–2** exactly; output differs mainly because of slightly refined breakdown definition of rooms factoring out SANITARY and WC elements completely (an artificial 'WC-SAN'-Axis would have to be introduced as a specific constraint instead of 'SANITARY * WC').

The model M variant reads as

T4.6m ←
    COMPUTE table; **OF** *n*; **ON** *room*; **FOR** #;
    WITH *enterprise*♦*accommodation type*@reduced(A):
                                    =gewerbliche Beherbergungsbetriebe;
    BROKEN DOWN BY *room-type* | Σ * Σ | (*sanitary * wc + heating*);
    OVER nation: =Austria_1 | Σ; **IN** =31May.1993 ⌐

Apparently, differences between T4.6s and T4.6m are rather marginal – a simple count function replaces the summing operation in the **OF** clause, and the QuClusters are turned, in *1:1*-correspondence, into ordinary QUality references. As to the *time* restriction in the **IN** clause, the remarks to table T4.6s apply fully again. The QUality *accommodation type* must, of course, be referenced explicitly since it is adjoined to the *room* FRAME; the **WITH**-clause of T4.6m excludes all *room* instances carrying enterprise labels not belonging to the specified constraint within the *enterprise* FRAME (note that, if a CONDEF – say: "Gewerblicher Beherberger" – equivalent to the **WITH**-clause's restriction was available, the clause could have been stated simply as '**WITH** *enterprise*♦Gewerblicher Beherberger'). Conversely, since *space* is inherited implicitly from *enterprise* by frame linkage definition, it is dealt with like a proper *room* QUality.

T4.7m presents a variant of table T4.6m showing percentages of absolute counts (that is, relative frequencies) of rooms available:

T4.7m ←
    COMPUTE table; **OF** *%(n)*; **ON** *room*; **FOR** #;
    WITH *enterprise*♦*accommodation type*@reduced(A):
                                      =gewerbliche Beherbergungsbetriebe;
    BROKEN DOWN BY *room-type* | $\Sigma * \Sigma$ | (*sanitary * wc + heating*);
    OVER nation: =Austria _1 | Σ; **IN** =31May.1993;
    BASED ON *enterprise*♦*accommodation type*@base(A) *
                                        space@country ↵

The non-native **BASED ON** clause of T4.7m defines the intermediary denominator table assigning 100 % to each combination of accommodation type and (Austrian) federal country, thus yielding percentage breakdowns of *room*s with respect to '*sanitary * wc*' and *heating*, respectively, within each denominator table compartment.

Bednight Aggregates

Regarding bednight statistics, **Tab. 4.2.2** has been used to illustrate a typical sample table. Annual summaries of bednights at community level can be derived simply with respect to model S by a specification template as follows:

T4.8s ←
    COMPUTE table; **OF** Σ *#.bednights*; **FOR** #;
    OVER community: # **with** nation: =Austria;
    IN year: =1986 .. 1993 ↵

In contrast to **Tab. 4.2.2**, T4.8s comprises communities as defined in the data model set-up, that is, communities are *not* sorted decreasingly by bednight counts automatically. To achieve a particular sort ordering, a corresponding table editing function must be applied such as

    T4.8s_sorted ←   T4.8s *sort: space*@community
                                mode: dec *on:* (*time*@year: =1993) ↵

Furthermore, **Tab. 4.2.2** in fact presents a time series of bednights for each included community whereas T4.8s reports a single bednight value summing up *all* bednights during 1986–1993. METASPEL's flexible specification syntax, however, permits a multitude of temporal breakdowns:

- 'IN year: 1986 .. 1993;' (that is, dropping the '=' sign) would reproduce exactly the temporal breakdown of **Tab. 4.2.2** ('IN year: {1986,1987,1988,1989, 1990,1991,1992,1993};' is, in fact, a completely synonymous specification although a bit lengthy);
- 'IN year: =1986 .. 1993 _1;' would generate both, annual breakdown and sum total, just as 'IN year: 1986 .. 1993 | Σ;' would.

In addition to this, temporal resolution may be increased by specifying, for instance, **IN** clauses such as

- '**IN** month: # **with** season: =summer.93;' asking for a monthly breakdown including 'May.93' to 'October.93';
- '**IN** month: {December, January}.1986 .. 1993;' permitting a comparison of December and January from 1986 through 1993;
- '**IN** month: (={December, January}).1986 .. 1993;' giving a comparison, with December and January totalled, for 1986 through 1993;
- '**IN** month: ({December, January} | Σ).1986 .. 1993;' adding a time series of bednights summed over December and January for each year;
- '**IN** month: # **with** season: {winter,summer}.88-89 .. 93;' resulting in a monthly time series starting in 'November.1988' and ending in 'October.1993'.

The model M variant of table T4.8s looks like

    T4.8m ←
        COMPUTE table; **OF** *n*; **ON** *bednight*; **FOR** #;
        OVER community: # **with** nation: =Austria;
        IN year: =1986 .. 1993 ↵

which is almost identical with T4.8s except that the **OF** clause and the additional **ON** clause necessary to reference a particular FRAME of model M have been modified.

Rather naturally, a little more information about the structure of bednights is of interest. T4.9s, with regard to model S, adds *accommodation type* as a further breakdown dimension (but excluding camping sites) using AGGLEvel '5types' as defined in **Tab. 4.4**:

T4.9s ←
    COMPUTE table; **OF** Σ #.*bednights*; **FOR** #;
    WITH *accommodation type*: **except** =camping site;
    BROKEN DOWN BY *accommodation type*@5types _1 | Σ;
    OVER *community*: # **with** *nation*: =Austria;
    IN *year*: =1993 ↵

For model M, the specification simply turns into

T4.9m ←
    COMPUTE table; **OF** *n*; **ON** *bednight*; **FOR** #;
    WITH *enterprise* ♦ *accommodation type*: **except** =camping site;
    BROKEN DOWN BY *enterprise* ♦ *accommodation type*@5types _1 | Σ;
    OVER *community*: # **with** *nation*: =Austria;
    IN *year*: =1993 ↵

Compared to **Tab. 4.2.2**, **Tab. 4.2.3** gives additional information about the shares of foreign tourists for tourism communities. Regarding model S, this amounts to a table specification as given for T4.10s:

T4.10s ←
    COMPUTE table; **OF** %(Σ #.*bednights*); **FOR** #;
    BROKEN DOWN BY RESIDENCE@Austria-3 ∗
                                      accommodation type@5types _1 | Σ;
    OVER *community*: # **with** *nation*: =Austria;
    IN *year*: =1993
    BASED ON *accommodation type*@5types _1 | Σ ↵

In T4.10s, 'RESIDENCE@Austria-3' generates the breakdown with respect to Austrian (=domestic) and non-Austrian (=foreign) tourists; for intermediary denominator table computations, of course, this subdivision is omitted. METASPEL does not support the specification of percentage tables covering only a part of table compartments; thus, in order to obtain **Tab. 4.2.2** a subsequent table editing function must be applied to cancel all compartments belonging to 'RESIDENCE@Austria-3: =Austrian' as follows:

T4.10s_foreign ←
    T4.10s *drop:* (RESIDENCE@Austria-3: =domestic-Austrian) ↵

For model M, essentially, the **OF** clause is replaced by '%(*n*)', QUCLUSTers change into ordinary QUalities, and adjoined *enterprise* QUalities must be referenced explicitly again:

.4 Information Delivery at the Management Level

    T4.10m ←
        COMPUTE table; **OF** %(n); **ON** *bednight*; **FOR** #;
        BROKEN DOWN BY RESIDENCE@Austria-3 ∗
                      enterprise ♦ *accommodation type*@5types _1 | Σ;
        OVER community: # **with** nation: =Austria;
        IN year: =1993
        BASED ON *enterprise* ♦ *accommodation type*@5types _1 | Σ ↵

If, for instance, guest shares for a particular month, say: February, shall be traced for several years, only the **IN** clause of T4.10m needs to be replaced by

        IN month: February.1983 .. 1993

Of course, all bednight-related tables can be re-computed for *arrivals* easily: only the FRAMELABel '*bednight*' has to be replaced with '*arrival*' in **OF** clauses (model S) or **ON** clauses (model M), respectively.

Destination Comparisons

Of primary interest to tourism managers are aggregates of arrivals and bednights directly comparing destinations; from the marketing point of view, furthermore, the mix of guest shares in destinations is of particular importance for obvious reasons. Hence, the following examples illustrate the specification of '*arrivals*' and '*bednights*' aggregates broken down primarily with respect to the RESIDENCE or *residence* DIMension, respectively. T4.11s is a straightforward (model S) approach:

    T4.11s ←
        COMPUTE table; **OF** Σ #.*bednights*; **FOR** #;
        BROKEN DOWN BY RESIDENCE@Austria-1;
        OVER community: # **with** nation: =Austria;
        IN season: =summer.93 ↵

However, since – by law – Austrian private accommodations and non-commercial enterprises report mesodata at 'Austria-2' level for RESIDENCE, the resulting table T4.11s will, in fact, return a breakdown of aggregation level 'RESIDENCE@Austria-2' only.

A similar table for health resorts in a couple of European countries (defined suitably as proper or improper AGGLEvel for *space*), including arrivals as well, but only discerning Austrian and non-Austrian guests, is T4.12s:

T4.12s ←
 COMPUTE table; **OF** [∑ #.*arrivals*, ∑ #.*bednights*]; **FOR** #;
 WITH *accommodation type*@reduced(A):
          =gewerbliche Beherbergungsbetriebe;
 BROKEN DOWN BY RESIDENCE@Austria-3;
 OVER health resort: #
    with country: {Bayern,Luxembourg,Tirol,Kärnten};
 IN season: =summer.93 ↵

The model M variant of T4.11s reads as:

T4.11m ←
 COMPUTE table; **OF** *n*; **ON** *bednight*; **FOR** #;
 BROKEN DOWN BY *residence*@Austria-1;
 OVER community: # **with** nation: =Austria;
 IN season: =summer.93 ↵

In model M, the combination of *arrival* and *bednight* aggregates is achieved simply by using METASPEL's folding principle with respect to the **ON** clause noting that both *arrival* and *bednight* FRAMEs make use of the same *residence* QUALITY:

T4.13m ←
 COMPUTE table; **OF** *n*; **ON** [*arrival,bednight*]; **FOR** #;
 BROKEN DOWN BY *residence*@Austria-1;
 OVER community: # **with** nation: =Austria;
 IN season: =summer.93 ↵

For individual decision makers as well as from a macro-economic point of view, bed-utilization rates are important indicators of prosperity and amortisation of invested capital. Arithmetically, bed-utilization is defined as the proportion

$$\frac{\text{bednights sold in some period of interest}}{\text{offered bedplaces } * \text{ days in this period}}$$

In this computation, typically, only regular *bedplace count*s are included (that is, spare beds are excluded).

 In terms of model S, T4.14s compares bed-utilization rates for Austrian commercial lodging enterprises in 4 federal countries for April 1993 and broken down by (Austrian) category levels, with a *time* unit of day used for *bedplace count* data understood implicitly:

T4.14s ← T4.14s.1 ÷ (30 * T4.14s.2) ↵

where the intermediary tables are obtained as

T4.14s.1 ←
  COMPUTE table; **OF** Σ #.*bednights*; **FOR** #;
  WITH *accommodation type*@reduced(A):
                  =gewerbliche Beherbergungsbetriebe;
  BROKEN DOWN BY *accommodation type*@base(A);
  OVER country: {Tirol,Kärnten,Salzburg,Wien};
  IN month: =April.1993 ⏎

T4.14s.2 ←
  COMPUTE table; **OF** Σ #. *bedplace count*; **FOR** #;
  WITH *accommodation type*@reduced(A):
                  =gewerbliche Beherbergungsbetriebe;
  BROKEN DOWN BY *accommodation type*@base(A);
  OVER country: {Tirol,Kärnten,Salzburg,Wien};
  IN month: =April.1993 ⏎

Since the TACONCEPTs of the intermediary tables T4.14s.1 and T4.14s.2 are identical, one is tempted to suggest an "**OF**" clause with a parameter phrase

$$(\Sigma \text{ \#.}bednights) \div (30 * \Sigma \text{ \#.}bedplace\ count)$$

instead of the META algebraic expression (this, however, would introduce a non-native variant better named differently). Note also that the summation 'Σ #.*bedplace count*' over *time* implies the application of a (standard) averaging or disaggregation procedure depending on the temporal resolution of reported source data (in Austria, semester data is reported whence source data would have to be converted to monthly data in deriving T4.14s.2).

Unfortunately, METASPEL cannot adapt automatically to changing periods (in the present case, this would imply the availability of some *time* unit counting function to be included in the table calculating formula). Thus, the following table reporting bed-utilization rates for the winter season 1992/93 requires a full restatement as follows where the value 181 has to be inserted manually (noting that 'winter.92-93' includes the months 'November.1992 .. April.1993' and, hence, counting 181 days):

  T4.15s ← T4.15s.1 ÷ (181 * T4.15s.2) ⏎

where, analogous to the intermediary tables in the derivation of T4.14s,

T4.15s.1 ←
   COMPUTE table; **OF** $\sum$#.*bednights*; **FOR** #;
   WITH *accommodation type*@reduced(A):
   =gewerbliche Beherbergungsbetriebe;
   BROKEN DOWN BY *accommodation type*@base(A);
   OVER country: {Tirol,Kärnten,Salzburg,Wien};
   IN season: =winter.92-93 ⌐

T4.15s.2 ←
   COMPUTE table; **OF** $\sum$#. *bedplace count*; **FOR** #;
   WITH *accommodation type*@reduced(A):
   =gewerbliche Beherbergungsbetriebe;
   BROKEN DOWN BY *accommodation type*@base(A);
   OVER country: {Tirol,Kärnten,Salzburg,Wien};
   IN season: =winter.92-93 ⌐

The specification of bed-utilization rates in model M is somewhat more cumbersome because enumerator and denominator of the bed-utilization formula now refer to different FRAMEs, viz. *bednight* and *enterprise*. Therefore, major parts of the specification of either intermediary table, though again largely identical, need to be repeated:

   T4.15m ← T4.15m.1 ÷ (181 ∗ T4.15m.2) ⌐

letting

T4.15m.1 ←
   COMPUTE table; **OF** *n*; **ON** *bednight*; **FOR** #;
   WITH *enterprise*♦*accommodation type*: 5* .. 1*;
   BROKEN DOWN BY *enterprise*♦*accommodation type*@base(A);
   OVER country: {Tirol,Kärnten,Salzburg,Wien};
   IN season: =winter.92-93 ⌐

T4.15m.2 ←
   COMPUTE table; **OF** $\sum$#.*bedplace count*; **ON** *enterprise*;
   FOR #;
   WITH *accommodation type*: 5* .. 1*;
   BROKEN DOWN BY *accommodation type*@base(A);
   OVER country: {Tirol,Kärnten,Salzburg,Wien};
   IN season: =winter.92-93 ⌐

Some More Complex Examples

These examples by no means exhaust the set of conceivable and reasonable tourism aggregates derivable from either of the discussed model variants. Particularly in the extended multi-frame model M, a lot of even more complex analyses becomes possible; for instance, T4.4.16m reports a breakdown of arrivals by *residence*, lodging category, and social status – expressed in terms of *companionship* and *age* of tourists – for the Tyrol in the winter season 1992/93:

> T4.16m ←
> **COMPUTE** table; **OF** *n*; **ON** *arrival*; **FOR** #;
> **WITH** *enterprise ♦ accommodation type*@reduced(A):
> {gewerbliche Beherbergungsbetriebe, Privatquartiere,
> andere Privatunterkünfte, Jugendgästehäuser,
> Jugendherbergen, bewirtschaftete Schutzhütten};
> **BROKEN DOWN BY** *residence*@Austria-2 | Σ *
> *enterprise ♦ accommodation type*@base(A) | Σ *
> *companionship* | Σ * *age*@5class | Σ;
> **OVER** country: =Tirol; **IN** season: =winter.92-93 ⏎

The resulting table T4.16m is cross-classified 4-dimensionally and comprises all derivable lower-dimensional marginals. Subdividing this table further by spatial and temporal breakdowns would give a rather huge 6-dimensional external array amenable to quite sophisticated statistical data analysis procedures. However, for principal reasons already discussed in Subsection 4.3.2 (in particular, cf. **Fig. 4.3.2–3**) it is impossible to compute the tourists' *duration of stay* with respect to a breakdown as specified for T4.16m since in the corresponding enumerator table (with '*bednight*' replacing the **ON** clause's '*arrival*' parameter) the breakdown dimensions

$$companionship \mid \Sigma * age@5class \mid \Sigma$$

are not supported in the *bednight* FRAME: because of missing FRAME linkages between weak entity types only breakdowns with respect to DImensions common to *all* weak entity types involved are admissible. Thus, a table of duration of stay cannot be derived unless *companionship* and *age* breakdowns are dropped from this METa specification.

In all previous examples, a strong FRAME's QUalities have been adjoined to weak FRAMEs during TARGET specifications. Of course, this adjunction is possible in the opposite direction as well. As an example, T4.17m is a table of lodging enterprises with at least 100 bednights during May 1993 broken down by a particular *arrival*-AGGLEvel 'arr.10':

T4.17m ←
    COMPUTE table; **OF** *n*; **ON** *enterprise*; **FOR** #;
    WITH *bednight*♦#: =100 .. ;
    BROKEN DOWN BY *arrival*♦#@arr.10;
    OVER nation: =Austria; **IN** month: =May.1993 ↵

As final example focusing on the linkage of *enterprise* and *destination* FRAMEs, T4.18m compares Tyrolean destinations in terms of arrivals broken down by residence, length of prepared alpine ski runs and number of ski-lifts, chair lifts and funiculars (comprised in a suitably defined 'LIFTS' QUCLUSTer); note also the special nesting of the table's marginals induced by thoughtfully setting parentheses around a sub-expression of the **BROKEN DOWN BY** clause of table T4.18m:

T4.18m ←
    COMPUTE table; **OF** *n*; **ON** *arrival*; **FOR** #;
    WITH *enterprise*♦*accommodation type*@reduced(A):
                                  =gewerbliche Beherbergungsbetriebe;
    BROKEN DOWN BY (*residence*@Austria-3 | ∑ *
                                destination♦LIFTS) | ∑ *
              destination♦*total length of ski runs*@8levels | ∑;
    OVER country: =Tirol; **IN** season: =winter.93-94 ↵

In this table specification, the references to the *destination* FRAME cause a grouping of enterprise labels (belonging to arrivals and, hence, enterprises located in the Tyrol) within the *arrival* FRAME according to the indicated *destination* breakdowns.

Electronic Tables On-line

Most of the presented tourism METAs referred to a particular printed statistical table as found typically in traditional statistics publications. These basic table designs have been slightly amended and extended in a few directions, too. However, the vast amount of conceivable combinations and breakdowns eventually derivable from a well-fed and carefully maintained METASTASYS ATIN-database must go unnoticed in a printed publication. In spite of this, the selection of examples should have conveyed at least a glimpse of what sort of information is derivable *on-line* in the ATIN and, thus, at the disposal of executive tourism management. Additionally, it should be kept in mind that output data is furnished amply with *statistical* metadata (cf. Section 1.5.2) to be used in subsequent statistical data analyses.

    In the longer run, of course, tourism data models will be less scanty than those discussed in Section 4.3. As more and more FRAMEs are established and accessible by ATIN clients, and with a growing supply also of secondary (that is, non-tourism) domestic and foreign data, extending the core ATIN domain data model by includ-

.4 Information Delivery at the Management Level

ing further (FRAME-mapped) internal marketing data becomes an increasingly rewarding endeavour for tourism marketing management. From such a client's point of view, the ATIN turns into a convenient "data window" to *external* tourism management data feedstocks. The crucial question, then, concerns the readiness of the economic market participants in the electronic tourism information market to support and maintain vigorously the data production infrastructure, providing a qualitatively high and market-oriented level of up-to-date information for tourism management. As with any market, of course, the thoughtful deployment of incurred costs and gained benefits only is capable to create the incentives to actually establish an information infrastructure like the ATIN.

Conclusion * Epilogue

Traditional statistical data management is organized "locally", obeying local constraints and serving local needs, depending on local contexts and milieux. As a natural consequence of this decentralized organization, data management procedures and statistical data systems differ considerably with respect to software, hardware, and "orgware". Although this will not – and, in fact, should not – change anyway, the increasing need to interlace statistical systems calls vigorously for the establishment of a kind of *interoperable* network "super-structure" connecting all of the local statistical systems comprised such that statistical data provided can be transferred easily and combined in a semantically sound way. As a matter of fact, analysis of current statistical data management practices nourished the suspicion that semantic data combination might still be a challenge even within local statistical data management systems. Moreover, the emergent demand for data integration is escorted by severely toughening requirements in speed and flexibility which must be addressed by statistical data systems lest they take the risk of vanishing into obsolescence in the not too distant future.

While these latter challenges have been recognized well by most data suppliers (which, in their majority, are national and international statistical offices or agencies) which are responding accordingly by increased utilization of information and communication technologies to assemble complete electronic data processing chains and to establish more efficient ways of data access and data dissemination (drawing on new media like CD-ROM, hypertext, Internet, etc.), the crucial aspect of *data integration* is not yet dealt with adequately, mainly because of either underestimating the importance of this subject or failing to grasp the potential of advanced symbol processing techniques. Apparently, the present phase of modernizing in statistical data management is concerned basically with improving local data production infrastructures, thus aiming at a re-engineering of local organizations and systems to gain efficiency in a time of strained budgets. This certainly is an inevitable, and worthwhile, endeavour which has already sharpened the view as to the fact that statistical data comes with a semantics quite different from commercial and other business data and, hence, requires rather specific data management tools – such as custom-tailored statistical data models – instead of borrowing well-maintained but badly-suited models, systems, and techniques from the general database domain (notably, the *relational* database domain). Fortunately, statistical data management has begun to develop a distinct outline although, at present, it can be hardly claimed to have arrived at a coherent body of theory yet. Entirely underrepresented, however, has been the topic of effective data integration – contrasting markedly with the apparent lack of integration which is felt, naturally, as most pressing by those in need of integrated data. Increasing economic and social interdependency as well as the general globalization tendency enforce an integration of data both within and between local systems and contexts to get hold of a more and more complex societal reality. For apparent reasons, more complex – that is to say: more adequate – planning and decision models safeguard competitive advantage at both micro- and macro-economic levels; to be effective in steering resource allocations, process control, and social engineering, these models must be fed, in turn, with more complex data, typically from various sources, obtained from more powerful – that is: comprehensive and flexible – data systems. For instance, multinational corporations (MNCs) depend vitally on advanced distributed information systems to control and manage internal logistics [DuBois and Carmel, 1994] as well as integrated statistical aggregates delivered by both internal and external data sources for deriving rationally tactical and strategic production, outsourcing, financing, retail management, materials procurement, and marketing decisions [Marshall, 1994]. This holds true even more so for governmental or supra-national policy making and monitoring, for instance in preparing the gradual transition from growth economies to sustainable ones [SUSTAIN, 1994] or in establishing the Single Market within the boundaries of the European Union. However, while industry and commerce seem to have realized quite swiftly the potentials of advanced information technology and its impact – for instance, in the tourism branch [Froeschl and Werthner, 1994] – and have started actually to take appropriate measures, the public sector lags behind to a perilous degree. Accordingly, the difficulty to come to terms with integrated statistics is a frequently recurring complaint. To cite just two (of numerous) examples, a recent authoritative publication of EUROSTAT [1995a] on "Tourism in Europe" states somewhat clumsily, in its introduction, that

"*[D]espite the current lack of harmonised statistical data on tourism, an attempt is made to present the relevant elements of Europe's tourism structure and to outline recent trends.*"

Similarly, "The Panelist" newsletter of EUROSTAT remarks in its first 1995 issue [1995b] that

"*[T]he statistics available at present, based on national references, are no longer sufficient to describe the activities of transnational corporations and their economic consequences.*"

While the geopolitical system is being extended to adapt its decision spaces to the extending societal and commercial spaces of disposition, the statistical infrastructures – notably the European statistical system – have failed to catch up technically in spite of their tremendous relevance for rational and responsible policy making; and this failure surely cannot be remedied without introducing more sophisticated data management and symbol processing techniques to (official) statistical information processing. Rather, supra-national statistical systems have commenced to streamline their operation in moderately innovative ways following the mainstream of telematics and digital network systems just like national or regional statistical systems do with varying eagerness, thus no doubt missing the prospective opportunity to make an anticipatory leap in statistical information processing destined to meet tomorrow's demands. After all, the hitherto envisaged improvements in statistical data management remain restricted merely to the *syntactic* level of symbol processing, leaving the semantic level – the meaning of data – still entirely to the data consumers' responsibility.

In this respect, the proposal of the integrated statistical information system architecture set out in Chapter 2 focuses on a practical, though theoretically rigorous, approach to achieve interoperability of established local statistical data production and supply systems by arranging an "umbrella" network system as a technically uniform and semantically coherent data interface to data consumers shielding logically the local details of data structures and data access organization. Basically, the METASTASYS framework has been conceived deliberately as a *toolbox* to set up effectively semantic distributed data integration systems, giving special emphasis to data integration in a narrower sense, viz. *data harmonization*: the semantic integration of (physically distributed, as a rule) statistical data pertaining to specific subject domains such as labour force statistics, tourism statistics, as well as domains of similar nature like, for instance, demography, family budget statistics, commodity production statistics, external trade statistics, balance of payments statistics, etc. Plainly speaking, METASTASYS aims at supporting the compilation of coherent topical statistical analyses like those of employment and unemployment (for instance, [EUROSTAT, 1991; 1993b]) or tourism statistics [EUROSTAT, 1993d; 1995a] such that the aggregates underlying socio-economic reasoning and forecasting can be derived directly and on-line from a federated database system encompassing all data sources of interest. In fact, as particularly the tourism statistics case highlights, profound statistical reasoning requires not only an effective *harmonization* of (national) tourism statistics but the *integration* of different statistics branches – like balance of payments statistics, currency exchange rate time series, etc. – as well; hence, the ultimate *statistical desktop* presupposes a single comprehensive statistical data model instead of a sectorally segregated set of partial models inhibiting effective data integration. Therefore, practically, data models as discussed in Chapter 3 and, briefly, in Chapter 4 in fact must be linked in the longer run which, in turn, requires formal data models prepared for a formal representation of this type of semantic linkages. Thus, from an symbol-analytical point of view, the arrangement of data warehouses supplying well-documented off-the-shelf data (as suggested, for instance, by [Fessey, 1989]) is but a first, and surely insufficient step on the way towards accomplishing interoperable statistical data management systems; quite on the contrary, to prevent apparent overload of data analysts facing a dozen data sources or so, each with its peculiar set of definitions and annotations, data integration and data harmonization cannot be brought about successfully without a maximum of mechanized support; and precisely this is the cardinal point of the METASTASYS proposal: to show the practical possibility of providing mechanized support in data harmonization by employing advanced symbol processing techniques, viz. *metadata management*, as a preparatory step for subsequent statistical modelling and analysis. It is of utmost importance to keep in mind that, in this conception, metadata is not just meta-information supplied to data consumers in connection with plain data – what it is anyway – but it attains a functional, or *operative*, role in fusing data of different origins in a semantically coherent fashion by virtue of metadata processing. As an appreciable side effect, the METASTASYS architecture still renders traditional statistical database services available though in a considerably more efficient, flexible, uniform, and time-saving manner.

In developing METASTASYS, much attention has been devoted to the socio-economic environment conditions and constraints which any such technical system of statistical information processing is subject to. Perhaps this approach may have traded formal rigour for additional complexity at the cost of mathematical elegance, but it should nevertheless compensate for the acute danger of all engineering masking out the unpleasant "disturbances" of reality. After all, the development of statistical information infrastructures amounts to a work of *design* rather than of formal proof, receiving its confirmation of validity by successful operation under real circumstances. Hence, first and foremost, the milieu determines the technical system (which definitely is by no means to deny the feedback of technical systems to their surrounding milieux). The design of METASTASYS seeks to strike a reasonable balance between formal simplicity and plasticity to embrace a large as possible set of established data sources and data production systems without forcing them into an all too rigid framework they most probably would be reluctant to adapt to for both economic and political reasons. Geared to the data consumers' view of statistical information processing, two functional system components have been identified, viz. an *interaction component* responsible for arranging a conceptual (that is: purely logical) data view, and a unifying basic *semantic data model* representing real observation data, at any processing stage, such that relative semantics of contributing data sources are captured formally in terms of operative metadata; a formal metadata processing system facilitates the mechanical translation of specific conceptual data views into data access and data harmonization/integration operations in a fully transparent way. To this end, of course, the relative semantics of data sources have to be expressed explicitly with respect to a common taxonomic reference frame whence, for the time being, the exposition of METASTASYS in Chapter 2 gave most emphasis to *taxonomic metadata*.

Although the presupposition of common taxonomic reference frames makes some feel not quite easy [Fessey, 1993], it nevertheless must be acknowledged that otherwise there is no sound possibility to harmonize statistical data at all. The common reference frame, furthermore, is a fictitious one in that no actual data source is forced to adopt it except for the sole purpose of creating the "semantic mappings" actually defining a data source's relative semantics. Data sources remain completely autonomous otherwise (especially with respect to data production methodology, data privacy standards, and data usage), and the federated database induced by the collection of arranged mappings in fact reduces to a data (and metadata) interchange *protocol* interconnecting the involved data hosts physically. Practically, there is no centralized computing either, as the conceptual data views are broken down into constituent parts which, as a rule, may be prepared locally before being shipped to the inquirers' sites, thus exploiting the readily available computing capacity of distributed network systems instead of imposing artificial

EPILOGUE

demands for expensive and inflexible supercomputing centres. Perhaps another advantage of the METASTASYS model is its deliberate attempt to reconcile sampling theory, stochastic modelling, and data integration in a unified statistical information processing framework.

As yet, METASTASYS has not been exposed to the vicissitudes of a real application field for two reasons mainly: first, the approach is too novel for having it introduced in a practical setting, and, secondly, as the proposal stands now it is barely more than paperwork leaving open a couple of issues needing clarification before the framework can be put to use actually. The experiences with the DÖS'chen prototype (cf. Section 2.7), however, have been quite encouraging despite the rather limited functionality of this premature, quick-and-dirty implementation piece of software. Major aspects not yet addressed adequately in the development of the METASTASYS model certainly are, inter alia:

- a method for handling discontinuities in FRAME definitions since it cannot be assumed tacitly that any future adaptation can be anticipated properly at initial FRAME set-up time – to this end, a formal mapping between FRAMES is conceivable to bridge taxonomic discontinuities at least within a moderate range of changes;
- likewise, sometimes it might be wishful to change FRAMEUNITs which could be accomplished by a formal FRAME mapping either;
- a special problem could occur in case of dynamic populations where – during the observation period(s) – population instances cease to exist, generate offspring, or become merged;
- in some cases, the semantics of observation settings may require yet different data structures (such as, for instance, in longitudinal sample designs [Froeschl and Grossmann, 1988]) calling for specific measures to preserve data context and sample structure representation;
- a similar remark applies to METAMOD's capabilities to embed, for instance, GIS data structures such as those described by Plank [1994], and to the coupling of GIS to METASTASYS for the generation of thematic maps for presenting spatial data distributions visually;
- the practical coupling of METASTASYS to established data production systems and databases such as, for instance, the data repositories of EUROSTAT [Knüppel and Platte, 1991] which is an economically critical factor in arranging the semantic mappings (at present, it is assumed simply that data production systems already provide, in machine-readable formats, *all* the metadata asked for);
- this is even more delicate, as in addition to taxonomic metadata concerned up till now also inferential-statistical metadata must be taken into account involving, for instance, information about sampling structures (stratification, periodicity, sampling rate, eligibility criteria, non-response, etc.), measuring devices and methods, aggregation methodology (in case of macrodata supplied), and so on;
- with respect to the *merge* operation the statistical methodology needs to be improved considerably lest METASTASYS's capabilities to combine data will remain limited severely, thereby sacrificing much of its appeal; and

- finally, METASTASYS would benefit enormously by including more or less advanced statistical analysis methods through the extension of the set of SUM-TYPES provided although this, of course, would blur the distinction between data supply and inferential statistics even further.

Altogether, this sums up to quite an impressive research programme which, in the end, leads up to the clear concept of a comprehensive statistical information system providing *manageable* (cf. Section 1.1) data images of a "positively" perceived world. Noting that, at present and even more so next time, information and the possession of powerful information processing tools are overtaking the roles investment capital and means of production have held during the 19th and 20th centuries, it can be foreseen at almost no risk that this type of statistical data management is about to become a cardinal practical building block of information economics as data and especially the flexible combination of data to gain information are the most valuable assets of information society and the most critical resource in global competition [Rifkin, 1995]. METASTASYS-like statistical data management systems will occupy a pivotal role in information economics basically because of their powerful, logically effected interoperability allowing to "zoom" reality, bringing the – digitized – world into reach, quite in the sense of the French trick shot pioneer Georges Méliès's „le monde à portée de la main"-metaphor [Storch, 1995], by means of formal manipulation (sic!) of data intensions [Bourgine, 1989], or metadata management, making possible a concrete discursive mental dealing with reality in terms of empirical data. It is hard to resist drawing a parallel to the development of literate society after the invention of book printing [Eisenstein, 1983] in order to get an impression of the possible impact of this next revolution in information technology. Books have changed the very notion of information and introduced novel ways to structure and digest it (segmentation of texts into chapters, sections, etc., indices, tables of contents, pagination, and so on), and just this episode is now happening again with electronic information – thinking of "knowledge bases", multi-media systems, hypertext, Internet, etc. – though replacing the static, mechanical view of the "Gutenberg era" (McLuhan) with the more dynamic, "cybernetic" one of today. As a matter of fact, contrary to books, electronic information dissemination covers both data and data processing methods as the latter can be encoded in data as well, thus creating added value of supplied data by facilitating actual data *usage* (note that CD-ROMs typically comprise both data and retrieval software at once).

Simply speaking, the digital symbol-mediated zooming of reality consists of two aspects, viz. fetching data here and combining data. The former is accomplished quite satisfactorily at a global scale due to the dense connectivity of telecommunication links; the latter, conversely, is a vast resource open for exploitation, and provides a fertile and promising area of value-adding business: *symbolic analysis* [Reich, 1993]. Analogous to the emergence of truly fanciful financial derivatives traded at stock exchanges, data combination and symbolic analysis is giving rise to *information derivatives* comprising a palette of "products" ranging from crude data through macrodata on demand up to custom-tailored information services and decision support systems converting input data to decisions. Apparently, and again strictly analogous to finance capital, the economic value of data is an inherent property of its instant availability and continuous circulation (a metaphor recurring in Western thinking since its inception by early scientific medical research, particularly of William Harvey's *De motu cordis* of 1628 [Sennet, 1994]) – its value coincides with the multiple use borne in the symbols stored and traded. This, in turn, presupposes a preservation of continuous interpretation frames to convert data into information proper. From an economic point of view, the creation and maintenance of data interpretation contexts – metadata – is but an investment to prevent premature depreciation of data stocks.

First and foremost, the structural change leading from industrial economies to a more and more uniform global information society means a transition to the collective business of designing the future. In former days, when priests and astronomers were the early "model fitters" and forecasters (and, thus, the timekeepers of society as well), the coming-true of predictions was seen as a fulfilling of mystic secrets governed by an elusive, or divine, plan arranged by a superior authority, and this belief and predictability entertained a feeling of eternal harmony and existential security in full agreement with nature. In our days, future is considered mouldable; while formerly *future* came to pass by striding ahead steadily, we essentially experience *presence* as yesterday's implemented future. This is to say that the business of shaping future *is* our sensed presence as we are concerned to make reality fit to our own forecasts (what is called *controlling* in management science); for instance, products more and more are not only designed and mounted electronically by computer aided design and flexible manufacturing, numerically controlled production systems but examined with respect to functionality in simulated application environments ("virtual robotics") and with respect to profitability in simulated consumer markets, too. Many goods, in fact, are no longer created physically before being sold by virtue of purely fictitious digital images of them which, of course, is the ultimate aim of electronic business management. Based on comprehensive data models feeding market models or even whole socio-economic system models, decision making aims to react in time and "appropriately" to market or system behaviour inferred from extrapolating recorded historical traces. It is everything but far-fetched to speak of a large-scale *digital illusion* which is maintained to engineer and implement effectively self-fulfilling prophecies using formal symbol-processing techniques as a tool of instrumental rationality, thus devoting absolute priority to the Platonic "reason" over "necessity" [Bolter, 1984] by stipulating that the *digital* future is the least affected by the annoying imperfections of reality. However, simulated digital future is traded for the ephemeral – though definite – presence, and that's how *we* make headway ...

APPENDIX A

METASTASYS Operation Definitions

Using the "abstract" FOOL language for defining functional computation structures (introduced briefly in Subsection 2.1.1), this appendix presents the most elementary METASTASYS functions covering the basic operations of METAGEN defined in Subsection 2.5.1. Though only worked out partly, the given definitions should elucidate sufficiently how higher-level operations, up to the METAGEN operators described in Subsection 2.5.2, can be assembled in a bottom-up fashion from these elementary functions.

In the following presentation, pure *pseudocode* is exhibited without any comments attached; if read top-down, however, the meaning of the specified functions can be grasped easily because, throughout, the symbols used reflect standard mathematical usage. In order to disclose the meaning of "built-in" messages, of course, the respective definitions given in Subsection 2.1.2 must be referred to. While studying the function definitions it must be kept in mind that not all of the referenced messages (functions) are stated in this sketchy account of METASTASYS operations (this would quickly exceed available space and, after all, amount to a rather exhausting exercise).

The material presented is divided into five small sections dealing, in turn, with identity, equality, and structure comparison predicates (A.1), Boolean operators (A.2), ordering relations (A.3), type casting operations (A.4), as well as a couple of basic sequence-based operations both for SETs and XSETs (A.5). Quite naturally, the last section is a bit more extended although no functions manipulating BLOCKs or higher-level structure are dealt with. Apparently, using FOOL's type polymorphism, most of the operations (function) symbols should be used over and over again for respective operations on all OBJECT types. To assure non-circularity of function definitions, of course, care must be exercised to prevail the *dag* structure (condition of *well-foundedness* [Manna and Waldinger, 1985]) of recursive function calling patterns.

Occasionally, functions generate new instances of OBJECT classes. As a shorthand notation, a new OBJECT of a particular <type> (cf. **Def. 2.1.2–10**) is created by sending the special message '*obtype:* <type>' to the generic OBJECT designated by symbol '☐'. In view of the implicit type resolution by type polymorphism, instead of writing $\subset_G, \prec_G, \approx_G$, etc., the symbols are introduced formally without any subscripts.

A.1 Identity, Equality, Structure Comparison

$=$ [infix: $(\mathcal{V},\mathcal{V}) \to$ BOOLE where $\mathcal{V} \in \{$VAL,XVAL$\}$]:
$\quad v = v \Rightarrow \top.$
$\quad \_ = \_ \Rightarrow \bot.$

$=$ [infix: $(\mathcal{S},\mathcal{S}) \to$ BOOLE where $\mathcal{S} \in \{$SET,XSET,XDOM$\}$]:
$\quad s_1 = s_2 \Rightarrow s_1 \text{ } \mathbf{id?} \text{ } s_2 \qquad | \text{ } s_1 \doteq s_2.$
$\quad \_ = \_ \Rightarrow \bot.$

$\doteq$ [infix: $(\mathcal{A},\mathcal{A}) \to$ BOOLE where
$\qquad\qquad\qquad \mathcal{A} \in \{$SET,XSET,XDOM,BLOCK,BLOCKCOLL,BOX$\}$]:
$\quad a_1 \doteq a_2 \Rightarrow \top \qquad\qquad | \text{ } (a_1 \text{ } \mathbf{base?}) \wedge (a_2 \text{ } \mathbf{base?}).$
$\quad \_ \doteq \_ \Rightarrow \bot.$

$\doteq$ [infix: $\mathcal{B} \to$ BOOLE
$\qquad\qquad$ where $\mathcal{B} \in \{($BLOCK,BLOCKCOLL$),($BLOCKCOLL,BLOCK$)\}$]:
$\quad b_1 \doteq b_2 \Rightarrow \top \qquad\qquad | \text{ } (b_1 \text{ } \mathbf{base?}) \wedge (b_2 \text{ } \mathbf{base?}).$
$\quad \_ \doteq \_ \Rightarrow \bot.$

$\hat{=}$ [infix: $(\mathcal{S},\mathcal{S}) \to$ BOOLE where $\mathcal{S} \in \{$SET,XSET,XDOM$\}$]:
$\quad s_1 \hat{=} s_2 \Rightarrow \top \qquad\qquad | \text{ } ((s_1 \text{ } \mathbf{mode?}) = (s_2 \text{ } \mathbf{mode?})) \wedge (s_1 \doteq s_2).$
$\quad \_ \hat{=} \_ \Rightarrow \bot.$

$\approx$ [infix: $(\mathcal{S},\mathcal{S}) \to$ BOOLE where $\mathcal{S} \in \{$SET,XSET$\}$]:
$\quad s_1 \approx s_2 \Rightarrow s_1 \text{ } \mathbf{match?} \text{ } s_2 \qquad | \text{ } s_1 \doteq s_2.$
$\quad \_ \approx \_ \Rightarrow \bot.$

$\cong$ [infix: $(\mathcal{S},\mathcal{S}) \to$ BOOLE where $\mathcal{S} \in \{$SET,XSET$\}$]:
$\quad s_1 \cong s_2 \Rightarrow s_1 \text{ } \mathbf{conf?} \text{ } s_2 \qquad | \text{ } s_1 \doteq s_2.$
$\quad \_ \cong \_ \Rightarrow \bot.$

$\prec$ [infix: $(\mathcal{S},\mathcal{S}) \to$ BOOLE where $\mathcal{S} \in \{$SET,XSET$\}$]:
$\quad s_1 \prec s_2 \Rightarrow s_1 \text{ } \mathbf{subsum?} \text{ } s_2 \qquad | \text{ } s_1 \doteq s_2.$
$\quad \_ \prec \_ \Rightarrow \bot.$

$\supset\subset$ [infix: $(\mathcal{A},\mathcal{A}) \to$ BOOLE where $\mathcal{A} \in \{$SET,XSET,BLOCK,BLOCKCOLL,BOX$\}$]:
$\quad a_1 \supset\subset a_2 \Rightarrow a_1 \text{ } \mathbf{disj?} \text{ } a_2 \qquad | \text{ } a_1 \doteq a_2.$
$\quad \_ \supset\subset \_ \Rightarrow \bot.$

.1 Identity, Equality, Structure comparison

⊃⊂ [infix: $\mathcal{B} \to$ Boole
              where $\mathcal{B} \in \{$(Block,BlockColl),(BlockColl,Block)$\}$]:
    $b_1 \supset\subset b_2 \Rightarrow b_1$ *disj?* b_2 | $b_1 \doteq b_2$.
     $\_ \supset\subset \_ \Rightarrow \bot$.

conf? [infix: (Set,Set) $\to$ Boole]:
    s_1 *conf?* $s_2 \Rightarrow \top$ | $(s_1$ *plain?*$) \wedge (s_2$ *plain?*$)$.
    s_1 *conf?* $s_2 \Rightarrow (s_1 \cap s_2)$ *eq?* $((s_1$ *plain!*$) \cap (s_2$ *plain!*$))$
                                              | $(s_1$ *grouped?*$) \wedge (s_2$ *plain?*$)$.
    $\_$ *conf?* $\_ \Rightarrow \bot$.

conf? [infix: (XSet,XSet) $\to$ Boole]:
    $\emptyset$ *conf?* $\emptyset \Rightarrow \top$.
    () *conf?* () $\Rightarrow \top$.
    (x_1,s_1) *conf?* $(x_2,s_2) \Rightarrow (x_1 \equiv x_2) \wedge (s_1$ *conf?* $s_2)$.
    () *conf?* $\_ \Rightarrow \bot$.
    $\_$ *conf?* () $\Rightarrow \bot$.

match? [infix: (Set,Set) $\to$ Boole]:
    s_1 *match?* $s_2 \Rightarrow \top$ | $(s_1$ *plain?*$) \wedge (s_2$ *plain?*$)$.
    s_1 *match?* $s_2 \Rightarrow s_1$ *mtchgrp?* s_2 | $(s_1$ *grouped?*$) \wedge (s_2$ *grouped?*$)$.
    s_1 *match?* $s_2 \Rightarrow s_1$ *mtchmxd?* s_2 | $(s_1$ *plain?*$) \wedge (s_2$ *grouped?*$)$.
    s_1 *match?* $s_2 \Rightarrow s_2$ *mtchmxd?* s_1 | $(s_1$ *grouped?*$) \wedge (s_2$ *plain?*$)$.

match? [infix: (XSet,XSet) $\to$ Boole]:
    $\emptyset$ *match?* $\emptyset \Rightarrow \top$.
    () *match?* () $\Rightarrow \top$.
    (x_1,s_1) *match?* $(x_2,s_2) \Rightarrow (x_1 \approx x_2) \wedge (s_1$ *match?* $s_2)$.
    () *match?* $\_ \Rightarrow \bot$.
    $\_$ *match?* () $\Rightarrow \bot$.

mtchgrp? [infix: (GruSet, GruSet) $\to$ Boole]:
    s_1 *mtchgrp?* $s_2 \Rightarrow (((s_1$ *plain!*$) \cap (s_2$ *plain!*$))$ *mtchmxd?* $s_1) \wedge$
                              $(((s_1$ *plain!*$) \cap (s_2$ *plain!*$))$ *mtchmxd?* $s_2)$.

mtchmxd? [infix: (PlaSet,GruSet) $\to$ Boole]:
    p *mtchmxd?* $\{\} \Rightarrow \top$.
    p_1 *mtchmxd?* $\langle p_2, g \rangle \Rightarrow ((p_2 \subseteq p_1) \vee (p_2 \subseteq (p_1$ *co!*$))) \wedge$
                                            $(p_1$ *mtchmxd?* $g)$.

disj? [infix: (Set,Set) $\to$ Boole]:
    s_1 *disj?* $s_2 \Rightarrow ((s_1$ *plain!*$) \cap (s_2$ *plain!*$)) = \langle\rangle$.

disj? [infix: (XSET,XSET) → BOOLE]:
  x_1 ***disj?*** $x_2 \Rightarrow ((x_1 \text{ \textbf{\textit{plain!}}}) \cap (x_2 \text{ \textbf{\textit{plain!}}})) = \varnothing$.

disj? [infix: (BLOCK,BLOCK) → BOOLE]:
  b_1 ***disj?*** $b_2 \Rightarrow (b_1 \text{ \textbf{\textit{=?}}})$ ***disj?*** $(b_2 \text{ \textbf{\textit{=?}}})$.

disj? [infix: (BLOCKCOLL,BLOCK) → BOOLE]:
  $\langle\rangle$ ***eq?*** $\_ \Rightarrow \top$.
  $\langle b_1, c \rangle$ ***disj?*** $b_2 \Rightarrow (b_1 \text{ \textbf{\textit{disj?}}} b_2) \wedge (c \text{ \textbf{\textit{disj?}}} b_2)$.

disj? [infix: (BLOCK,BLOCKCOLL) → BOOLE]:
  b ***disj?*** $c \Rightarrow c$ ***disj?*** b.

disj? [infix: (BLOCKCOLL,BLOCKCOLL) → BOOLE]:
  $\_$ ***eq?*** $\langle\rangle \Rightarrow \top$.
  c_1 ***disj?*** $\langle b, c_2 \rangle \Rightarrow (c_1 \text{ \textbf{\textit{disj?}}} b) \wedge (c_1 \text{ \textbf{\textit{disj?}}} c_2)$.

disj? [infix: (BOX,BOX) → BOOLE]:
  b_1 ***disj?*** $b_2 \Rightarrow (b_1 \text{ \textbf{\textit{text?}}})$ ***disj?*** $(b_2 \text{ \textbf{\textit{text?}}})$.

eq? [infix: ($\mathcal{S},\mathcal{S}$) → BOOLE where $\mathcal{S} \in \{\text{SET,XSET,XDOM}\}$]:
  s ***eq?*** $s \Rightarrow \top$.
  s_1 ***eq?*** $s_2 \Rightarrow (s_1 \text{ \textbf{\textit{plain!}}})$ ***id?*** $(s_2 \text{ \textbf{\textit{plain!}}})$.

eq? [infix: $\mathcal{D}$ → BOOLE where $\mathcal{D} \in \{(\text{BASET,SET}),(\text{BAXDOM,XDOM})\}$]:
  $\langle\rangle$ ***eq?*** $\_ \Rightarrow \top$.
  $\langle e, b \rangle$ ***eq?*** $d \Rightarrow (e \in (d \text{ \textbf{\textit{plain!}}})) \wedge (b \text{ \textbf{\textit{eq?}}} d)$.
  b ***eq?*** $d \Rightarrow \bot$.

eq? [infix: $\mathcal{D}$ → BOOLE where $\mathcal{D} \in \{(\text{SET,BASET}),(\text{XDOM,BAXDOM})\}$]:
  d ***eq?*** $b \Rightarrow b$ ***eq?*** d.

full? [postfix: $\mathcal{S}$ → BOOLE where $\mathcal{S} \in \{\text{SET,XSET,XDOM}\}$]:
  s ***full?*** $\Rightarrow \top \quad | \quad s$ ***eq?*** $(s \text{ \textbf{\textit{base?}}})$.
  s ***full?*** $\Rightarrow \bot$.

grouped? [postfix: $\mathcal{S}$ → BOOLE where $\mathcal{S} \in \{\text{SET,XSET,XDOM}\}$]:
  s ***grouped?*** $\Rightarrow (s \text{ \textbf{\textit{mode?}}}) = \text{GROUPED}$.

nonempty? [postfix: $\mathcal{D}$ → BOOLE where $\mathcal{D} \in \{\text{SET,XDOM}\}$]:
  $\{\}$ ***nonempty?*** $\Rightarrow \bot$.
  $\_$ ***nonempty?*** $\Rightarrow \top$.

nonempty? [postfix: XSET → BOOLE]:
$\quad\quad$ ∅ *nonempty?* ⇒ ⌊.
$\quad\quad$ _ *nonempty?* ⇒ ⌉.

id? [infix: S → BOOLE where S ∈ {SET,XSET,XDOM}]:
$\quad\quad$ *s id? s* ⇒ ⌉.
$\quad\quad$ _ *id?* _ ⇒ ⌊.

plain? [postfix: S → BOOLE where S ∈ {SET,XSET,XDOM}]:
$\quad\quad$ *s plain?* ⇒ (*s mode?*) = PLAIN.

subsum? [infix: (GRUSET, GRUSET) → BOOLE]:
$\quad\quad$ g_1 *subsum?* g_2 ⇒ (((g_1 ∪ g_2) * g_2) ∩ (g_1 *plain!*)) = g_2.

tail [postfix: S →$_{\text{BASE}}$ S where S ∈ {SET,XDOM,BLOCKCOLL}]:
$\quad\quad$ {} *tail* ⇒ ⊥.
$\quad\quad$ ⟨e_1, e_2, ..., e_n⟩ *tail* ⇒ ⟨e_2, ..., e_n⟩.

top [postfix: S → VAL where S ∈ {SET,XDOM,BLOCKCOLL}]:
$\quad\quad$ {} *top* ⇒ ⊥.
$\quad\quad$ ⟨e_1, e_2, ..., e_n⟩ *top* ⇒ e_1.

A.2 Boolean Operators

∧ [infix: (BOOLE, BOOLE) → BOOLE]:
$\quad\quad$ ⌉∧⌉ ⇒ ⌉.
$\quad\quad$ _ ∧ _ ⇒ ⌊.

∨ [infix: (BOOLE, BOOLE) → BOOLE]:
$\quad\quad$ ⌊∨⌊ ⇒ ⌊.
$\quad\quad$ _ ∨ _ ⇒ ⌉.

¬ [infix: (BOOLE, BOOLE) → BOOLE]:
$\quad\quad$ ⌊¬ ⇒ ⌉.
$\quad\quad$ ⌉¬ ⇒ ⌊.

A.3 Ordering Relations

◁ [infix: $\mathcal{P} \to$ BOOLE where $\mathcal{P} \in \{$PLASET,PLAXSET$\}$]:
    $p_1 \triangleleft p_2 \Rightarrow p_1$ ***before:*** p_2 $| p_1 \doteq p_2.$
    $\_ \triangleleft \_ \Rightarrow \bot.$

before: [arg: (PLASET;PLASET) $\to$ BOOLE]:
    $\langle\rangle$ ***before:*** $\_ \Rightarrow \top.$
    $\langle e, p_1 \rangle$ ***before:*** $\langle e, p_2 \rangle \Rightarrow p_1$ ***before:*** $p_2.$
    $\langle e, p_1 \rangle$ ***before:*** $\langle e, p_2 \rangle \Rightarrow e_1$ ***before:*** e_2 ***in:*** $(p_1$ ***base?***$).$
    $\_$ ***before:*** $\_ \Rightarrow \bot.$

before: [arg: (PLAXSET;PLAXSET) $\to$ BOOLE]:
    $\varnothing$ ***before:*** $\_ \Rightarrow \top.$
    $()$ ***before:*** $() \Rightarrow \top.$
    $\langle p_1, \_ \rangle$ ***before:*** $\langle p_2, \_ \rangle \Rightarrow p_1 \triangleleft p_2.$
    $\langle p, x_1 \rangle$ ***before:*** $\langle p, x_2 \rangle \Rightarrow x_1$ ***before:*** $x_2.$
    $\_$ ***before:*** $\_ \Rightarrow \bot.$

before: in: [arg: $\mathcal{D} \to$ BOOLE where $\mathcal{D} \in \{$(VAL;VAL,BASET),
                                                                      (XVAL;XVAL,BAXDOM)$\}$]:
    v_1 ***before:*** v_2 ***in:*** $\langle v_1, b \rangle \Rightarrow \top$ $| v_2 \in b.$
    v_1 ***before:*** v_2 ***in:*** $\langle e, b \rangle \Rightarrow v_1$ ***before:*** v_2 ***in:*** $b | (v_2 = e)\neg.$
    v_1 ***before:*** v_2 ***in:*** $b \Rightarrow \bot$ $| (v_1 \in b) \wedge (v_2 \in b).$
    $\_$ ***before:*** $\_$ ***in:*** $\_ \Rightarrow \bot.$

before: in: [arg: (XVAL;XVAL,BAXSET) $\to$ BOOLE]:
    $()$ ***before:*** $()$ ***in:*** $\_ \Rightarrow \top.$
    (v_1, x_1) ***before:*** (v_2, x_2) ***in:*** $(b, s) \Rightarrow (v_1$ ***before:*** v_2 ***in:*** $b) \wedge$
                                                                                $(x_1 \in s) \wedge (x_2 \in s).$
    (v, x_1) ***before:*** (v, x_2) ***in:*** $(\_, s) \Rightarrow (x_1$ ***before:*** x_2 ***in:*** $s).$
    x_1 ***before:*** x_2 ***in:*** $b \Rightarrow \bot$ $| (x_1 \in b) \wedge (x_2 \in b).$
    $\_$ ***before:*** $\_$ ***in:*** $\_ \Rightarrow \bot.$

A.4 Type Casting Operations

box! [postfix: BLOCKCOLL $\to_{\text{BASE}}$ PLABOX]:
  $\langle\rangle$ ***box!*** $\Rightarrow \bot$.
  $\langle b \rangle$ ***box!*** $\Rightarrow$ ($\Box$ ***obtype:*** PLABOX) ***text:*** (b ***xset!***) ***edits:*** $\langle\rangle$.
  $\langle b,c \rangle$ ***box!*** $\Rightarrow$ (c ***box!***) + b.

block! [postfix: PLAXSET $\to_{\text{BASE}}$ BLOCK]:
  $\varnothing$ ***block!*** $\Rightarrow \bot$.
  p ***block!*** $\Rightarrow$ ($\Box$ ***obtype:*** BLOCK) =: p ***tags:*** $\langle\rangle$.

block! [postfix: GRUSET $\to_{\text{BASE}}$ BLOCKCOLL]:
  [ ] ***block!*** $\Rightarrow \langle\rangle$.
  $\langle p,g \rangle$ ***block!*** $\Rightarrow \langle (p),g$ ***block!***$\rangle$.

block! [postfix: GRUXSET $\to_{\text{BASE}}$ BLOCKCOLL]:
  $\varnothing$ ***block!*** $\Rightarrow \bot$.
  ( ) ***block!*** $\Rightarrow \langle\rangle$.
  (g,( )) ***block!*** $\Rightarrow g$ ***block!***.
  ($\langle g,s \rangle$,x) ***block!*** $\Rightarrow$ ($g \times (x$ ***block!***)) $\circ$ ((s,x) ***block!***).

dom! [postfix: $\mathcal{V} \to$ PLAXSET where $\mathcal{V} \in$ {BASVAL,COMVAL}]:
  v ***dom!*** $\Rightarrow (v)$.

dom! [postfix: $\mathcal{M}$ where $\mathcal{M} \in$ {PLAXSET $\to_{\text{BASE}}$ PLAXDOM,
                             GRUXSET $\to_{\text{BASE}}$ GRUXDOM}]:
  $\varnothing$ ***dom!*** $\Rightarrow$ { }.
  (s,( )) ***dom!*** $\Rightarrow s$ ***dom!***.
  ({ },_) ***dom!*** $\Rightarrow$ { }.
  ($\langle e,x \rangle$,d) ***dom!*** $\Rightarrow$ ($e \times (d$ ***dom!***)) $\circ$ ((x,d) ***dom!***).

dom! [postfix: PLASET $\to_{\text{BASE}}$ PLAXDOM]:
  $\langle\rangle$ ***dom!*** $\Rightarrow \langle\rangle$.
  $\langle e,p \rangle$ ***dom!*** $\Rightarrow \langle (e),p$ ***dom!***$\rangle$.

dom! [postfix: GRUSET $\to_{\text{BASE}}$ GRUXDOM]:
  [ ] ***dom!*** $\Rightarrow$ [ ].
  $\langle p,g \rangle$ ***dom!*** $\Rightarrow \langle p$ ***dom!***$,g$ ***dom!***$\rangle$.

dom! [postfix: BAXSET → BAXDOM]:
  $()$ ***dom!*** ⇒ ⊥.
  b ***dom!*** ⇒ (b ***xdom!***) ***base:*** b.

group! [postfix: PLASET →$_{\text{BASE}}$ GRUSET]:
  $\langle\rangle$ ***group!*** ⇒ [].
  $\langle e,p \rangle$ ***group!*** ⇒ $\langle\langle e\rangle, p$ ***group!***$\rangle$.

plain! [postfix: $\mathcal{M}$ where $\mathcal{M} \in$ {SET →$_{\text{BASE}}$ PLASET, XDOM →$_{\text{BASE}}$ PLAXDOM}]:
  s ***plain!*** ⇒ s | s ***plain?***.
  [] ***plain!*** ⇒ $\langle\rangle$.
  $\langle e,s \rangle$ ***plain!*** ⇒ $e \cup (s$ ***plain!***$)$.

plain! [postfix: XSET →$_{\text{BASE}}$ PLAXSET]:
  $\emptyset$ ***plain!*** ⇒ $\emptyset$.
  $()$ ***plain!*** ⇒ $()$.
  (e,x) ***plain!*** ⇒ $(e$ ***plain!***$, x$ ***plain!***$)$.

xset! [postfix: BLOCK →$_{\text{BASE}}$ PLAXSET]:
  b ***xset!*** ⇒ $b =?$.

xdom! [postfix: BAXSET →$_{\text{BASE}}$ BAXDOM]:
  $(\langle\rangle,\_)$ ***xdom!*** ⇒ $\langle\rangle$.
  $(\langle e,p\rangle,())$ ***xdom!*** ⇒ $\langle(e),(p,())$ ***xdom!***$\rangle$.
  $(\langle e,p\rangle,b)$ ***xdom!*** ⇒ $(e \times (b$ ***xdom!***$)) \circ ((p,b)$ ***xdom!***$)$.

A.5 Set Operations

$\circ$ [infix: (BAXDOM, BAXDOM) → BAXDOM]:
  $\langle\rangle \circ b$ ⇒ b.
  $\langle e,b_1\rangle \circ b_2$ ⇒ $\langle e, b_1 \circ b_2\rangle$.

$\circ$ [infix: $(\mathcal{D},\mathcal{D})$ →$_{\text{BASE}}$ $\mathcal{D}$ where $\mathcal{D} \in$ {SET, XDOM}]:
  $\{\} \circ x$ ⇒ x.
  $\langle e,x_1\rangle \circ x_2$ ⇒ $\langle e, x_1 \circ x_2\rangle$.

.5 SET OPERATIONS

× [infix: $(\mathcal{V},\mathcal{D}) \to \mathcal{D}$ where $\mathcal{V} \in \{\text{BASVAL},\text{COMVAL}\}$,
$\mathcal{D} \in \{\text{BAXDOM, PLAXDOM}\}$]:
$v \times \langle\rangle \Rightarrow \langle\rangle$.
$v \times \langle e,d \rangle \Rightarrow \langle (v,e), v \times d \rangle$.

× [infix: (PLASET,PLAXDOM) $\to_{\text{BASE}}$ PLAXDOM]:
$\langle\rangle \times \_ \Rightarrow \langle\rangle$.
$\langle e,p \rangle \times g \Rightarrow (e \times g) \circ (p \times g)$.

× [infix: (PLASET,GRUXDOM) $\to_{\text{BASE}}$ GRUXDOM]:
$\_ \times [] \Rightarrow []$.
$p \times \langle g,d \rangle \Rightarrow \langle p \times g, p \times d \rangle$.

× [infix: (PLASET,BLOCKCOLL) $\to_{\text{BASE}}$ BLOCKCOLL]:
$p \times \langle\rangle \Rightarrow \langle\rangle$.
$p \times \langle b,c \rangle \Rightarrow \langle (p,b), p \times c \rangle$.

∈ [infix: $(\mathcal{V},\text{BASET}) \to$ BOOLE where $\mathcal{V} \in \{\text{BASVAL},\text{COMVAL}\}$]:
$\_ \in \langle\rangle \Rightarrow \mathsf{L}$.
$v \in \langle v,\_ \rangle \Rightarrow \mathsf{T}$.
$v \in \langle \_,b \rangle \Rightarrow v \in b$.

∈ [infix: (XVAL,BAXSET) $\to$ BOOLE]:
$() \in () \Rightarrow \mathsf{T}$.
$(e,v) \in \langle b,x \rangle \Rightarrow (e \in b) \wedge (v \in x)$.
$\_ \in () \Rightarrow \bot$.
$() \in \_ \Rightarrow \bot$.

∈ [infix: (XVAL,BASET) $\to$ BOOLE]:
$\_ \in \langle\rangle \Rightarrow \mathsf{L}$.
$v \in \langle v,\_ \rangle \Rightarrow \mathsf{T}$.
$v \in \langle \_,b \rangle \Rightarrow v \in b$.

∈ [infix: $(\mathcal{V},\text{PLASET}) \to$ BOOLE where $\mathcal{V} \in \{\text{BASVAL},\text{COMVAL}\}$]:
$v \in p \Rightarrow v \text{ in? } p$ | $v \in (p \text{ **base?**})$.
$\_ \in \_ \Rightarrow \bot$.

∈ [infix: $(\text{XVAL},\mathcal{X}) \to$ BOOLE where $\mathcal{X} \in \{\text{PLAXSET},\text{PLAXDOM}\}$]:
$x \in p \Rightarrow x \text{ in? } p$ | $x \in (p \text{ **base?**})$.
$\_ \in \_ \Rightarrow \bot$.

∈ [infix: $\mathcal{D} \to$ BOOLE
　　　　where $\mathcal{D} \in \{$(PLASET,GRUSET), (PLAXSET,GRUXSET),
　　　　　　　　　　　　　　　　　　　　(PLAXDOM,GRUXDOM)$\}$]:
　　$p \in g \Rightarrow p$ *in?* g　　　　　$| p \doteq g.$
　　$\_ \in \_ \Rightarrow \perp.$

$+$ [infix: $\mathcal{M}$ where $\mathcal{M} \in \{$(PLASET,VAL) $\to_{\text{BASE}}$ PLASET,
　　　　　　　　　　　　　　(PLAXDOM,XVAL) $\to_{\text{BASE}}$ PLAXDOM$\}$]:
　　$p + v \Rightarrow p$ *add:* v　　　　　$| v \in (p$ *base?*$).$
　　$\_ + \_ \Rightarrow \perp.$

$+$ [infix: (GRUSET,PLASET) $\to_{\text{BASE}}$ GRUSET]:
　　$g + p \Rightarrow g$ *add:* p　　　　　$| (g \doteq p) \wedge (g$ *disj?* $p).$
　　$g + p \Rightarrow g$　　　　　　　　　$| g \doteq p.$
　　$\_ + \_ \Rightarrow \perp.$

$-$ [infix: $\mathcal{M}$ where $\mathcal{M} \in \{$(PLASET,VAL) $\to_{\text{BASE}}$ PLASET,
　　　　　　　　　　　　　　(PLAXDOM,XVAL) $\to_{\text{BASE}}$ PLAXDOM$\}$]:
　　$p - v \Rightarrow p$ *remove:* v　　　　$| v \in (p$ *base?*$).$
　　$\_ - \_ \Rightarrow \perp.$

$-$ [infix: (GRUSET,PLASET) $\to_{\text{BASE}}$ GRUSET]:
　　$g - p \Rightarrow g$ *remove:* p　　　　$| g \doteq p.$
　　$\_ - \_ \Rightarrow \perp.$

$-$ [infix: (PLAXSET,XVAL) $\to_{\text{BASE}}$ PLAXSET]:
　　$p - x \Rightarrow p$ *remove:* x　　　　$| x \in p.$
　　$\_ - \_ \Rightarrow \perp.$

$\subset$ [infix: $(\mathcal{S},\mathcal{S}) \to$ BOOLE where $\mathcal{S} \in \{$SET,XSET,XDOM$\}$]:
　　$s_1 \subset s_2 \Rightarrow s_1$ *subset?* s_2　　$| s_1 \doteq s_2.$
　　$s_1 \subset s_2 \Rightarrow (s_1$ *plain!*$)$ *subset?* s_2　$| (s_1 \doteq s_2) \wedge$
　　　　　　　　　　　　　　　　　　　　　　$(s_1$ *grouped?* $) \wedge (s_2$ *plain?*$).$
　　$s_1 \subset s_2 \Rightarrow \perp$　　　　　　$| s_1 \doteq s_2.$
　　$\_ \subset \_ \Rightarrow \perp.$

$\subseteq$ [infix: $(\mathcal{S},\mathcal{S}) \to$ BOOLE where $\mathcal{S} \in \{$SET,XSET,XDOM$\}$]:
　　$s_1 \subseteq s_2 \Rightarrow (s_1 = s_2) \vee (s_1 \subset s_2).$

.5 Set Operations

∪ [infix: (SET,SET) →$_{BASE}$ SET]:
    $s_1 \cup s_2 \Rightarrow s_1$ **union** s_2 | $s_1 \doteq s_2$.
    _ ∪ _ ⇒ ⊥.

∩ [infix: (𝑆,𝑆) →$_{BASE}$ 𝑆 where 𝑆 ∈ {SET,XSET}]:
    $s_1 \cap s_2 \Rightarrow s_1$ **intersect** s_2 | $s_1 \doteq s_2$.
    _ ∩ _ ⇒ ⊥.

\ [infix: (SET,SET) →$_{BASE}$ SET]:
    $s_1 \setminus s_2 \Rightarrow s_1$ **minus** s_2 | $s_1 \doteq s_2$.
    _ \ _ ⇒ ⊥.

[infix: (SET,SET) →$_{BASE}$ GRUSET]:
    $s_1 \# s_2 \Rightarrow s_1$ **tile** s_2 | $s_1 \doteq s_2$.
    _ # _ ⇒ ⊥.

[infix: (PLAXSET, PLAXSET) →$_{BASE}$ GRUXSET]:
    $s_1 \# s_2 \Rightarrow s_1$ **tile** s_2 | $s_1 \doteq s_2$.
    _ # _ ⇒ ⊥.

• [infix: (GRUSET,GRUSET) →$_{BASE}$ GRUSET]:
    $s_1 \bullet s_2 \Rightarrow s_1$ **restrict** s_2 | $s_1 \doteq s_2$.
    _ • _ ⇒ ⊥.

* [infix: (SET,SET) →$_{BASE}$ SET]:
    $s_1 * s_2 \Rightarrow s_1$ **select** s_2 | $s_1 \doteq s_2$.
    _ * _ ⇒ ⊥.

add: [arg: 𝓜 where 𝓜 ∈ {(PLASET,VAL) →$_{BASE}$ PLASET,
                                      (PLAXDOM,XVAL) →$_{BASE}$ PLAXDOM}]:
    ⟨⟩ **add:** $v \Rightarrow \langle v \rangle$.
    ⟨v,p⟩ **add:** $v \Rightarrow \langle v,p \rangle$.
    ⟨e,p⟩ **add:** $v \Rightarrow \langle e,p$ **add:** $v \rangle$ | e **before:** v **in:** (⟨e,p⟩ **base?**).
    p **add:** $v \Rightarrow \langle v,p \rangle$.

add: [arg: (GRUSET;PLASET) →$_{BASE}$ GRUSET]:
    [] **add:** $p \Rightarrow [p]$.
    ⟨p_1,g⟩ **add:** $p_2 \Rightarrow \langle p_1,g$ **add:** $p_2 \rangle$ | $p_1 \triangleleft p_2$.
    g **add:**$p \Rightarrow \langle p,g \rangle$.

co! [postfix: SET →$_{BASE}$ PLASET}]:
    s co! ⇒ (☐ *obtype:* PLASET =: (*s base?*)) \ (*s plain!*).

enc? [infix: (S,S) → BOOLE where S ∈ {SET,XSET,XDOM}]:
    s_1 *enc?* s_2 ⇒ (s_1 *plain!*) ⊆ (s_2 *plain!*).

in? [infix: D → BOOLE where D ∈ {(VAL,PLASET),(PLASET,GRUSET),
                    (XVAL,PLAXDOM),(PLAXDOM,GRUXDOM)}]:
    e in? {} ⇒ ⊥.
    e in? d ⇒ (*e* = (*d top*)) ∨ (*e in?* (*d tail*)).

in? [infix: D → BOOLE where D ∈ {(XVAL,PLAXSET),(PLAXSET,GRUXSET)}]:
    () *in?* () ⇒ ⊤.
    (*e,s*) *in?* (*x,t*) ⇒ (*e in? x*) ∧ (*s in? t*).
    _ *in?* () ⇒ ⊥.
    () *in?* _ ⇒ ⊥.

intersect [infix: (PLASET,PLASET) →$_{BASE}$ PLASET]:
    ⟨⟩ *intersect* _ ⇒ ⟨⟩.
    ⟨*e,p$_1$*⟩ *intersect* p_2 ⇒ ⟨*e,p$_1$ intersect p_2*⟩ | *e* ∈ p_2.
    ⟨_,p_1⟩ *intersect* p_2 ⇒ p_1 *intersect* p_2.

intersect [infix: (GRUSET,SET) →$_{BASE}$ GRUSET]:
    [] *intersect* _ ⇒ [].
    ⟨*p,g*⟩ *intersect s* ⇒ (*p intersect s*) ∪ (*g intersect s*).

intersect [infix: (PLASET,GRUSET) →$_{BASE}$ GRUSET]:
    p intersect g ⇒ *g intersect p*.

intersect [infix: (XSET,XSET) →$_{BASE}$ XSET]:
    ∅ *intersect* _ ⇒ ∅.
    _ *intersect* ∅ ⇒ ∅.
    x intersect x ⇒ *x*.
    (s_1,x_1) *intersect* (s_2,x_2) ⇒ ∅ | (s_1 *disj?* s_2) ∨ (x_1 *disj?* x_2).
    (s_1,x_1) *intersect* (s_2,x_2) ⇒ ($s_1 ∩ s_2, x_1$ *intersect* x_2).
    () *intersect* _ ⇒ ⊥.
    _ *intersect* () ⇒ ⊥.

.5 Set Operations

minus [infix: (PLASET,PLASET) →$_{BASE}$ PLASET]:
      p ***minus*** $\langle\rangle \Rightarrow p$.
      $\langle\rangle$ ***minus*** $\_ \Rightarrow \langle\rangle$.
      $\langle e,p_1\rangle$ ***minus*** $\langle e,p_2\rangle \Rightarrow p_1$ ***minus*** p_2.
      $\langle e_1,p_1\rangle$ ***minus*** $\langle e_2,p_2\rangle \Rightarrow \langle e_1,p_1$ ***minus*** $\langle e_2,p_2\rangle\rangle$ | $e_2 \in p_1$.
      p_1 ***minus*** $\langle \_,p_2\rangle \Rightarrow p_1$ ***minus*** p_2.

minus [infix: (GRUSET,GRUSET) →$_{BASE}$ GRUSET]:
      g ***minus*** $[] \Rightarrow g$.
      $[]$ ***minus*** $\_ \Rightarrow []$.
      g_1 ***minus*** $\langle p,g_2\rangle \Rightarrow (g_1$ ***minus*** $p)$ ***minus*** g_2.

minus [infix: (GRUSET,PLASET) →$_{BASE}$ GRUSET]:
      $[]$ ***minus*** $\_ \Rightarrow []$.
      $\langle p_1,g\rangle$ ***minus*** $p_2 \Rightarrow \langle p_1,g$ ***minus*** $p_2\rangle$ | $p_1 \supset\subset p_2$.
      $\langle \_,g\rangle$ ***minus*** $p \Rightarrow g$ ***minus*** p.

minus [infix: (PLASET,GRUSET) →$_{BASE}$ GRUSET]:
      p ***minus*** $g \Rightarrow (p$ ***group!***$)$ ***minus*** g.

remove: [arg: $\mathcal{M}$ where $\mathcal{M} \in$ {(PLASET,VAL) →$_{BASE}$ PLASET,
                             (PLAXDOM,XVAL) →$_{BASE}$ PLAXDOM}]:
      $\langle\rangle$ ***remove:*** $\_ \Rightarrow \langle\rangle$.
      $\langle e,p\rangle$ ***remove:*** $e \Rightarrow p$.
      $\langle e,p\rangle$ ***remove:*** $v \Rightarrow \langle e,p$ ***remove:*** $v\rangle$.

remove: [arg: (GRUSET; PLASET) →$_{BASE}$ GRUSET]:
      $[]$ ***remove:*** $\_ \Rightarrow []$.
      $\langle p,g\rangle$ ***remove:*** $p \Rightarrow g$.
      $\langle p_1,g\rangle$ ***remove:*** $p_2 \Rightarrow \langle p_1,g$ ***remove:*** $p_2\rangle$.

remove: [arg: (PLAXSET;XVAL) →$_{BASE}$ PLAXSET]:
      $\emptyset$ ***remove:*** $\_ \Rightarrow \emptyset$.
      $()$ ***remove:*** $() \Rightarrow ()$.
      (b,x) ***remove:*** $(e,t) \Rightarrow (b$ ***remove:*** e,x ***remove:*** $t)$.

restrict [infix: (GRUSET,GRUSET) →$_{BASE}$ GRUSET]:
      $[]$ ***restrict*** $\_ \Rightarrow []$.
      $\langle p,g_1\rangle$ ***restrict*** $g_2 \Rightarrow g_1$ ***restrict*** g_2 | p ***disj?*** g_2.
      $\langle p,g_1\rangle$ ***restrict*** $g_2 \Rightarrow (g_1$ ***restrict*** $g_2) + p$.

select [infix: (PLASET,SET) $\rightarrow_{\text{BASE}}$ SET]:
  p *select* $s \Rightarrow p \cap s$.

select [infix: (GRUSET,SET) $\rightarrow_{\text{BASE}}$ GRUSET]:
  g *select* $s \Rightarrow ((g \cup s) \cap (g \, \textit{plain!})) \bullet (g \cap s)$.

subset? [infix: ($\mathcal{D},\mathcal{D}$) $\rightarrow$ BOOLE where $\mathcal{D} \in$ {PLASET,GRUSET,
                                                                                  PLAXDOM,GRUXDOM}]:
  $\{\}$ *subset?* _ $\Rightarrow \top$.
  $\langle e, t_1 \rangle$ *subset?* $t_2 \Rightarrow (e \in t_2) \wedge (t_1 \, \textit{subset?} \, t_2)$.

subset? [infix: ($\mathcal{X},\mathcal{X}$) $\rightarrow$ BOOLE where $\mathcal{X} \in$ {PLAXSET,GRUXSET}]:
  $\emptyset$ *subset?* _ $\Rightarrow \top$.
  _ *subset?* $\emptyset \Rightarrow \bot$.
  x *subset?* $x \Rightarrow \top$.
  $\langle s_1, x_1 \rangle$ *subset?* $\langle s_2, x_2 \rangle \Rightarrow (s_1 \subset s_2) \wedge (x_1 \, \textit{subset?} \, x_2)$.
  $()$ *subset?* _ $\Rightarrow \bot$.
  _ *subset?* $() \Rightarrow \bot$.

tile [infix: (SET,SET) $\rightarrow_{\text{BASE}}$ GRUSET]:
  s_1 *tile* $s_2 \Rightarrow \langle s_1 \cap s_2 \rangle \cup (\langle (s_1 \, \textit{co!}) \cap s_2 \rangle \cup \langle s_1 \cap (s_2 \, \textit{co!}) \rangle)$
                                                          $| \, (s_1 \, \textit{plain?}) \wedge (s_2 \, \textit{plain?})$.
  s_1 *tile* $s_2 \Rightarrow ((s_1 \cap s_2) \cup ((s_1 \, \textit{co!}) \cap s_2) \cup ((s_1 \cap (s_2 \, \textit{co!}))) \, \textit{group!})$
                                                          $| \, (s_1 \, \textit{plain?}) \wedge (s_2 \, \textit{grouped?})$.
  s_1 *tile* $s_2 \Rightarrow (s_1 \cap s_2) \cup (((s_1 \, \textit{co!}) \cap s_2) \cup (s_1 \cap (s_2 \, \textit{co!})))$
                                                          $| \, (s_1 \, \textit{grouped?}) \wedge (s_2 \, \textit{grouped?})$.
  s_1 *tile* $s_2 \Rightarrow s_2$ *tile* s_1 $| \, (s_1 \, \textit{grouped?}) \wedge (s_2 \, \textit{plain?})$.

tile [infix: (PLAXSET, PLAXSET) $\rightarrow_{\text{BASE}}$ GRUXSET]:
  $()$ *tile* $() \Rightarrow ()$.
  (s_1, x_1) *tile* $(s_2, x_2) \Rightarrow (s_1 \# s_2, x_1 \, \textit{tile} \, x_2)$.

.5 Set Operations

union [infix: $(\mathcal{P},\mathcal{P}) \to_{\text{BASE}} \mathcal{P} \in \{\text{PLASET},\text{PLAXDOM}\}$]:
    $\langle\rangle$ *union* $p \Rightarrow p$.
    $\langle e, p_1 \rangle$ *union* $p_2 \Rightarrow p_1$ *union* p_2 | $e \in p_2$.
    $\langle e, p_1 \rangle$ *union* $p_2 \Rightarrow \langle e, p_1$ *union* $p_2 \rangle$
                                         | e *before*: $(p_2$ *top*) *in*: $(p_2$ *base*?).
    p_1 *union* $\langle e, p_2 \rangle \Rightarrow \langle e, p_1$ *union* $p_2 \rangle$.

union [infix: (GRUSET,GRUSET) $\to_{\text{BASE}}$ GRUSET]:
    [] *union* $g \Rightarrow g$.
    $\langle p, g_1 \rangle$ *union* $g_2 \Rightarrow g_1$ *union* $((p$ *unioneco* $g_2) + (p \cup (p$ *unione* $g_2)))$.

union [infix: (GRUSET,PLASET) $\to_{\text{BASE}}$ GRUSET]:
    g *union* $p \Rightarrow (((g$ *co!*$) \cap p)$ *unionmxd*) *union* g.

union [infix: (PLAXSET, PLAXSET) $\to_{\text{BASE}}$ PLAXSET]:
    $\emptyset$ *union* $p \Rightarrow p$.
    p *union* $\emptyset \Rightarrow p$.
    p *union* $p \Rightarrow p$.
    (p_1, x_1) *union* $(p_2, x_2) \Rightarrow (p_1 \cup p_2, x_1$ *union* $x_2)$.
    () *union* _ $\Rightarrow \bot$.
    _ *union* () $\Rightarrow \bot$.

union [infix: (PLASET,GRUSET) $\to_{\text{BASE}}$ GRUSET]:
    p *union* $g \Rightarrow g$ *union* p.

unione [infix: (PLASET,GRUSET) $\to_{\text{BASE}}$ PLASET]:
    _ *unione* [] $\Rightarrow \langle\rangle$.
    p *unione* $\langle e, g \rangle \Rightarrow e$ *union* $(p$ *unione* $g)$ | $(p \cap e)$ *nonempty*?.
    p *unione* $\langle \_, g \rangle \Rightarrow p$ *unione* g.

unioneco [infix: (PLASET,GRUSET) $\to_{\text{BASE}}$ GRUSET]:
    _ *unioneco* [] $\Rightarrow$ [].
    p *unioneco* $\langle e, g \rangle \Rightarrow (p$ *unione* $g) + e$ | $(p \cap e)$ *nonempty*?.
    p *unioneco* $\langle \_, g \rangle \Rightarrow p$ *unioneco* g.

unionmxd [postfix: PLASET $\to_{\text{BASE}}$ GRUSET]:
    $\langle\rangle$ *unionmxd* $\Rightarrow$ [].
    $\langle e, p \rangle$ *unionmxd* $\Rightarrow [e]$ *union* $(p$ *unionmxd*$)$.

APPENDIX B

Dependency Edit Processing

In Section 2.3 it has been assumed simply that source edits are defined appropriately without making any provision for deriving them actually. In fact, in case of sporadic edits only a manual definition of edits is possible; structural edits (such as those generated by conditional questions in a survey questionnaire), however, can be derived in a systematic way as sketched below. In contrast to previous work spent on the analysis of edits (particularly with respect to correction and imputation tasks in survey processing; cf. [Felligi and Holt, 1976; Barcaroli and Di Pace, 1992]), the present expositions aims at the derivation of SoEDits starting from a *dependency graph* capturing the conditional relationships (route information) between a data source's variables. Starting point of this transformation procedure is a set of pre-processed SoVARiables according to the rules specified in Subsection 2.3.1. Essentially, the dependency graph of a source's SoVARiables is translated into a canonical representation from which the SoEDits can be gained by a couple of simple algebraic combination rules. Prerequisites of the outlined procedure are (i) a given dependency graph and (ii) a specification of μ with respect to at least the variable mapping component (telling to which IMAxes the SoVARiables actually map).

In the following, V_i denotes the i-th SoVARiable and A_j the j-th Axis of the associated IMAGE. If V_c is a *conditioning* SoVARiable with SCALEDOMain $D^{(c)} =_{def} D(V_c)$, $C_+^{(c)} \subseteq D^{(c)}$ denotes the *condition set*, that is the subset of SoVALues the conditional variable V_d is dependent upon; $C_-^{(c)}$ is the complement of $C_+^{(c)}$ with respect to $D^{(c)}$. Additionally, deviating from previous usage, let $D_0^{(d)} =_{def} D^{(d)} \cup \langle \phi^{(d)} \rangle$ for a conditional SoVAR V_d, that is the SCALEDOMain augmented with the 'NULL' value. This notation applies to the A_j as well, letting context decide if SoVARiables or Axis are referenced. Likewise, $\phi^{(d)}$ may denote either V_d's 'NULL' value or $\langle \phi^{(d)} \rangle$ as appropriate; in particular, deviating from definition in Subsection 2.4.1, $\mathcal{C}^P(\phi^{(d)}) =_{def} \mathcal{C}^P_{d:\langle \phi \rangle}$. According to **Def. 2.4.1–2**, '$\otimes$' denotes the *primitive concept product*.

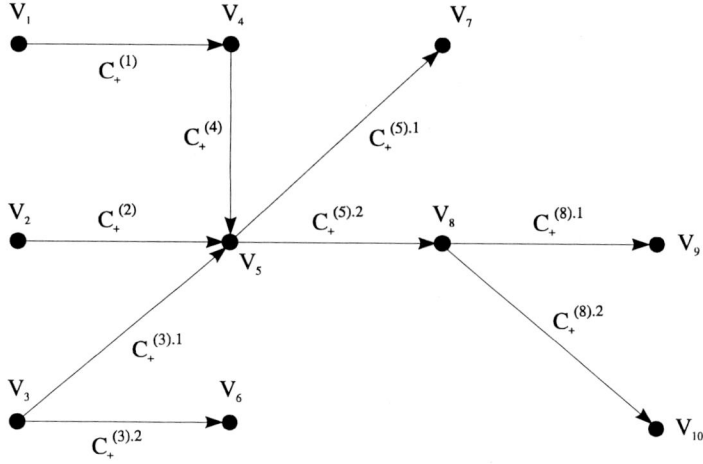

Figure B–1: A Sample Dependency Graph

The derivation of dependency edits is accomplished in a two stage procedure. In stage 1, a source's dependency graph is reduced by (i) eliminating conditionally nested variables (if any) and (ii) eliminating downstream branches. A variable is *nested conditionally*, if it replaces a measurement value of the conditioning variable provided that an additional condition, expressed in terms of another variable, holds; strictly speaking, this implies that a set of SOVALues is clustered unless an additional condition is fulfilled whence the SOVALue cluster is resolved into its constituent elements (note that unconditionally nested variables have been eliminated in a pre-processing step already by adapting the SCALEDOMain of the conditioning variable accordingly). A *downstream branch* of a dependency graph, which is always a *dag* (directed acyclic graph) by its very nature, occurs whenever a variable is conditioning more than one dependent variable, that is there are several conditional variables depending on the *same* conditioning variable. After reducing the dependency graph, stage 2 commences actually composing the set of SOEDits implied by the variables' dependency structure. In order to illustrate the derivation procedure, **Fig. B–1** shows a (rather contrived) dependency graph comprising 10 variables; $\{V_1, V_2, V_3\}$ is the set of *roots* (unconditional variables), and $\{V_6, V_7, V_9, V_{10}\}$ the set of *terminals*. Evidently, all non-root variables represent conditional variables. Each variable is represented by a node carrying the variable designation as its label; a directed edge between two nodes connects a conditioning variable to a conditional variable (pointing to the latter). Edges are labelled with the condition set admitting a transition from the conditioning to the conditional

B Dependency Edit Processing

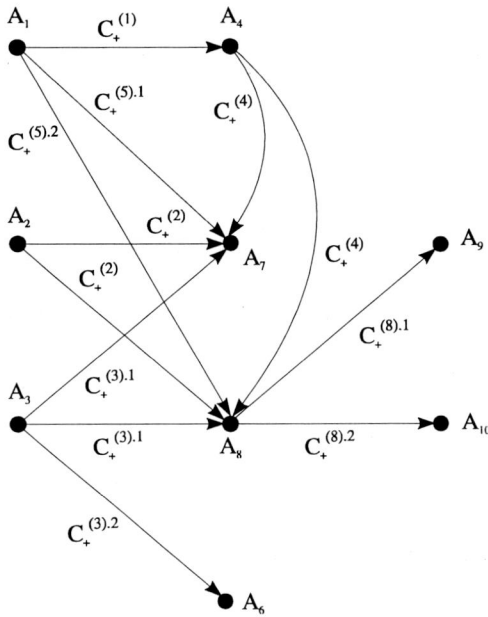

Figure B–2: Sample Dependency Graph after Elimination of Conditionally Nested Variable

variable. In this example, it is assumed furthermore that V_5 represents a conditionally nested variable with respect to V_1. As a consequence, consistency of graph definition requires that $C_+^{(5).1} \subset C_+^{(1)}$ and $C_+^{(5).2} \subset C_+^{(1)}$. In turn, both V_1 and V_5 map to the same SOFRAME AXis, say A_1; for the sake of simplicity it is assumed that V_k maps to A_k, for all k except $k = 5$.

In stage 1 of the edit derivation procedure, conditionally nested variables are spotted and eliminated first. This is achieved as follows:

For each conditionally nested variable V mapping to Axis A, its most upstream parent (predecessor) is sought; let this node be called P. Connect all outgoing edges of V, $E^+(V)$, to node P. For each ingoing edge e to V, $e \in E^-(V)$, e is redirected and, if necessary, replicated such that an instance of e points to each of the successor nodes of V, $N^+(V)$, in the original dependency graph. Now, node V is dropped and SOVAR labels are (provisionally) exchanged by AXis labels.

The result of applying this transformation to the dependency graph of **Fig. B–1** is shown in **Fig. B–2**. Note also the rearrangement of edge labels induced by edge redirection.

Still in stage 1, the elimination of downstream branches proceeds as follows:

Among all nodes in the graph, the set of *splitting nodes* is determined; a node V is a splitting node if $|E^+(V)| > 1$. For each splitting node V, the graph is rearranged such that (i) V is replicated $|E^+(V)|$ times, (ii) each replication is made incident with an edge from $E^+(V)$, and (iii) each replication of V inherits $E^-(V)$, the set of ingoing edges of V, which, of course needs to be replicated either (thus inducing a growth in predecessors' outgoing edges sets).

Since node splitting depends on local conditions only, it can be done in parallel and, hence, is an associative operation. Upon termination of stage 1, a reduced dependency graph is returned consisting, in general, of more than one component (cf. [Harary, 1969] with respect to terminology). Considering the running example, **Fig. B–3** illustrates the reduced state of **Fig. B–1**'s dependency graph.

Apparently, a reduced dependency graph (RDG, for short) is composed of (cascaded) linear or upstream branching dependencies only. In fact, upstream branches may become complicated further by replacing the simple edges hitherto considered with *multiple* edges (which, as little consideration reveals, does not affect RDG derivation). Thus, the elementary cases to be concerned are the following ones:

- A *simple* dependency connects a pair of a conditioning and a conditional node; symbolically $A_c \rightarrow A_d$.
- A *multiple* dependency connects a (non-singleton) set of conditioning nodes with a conditional node; symbolically $A_{c_1}, \ldots, A_{c_l} \rightarrow A_d$, $l > 1$.
- A *composite* dependency connects a (non-singleton) set of conditioning nodes with a conditional node; symbolically $A_{c_1}, \ldots, A_{c_l} \Rightarrow A_d$, $l > 1$, such that, for each j, $1 \leq j \leq l$, A_{c_j} and A_d are connected by a multiple edge. In a composite dependency, from each multiple edge exactly one variant is chosen to be combined to a multiple dependency; hence, for each j, $1 \leq j \leq l$, there is the same number of variants in a multiple edge.

In each of these elementary dependencies, the left-hand side nodes of symbolic representations are called *entries*.

Taking $u = 3$, **Fig. B–4** (overleaf) exhibits a sample composite dependency with three variant combinations of multiple edges. Note that the conditioning sets attached to the variants of a multiple edge need not be disjoint. If multiple and single edges occur jointly, it apparently is always possible to replicate single edges by a corresponding factor until the necessary number of variants is obtained (other combinations are not consistent for obvious reasons).

The actual composition of SoEDits is now brought about algebraically by combining the *edit generating rules* associated with each of the three discerned elemen-

B Dependency Edit Processing

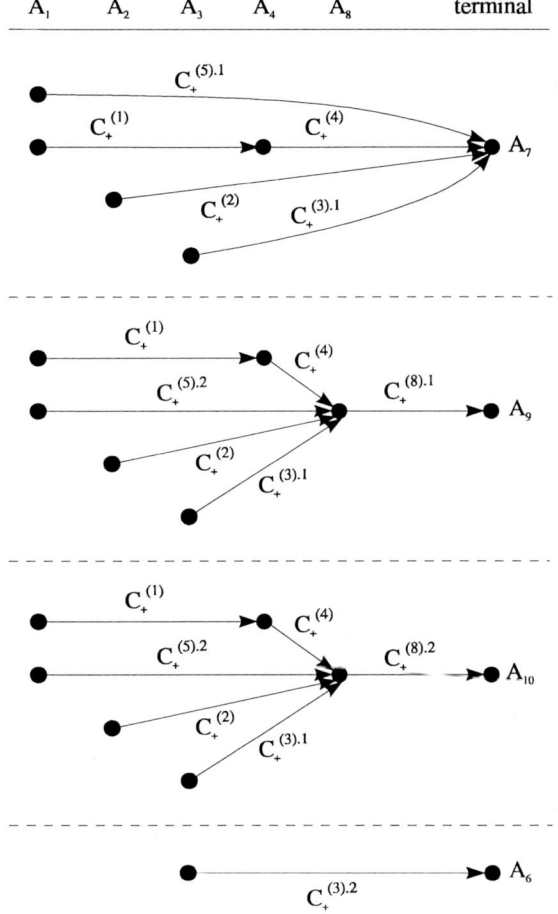

Figure B–3: A Sample RDG

tary dependency cases described above. These rules can be stated in a general way as follows (making use of the notation introduced in **Def. 2.4.1–2**; cf. [Barcaroli and Di Pace, 1992], too):

- A *simple* dependency induces the edit

$$e^P\left(C_-^{(c)}\right) \otimes e^P\left(D^{(d)}\right)$$

and its "dual",

$$e^P\left(C_+^{(c)}\right) \otimes e^P\left(\phi^{(d)}\right).$$

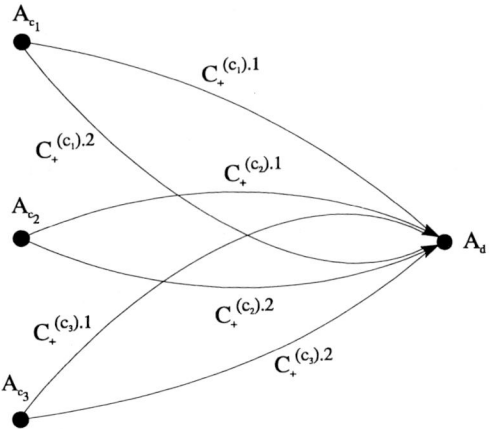

Figure B–4: A Sample Composite Dependency

- A *multiple* dependency induces the edits

$$e_1^P \otimes \cdots \otimes e_l^P \otimes e^P\left(D^{(d)}\right)$$

such that $\exists j.1 \leq j \leq l : e_j^P = e^P\left(C_-^{(c_j)}\right)$, letting, for $k \neq j$, $e_k^P = e^P\left(C_+^{(c_k)}\right)$ or $e_k^P = e^P\left(C_-^{(c_k)}\right)$,

as well as the dual edit

$$e^P\left(C_+^{(c_1)}\right) \otimes \cdots \otimes e^P\left(C_+^{(c_l)}\right) \otimes e^P\left(\phi^{(d)}\right).$$

- A *composite* dependency induces the edits

$$\left(\bigcap_u e_u\right) \otimes e^P\left(D^{(d)}\right)$$

where $e_u = e_{1.u}^P \otimes \cdots \otimes e_{l.u}^P$, and their duals

$$\left(\bigcup_u e_u^+\right) \otimes e^P\left(\phi^{(d)}\right)$$

B DEPENDENCY EDIT PROCESSING

where $e_u^+ = e^P\left(C_+^{(c_1).u}\right) \otimes \cdots \otimes e^P\left(C_+^{(c_l).u}\right)$, letting u range over the variant combinations of multiple edges.

Using these edit generating rules, the SOEDits of each RDG component are composed simply by collecting the respective edit generating rules instantiated appropriately with condition sets and complements thereof; furthermore, additional edits are induced for downstream-adjacent dependencies – that is, the immediate successor dependency of the one presently considered – as follows:

For the successor's *conditional* entries, $A_{e_1}, \ldots, A_{e_l}$, $l \geq 1$, such that $\phi^{(e_j)} \in D_0^{(e_j)}$ for $1 \leq j \leq l$, and the sequence of all downstream (dependent) Axis nodes, $A_{d_1}, \ldots, A_{d_m}$, $m \geq 1$, of the (elementary or already compound) successor dependency, edits are generated according to the *sequencing rule*

$$e_1^P \otimes \cdots \otimes e_l^P \otimes e_{l+1}^P \otimes \cdots \otimes e_{l+m}^P$$

where $\exists j.1 \leq j \leq l : e_j^P = e^P\left(\phi^{(e_j)}\right)$, letting, for $k \neq j$, $e_k^P = e^P\left(D^{(e_k)}\right)$ or $e_k^P = e^P\left(\phi^{(k)}\right)$ with $\phi^{(k)} \in D_0^{(e_k)}$, as well as

$\exists j.1 \leq j \leq m : e_{l+j}^P = e^P\left(D^{(d_j)}\right)$, letting, for $k \neq j$, $e_{l+k}^P = e^P\left(D^{(d_k)}\right)$ or $e_{l+k}^P = e^P\left(\phi^{(k)}\right)$ with $\phi^{(k)} \in D_0^{(d_k)}$.

Frequently, of course, the edits generated by the sequencing rule could be collapsed, thus rendering fewer factors. Although this sequencing rule ensures full associativity of edit generation, it is advisable to process RDGs from left to right; in this case the sequencing rules always induce the simplest representation of edits. For instance, processing the component with terminal A_{10} of the RDG shown in **Fig. B–3** generates the set of SOEDits presented in **Tab. B**.

Table B: Sample Edits

simple dependency $A_1 \rightarrow A_4$:
$e^P\left(C_-^{(1)}\right) \otimes e^P\left(D^{(4)}\right)$
$e^P\left(C_+^{(1)}\right) \otimes e^P\left(\phi^{(4)}\right)$ (dual edit)

Table B: Sample Edits (cont'd)

multiple dependency $A_1, A_2, A_3, A_4 \rightarrow A_8$ with conditional entry A_4:

$$e^P\left(C_-^{(4)}\right) \otimes e^P\left(C_+^{(5).2}\right) \otimes e^P\left(C_+^{(2)}\right) \otimes e^P\left(C_+^{(3).1}\right) \otimes e^P\left(D^{(8)}\right)$$

$$e^P\left(C_+^{(4)}\right) \otimes e^P\left(C_-^{(5).2}\right) \otimes e^P\left(C_+^{(2)}\right) \otimes e^P\left(C_+^{(3).1}\right) \otimes e^P\left(D^{(8)}\right)$$

$$e^P\left(C_+^{(4)}\right) \otimes e^P\left(C_+^{(5).2}\right) \otimes e^P\left(C_-^{(2)}\right) \otimes e^P\left(C_+^{(3).1}\right) \otimes e^P\left(D^{(8)}\right)$$

$$e^P\left(C_+^{(4)}\right) \otimes e^P\left(C_+^{(5).2}\right) \otimes e^P\left(C_+^{(2)}\right) \otimes e^P\left(C_-^{(3).1}\right) \otimes e^P\left(D^{(8)}\right)$$

$$e^P\left(C_-^{(4)}\right) \otimes e^P\left(C_-^{(5).2}\right) \otimes e^P\left(C_+^{(2)}\right) \otimes e^P\left(C_+^{(3).1}\right) \otimes e^P\left(D^{(8)}\right)$$

$$e^P\left(C_-^{(4)}\right) \otimes e^P\left(C_+^{(5).2}\right) \otimes e^P\left(C_-^{(2)}\right) \otimes e^P\left(C_+^{(3).1}\right) \otimes e^P\left(D^{(8)}\right)$$

$$e^P\left(C_-^{(4)}\right) \otimes e^P\left(C_+^{(5).2}\right) \otimes e^P\left(C_+^{(2)}\right) \otimes e^P\left(C_-^{(3).1}\right) \otimes e^P\left(D^{(8)}\right)$$

$$e^P\left(C_+^{(4)}\right) \otimes e^P\left(C_-^{(5).2}\right) \otimes e^P\left(C_-^{(2)}\right) \otimes e^P\left(C_+^{(3).1}\right) \otimes e^P\left(D^{(8)}\right)$$

$$e^P\left(C_+^{(4)}\right) \otimes e^P\left(C_-^{(5).2}\right) \otimes e^P\left(C_+^{(2)}\right) \otimes e^P\left(C_-^{(3).1}\right) \otimes e^P\left(D^{(8)}\right)$$

$$e^P\left(C_+^{(4)}\right) \otimes e^P\left(C_+^{(5).2}\right) \otimes e^P\left(C_-^{(2)}\right) \otimes e^P\left(C_-^{(3).1}\right) \otimes e^P\left(D^{(8)}\right)$$

$$e^P\left(C_-^{(4)}\right) \otimes e^P\left(C_-^{(5).2}\right) \otimes e^P\left(C_-^{(2)}\right) \otimes e^P\left(C_+^{(3).1}\right) \otimes e^P\left(D^{(8)}\right)$$

$$e^P\left(C_-^{(4)}\right) \otimes e^P\left(C_-^{(5).2}\right) \otimes e^P\left(C_+^{(2)}\right) \otimes e^P\left(C_-^{(3).1}\right) \otimes e^P\left(D^{(8)}\right)$$

$$e^P\left(C_-^{(4)}\right) \otimes e^P\left(C_+^{(5).2}\right) \otimes e^P\left(C_-^{(2)}\right) \otimes e^P\left(C_-^{(3).1}\right) \otimes e^P\left(D^{(8)}\right)$$

$$e^P\left(C_+^{(4)}\right) \otimes e^P\left(C_-^{(5).2}\right) \otimes e^P\left(C_-^{(2)}\right) \otimes e^P\left(C_-^{(3).1}\right) \otimes e^P\left(D^{(8)}\right)$$

$$e^P\left(C_-^{(4)}\right) \otimes e^P\left(C_-^{(5).2}\right) \otimes e^P\left(C_-^{(2)}\right) \otimes e^P\left(C_-^{(3).1}\right) \otimes e^P\left(D^{(8)}\right)$$

$$e^P\left(C_+^{(4)}\right) \otimes e^P\left(C_+^{(5).2}\right) \otimes e^P\left(C_+^{(2)}\right) \otimes e^P\left(C_+^{(3).1}\right) \otimes e^P\left(\phi^{(8)}\right)$$
(dual edit)

$$e^P\left(\phi^{(4)}\right) \otimes e^P\left(D^{(8)}\right) \quad \text{(sequencing rule)}$$

simple dependency $A_8 \rightarrow A_{10}$ with conditional entry A_8:

$$e^P\left(C_-^{(8).2}\right) \otimes e^P\left(D^{(10)}\right)$$

$$e^P\left(C_+^{(8).2}\right) \otimes e^P\left(\phi^{(10)}\right) \quad \text{(dual edit)}$$

$$e^P\left(\phi^{(8)}\right) \otimes e^P\left(D^{(10)}\right) \quad \text{(sequencing rule)}$$

B Dependency Edit Processing

The sum total of all SOEDits is obtained by collecting the edits derived for each RDG component (since, formally, these edits are the complements of *jointly* applying conditions expressed by edit rules, the generated edits are combined disjunctively which, however, simply amounts to gathering the – conjunctive – terms induced by the edit generating and sequencing rules) and dropping duplicates. The edit generation process can be speeded up by exploiting the fact that elimination of downstream branches in stage 1 leads to identical sub-edits which need be derived only once.

APPENDIX C

Labour Force Metadata

In Chapter 3, the Labour Force (LF, for short) domain metadata structures have been discussed in rather abstract terms using METAMODel (Sections 2.2 and 2.3) notation conventions; however, to be of practical value in a computational setting, a translation of both, the nomenclature used and the mappings arranged formally, into METASTASYS code is indispensable. Although quite lengthy, this appendix gives an abbreviated account of LF metadata structures as they have been compiled particularly for the purpose of demonstrating the DÖS'chen software prototype described in the final section (2.7) of Chapter 2. Specifically, these excerpts of LF metadata are selected such that the screenshots taken from the prototype program user interface can be related directly to annotated parts of the following transcripts obtained from program printouts which, for the sake of better readability, have been reformatted and edited slightly.

The material presented in this appendix is organized as follows. In Section C.1 a sample of LF FRAMEQUalities is reproduced, followed by a selection of FRAM-EDITS (Section C.2). On top of the LF FRAME, several labelled LF CONDEFs (cf. Subsection 2.4.1 and, specifically, **Def. 2.2.3–1**) have been defined, some of which are reproduced in Section C.3. Although the defined CONDEFs refer to LF concepts valid up to the early 1990ies only (noting that research and development of DÖS'chen commenced within an earlier metadata project – cf. Section 1.4 – starting in 1989) it is especially interesting to recognize the variety of definitions being in use concurrently in different countries; despite on-going efforts to stand-ardize population concepts and the measurement of (un-)employment, this situation will most likely prevail due to continuing socio-economic changes taking place with different pace and against heterogeneous structural backgrounds in different regions or nations (cf. Section 3.2). Section C.4 is devoted to an illustration of survey mappings; more specifically, several variable and value mappings for three distinct LF surveys are reproduced focusing on the cardinal LF QUalities, viz. *economic activity status* and *status in employment*. The final Section C.5 sketches a DÖS'chen client-server protocol record reproduced from corresponding query processing log files.

The transcripts are making use of a specific notation which, essentially, is self-explanatory. Generally, transcripts are segmented in paragraphs beginning with a '@' sign followed by the paragraph header (keyword). More specific punctuations are explained where appropriate.

C.1 Labour Force Quality Definitions

As discussed in Chapter 3, LF surveys comprise a rather extensive set of queries (SoVARiables; cf. Subsection 2.3.1) leading to an even higher number of FRAME-QUalities. For the sake of brevity, only a few typical and illustrative examples of QUality definitions are shown here. Furthermore, SoVARiables involving trivial, though rather tedious, mappings (such as *age, usual working time, actual working time*, etc.) are excluded from the presentation; either these mappings are associating SoVALues with MoDalities simply *1:1* or, as in the *age* case, require an intermediary conversion of birth dates first. Somewhat special are the mappings to the universal QUalities, *space* and *time*, which are coded usually as implicit variables in questionnaire records and, hence, need some pre-processing either before mapping can take place actually.

In the following transcript, two simple QUalities are stated first, viz. *sex* and *marital status*. Note that, within DÖS'chen, FRAMEQUalities are assigned QULABs (cf. **Def. 2.2.1–10**) written upper-case throughout for reasons of optical discernment from survey variables; QULABs are displayed in the leftmost pane of DÖS'chen's *GME*; cf. **Fig. 2.7–1**. In the transcript, each definition starts with a Frame-Quality paragraph and lists the MoDalities (ordered implicitly by the listing sequence) in the BaseValues paragraph which always terminates with a '#' symbol; in between there is an optional Info paragraph holding unformatted text. The QUality *marital status* has attached a further MapGroups paragraph used for the definition of VALMAPpings (cf. **Def. 2.3.3–5**) as far as SoVALue-MoDality associations *are not of* 1:1 *type* (with respect to VARMAPs, see Section C.4): each *map group* consists of a label followed by '<<' and a list of associated MoDalities enclosed in set braces '{' and '}', thus encoding part of $\mu_{lm}^{(s)}$.

```
@FrameQuality
SEX

@Info
[ DOMAIN GENERAL ((internal code: G-2)) ]    Frame unit: PERSON

(self-explanatory)

@BaseValues
  male
  female
#

@FrameQuality
MARITAL STATUS
[ GENERAL DOMAIN ((internal code: G-3)) ]    Frame unit: PERSON

@BaseValues
```

.1 LABOUR FORCE QUALITY DEFINITIONS

```
    single
    married
    widowed
    divorced
    living in common household
    separated
#
@MapGroups
    verheiratet (A)
        << {married,separated}
    living separated
        << {divorced,separated}
    living together
        << {married,living in common household}
#
```

Another example of a categorical QUality is *relation to head of household*; in addition to the paragraphs already mentioned there is also an AggLevels paragraph used to specify aggregation levels (cf. **Def. 2.2.1–9**) syntactically similar to map group definition; before AGGDOM values are defined, an AGGLABel is stated followed by '<<' and, in brackets, the AGGBASE (in case of AGGBASE coinciding with the bottom SCALEDOM of the referenced FRAMEQUality, '[ ]' is stated). Ordering of AGGDOM values is significant. Each AGGDOM enumeration terminates with a '#' symbol.

```
@FrameQuality
RELATION TO HEAD OF HOUSEHOLD

@Info
[ GENERAL DOMAIN ((internal code: G-6)) ]   Frame unit: PERSON

@BaseValues
    head of household
    spouse
    child (of head of household)
    child (of spouse)
    mother/father of head of household
    mother/father of spouse
    daughter or son in law
    other relative of head of household or spouse
    other relation
#
@AggLevels
*3*     << []:
    head of household
        << {head of household}
    spouse or relatives
        << {spouse,child (of head of household),child (of spouse),
            mother/father of head of household,
            mother/father of spouse,daughter or son in law,
            other relative of head of household or spouse}
    other
        << {other relation}
#
@MapGroups
    ascendent relative (EU)
```

```
            << {mother/father of head of household,
                mother/father of spouse}
      other relative (EU)
            << {daughter or son in law,
                other relative of head of household or spouse}
      Sonstige Person (A)
            << {child (of spouse),mother/father of spouse,
                daughter or son in law,
                other relative of head of household or spouse,
                other relation}
      Kinder (FRG)
            << {child (of head of household),child (of spouse)}
 #
```

The most important FRAMEQUalities of the LF FRAME certainly are *economic activity status* and *status in employment* stated below. In view of their pivotal role, both definitions comprise a variety of map groups as well as hierarchically layered AGGLEvels. **Fig. 2.7–2** presents DÖS'chen's *GME* view on QUality ECONOMIC ACTIVITY STATUS; **Fig. 2.7–3** illustrates the step of defining a MODality 'registered unemployed' of AGGLEvel '*6*'.

```
      @FrameQuality
      ECONOMIC ACTIVITY STATUS

      @Info
      [ DOMAIN SPECIFIC ((internal code: D-1)) ]   Frame unit: PERSON

      main distinction of type of participation in the labour force
      (w.r.t. the reference period)

      @BaseValues
        work for pay or profit
        not active because of bad weather
        not active because of slack work
        not active because of mechanical breakdown
        not active because of labour dispute
        not active because of education/formation
        not active because of accident of work or professional
                                                           affection
        not active because of maternity leave
        not active because of vacation
        not active - starts new job later
        not active - other reasons
        military or substitute service (conscripts)
        registered at employment exchange
        (registered) unemployed participating in employment
                                                  promotion scheme
        registered lay-off
        not registered lay-off
        retired and receiving a pension
        housewife / domestic family responsibilities
        in education/formation
        other supported person
        other not supported person
      #
      @AggLevels
      *6*     << []:
```

.1 LABOUR FORCE QUALITY DEFINITIONS

```
active
    << {work for pay and profit}
currently not active
    << {not active because of bad weather,
        not active because of slack work,
        not active because of mechanical breakdown,
        not active because of labour dispute,
        not active because of education/formation,
        not active because of accident of work or
                              professional affection,
        not active because of maternity leave,
        not active because of vacation,
        not active - other reasons}
not active
    << {military or substitute service (conscripts),
        retired and receiving a pension
        housewife / domestic family responsibilities
        in education/formation}
lay-off/new job
    << {not active - starts new job later,
        registered lay-off,not registered lay-off}
registered unemployed
    << {registered at employment exchange,
        (registered) unemployed participating
                              in employment promotion scheme}
other person
    << {other supported person,other not supported person}
#
*3*     << [ *6* ]:
busy
    << {active,currently not active}
unbusy
    << {registered unemployed,lay-off/new job}
other
    << {not active,other person}
#
*2*     << [ *3* ]:
working
    << {busy}
other
    << {unbusy,other}
#
@MapGroups
  active person currently not working (EU)
    << {not active because of bad weather,
        not active because of slack work,
        not active because of mechanical breakdown,
        not active because of labour dispute,
        not active because of education/formation,
        not active because of accident of work or
                              professional affection,
        not active because of maternity leave,
        not active because of vacation,
        not active - starts new job later,
        not active - other reasons}
  registrierter Arbeitsloser (A)
    << {registered at employment exchange,
        (registered) unemployed participating in
                              employment promotion scheme,
        registered lay-off}
  Null Stunden (A)
    << {not active because of bad weather,
        not active because of slack work,
```

```
           not active because of mechanical breakdown,
           not active because of labour dispute,
           not active because of education/formation,
           not active because of accident of work or
                                       professional affection,
           not active because of vacation,
           not active - starts new job later,
           not active - other reasons}
    sonstige Person (A)
       << {other supported person,other not supported person}
    Präsenzdiener/Karenzurlauberin (A)
       << {military or substitute service (conscripts),
           not active because of maternity leave}
    slack work/mechanical breakdown
       << {not active because of slack work,
           not active because of mechanical breakdown}
    lay-off
           @Info  referring to persons over 13 years
       << {registered lay-off,
           not registered lay-off}
    inactive (EU)
       << {registered at employment exchange,
           retired and receiving a pension,
           housewife / domestic family responsibilities,
           in education/formation,
           other supported person,other not supported person}
    other person (over 13 years) neither employed nor active
       << {registered at employment exchange,
           (registered) unemployed participating in
                                       employment promotion scheme,
           retired and receiving a pension,
           housewife / domestic family responsibilities,
           in education/formation,
           other supported person,
           other not supported person}
    registered unemployed (P)
       << {registered at employment exchange,
           (registered) unemployed participating in
                                       employment promotion scheme}
#

@FrameQuality
STATUS IN EMPLOYMENT

@Info
[ DOMAIN SPECIFIC ((internal code: D-2)) ]   Frame unit: PERSON

first/primary job only

@BaseValues
    own-account worker (self-employed without paid employees)
    self-employed with paid employees (small independent)
    employer (many employees)
    employee - private sector (manual work)
    employee - private sector (clerical work)
    employee - private sector (managerial work)
    employee - private sector (domestic servant)
    employee - public sector (manual work)
    employee - public sector (clerical work)
    employee - public sector (managerial work)
    assisting relative (unpaid family worker)
    assistant (unpaid worker)
    member of producers' cooperative
```

.1 Labour Force Quality Definitions

```
      doing unpaid community or social work
      unemployed person never employed before
      soldier
#
@AggLevels
ICSE     << []:
  employer
      << {employer (many employees)}
  self-employed
      << {own-account worker (self-employed without paid
                                                  employees),
          self-employed with paid employees (small independent)}
  employee
      << {employee - private sector (manual work),
          employee - private sector (clerical work),
          employee - private sector (managerial work),
          employee - private sector (domestic servant),
          employee - public sector (manual work),
          employee - public sector (clerical work),
          employee - public sector (managerial)}
  employee (P)
      << {employee - private sector (manual work),
          employee - private sector (clerical work),
          employee - private sector (managerial work)}
  assisting relative (unpaid family worker)
      << {assisting relative (unpaid family worker)}
  member of producers' cooperative
      << {member of producers' cooperative}
#
EUROSTAT    << []:
  self-employed without employees
      << {own-account worker (self-employed without paid
                                                  employees)}
  self-employed with employees
      << {self-employed with paid employees (small independent),
          employer (many employees)}
  employee
      << {employee - private sector (manual work),
          employee - private sector (clerical work),
          employee - private sector (managerial work),
          employee - public sector (manual work),
          employee - public sector (clerical work),
          employee - public sector (managerial)}
  family worker
      << {assisting relative (unpaid family worker)}
  other
      << {member of producers' cooperative,
          soldier,
          unemployed person never employed before,
          doing unpaid community or social work}
#
*4*      << []:
  self-employed
      << {own-account worker (self-employed without paid
                                                  employees),
          self-employed with paid employees (small independent),
          employer (many employees)}
  employee
      << {employee - private sector (manual work),
          employee - private sector (clerical work),
          employee - private sector (managerial work),
          employee - private sector (domestic servant),
          employee - public sector (manual work),
```

```
            employee - public sector (clerical work),
            employee - public sector (managerial)}
    unpaid assisting person
        << {assisting relative (unpaid family worker),
            assistant (unpaid worker)}
    other
        << {member of producers' cooperative,
            soldier,
            unemployed person never employed before,
            doing unpaid community or social work}
#
Austria-3    << [ EUROSTAT ]:
    Selbständige
        << {self-employed without employees,
            self-employed with employees}
    Mithelfende Familienangehörige
        << {family worker}
    Unselbständige
        << {employee}
#
@MapGroups
    employee
        << {employee - private sector (manual work),
            employee - private sector (clerical work),
            employee - private sector (managerial work),
            employee - private sector (domestic servant),
            employee - public sector (manual work),
            employee - public sector (clerical work),
            employee - public sector (managerial)}
    employer
        << {self-employed with paid employees (small independent),
            employer (many employees)}
    own-account-worker (NL)
        << {own-account worker (self-employed without paid
                                                    employees),
            self-employed with paid employees (small independent)}
    other (P)
        << {assistant (unpaid worker),
            doing unpaid community or social work,
            unemployed person never employed before}
    other (FRG)
        << {member of producers' cooperative,
            doing unpaid community or social work,
            unemployed person never employed before}
#
```

Another key QUality relevant especially for measuring full-time employment and underemployment is *employment type* (termed *full-time/part-time distinction* in LF surveys); in this example, an explanatory remark has been added to the `Info` paragraph.

```
@FrameQuality
EMPLOYMENT TYPE (TIME)

@Info
[ DOMAIN SPECIFIC ((internal code: D-8)) ]  Frame unit: PERSON
```

.1 Labour Force Quality Definitions

```
  this criterion refers to the first, or main, activity only
  (note: it does NOT depend on usual or actual hours worked per
  week; in case of part-time most important reason is stated)

@BaseValues
  working full-time
  working part-time due to education
  working part-time due to illness
  working part-time because could not find full-time job
  working part-time because did not want full-time job
  working part-time because of probationary period
  working part-time due to other reasons
  home/tele-worker
  working occasionally
#
@AggLevels
full_part   << []:
  part-time
      << {working part-time due to education,
          working part-time due to illness,
          working part-time because could not find full-time job,
          working part-time because did not want full-time job,
          working part-time because of probationary period,
          working part-time due to other reasons}
  full-time
      << {working full-time}
  other
      << {home/tele-worker,working occasionally}
#
```

As a final example of DÖS'chen QUality definition, *employment seeking* has been specified in DÖS'chen as follows:

```
@FrameQuality
EMPLOYMENT SEEKING

@Info
[ DOMAIN SPECIFIC ((internal code: D-17)) ]   Frame unit: PERSON

type of job sought by unemployed person, or reason why not seeking
any job

@BaseValues
  seeking (dependent) employment
  seeking apprenticeship
  seeking self-employment
  not seeking - job already found
  not seeking - apprenticeship already found
  not seeking because lay-off
  not seeking since having no hope to find a job
  not seeking for other reasons
  not seeking without telling reason
#
@AggLevels
  *4*    << []:
    seeking work as employee
        << {seeking (dependent) employment,seeking apprenticeship}
    starting job later
        << {not seeking - job already found,
```

```
                 not seeking - apprenticeship already found}
      does not seek
         << {not seeking because lay-off,
             not seeking since having no hope to find a job,
             not seeking for other reasons,
             not seeking without telling reason}
      seeking self-employment
         << {seeking self-employment}
   #
   @MapGroups
      sucht nicht (A)
         @ Info
            alle Fälle, in denen nicht aktiv Arbeit gesucht wird
         << {not seeking - job already found,
             not seeking - apprenticeship already found,
             not seeking because lay-off,
             not seeking since having no hope to find a job,
             not seeking for other reasons,
             not seeking without telling reason}
      seek_work (EU)
         << {seeking (dependent) employment,seeking apprenticeship,
             seeking self-employment}
      later_work (EU)
         << {not seeking - job already found,
             not seeking - apprenticeship already found}
   #
```

C.2 Labour Force Frame Edits

The LF FRAME is rich in FRAMEDits because of the manifold dependencies between FRAMEQUalities; in particular, many relevant measurement dimensions can be asked persons over a certain *age* only. Additionally, there is a couple of mutually incompatible MOdality combinations involving two or more QUalities all of which ought to be captured formally as FRAMEDits as far as they are foreseen to hold generally in all LF surveys conceivable (cf. Subsection 3.3.2).

Technically, in DÖS'chen FRAMEDits are composed like CONDEFs (that is, composed of REGCONs; cf. **Def. 2.4.1–2**); the definition of constituent REGCONs always terminates with a '#' symbol. Implicitly, all FRAMEQUalities not stated in a REGCON participate with their full range of MOdalities as defined in the QUalities' DIMDOMs (cf. **Def. 2.2.1–11**). As a shorthand notation, '{==ALL VALUES==}' is used to designate a QUality's DIMDOM *except* SPECVALues. For convenience, defined AGGLEVels may be used to state a FRAMEQUality's actual FRAMEDit range; to this end, a suitable AGGLABel *<agglab>* is inserted in the 'at [*<agglab>*]:' string following the respective QULABel (otherwise, *<agglab>* is simply omitted).

Each FRAMEDit is given a label (contrary to METASTASYS conventions which would treat the constituent REGCONs as genuine FRAMEDits instead) visible in the rightmost pane of DÖS'chen's *GME* interface (**Fig. 2.7–1**); this label is stated in the Edit paragraph opening each FRAMEDit definition which is actually given in a subsequent EditDef paragraph.

.2 LABOUR FORCE FRAME EDITS

The first couple of sample FRAMEDit definitions refers to *age*-dependent incompatibilities assuming, for instance, that it is ubiquitously infeasible for a person under 13 years to have a *marital status* other than 'single':

```
@Edit
General edits relating to general domain qualities

@EditDef
AGE   at []:     {0,1,2,3,4,5,6,7,8,9,10,11,12}
MARITAL STATUS   at []:
    {married,widowed,divorced,living in common household,separated}
#

...

#
AGE   at []:     {0,1,2,3,4,5,6,7,8,9,10,11,12,13,14,17}
RELATION TO HEAD OF HOUSEHOLD   at []:
    {head of household,mother/father of head of household,
     mother/father of spouse}
#

@Edit
Age-dependent LFS edits

@EditDef
AGE   at []:
    {0,1,2,3,4,5,6,7,8,9,10,11}
ECONOMIC ACTIVITY STATUS   at []:      {==ALL VALUES==}
STATUS IN EMPLOYMENT   at []:      {==ALL VALUES==}
SITE OF EMPLOYMENT   at []:      {==ALL VALUES==}
REGULARITY OF ECONOMIC ACTIVITY   at []:      {==ALL VALUES==}
TIMING OF EMPLOYMENT CONTRACT   at []:      {==ALL VALUES==}
EMPLOYMENT TYPE (TIME)   at []:      {==ALL VALUES==}
USUAL WORKING TIME   at []:      {==ALL VALUES==}
ACTUAL WORKING TIME   at []:      {==ALL VALUES==}
REASON FOR DIFFERENCE BETWEEN USUAL AND ACTUAL WORK HOURS   at []:
    {==ALL VALUES==}
NUMBER OF CURRENT JOBS   at []:      {==ALL VALUES==}
EMPLOYMENT SITUATION A YEAR BEFORE   at []:      {==ALL VALUES==}
STATUS IN EMPLOYMENT A YEAR BEFORE   at []:      {==ALL VALUES==}
LOOKING FOR OTHER JOB   at []:      {==ALL VALUES==}
REASON FOR TERMINATING EMPLOYMENT   at []:      {==ALL VALUES==}
EMPLOYMENT SEEKING   at []:      {==ALL VALUES==}
SITUATION BEFORE SEEKING EMPLOYMENT   at []:      {==ALL VALUES==}
TYPE OF EMPLOYMENT SOUGHT   at []:      {==ALL VALUES==}
AVAILABILITY (IMMEDIATE)   at []:      {==ALL VALUES==}
AVAILABILITY (TIME HORIZON)   at []:      {==ALL VALUES==}
REGISTRATION TYPE   at []:      {==ALL VALUES==}
DURATION SINCE LAST EMPLOYMENT   at []:      {==ALL VALUES==}
DURATION OF JOB SEARCH   at []:      {==ALL VALUES==}
FINANCIAL ASSISTANCE (UNEMPLOYMENT BENEFITS)   at []:
    {==ALL VALUES==}
#

@Edit
AGE / ECONOMIC ACTIVITY STATUS edit

@EditDef
```

```
AGE   at []:      {12,13,14,15,16,17}
ECONOMIC ACTIVITY STATUS   at []:
  {military or substitute service (conscripts),
   retired and receiving a pension}
#
```

Many FRAMEDits between domain specific QUalities involve QUality *pairs*; triplets or higher-order interactions are rare (and difficult to grasp, after all). The FRAM-EDit labelled 'D-1 / D-10' (not a very illuminating name, in fact) corresponds to the screenshot exhibited in **Fig. 2.7–4**. Since the Info paragraph is empty (and, hence, dropped from the transcript, the respective pane on the *GME* screen simply displays '-NOINFO-'.

```
@Edit
D-1 / D-2

@EditDef
ECONOMIC ACTIVITY STATUS   at [ *3* ]:
  {busy}
STATUS IN EMPLOYMENT   at []:
  {assisting relative (unpaid family worker),
   assistant (unpaid worker),
   unemployed person never employed before,
   doing unpaid community or social work,soldier}
#
ECONOMIC ACTIVITY STATUS   at []:
  {not active - starts new job later,
   registered lay-off,not registered lay-off}
STATUS IN EMPLOYMENT   at []:
  {unemployed person never employed before,soldier}
#
ECONOMIC ACTIVITY STATUS   at []:
  {military or substitute service (conscripts),
   in education/formation,
   housewife / domestic family responsibilities}
STATUS IN EMPLOYMENT   at []:     {==ALL VALUES==}
#
ECONOMIC ACTIVITY STATUS   at []:
  {retired and receiving a pension,
   other supported person,other not supported person}
STATUS IN EMPLOYMENT   at [ *ICSE* ]:
  {employer,self-employed,employee,
   member of producers' cooperative}
#

@Edit
D-1, D-6, D-14

@EditDef
ECONOMIC ACTIVITY STATUS    at [ *3* ]:    {other}
REGULARITY OF ECONOMIC ACTIVITY   at []:   {==ALL VALUES==}
STATUS IN EMPLOYMENT A YEAR BEFORE   at []:   {==ALL VALUES==}
#

@Edit
D-1 / D-10
```

.2 Labour Force Frame Edits

```
@EditDef
ECONOMIC ACTIVITY STATUS   at []:     {work for pay and profit}
ACTUAL WORKING TIME   at []:     {did not work at all}
#
ECONOMIC ACTIVITY STATUS   at [ *6* ]:     {currently not active}
ACTUAL WORKING TIME   at [ ILO - active/inactive ]:
   {worked at least 1 hour}
#

@Edit
Edits depending on ECONOMIC ACTIVITY STATUS

@EditDef
ECONOMIC ACTIVITY STATUS   at [ *2* ]:     {other}
SITE OF EMPLOYMENT   at []:     {==ALL VALUES==}
TIMING OF EMPLOYMENT CONTRACT   at []:     {==ALL VALUES==}
EMPLOYMENT TYPE (TIME)   at []:     {==ALL VALUES==}
USUAL WORKING TIME   at []:     {==ALL VALUES==}
ACTUAL WORKING TIME   at []:     {==ALL VALUES==}
REASON FOR DIFFERENCE BETWEEN USUAL AND ACTUAL WORK HOURS   at []:
   {==ALL VALUES==}
NUMBER OF CURRENT JOBS   at []:     {==ALL VALUES==}
#

...

@Edit
D-1 / D-16

@EditDef
ECONOMIC ACTIVITY STATUS   at [ *3* ]:     {busy}
REASON FOR TERMINATING EMPLOYMENT   at []:     {==ALL VALUES==}
#

@Edit
D-1 *working* excluded

@EditDef
ECONOMIC ACTIVITY STATUS   at [ *2* ]:     {working}
EMPLOYMENT SEEKING   at []:     {==ALL VALUES==}
TYPE OF EMPLOYMENT SOUGHT   at []:     {==ALL VALUES==}
AVAILABILITY (IMMEDIATE)   at []:     {==ALL VALUES==}
AVAILABILITY (TIME HORIZON)   at []:     {==ALL VALUES==}
#

...

@Edit
Working time edit

@EditDef
USUAL WORKING TIME   at []:     {==ALL VALUES==}
ACTUAL WORKING TIME   at []:     {did not work at all}
REASON FOR DIFFERENCE BETWEEN USUAL AND ACTUAL WORK HOURS   at []:
   {==ALL VALUES==}
#
```

C.3 Labour Force Concept Definitions

Basically, there are two types of LF subject-matter concepts of interest, viz. *harmonized* concepts expressing a data source-independent view and "local" concepts according to which, as a rule, the local LF surveys have been designed and implemented. The METASTASYS data integration approach, of course, enables a rather ad lib application of more or less locally inclined LF concepts to any available source of LF data whatsoever, bearing the obvious risk of queried concepts and survey-supported concepts matching badly and, hence, inhibiting successful table derivation.

The following CONDEFs (which are partial – cf. Subsection 2.4.1 – as well as generic – cf. Subsection 2.4.3 – CONDEFs assuming that FRAMEDITS are handled implicitly and *space* and *time* AXes are specified at REQUEST submitting time) illustrate a variety of relevant LF concepts as used traditionally in several European countries. The CONDEFs including an '(EU)' suffix in their label conform to the European CLFS as enacted by the EC (cf. Subsection 3.2.2). The transcript includes also LF concepts originating from Austrian, Belgian, and Portuguese LF statistics; in each case, the LF concepts shown are related hierarchically. Some of the presented CONDEFs have been valid and in use during past periods only (these temporal restrictions, of course, could be expressed formally by adding respective temporal restrictions accordingly, if so desired); the different concept versions reflect the changing demands LF statistics is expected to serve. However, as the main purpose of these CONDEFs has been to prove DÖS'chen's representation capabilities, no validity with regard to contents is warranted.

Each CONDEF bears a CONLAB (**Def. 2.2.3–4**) contained in the opening `Concept` paragraph; next, in the `SuperConcept` paragraph, comes the reference to a super-concept defined separately which the present CONDEF is going to refine in terms of a sub-concept. If a concept's superconcept coincides with a FRAME's *ur-concept* (cf. Subsection 2.4.1), '`*FRAME*`' is stated in the `SuperConcept` paragraph. After an optional `Info` paragraph, the `ConceptDef` paragraph provides the actual CONDEF definition structurally identical to a FRAMEDit definition.

The prepared CONDEFs are indicated by CONLAB in DÖS'chen's *GME* interface in the second from right pane. The first CONDEF shown in the following transcript corresponds to **Fig. 2.7–5** highlighting the second REGCON of its definition.

```
@Concept
Erwerbspersonen (A/87)

@SuperConcept
*FRAME*
```

.3 Labour Force Concept Definitions

@Info
Definition of Active Population according to Austrian Mikrozensus as of 1987 (covering the whole of Austria's resident population): in contrast to its predecessor this definition now includes unemployed persons not being registered but seeking employment and being available within a month after the reference week.

@ConceptDef
AGE   at [half-decades]:
   {15-19,20-24,25-29,30-34,35-39,40-44,45-49,50-54,55-59,60-64,
    65-69,70-74,75 and over}
USUAL WORKING TIME   at [*6*]:
   {1-12 hours,13-20 hours,21-30 hours,31-40 hours,over 40 hours}
ECONOMIC ACTIVITY STATUS   at []:
   {work for pay and profit}
#
AGE   at [half-decades]:
   {15-19,20-24,25-29,30-34,35-39,40-44,45-49,50-54,55-59,60-64,
    65-69,70-74,75 and over}
ECONOMIC ACTIVITY STATUS   at []:
   {not active because of bad weather,
    not active because of slack work,
    not active because of mechanical breakdown,
    not active because of labour dispute,
    not active because of education/formation,
    not active because of accident of work or
                                       professional affection,
    not active because of maternity leave,
    not active because of vacation,
    not active - starts new job later,
    not active - other reasons,
    military or substitute service (conscripts),
    registered at employment exchange,
    (registered) unemployed participating in
                                employment promotion scheme,
    registered lay-off}
#
AGE   at [half-decades]:
   {15-19,20-24,25-29,30-34,35-39,40-44,45-49,50-54,55-59,60-64,
    65-69,70-74,75 and over}
EMPLOYMENT SEEKING   at []:
   {seeking (dependent) employment,seeking apprenticeship}
ECONOMIC ACTIVITY STATUS   at []:
   {retired and receiving a pension,
    housewife / domestic family responsibilities,
    in education/formation,
    other supported person}
AVAILABILITY (TIME HORIZON)   at []:
   {within next week,in second week from now,
    in third week from now,in fourth week from now}
#

@Concept
Nichterwerbspersonen (A/87)

@SuperConcept
FRAME

@Info
Definition of Inactive Population according to Austrian Mikrozensus as of 1987

```
@ConceptDef
ECONOMIC ACTIVITY STATUS   at []:
  {retired and receiving a pension,
   housewife / domestic family responsibilities,
   in education/formation,other supported person}
EMPLOYMENT SEEKING    at [ *4* ]:
  {does not seek}
#
ECONOMIC ACTIVITY STATUS   at []:
  {retired and receiving a pension,
   housewife / domestic family responsibilities,
   in education/formation,other supported person}
AVAILABILITY (TIME HORIZON)   at []:
  {in 5th to 8th week from now,in 9th to 12th week from now,
   after 12 weeks from now}
#

@Concept
Erwerbspersonen (A/84)

@SuperConcept
*FRAME*

@Info
Definition of Active Population according to Austrian Mikrozensus
as of 1984 (covering the whole of Austria's resident population):
women on maternity leave are now included in the labour force
(i.e. they are no longer counted as unemployed)

@ConceptDef
AGE   at [ half-decades ]:
  {15-19,20-24,25-29,30-34,35-39,40-44,45-49,50-54,55-59,60-64,
   65-69,70-74,75 and over}
USUAL WORKING TIME    at [ *6* ]:
  {1-12 hours,13-20 hours,21-30 hours,31-40 hours, over 40 hours}
ECONOMIC ACTIVITY STATUS    at []:
  {work for pay and profit}
#
AGE   at [ half-decades ]:
  {15-19,20-24,25-29,30-34,35-39,40-44,45-49,50-54,55-59,60-64,
   65-69,70-74,75 and over}
ECONOMIC ACTIVITY STATUS    at []:
  {not active because of bad weather,
   not active because of slack work,
   not active because of mechanical breakdown,
   not active because of labour dispute,
   not active because of education/formation,
   not active because of accident of work or
                                          professional affection,
   not active because of maternity leave,
   not active because of vacation,
   not active - starts new job later,
   not active - other reasons,
   military or substitute service (conscripts),
   registered at employment exchange,
   (registered) unemployed participating in
                                 employment promotion scheme,
   registered lay-off}
#

@Concept
Nichterwerbspersonen (A/84)
```

.3 Labour Force Concept Definitions

```
@SuperConcept
*FRAME*

@Info
Definition of Inactive Population according to Austrian Mikrozen-
sus as of 1984

@ConceptDef
ECONOMIC ACTIVITY STATUS   at []:
  {not registered lay-off,
   retired and receiving a pension,
   housewife / domestic family responsibilities,
   in education/formation,other supported person}
#

@Concept
Arbeitslose (A/87)

@SuperConcept
Erwerbspersonen (A/87)

@Info
Unemployed persons according to the definition used in Austrian
Mikrozensus as of 1987

@ConceptDef
ECONOMIC ACTIVITY STATUS   at []:
  {registered at employment exchange,
    (registered) unemployed participating in
                               employment promotion scheme,
   registered lay-off,not registered lay-off}
#

@Concept
Erwerbstätige (A/87)

@SuperConcept
Erwerbspersonen (A/87)

@Info
Persons in employment according to the definition used in Austrian
Mikrozensus as of 1987

@ConceptDef
ECONOMIC ACTIVITY STATUS   at []:
  {working for pay or profit,
   not active because of bad weather,
   not active because of slack work,
   not active because of labour dispute,
   not active because of mechanical breakdown,
   not active because of education/formation,
   not active because of accident of work or
                                       professional affection,
   not active because of vacation,
   not active - starts new job later,
   not active - other reasons}
#

@Concept
Arbeitslose (A/84)

@SuperConcept
Erwerbspersonen (A/84)
```

@Info
Unemployed persons according to the definition used in Austrian
Mikrozensus as of 1984

@ConceptDef
ECONOMIC ACTIVITY STATUS   at []:
  {registered at employment exchange, registered lay-off,
   not registered lay-off}
#

@Concept
Erwerbstätige (A/84)

@SuperConcept
Erwerbspersonen (A/84)

@Info
Persons in employment according to the definition used in Austrian
Mikrozensus as of 1984

@ConceptDef
ECONOMIC ACTIVITY STATUS   at []:
  {work for pay or profit,
   not active because of bad weather,
   not active because of slack work,
   not active because of labour dispute,
   not active because of mechanical breakdown,
   not active because of education/formation,
   not active because of accident of work or
                                            professional affection,
   not active because of vacation,
   not active - starts new job later,
   not active - other reasons}
#

@Concept
Beschäftigte (A/84)

@SuperConcept
Erwerbstätige (A/84)

@Info
Actually working persons (Labour Force) being employed continuously according to definition in Austrian Mikrozensus as of 1984

@ConceptDef
REGULARITY OF ECONOMIC ACTIVITY   at []:
  {permanent}
#

@Concept
Resident Population over 13 years (EU)

@SuperConcept
FRAME

@Info
Note: implicitly, this definition is not restricted to a particular political area (i.e. it is "generic" in the geographical
sense)

.3 Labour Force Concept Definitions

```
@ConceptDef
AGE   at []:
  {14,15,16,17,18,19,20,21,22,23,24,25,26,27,28,29,30,31,32,
   33,34,35,36,37,38,39,40,41,42,43,44,45,46,47,48,49,50,51,
   52,53,54,55,56,57,58,59,60,61,62,63,64,65,66,67,68,69,70,
   71,72,73,74,75+}
#


@Concept
Active Population (EU)

@SuperConcept
Resident Population over 13 years (EU)

@ConceptDef
ECONOMIC ACTIVITY STATUS   at []:
  {work for pay and profit}
ACTUAL WORKING TIME   at [ ILO - active/inactive ]:
  {worked at least 1 hour}
#
ECONOMIC ACTIVITY STATUS   at []:
  {not active because of bad weather,
   not active because of slack work,
   not active because of mechanical breakdown,
   not active because of labour dispute,
   not active because of education/formation,
   not active because of accident of work or
                                      professional affection,
   not active because of maternity leave,
   not active because of vacation,
   not active - starts new job later,
   not active - other reasons,
   military or substitute service (conscripts),
   registered at employment exchange,
   (registered) unemployed participating in
                                  employment promotion scheme,
   registered lay-off,not registered lay-off}
#
ECONOMIC ACTIVITY STATUS   at []:
  {retired and receiving a pension,
   housewife / domestic family responsibilities,
   in education/formation,
   other supported person,other not supported person}
EMPLOYMENT SEEKING   at [ *4* ]:
  {seeking self-employment,seeking work as employee}
AVAILABILITY (IMMEDIATE)   at [ within 2 weeks ]:
  {yes}
#


@Concept
Inactive Population (EU)

@SuperConcept
Resident Population over 13 years (EU)

@ConceptDef
ECONOMIC ACTIVITY STATUS   at []:
  {retired and receiving a pension,
   housewife / domestic family responsibilities,
   in education/formation,
   other supported person,other not supported person}
EMPLOYMENT SEEKING   at [ *4* ]:
```

```
  {does not seek}
#
ECONOMIC ACTIVITY STATUS   at []:
  {retired and receiving a pension,
   housewife / domestic family responsibilities,
   in education/formation,
   other supported person,other not supported person}
EMPLOYMENT SEEKING    at [ *4* ]:
  {seeking self-employment,seeking work as employee}
AVAILABILITY (IMMEDIATE)   at [ within 2 weeks ]:
  {no}
#

@Concept
Unemployed persons (EU)

@SuperConcept
Active Population (EU)

@ConceptDef
ECONOMIC ACTIVITY STATUS   at []:
  {not active - starts new job later,
   registered at employment exchange,
   (registered) unemployed participating in
                            employment promotion scheme,
   registered lay-off,not registered lay-off}
#
ECONOMIC ACTIVITY STATUS   at []:
  {retired and receiving a pension,
   housewife / domestic family responsibilities,
   in education/formation,
   other supported person,other not supported person}
EMPLOYMENT SEEKING    at [ *4* ]:
  {seeking self-employment,seeking work as employee}
#

@Concept
Persons in employment (EU)

@SuperConcept
Active Population (EU)

@ConceptDef
ECONOMIC ACTIVITY STATUS   at []:
  {work for pay and profit,
   not active because of bad weather,
   not active because of slack work,
   not active because of mechanical breakdown,
   not active because of labour dispute,
   not active because of education/formation,
   not active because of accident of work or
                            professional affection,
   not active because of maternity leave,
   not active because of vacation,
   not active - starts new job later,
   not active - other reasons,
   military or substitute service (conscripts)}
#
```

@Concept
Active Resident Population (B/85)

.3 LABOUR FORCE CONCEPT DEFINITIONS

```
@SuperConcept
Resident Population over 13 years (EU)

@Info
Note: Belgium's Active Population excludes foreign persons in
      contrast to the Labour Force concept commonly used (which
      includes foreign persons residing and working within the
      country conducting the survey)

@ConceptDef
NATIONALITY   at []:
  {Belgium}
#

@Concept
Active Population (B/85)

@SuperConcept
Active Resident Population (B/85)

@Info
Definition as of 1985 excluding unemployed persons over 55 receiv-
ing official unemployment benefits not seeking work anymore

@ConceptDef
ECONOMIC ACTIVITY STATUS   at []:
  {work for pay and profit}
ACTUAL WORKING TIME   at [ ILO - active/inactive ]:
  {worked at least 1 hour}
#
ECONOMIC ACTIVITY STATUS   at []:
  {not active because of bad weather,
   not active because of slack work,
   not active because of mechanical breakdown,
   not active because of labour dispute,
   not active because of education/formation,
   not active because of accident of work or
                                       professional affection,
   not active because of maternity leave,
   not active because of vacation,
   not active - starts new job later,
   not active because of other reasons}
ACTUAL WORKING TIME   at []:
  {did not work at all}
#
ECONOMIC ACTIVITY STATUS   at []:
  {other supported person}
AGE   at []:
  {14,15,16,17,18,19,20,21,22,23,24,25,26,27,28,29,30,31,32,33,
   34,35,36,37,38,39,40,41,42,43,44,45,46,47,48,49,50,51,52,53,
   54,55}
FINANCIAL ASSISTANCE (UNEMPLOYMENT BENEFITS)   at []:
  {granted by government/official unemployment care,
   receiving no assistance}
#
ECONOMIC ACTIVITY STATUS   at []:
  {other supported person}
AGE   at []:
  {55,56,57,58,59,60,61,62,63,64,65,66,67,68,69,70,71,72,73,
   74,75+}
FINANCIAL ASSISTANCE (UNEMPLOYMENT BENEFITS)   at []:
  {granted by government/official unemployment care,
   receiving no assistance}
```

```
EMPLOYMENT SEEKING   at [ *4* ]:
  {seeking work as employee,seeking self-employment}
#
ECONOMIC ACTIVITY STATUS   at []:
  {registered at employment exchange}
FINANCIAL ASSISTANCE (UNEMPLOYMENT BENEFITS)   at []:
  {receiving no assistance}
#
ECONOMIC ACTIVITY STATUS   at []:
  {other not supported person}
AGE   at []:
  {14,15,16,17,18,19,20,21,22,23,24,25,26,27,28,29,30,31,32,33,
   34,35,36,37,38,39,40,41,42,43,44,45,46,47,48,49,50,51,52,53,
   54,55}
REGISTRATION TYPE   at []:
  {currently not registered but forced to do so by (local) law}
#
ECONOMIC ACTIVITY STATUS   at []:
  {other not supported person}
REGISTRATION TYPE   at []:
  {none/not registered at all}
EMPLOYMENT SEEKING   at [ *4* ]:
  {seeking work as employee,seeking self-employment}
#

@Concept
Persons in employment (B/85)

@SuperConcept
Resident Population over 13 years (EU)

@ConceptDef
ECONOMIC ACTIVITY STATUS   at []:
  {work for pay and profit}
USUAL WORKING TIME   at [ *6* ]:
  {1-12 hours,13-20 hours,21-30 hours,31-40 hours,over 40 hours}
STATUS IN EMPLOYMENT   at []:
  {employee,helper}
#
ECONOMIC ACTIVITY STATUS   at []:
  {in education/formation}
AGE   at []:
  {16,17,18,19,20,21,22,23,24,25,26,27,28,29,30,31,32,33,34,35,
   36,37,38,39,40,41,42,43,44,45,46,47,48,49,50,51,52,53,54,55,
   56,57,58,59,60,61,62,63,64,65,66,67,68,69,70,71,72,73,74,75+}
REGULARITY OF ECONOMIC ACTIVITY   at []:
  {worked (continuously) more than one
                       month during July-September only}
STATUS IN EMPLOYMENT   at [ *4* ]:
  {employee,unpaid assisting worker}
#
ECONOMIC ACTIVITY STATUS   at []:
  {work for pay and profit}
USUAL WORKING TIME   at [ *6* ]:
  {1-12 hours,13-20 hours,21-30 hours,31-40 hours,over 40 hours}
STATUS IN EMPLOYMENT   at [ *4* ]:
  {self-employed}
NUMBER OF CURRENT JOBS   at []:
  {one job,2nd job (other),more than one other job
                       (but none as dependent employee)}
#
```

.3 LABOUR FORCE CONCEPT DEFINITIONS

```
@Concept
Active Resident Population (P/83)

@SuperConcept
*FRAME*

@Info
... excluded from the Active Population are persons living
continuously in a foreign household if they reside less than
3 month in a domestic Portuguese dwelling; also excluded are
persons living in institutional organizations and mobile
housing units

@ConceptDef
AGE   at []:
   {12,13,14,15,16,17,18,19,20,21,22,23,24,25,26,27,28,29,30,31,
    32,33,34,35,36,37,38,39,40,41,42,43,44,45,46,47,48,49,50,51,
    52,53,54,55,56,57,58,59,60,61,62,63,64,65,66,67,68,69,70,71,
    72,73,74,75+}
DURATION OF RESIDENCE IN USUAL HOUSEHOLD   at []:
   {3-6 months,7-12 months,1-2 years,more than two years}
#
AGE   at []:
   {12,13,14,15,16,17,18,19,20,21,22,23,24,25,26,27,28,29,30,31,
    32,33,34,35,36,37,38,39,40,41,42,43,44,45,46,47,48,49,50,51,
    52,53,54,55,56,57,58,59,60,61,62,63,64,65,66,67,68,69,70,71,
    72,73,74,75+}
DURATION OF RESIDENCE IN USUAL HOUSEHOLD   at []:
   {up to one month,less than 3 months}
TYPE OF HOUSEHOLD (USUAL)   at []:
   {private domestic household (dwelling)}
#

@Concept
Active Population (P/83)

@SuperConcept
Active Resident Population (P/83)

@ConceptDef
ECONOMIC ACTIVITY STATUS   at []:
   {work for pay and profit}
ACTUAL WORKING TIME   at [ ILO - active/inactive ]:
   {worked at least 1 hour}
#
ECONOMIC ACTIVITY STATUS   at []:
   {not active because of bad weather,
    not active because of slack work,
    not active because of mechanical breakdown,
    not active because of labour dispute,
    not active because of education/formation,
    not active because of accident of work or
                                      professional affection,
    not active because of maternity leave,
    not active because of vacation,
    not active - starts new job later,
    not active because of other reasons,
    registered at employment exchange,
    (registered) unemployed participating in
                                employment promotion scheme,
    retired and receiving a pension}
#
ECONOMIC ACTIVITY STATUS   at []:
```

```
  {registered lay-off,not registered lay-off}
EMPLOYMENT SEEKING   at []:
  {not seeking because lay-off}
DURATION SINCE LAST EMPLOYMENT   at []:
  {less than a month,1-2 months,3-5 months,6-11 months}
#
ECONOMIC ACTIVITY STATUS   at []:
  {in education/formation,other supported person,
   other unsupported person}
EMPLOYMENT SEEKING   at [ *4* ]:
  {seeking work as employee,seeking self-employment"}
AVAILABILITY (TIME HORIZON)   at []:
  {within a week,in second week from now}
ACTUAL WORKING TIME   at []:
  {did not work at all}
#

@Concept
Active Population (P/74)

@SuperConcept
*FRAME*

@Info
Definition as of 1974 (valid through 1982)

@ConceptDef
ECONOMIC ACTIVITY STATUS   at []:
  {work for pay and profit}
ACTUAL WORKING TIME   at []:
  {15,16,17,18,19,20,21,22,23,24,25,26,27,28,29,30,31,32,33,34,
   35,36,37,38,39,40,41,42,43,44,45,46,47,48,49,50,51,52,53,54,
   55,56,57,58,59,60,61,62,63,64,65,66,67,68,69,70,71,72,73,74,
   75,76,77,78,79,80,81,82,83,84,85,86,87,88,89,90,91,92,93,94,
   95,96,97,98}
AGE   at []:
  {11,12,13,14,15,16,17,18,19,20,21,22,23,24,25,26,27,28,29,30,
   31,32,33,34,35,36,37,38,39,40,41,42,43,44,45,46,47,48,49,50,
   51,52,53,54,55,56,57,58,59,60,61,62,63,64,65,66,67,68,69,70,
   71,72,73,74,75+}
#
ECONOMIC ACTIVITY STATUS   at []:
  {not active because of bad weather,
   not active because of slack work,
   not active because of mechanical breakdown,
   not active because of labour dispute,
   not active because of education/formation,
   not active because of accident of work or
                                          professional affection,
   not active because of maternity leave,
   not active because of vacation,
   not active - starts new job later,
   not active because of other reasons}
USUAL WORKING TIME   at []:
  {15,16,17,18,19,20,21,22,23,24,25,26,27,28,29,30,31,32,33,34,
   35,36,37,38,39,40,41,42,43,44,45,46,47,48,49,50,51,52,53,54,
   55,56,57,58,59,60,61,62,63,64,65,66,67,68,69,70,71,72,73,74,
   75,76,77,78,79,80,81,82,83,84,85,86,87,88,89,90,91,92,93,94,
   95,96,97,98}
AGE   at []:
  {11,12,13,14,15,16,17,18,19,20,21,22,23,24,25,26,27,28,29,30,
   31,32,33,34,35,36,37,38,39,40,41,42,43,44,45,46,47,48,49,50,
   51,52,53,54,55,56,57,58,59,60,61,62,63,64,65,66,67,68,69,70,
```

.3 LABOUR FORCE CONCEPT DEFINITIONS

```
      71,72,73,74,75+}
#
ECONOMIC ACTIVITY STATUS    at []:
  {registered at employment exchange,
   registered lay-off,not registered lay-off}
AGE    at []:
  {11,12,13,14,15,16,17,18,19,20,21,22,23,24,25,26,27,28,29,30,
   31,32,33,34,35,36,37,38,39,40,41,42,43,44,45,46,47,48,49,50,
   51,52,53,54,55,56,57,58,59,60,61,62,63,64,65,66,67,68,69,70,
   71,72,73,74,75+}
#

@Concept
Unemployed persons (P/83)

@SuperConcept
Active Population (P/83)

@ConceptDef
ECONOMIC ACTIVITY STATUS    at []:
  {registered at employment exchange,
     (registered) unemployed participating in
                                 employment promotion scheme,
   registered lay-off,not registered lay-off,
   in education/formation,other unsupported person}
#

@Concept
Underemployed persons (P/83)

@SuperConcept
Active Population (P/83)

@ConceptDef
REASON FOR DIFFERENCE BETWEEN USUAL AND ACTUAL WORK HOURS   at []:
  {less hours due to slack work,
   less hours due to labour dispute,
   less hours due to mechanical breakdown,
   less hours due to variable hours",
   less hours due to  other reasons}
AVAILABILITY (IMMEDIATE)    at []:
  {at survey time (reference period)}
#
REASON FOR DIFFERENCE BETWEEN USUAL AND ACTUAL WORK HOURS   at []:
  {less hours due to slack work,
   less hours due to labour dispute,
   less hours due to mechanical breakdown,
   less hours due to variable hours,
   less hours due to  other reasons}
EMPLOYMENT SEEKING    at [ *4* ]:
  {seeking work as employee,seeking self-employment"}
#

@Concept
Harmonized Statistics of Dependent Employed (P)

@SuperConcept
*FRAME*

@Info
[this is an example of a concept pertaining to a different data
source; as a consequence, ECONOMIC ACTIVITY STATUS is introduced
artificially to conform to frame structure ! ]
```

```
@ConceptDef
ECONOMIC ACTIVITY STATUS   at []:
  {work for pay or profit,
   not active because of bad weather,
   not active because of slack work,
   not active because of mechanical breakdown,
   not active because of labour dispute,
   not active because of education/formation,
   not active because of accident of
                              work or professional affection,
   not active because of maternity leave,
   not active because of vacation}
STATUS IN EMPLOYMENT   at []:
  {employee - private sector (manual work),
   employee - private sector (clerical work),
   employee - private sector (managerial work)}
EMPLOYMENT TYPE (TIME)   at [ full_part ]:
  {full-time,part-time}
#
ECONOMIC ACTIVITY STATUS   at []:
  {in education/formation}
AGE   at [ half-decades ]:
  {15-19,20-24,25-29,30-34,35-39,40-44,45-49,50-54,55-59,
   60-64,65-69,70-74,75 and over}
#
ECONOMIC ACTIVITY STATUS   at []:
  {work for pay or profit}
STATUS IN EMPLOYMENT   at []:
  {soldier}
TIMING OF EMPLOYMENT CONTRACT   at [ *2* ]:
  {timed}
#
```

C.4 Labour Force Survey Mappings

For demonstrating DÖS'chen, only a small set of LF surveys has been analyzed and the respective metadata encoded formally. Furthermore, with exception of the Austrian *Mikrozensus* and the European *Labour Force Survey* (cf. Section 3.2), a small part of the Portuguese *Inquérito ao Emprego* [Gomes and Miranda, 1991] was translated into DÖS'chen data structures (cf. the 'Surveys' pane of the top-level *GME* screen shown in **Fig. 2.7–1**). Despite these limitations, the chosen sample mappings highlight vividly the enormous range of variability in survey structures even in a rather narrow subject domain area such as LF statistics. Since, by legal commitment, the CLFS is the declared common yardstick of European harmonized LF statistics among all member states, it is little surprising that CLFS mappings tend to be somewhat simpler in structure compared to those of the "home-grown" and historically rooted national LF surveys.

The first part of the following transcript refers to an excerpt of the Austrian *Mikrozensus* survey; some of its SoVars appear in the 'SurveyVariables' pane of the *GME* screenshot reproduced in **Fig. 2.7–6**. In DÖS'chen, a mapping associates a

.4 LABOUR FORCE SURVEY MAPPINGS

SoVar with a specific FrameQuality and maps each SoValue of the respective SoVar to either a MODality of this FrameQuality or a map group defined for it. After opening such a mapping by a `Survey` paragraph, followed by an optional `Info` paragraph, each SoVar mapping starts with a `SurveyVariable` paragraph stating the mapped SoVar's label, SoVarLABel (cf. **Def. 2.3.1–1**); this is followed by an `AssociatedFrameQuality` paragraph indicating the respective FrameQuality. After another optional `Info` paragraph, the `ValueMappings` paragraph encodes the ValMap in a triplet notation: in each triplet, first the code used in the survey is listed, in brackets, followed by the SoVal label; after '|->' either the assigned MODality, a map group, or the reference to a nested SoVar is stated. A trivial example is the mapping of the *Mikrozensus* variable »*Geschlecht*« to *sex*; the mapping of variable »*Stellung im Haushalt*« to *relation to head of household* makes use of a map group named 'Sonstige Person (A)' included in the transcript comprising FrameQuality definitions (cf. Section C.1). Both *Mikrozensus* variables »*Teilnahme am Erwerbsleben*« as well as »*1_Stunde*« (which is an artificial label) map to the FrameQuality *economic activity status* such that the latter refines the SoVal ›beschäftigt‹; this nested SoVar optionally states also the SoVal labels to be used in the *GME* interface after '>>'. A view on the resulting composite mapping is exhibited in the screenshot reproduced in **Fig. 2.7–7**.

```
@Survey
Mikrozensus (Austria)

@Info
Grundstruktur der Variablenabbildung vom österreichischen Mikro-
zensus - Fragebogen (Personenblatt B)

...

@SurveyVariable
Stellung im Haushalt

@AssociatedFrameQuality
RELATION TO HEAD OF HOUSEHOLD

@Info
Frage 5

@ValueMapping
[0]     Haushaltsvorstand
            |-> {head of household}
[1]     Ehegatte des HV
            |-> {spouse}
[2]     Kind des HV
            |-> {child (of head of household)}
[3]     Mutter, Vater des HV
            |-> {mother/father of head of household}
[4]     Sonstige Person
            |-> $MapGroup   Sonstige Person (A)
#
```

```
@SurveyVariable
Geschlecht

@AssociatedFrameQuality
SEX

@Info
Frage 6

@ValueMapping
[0]    männlich
          |-> {male}
[1]    weiblich
          |-> {female}
#


...


@SurveyVariable
Familienstand

@AssociatedFrameQuality
MARITAL STATUS

@Info
Frage 8

@ValueMapping
[0]    ledig
          |-> {single}
[1]    verheiratet
          |-> $MapGroup   verheiratet (A)
[2]    verwitwet
          |-> {widowed}
[3]    geschieden
          |-> {divorced}
#


...


@SurveyVariable
Teilnahme am Erwerbsleben / 1_Stunde

@AssociatedFrameQuality
ECONOMIC ACTIVITY STATUS

@Info
Frage 9
Anmerkung: "Kind im Vorschulalter" wird zu "Sonstige erhaltene
Person" gerechnet (diese Unterscheidung geht also verloren !)
Anmerkung: Frage 9 (Code [1]) wird gemeinsam mit Frage 20 abge-
bildet

@ValueMapping
[1]    beschäftigt
          |-> $NestedVariable  1_Stunde
[2]    arbeitslos (Arbeitsamt vorgemerkt)
          |-> $MapGroup   registrierter Arbeitsloser (A)
[3]    Pensionist, Rentner
          |-> {retired and receiving a pension}
[4]    Nicht berufstätige Hausfrau
```

.4 Labour Force Survey Mappings

```
                    |-> {housewife / domestic family responsibilities}
        [5]     Student, Schüler
                    |-> {in education/formation}
        [67]    Sonstige erhaltene Person
                    |-> {other supported person}
        [8]     Präsenzdiener, Karenzurlauberin
                    |-> $MapGroup   Präsenzdiener/Karenzurlauberin (A)
#

@SurveyVariable
Status im Erwerbsleben

@AssociatedFrameQuality
STATUS IN EMPLOYMENT

@Info
Frage 10
Anmerkung: in einem Vorverarbeitungsschritt werden die insgesamt
47 unterschiedenen Ausprägungen auf 11 Werte aggregiert und dann
wie angegeben 1:1 den Modalitäten von STATUS IN EMPLOYMENT
zugeordnet

@ValueMapping
[01]    Selbständiger ohne bezahlte Arbeitnehmer (Code 10)
            |-> own-account worker (self-employed without paid
                                                      employees)
[02]    Selbständiger mit kleinerem Betrieb (ggf. mit bis zu 9
              bezahlten Arbeitnehmern; Codes 1,2,3,11,12,20,21)
            |-> self-employed with paid employees (small
                                                independent)
[03]    Mithelfender Familienangehöriger
                                (Codes 5,6,7,15,16,17,18,25,26)
            |-> assisting relative (unpaid family worker)
[04]    Arbeitgeber (Code 13)
            |-> employer (many employees)
[05]    Arbeitnehmer (Arbeiter) im privatrechtl. Dienstverhältnis
          (einschl. privatrechtl. Dienstnehmer im öffentl. Dienst;
                                Codes 30,40,41,42,43,44,70,71)
            |-> employee - private sector (manual work)
[06]    Arbeitnehmer (Angestellter) im privatrechtl.
                                                 Dienstverhältnis
          (einschl. privatrechtl. Dienstnehmer im öffentl. Dienst;
                                Codes 31,50,51,52,53,72,73,74,75)
            |-> employee - private sector (clerical work)
[07]    Arbeitnehmer (Leitungsfkt.) im privatrechtl.
                    Dienstverhältnis (einschl. privatrechtl.
                    Dienstnehmer im öffentl. Dienst; Codes 54,55,76)
            |-> employee - private sector (managerial work)
[08]    Arbeitnehmer (Arbeiter) im öffentl.rechtl. Dienstverhältnis
                                                  (Codes 60,61)
            |-> employee - public sector (manual work)
[09]    Arbeitnehmer (Angestellter) im öffentl.rechtl.
                         Dienstverhältnis (Codes 62,63,64,65)
            |-> employee - public sector (clerical work)
[10]    Arbeitnehmer (Leitungsfkt.) im öffentl.rechtl.
                                Dienstverhältnis (Codes 66,67)
            |-> employee - public sector (managerial work)
[11]    Nie berufstätig gewesen (Code X)
            |-> unemployed person never employed before
#

...
```

```
@SurveyVariable
Arbeitssuche

@AssociatedFrameQuality
EMPLOYMENT SEEKING

@Info
Frage 15
"In den letzten vier Wochen aktiv einen Arbeitsplatz gesucht ?"

@ValueMapping
[0]    Arbeitsplatz
          |-> {seeking (dependent) employment}
[1]    Lehrstelle
          |-> {seeking apprenticeship}
[2]    Nein
          |-> $MapGroup  sucht nicht (A)
#


...


@SurveyVariable
1_Stunde

@AssociatedFrameQuality
ECONOMIC ACTIVITY STATUS

@ConditionVariable
TEILNAHME AM ERWERBSLEBEN

@ConditionValue
[1]    beschäftigt

@Info
Frage 20
Anmerkung: diese Frage wird mit Frage 9 (TEILNAHME AM ERWERBSLE-
BEN, Code [1]) gemeinsam ECONOMIC ACTIVITY STATUS zugeordnet;
außerdem werden die Ausprägungen "Bezahlt" und
"Mitgearbeitet" in einem Vorverarbeitungsschritt zusammengefaßt.

Person ist beschäftigt und hat "In der letzten Woche mind. 1
Stunde gegen Bezahlung gearbeitet od. im Familienbetrieb mitgear-
beitet ?"

@ValueMapping
[01]   ja        >> beschäftigt .. mind. 1 Stunde pro Woche
          |-> {work for pay and profit}
[2]    nein      >> beschäftigt .. nicht mind. 1 Stunde pro Woche
          |-> $MapGroup  Null Stunden (A)
```

The CLFS makes use of SoVar nestings either, for instance, in mapping both variables, »*Economic activity*« and »*Reason of current inactivity*«, to FrameQuality *economic activity status*:

```
@Survey
LABOUR FORCE SURVEY of the European Community (1983-91)
```

.4 Labour Force Survey Mappings

...

```
@SurveyVariable
Economic activity

@AssociatedFrameQuality
ECONOMIC ACTIVITY STATUS

@Info
Position 25
Note: this variable is mapped in conjunction with position 26
(code [2])

@ValueMapping
[1]    worked (incl. family workers)
           |-> {work for pay and profit}
[2]    did not work
           |-> $NestedVariable  Reason of current inactivity
[3]    did not work because of lay-off
           |-> $MapGroup  lay-off
[4]    military service (conscript)
           |-> {military or substitute service (conscripts)}
[5]    other person (over 13 years)
           |-> $MapGroup  other person (over 13 years) neither
                                                 employed nor active
#

@SurveyVariable
Reason of current inactivity

@AssociatedFrameQuality
ECONOMIC ACTIVITY STATUS

@ConditionVariable
Economic activity

@ConditionValue
[2]    did not work

@Info
Position 26

@ValueMapping
[0]    bad weather
              >> did not work because of bad weather
           |-> {not active because of bad weather}
[1]    slack work
              >> did not work because of slack work
           |-> $MapGroup  slack work/mechanical breakdown
[2]    labour dispute
              >> did not work because of labour dispute
           |-> {not active because of labour dispute}
[3]    education/vocational training
              >> did not work because of
                              education/vocational training
           |-> {not active because of education/formation}
[4]    illness, accident, temporary disability
              >> did not work because of illness, accident, or
                                            temporary disability
           |-> {not active because of accident of work or
                                       professional affection}
```

```
       [5]    maternity leave
                      >> did not work because of maternity leave
                |-> {not active because of maternity leave}
       [6]    vacation
                      >> did not work because of vacation
                |-> {not active because of vacation}
       [7]    new job starting later
                      >> did not work because of new job starting later
                |-> {not active - starts new job later}
       [8]    other reason (personal, familiar responsibilities)
                      >> did not work because of other reasons
                |-> {not active - other reasons}
#

@SurveyVariable
Seeking Employment

@AssociatedFrameQuality
EMPLOYMENT SEEKING

@Info
Position 49

@ValueMapping
       [1]    seeking employment
                |-> $MapGroup  seek_work (EU)
       [2]    found employment starting later
                |-> $MapGroup  later_work (EU)
       [3]    not seeking because lay-off
                |-> {not seeking because lay-off}
       [4]    not seeking because does not expect to find work
                              or does not know where to look for work
                |-> {not seeking since having no hope to find job}
       [5]    not seeking because of other reasons
                |-> {not seeking for other reasons}
       [6]    not seeking without stating any reason
                |-> {not seeking without telling reason}
#
```

Unlike other questionnaires, the Portuguese LF survey, *Inquérito ao Emprego* breaks down the variables contributing to FRAMEQUality *economic activity status* into an extended hierarchy of nested binary questions listed after a couple of more elementary mappings:

```
@Survey
Inquérito ao Emprego (Portugal)

@Info
for the sake of comprehension, Portuguese texts are stated in
plain English; source of reference is Deliverable 1.3 "Labour
Force Survey" of INE (National Statistical Institute of Portugal)
of DOSES Project No B 41

   ...

@SurveyVariable
Marital Status
```

.4 Labour Force Survey Mappings

```
@AssociatedFrameQuality
MARITAL STATUS

@Info
Query 12

@ValueMapping
[1]     Single
        |-> {single}
[2]     Married or living together
        |-> $MapGroup   living together
[3]     Widower
        |-> {widowed}
[4]     Divorced or legally separated
        |-> $MapGroup   living separated
#

@SurveyVariable
Affinity with the representative of the family

@AssociatedFrameQuality
RELATION TO HEAD OF HOUSEHOLD

@Info
Query 13

@ValueMapping
[1]     Head of Household
        |-> {head of household}
[2]     Spouse of head of household
        |-> {spouse}
[3]     Child of head of household (or his/her spouse)
        |-> {child (of head of household or spouse)}
[4]     Ascendant of head of household (or his/her spouse)
        |-> $MapGroup   ascendant relative (EU)
[5]     Daughter in law or son-in-law from the family's
                            representative or from his consort
        |-> {daughter or son in law}
[6]     Other relative
        |-> {other relative of head of household or spouse}
[7]     Non relative
        |-> {other relation}
#

@SurveyVariable
Military service

@AssociatedFrameQuality
ECONOMIC ACTIVITY STATUS

@Info
Query 22
"Are you making the compulsory military service ?"
Note: this query is processed together with queries 23-29, 133
Note: this query are asked males over 15 and under 36 years
(which are not handicapped) only

@ValueMapping
[1]     Yes     >> Military/substitute service
        |-> {military or substitute service (conscripts)}
[2]     No
        |-> $NestedVariable   1_hour work
#
```

```
@SurveyVariable
1_hour work

@AssociatedFrameQuality
ECONOMIC ACTIVITY STATUS

@ConditionVariable
Military service

@ConditionValue
[2]    No

@Info
Query 23
"Did you work at least one hour last week, with the objective of a
salary or a benefit ?"
Note: this query is processed together with queries 22, 24-29, and
133

@ValueMapping
[1]    Yes      >> worked at least 1 hour for pay/profit
                |-> {work for pay and profit}
[2]    No
                |-> $NestedVariable  Job, but inactive
#

@SurveyVariable
Job, but inactive

@AssociatedFrameQuality
ECONOMIC ACTIVITY STATUS

@ConditionVariable
1_hour work

@ConditionValue
[2]    No

@Info
Query 24
(asked only in case of query 23, code [2])
"Though not having worked, is there any job you have been out tem-
porarily (holidays, for example, etc.) ?"
Note: this query is processed together with queries 22-23, 25-29,
and 133

@ValueMapping
[1]    Yes      >> currently inactive though having regular job
                |-> $MapGroup   active person currently not working (EU)
[2]    No
                |-> $NestedVariable  Contract interrupted
#

@SurveyVariable
Contract interrupted

@AssociatedFrameQuality
ECONOMIC ACTIVITY STATUS

@ConditionVariable
Job, but inactive
```

.4 LABOUR FORCE SURVEY MAPPINGS

```
@ConditionValue
[2]    No

@Info
Query 25
(asked only in case of query 24, code [2])
"Didn't you work because your contract was temporarily interrupted
?"
Note: this query is processed together with queries 22-24, 26-29,
and 133

@ValueMapping
[1]    Yes      >> work contract temporarily interrupted
           |-> $MapGroup   lay-off
[2]    No
           |-> $NestedVariable   Education
#

@SurveyVariable
Education

@AssociatedFrameQuality
ECONOMIC ACTIVITY STATUS

@ConditionVariable
Contract interrupted

@ConditionValue
[2]    No

@Info
Query 26
(asked only in case of query 25, code [2])
"Were you studying ?"
Note: this query is processed together with queries 22-25, 27-29,
and 133

@ValueMapping
[1]    Yes      >> studying
           |-> {in education/formation}
[2]    No
           |-> $NestedVariable   Domestic task
#

@SurveyVariable
Domestic task

@AssociatedFrameQuality
ECONOMIC ACTIVITY STATUS

@ConditionVariable
Education

@ConditionValue
[2]    No

@Info
Query 27
(asked only in case of query 26, code [2])
"Did you do domestic tasks at your own home ?"
Note: this query is processed together with queries 22-26, 28-29,
and 133
```

```
@ValueMapping
[1]    Yes      >> domestic task
                |-> {housewife / domestic family responsibilities}
[2]    No
                |-> $NestedVariable   Unemployed
#

@SurveyVariable
Unemployed

@AssociatedFrameQuality
ECONOMIC ACTIVITY STATUS

@ConditionVariable
Domestic task

@ConditionValue
[2]    No

@Info
Query 133
(asked in all cases, but relevant only in case of query 27,
code [2]; supposedly)
"Are you inscribed at an employment centre ?"
Note: this query is processed together with queries 22-29

@ValueMapping
[1]    Yes      >> inscribed at employment centre
                |-> $MapGroup   registered unemployed (P)
[2]    No
                |-> $NestedVariable   Support
#

@SurveyVariable
Support

@AssociatedFrameQuality
ECONOMIC ACTIVITY STATUS
t
@ConditionVariable
Unemployed

@ConditionValue
[2]    No

@Info
Query 29
"Have you some oldness, or survival periodical allowance or any
kind of pension ?"
Note: this query is processed together with queries 22-28, and 133

@ValueMapping
[1]    Yes
                |-> $NestedVariable   Retired
[2]    No       >> no financial support
                |-> {other not supported person}
#

@SurveyVariable
Retired

@AssociatedFrameQuality
ECONOMIC ACTIVITY STATUS
```

```
@ConditionVariable
Support

@ConditionValue
[1]    Yes

@Info
Query 28
(asked only in case of query 27, code [1], supposedly; further-
more, persons not handicapped and over 35 years are asked only)
"Are you a pensioner retired from any activity ?"
Note: this query is processed together with queries 22-27, 29, and
133

@ValueMapping
[1]    Yes      >> pensioner/retired
               |-> {retired and receiving a pension}
[2]    No       >> financially supported
               |-> {other supported person}
#

@SurveyVariable
Status in employment (pre-processed auxiliary)

@AssociatedFrameQuality
STATUS IN EMPLOYMENT

@Info
auxiliary variable resulting of pre-processing of queries 33 to
36a in Portuguese Labour Force survey

@ValueMapping
[331]  self-employed with employees
            |-> $MapGroup   employer
[341]  self-employed without employees
            |-> {own-account worker (self-employed without paid
                                                    employees)}
[351]  employee
            |-> $MapGroup   employee (P)
[361]  unpaid family worker
            |-> {assisting relative (unpaid family worker)}
[001]  cooperative member
            |-> {member of producers' cooperative}
[002]  other
            |-> $MapGroup   other (P)
#
```

C.5 Sample Request Processing Minutes

The following transcript excerpts are assembled from log files of DÖS'chen client and server processes executing a REQUEST submitted as outlined in Section 2.7. To this end, such a REQUEST needs to be composed using DÖS'chen's *AGen* (statistical Aggregate Generator interface) first. REQUEST composition is initiated by selecting a (generic) CONDEF; referring to **Fig. 2.7–8**, 'Erwerbstätige (A/87)' is chosen resulting in the log file transcript stated below:

```
@ConceptToStartWith
Erwerbstätige (A/87)

@ConceptInfo
Persons in employment according to the definition used in Austrian
Mikrozensus as of 1987

@ConceptValue
AGE at [ half-decades ]:
   {15-19,20-24,25-29,30-34,35-39,40-44,45-49,50-54,
    55-59,60-64,65-69,70-74,75 and over}
ECONOMIC ACTIVITY STATUS at []:
   {work for pay or profit}
USUAL WORKING TIME at [ *6* ]:
   {1-12 hours,13-20 hours,21-30 hours,31-40 hours, over 40 hours}
#
AGE at [ half-decades ]:
   {15-19,20-24,25-29,30-34,35-39,40-44,45-49,50-54,
    55-59,60-64,65-69,70-74,75 and over}
ECONOMIC ACTIVITY STATUS at []:
   {not active because of bad weather,
    not active because of slack work,
    not active because of mechanical breakdown,
    not active because of labour dispute,
    not active because of education/formation,
    not active because of accident of work or professional
                                                affection,
    not active because of vacation,
    not active - starts new job later,
    not active - other reasons}
#
```

Note that, since 'Erwerbstätige (A/87)' is a sub-concept (cf. Section C.3) of 'Erwerbspersonen (A/87)', @ConceptValue in fact represents the composed (viz., intersected) CONDEF. Converting this generic CONDEF internally into a BOX structure yields a TEXTure (comprised in a ConceptContainment paragraph) as well as an EDITS component:

.5 Sample Request Processing Minutes

```
@ConceptContainment
AGE at [ half-decades]:
  {15-19,20-24,25-29,30-34,35-39,40-44,45-49,50-54,55-59,60-64,
   65-69,70-74,75 and over}
ECONOMIC ACTIVITY STATUS at []:
  {work for pay and profit,
   not active because of bad weather,
   not active because of slack work,
   not active because of mechanical breakdown,
   not active because of labour dispute,
   not active because of education/formation,
   not active because of accident of work or professional
                                               affection,
   not active because of vacation,
   not active - starts new job later,
   not active - other reasons}
#

@ConceptEdits
ECONOMIC ACTIVITY STATUS at []:
  {work for pay and profit}
USUAL WORKING TIME at [ *6* ]:
  {varying hours}
#
```

Restricting now the generic CONDEF (cf. **Fig. 2.7–9**) yields a further log file entry:

```
@ConceptRestriction
SPACE at [ NUTS  I ]:
  {Austria-East/RD 1}
TIME at []:
  {09-90}
SEX at []:
  {female}
MARITAL STATUS at []:
  {married,divorced,separated}
#
```

Adding a breakdown (cf. **Fig. 2.7–10**) leads to a grouped (generic) TARGET (**Def. 2.4.1–9**) as shown in the DÖS'chen screenshot of the *AGen* in **Fig. 2.7–11** just before submitting the composed TARGET to the evaluation stage. Internally, the log file records the TARGET box indicating both, the original as well as the demanded and attained AGGLEvel of MODality groupings (the latter are stated after '>>'; a '*' indicates that all MODalities listed go in one group, turning this AXis into a reference dimension):

```
@Target
SPACE at [ NUTS  I ] >> *:
  {Hollabrunn,Mistelbach,Gmünd,Waidhofen/Thaya,Zwettl,Horn,Krems,
   Melk,Amstetten,Waidhofen/Ybbs,Scheibbs,St. Pölten,Baden,
   Lilienfeld,Neunkirchen,Wr. Neustadt,Tulln,Korneuburg,
   Wien-Umgebung,Mödling,Bruck an der Leitha,Gänserndorf,Wien,
```

```
          Eisenstadt,Güssing,Jennersdorf,Mattersburg,Neusiedl am See,
             Oberpullendorf,Oberwart} #
       TIME at [] >> *:
          {09-90} #
       AGE at [ half-decades ] >> [ 7 phases ]:
          {15}   {16,17,18,19}
          {20,21,22,23,24,25,26,27,28,29,30,31,32,33,34,35,36,37,38,39,
             40,41,42,43,44,45,46,47,48,49,50,51,52,53,54,55,56,57,58,59,
             60}
          {61,62,63,64,65}   {66,67,68,69,70,71,72,73,74,75 and over} #
       SEX at [] >> *:
          {female} #
       MARITAL STATUS at [] >> *:
          {married,divorced,separated} #
       EDUCATION LEVEL at [] >> [ Austria-4 ]:
          {second level (first stage - general education/
                                                    secondary school)}
          {second level (second stage - training in dual
                                      school/establishment mode),
             second level (second stage - training in school
                                           with specific subject)}
          {second level (third stage - general intermediary level),
             second level (third stage - intermediary level
                                           with specific subject)}
          {third level (non-university),third level (university),
             third level (post-graduate)} #
       ECONOMIC ACTIVITY STATUS at [] >> *:
          {work for pay or profit,not active because of bad weather,not
             active because of slack work,not active because of mechanical
             breakdown,not active because of labour dispute,not active
             because of education/formation,not active because of accident
             of work or professional affection,not active because of
             vacation,not active - starts new job later,
             not active - other reasons} #
       STATUS IN EMPLOYMENT at [] >> [ Austria-3 ]:
          {own-account worker (self-employed without paid employees),
             self-employed with paid employees (small independent),employer
             (many employees)}
          {employee - private sector (manual work),employee - private
             sector (clerical work),employee - private sector (managerial
             work),employee - public sector (manual work),employee - public
             sector (clerical work), employee - public sector (managerial)}
          {assisting relative (unpaid family worker)} #
```

At this point, the client process running the REQUEST makes a round of calls to all active servers of the network (the testing installation comprised two such servers — called "pools" — only). In the following transcript excerpt related to pool 'AUSTRIA', the first paragraphs record server start-up data; the @Client-Request paragraph registers that a connection has been established to a client and the present pool containment sent to it. The actual communication between client and server regarding the REQUEST under consideration is recorded in the @DialogSession paragraph coming next; in the specific example chosen, three of the four available candidate data sets are rejected — only the data set named 'A0990' is kept and, upon request, shipped to the client as recorded in the @SendRequest paragraph. After this action, the client terminates the server link which prompts the server to hang up (@DialogShutdown paragraph).

.5 Sample Request Processing Minutes

```
@PoolStartup
... this is Pool **AUSTRIA**
    > IP-Address:   143.130.16.21
    > TablePath:    <SOURCE>PubData>BaseTable>*.lmd
    > CostRatings (local): sel=1.1, agg=2.2, grp=3.3
... OK
@LoadingTables
... **A0390**: loading ... OK, indexing ... OK
... **A0690**: loading ... OK, indexing ... OK
... **A0990**: loading ... OK, indexing ... OK
... **A1290**: loading ... OK, indexing ... OK
    > found -4- BaseTables
@CreatePoolContainment
... OK; PoolServer up & listening ...

@ClientRequest
... request from  143.130.16.35  via port 555
    > transmitting containment to client   ... OK


@DialogSession
... starting dialog no.  1
         with client: 143.130.16.35  via port 555
... receiving CostRating: central (sel=1.1,agg=1.7,grp=2.3)
... receiving SearchMode: (MultComb,AllowAdapt,AllowRestrict)
... starting PoolSearch
    > BaseTable  **A0390**: disjoint; dropped !
    > BaseTable  **A0690**: disjoint; dropped !
    > BaseTable  **A0990**: restricting target ...
                 hit (modified) ... at cost:   30.5;  --- kept ---
    > BaseTable  **A1290**: disjoint; dropped !
... found:   0 FullHit(s);   1 ModifiedHit(s);   0 PartialHit(s)
... transmit response to client   ... OK


@SendRequest
... dialog session no.  1
         with client: 143.130.16.35  via port 555
... client SendRequest for BaseTable: **A0990**
    > BaseTable **A0990** preparing & shipping  ... OK

@DialogShutdown
... dialog session no.  1
         with client: 143.130.16.35  via port 555
... closing session   ... OK
```

Before exhibiting this dialog viewed from the client side, some parts of the pool containment delivered to the client by pool 'AUSTRIA' are reproduced in the following transcript:

```
@PoolContainment
-- Pool: **AUSTRIA**              -- Frame: **PERSON**

@PoolDimensions
SPACE
     {Wien,Bregenz,Bludenz,Dornbirn,Feldkirch,Eisenstadt,Güssing,
```

```
          Jennersdorf,Mattersburg,Neusiedl am See,Oberpullendorf,
    ...
    .....Krems,Melk,Amstetten,Waidhofen/Ybbs,Scheibbs,
          St. Pölten,Baden,Lilienfeld,Neunkirchen,
          Wr. Neustadt,Tulln,Korneuburg,Wien-Umgebung,Mödling,
          Bruck an der Leitha,Gänserndorf}
TIME
          {03-90,06-90,09-90,12-90}
AGE
          {0,1,2,3,4,5,6,7,8,9,10,11,12,13,14,15,16,17,18,19,20,21,22,
    ...
          60,61,62,63,64,65,66,67,68,69,70,71,72,73,74,75 and over}
SEX
          {male,female}
MARITAL STATUS
          {single,married,separated,widowed,divorced}

    ...

ECONOMIC ACTIVITY STATUS
          {work for pay or profit,not active because of bad weather,
          not active because of slack work,not active because of
          mechanical breakdown,not active because of labour dispute,
          not active because of education/formation,not active because
          of accident of work or professional affection,not active
          because of vacation,not active - starts new job later,not
          active - other reasons,not active because of maternity
          leave, military or substitute service conscripts),
          registered at employment exchange,(registered) unemployed
          participating in employment promotion scheme,
          registered lay-off,retired and receiving a pension,
          housewife / domestic family responsibilities,
          in education/formation,other supported person}

    ...

EMPLOYMENT SEEKING
          {seeking (dependent) employment,seeking apprenticeship,not
          seeking - job already found,not seeking - apprenticeship
          already found,not seeking because lay-off,not seeking since
          having no hope to find a job,not seeking for other reasons,
          not seeking without telling reason}

    ...

@EndPoolContainment
```

At client side, after requesting the pool containments from all active servers reachable in the network, the received pool containments are checked, one by one, against the TARGET of the current REQUEST; in the testing installation the client detects two pools ('AUSTRIA' and 'GERMANY') but drops the latter since its pool containment does not overlap the TARGET; this is registered in the transcript's @CheckTargetCoverage paragraph.

Next, the client initiates the dialogue with server 'AUSTRIA' in the @OpenServerDialog paragraph and transmits the search configuration to the attached pool in the @SendingSearchRequest paragraph. This inquiry's result is recorded in the following @PoolResponses paragraph; next to each FRAMEQual-

.5 Sample Request Processing Minutes

ity stated the operation types necessary for transforming source data into TARGET format are indicated (`sel`=selection/restriction, `grp`=re-grouping, `agg`=marginalizing).

The `@ResponseTableRequest` paragraph registers the 'A0990' data set fetched successfully. This finishes the dialogue sessions with all servers involved.

```
@ContainmentRequestToPools
... from Pool **AUSTRIA**:   received
... from Pool **GERMANY**:   received
    > found -2- Pools
@CheckPoolContainments
... PoolGroup   1 : [ **AUSTRIA**   ]
... PoolGroup   2 : [ **GERMANY**   ]
    > arranged -2- PoolGroups
@CheckTargetCoverage
... PoolGroup   1 :
    > **AUSTRIA**: target restriction ...
      full cover; Pool kept
... PoolGroup   2 :
    > **GERMANY**: disjoint; Pool dropped
#
@RelevantPools
... PoolGroups  1 : [ **AUSTRIA**   ]
#
@OpenServerDialogs
... PoolGroup   1
    > Pool **AUSTRIA**
         IP-Address 143.130.16.21 (port 555)   ... OK
#
@SendingSearchRequest
... Config [CostRating: central (sel=1.1,agg=1.7,grp=2.3);
           SearchMode: (MultComb,AllowAdapt,AllowRestrict)]
    > transmit to Pool **AUSTRIA**   ... OK
#

@PoolResponses
... Pool **AUSTRIA**: hit(s) covering target found
    > BaseTable: **A0990** at cost:  6.6 + 17.0 + 6.9 = 30.5
      [ SPACE: sel,agg; AGE: sel,grp; SEX: sel;
        MARITAL STATUS: sel,agg; NATIONALITY: agg;
        EDUCATION LEVEL: grp; RELATION TO HEAD OF HOUSEHOLD: agg;
        ECONOMIC ACTIVITY STATUS: sel,agg;
        STATUS IN EMPLOYMENT: sel,grp; USUAL WORKING TIME: agg;
        ACTUAL WORKING TIME: agg; EMPLOYMENT SEEKING: agg;
        AVAILABILITY (TIME HORIZON): agg;
        DURATION OF JOB SEARCH: agg ]
#

@ResponseTableRequest
... from Pool **AUSTRIA**: BaseTable **A0990**
    > mode: processed   ... OK
#
@ClosePoolDialogs
... with Pool **AUSTRIA**   ... OK
#
```

The containment of the work table derived in this REQUEST example is shown partly in a final transcript excerpt comprising particularly the FRAMEQUalities for which cross-classifications have been demanded, viz. *age, education level*, and *status in employment*.

```
@OutputWorkTable
SPACE
  {Hollabrunn,Mistelbach,Gmünd,Waidhofen/Thaya,Zwettl,Horn,Krems,
   Melk,Amstetten,Waidhofen/Ybbs,Scheibbs,St. Pölten,Baden,
   Lilienfeld,Neunkirchen,Wr. Neustadt,Tulln,Korneuburg,
   Wien-Umgebung,Mödling,Bruck an der Leitha,Gänserndorf,Wien,
   Eisenstadt,Güssing,Jennersdorf,Mattersburg,Neusiedl am See,
   Oberpullendorf,Oberwart} #
TIME
  {09-90} #
AGE
  {15}   {16,17,18,19}
  {20,21,22,23,24,25,26,27,28,29,30,31,32,33,34,35,36,37,38,39,
   40,41,42,43,44,45,46,47,48,49,50,51,52,53,54,55,56,57,58,
   59,60}
  {61,62,63,64,65}   {66,67,68,69,70,71,72,73,74,75 and over} #
SEX
  {female} #
MARITAL STATUS
  {married,divorced,separated} #
EDUCATION LEVEL
  {second level (first stage -
        general education/secondary school)}
  {second level (second stage -
        training in dual school/establishment mode),
   second level (second stage -
        training in school with specific subject)}
  {second level (third stage - general intermediary level),
   second level (third stage -
        intermediary level with specific subject)}
  {third level (non-university),third level (university),
   third level (post-graduate)} #
ECONOMIC ACTIVITY STATUS
  {work for pay or profit,not active because of bad weather,
   not active because of slack work,not active because of
   mechanical breakdown,not active because of labour dispute,
   not active because of education/formation,
   not active because of accident of work or professional
   affection,not active because of vacation,not active - starts
   new job later,not active - other reasons} #
STATUS IN EMPLOYMENT
  {own-account worker (self-employed without paid employees),
   self-employed with paid employees (small independent),
   employer (many employees)}
  {employee - private sector (manual work),employee - private
   sector (clerical work),employee - private sector (managerial
   work),employee - public sector (manual work),employee - public
   sector (clerical work),employee - public sector (managerial)}
  {assisting relative (unpaid family worker)} #
```

LITERATURE

Ahn T.H. et al. (1990) Temporal Summary Table Management and Graphic Interface. In: Proc. *Statistical and Scientific Database Management* (5th SSDBM; Michalewicz Z., ed.), Berlin et al.: Springer (Lecture Notes in Computer Science 420), pp. 112–130.

Aho A.V., Hopcroft J.E., Ullman J.D. (1983) *Data Structures and Algorithms*. Reading, Ma. et al.: Addison-Wesley.

Aho A.V., **Ullman** J.D. (1992) *Foundations of Computer Science*. New York: Computer Science Press.

Aluja-Banet T. et al. (1993) The EDA/System – A System for the Production and Analysis of Summary Statistics. In: Proc. *Statistical Meta-Information Systems* (EUROSTAT, ed.), Luxembourg: Office for Official Publications, pp. 81–90.

Angelaccio M., Catarci T., Santuconei G. (1990) Query by Diagram: A Fully Visual Query System. *Journal of Visual Languages and Computing* **1**, pp. 255–273.

Appel G. (1993a) A Metadata Driven Statistical Information System. In: Proc. *Statistical Meta-Information Systems* (EUROSTAT, ed.), Luxembourg: Office for Official Publications, pp. 291–309.

—— (1993b) Zum Entwurf eines metadatenbasierten statistischen Informationssystems (in German). *Allg. Statistisches Archiv* **77**, pp. 68–91.

Archdale G. (1994) Non-European Initiatives and Systems. In: Proc. *Information and Communications Technologies in Tourism* (Proc. ENTER '94; Schertler W. et al., eds), Wien–New York: Springer, pp. 56–63.

Augendre H., **Hatabian** G. (1992) Inside ESIA: An Open and Self-Consistent Knowledge-based System. In: Proc. *COMPSTAT '92*, Vol. 2 (Dodge Y., Whittaker J., eds), Heidelberg–New York: Physica, pp. 33–38.

Austin W.J. et al. (1994) Processing Travel Queries in a Multimedia Information System. In: Proc. *Information and Communications Technologies in Tourism* (Proc. ENTER '94; Schertler W. et al., eds), Wien–New York: Springer, pp. 64–71.

Backus J. (1981) The History of Fortran I, II, III. In: Wexelblat R. (ed.) *History of Programming Languages*. New York–London: Academic Press, pp. 25–45.

Baron R.J. (1987) *The Cerebral Computer*. Hillsdale, N.J.–London: Lawrence Erlbaum Ass.

Barcaroli G., **Di Pace** L. (1992) The Automatic Generation of Statistical Incompatibility Rules from Entity-Relationship Schemes. In: Proc. *New Techniques and Technologies for Statistics* (Bonn 1992; EUROSTAT, ed.), Luxembourg: Office for Official Publications, pp. 226–236.

Bartunek E. (1994) Berechnung von Arbeitslosenquoten in Österreich (in German). *Österr. Zeitschrift für Statistik und Informatik* **23** (1), pp. 13–20.

—— (1995) Personal communication.
Basili C., **Meo-Evoli** L. (1992) A Deductive Query Processor for Statistical Databases. In: Proc. *Database and Expert System Applications* (*DEXA '92*; Tjoa A M., Ramos I., eds), Wien-New York: Springer, pp. 390–395.
  See also: —— (1992) StEM: A Deductive Query Processor for Statistical Databases. In: Computational Statistics, Vol. 2 (Proc. *COMPSTAT '92*; Dodge, Y. Whittaker, J., eds), Heidelberg: Physica, pp. 409–414.
Becker R.A., Chambers J.M., Wilks A. (1988) *The New S Language.* Belmont, Ca.: Wadsworth & Brooks (Cole Computer Science Series).
Becker R.A. (1994) A Brief History of S. In: Dirschedl P., Ostermann R. (eds) *Computational Statistics.* Heidelberg: Physica, pp. 81–110.
Beniger J.R. (1986) *The Control Revolution – Technological and Economic Origins of the Information Society.* Cambridge, Ma.–London: Harvard University Press.
Bethlehem J.G., **Hundepool** A.J. (1992) Integrated Statistical Information Processing on Microcomputers. In: Proc. *New Techniques and Technologies for Statistics* (Bonn 1992; EUROSTAT, ed.), Luxembourg: Office for Official Publications, pp. 7–17.
Bezenchek A., Massari F., Rafanelli M. (1994) STORM+: Statistical Data Storage and Manipulation System. In: Proc. *COMPSTAT '94* (Dutter R., Grossmann W., eds), Heidelberg: Physica, pp. 351–356.
Biffl G. (1994) Eine national und eine internationale Arbeitslosenquote: der Stein der Weisen ? (in German). *Österr. Zeitschrift für Statistik und Informatik* **23** (1), pp. 3–7.
Bischoff J.P. (1990) *Versuch einer Geschichte der Rechenmaschine* (in German). München: Systhema (reprint, Weiß St., ed.; orig. edn. Ansbach 1804).
Bishop Y.M.M., Fienberg S.E., Holland P.W. (1975) *Discrete Multivariate Analysis: Theory and Practice.* Cambridge, Ma.: MIT Press.
Bisdorff R. (1992) The Conceptual Model of a Documented Statistical Database. In: Proc. *New Techniques and Technologies for Statistics* (Bonn 1992; EUROSTAT, ed.), Luxembourg: Office for Official Publications, pp. 310–319.
Blanpain R., **Sadowski** D. (1994) *Habe ich morgen noch einen Job? – Die Zukunft der Arbeit in Europa* (in German). München: Beck.
Blencke R. (1994) Standard Specifications as the Basis for Salesoriented Tourism. In: Proc. *Information and Communications Technologies in Tourism* (Proc. *ENTER '94*; Schertler W. *et al.*, eds), Wien–New York: Springer, pp. 46–49.
Bolter J.D. (1984) *Turing's Man – Western Culture in the Computer Age.* Chapel Hill: The University of North Carolina Press.
Bourgine P. (1989) The Role of Intentionality in the Design and Consultation of Statistical Data. In: Proc. *Development of Statistical Expert Systems* (Luxembourg 1987, EUROSTAT, ed.), Luxembourg: Office for Official Publications, pp. 202–209.
Brachman R.J., **Schmolze** J.G. (1985) An Overview of the KL-ONE Knowledge Representation System. *Cognitive Science* **9** (2), pp. 171–216.

Bretherton F.P. (1994) A Reference Model for Metadata – A Strawman. Report, University of Wisconsin, 17pp. *See also:* Bretherton F.P., Singley P.T. (1994) Metadata: A User's View. In: Proc. *Statistical and Scientific Database Management* (7th SSDBM; French J.C., Hinterberger H., eds), Los Alamitos, Ca.: IEEE Computer Society Press, pp. 166–174.

Brillouin L. (1962) *Science and Information Theory.* New York: Academic Press, 2nd. edn.

Bukhres O., **Elmagarmid** A., eds (1996) *Object-Oriented Multidatabase Systems – A Solution for Advanced Applications.* Englewood Cliffs, N.J.: Prentice-Hall.

Burroughs P.A. (1989) *Principles of Geographic Information Systems.* Oxford: Science Publications.

Byerley P.F. *et al.* (1995) A European Electronic Market Place for Tourism. In: Proc. *Information and Communications Technologies in Tourism* (Proc. *ENTER '95*; Schertler W. *et al.*, eds), Wien–New York: Springer, pp. 22–32.

Catarci T., D'Angiolini G., Lenzerini M. (1990) A Structured Language for Modelling Statistical Data. In: Proc. *COMPSTAT '90* (Momirovic K., Mildner V., eds), Heidelberg–New York: Physica, pp. 237–242.

Catarci T., **Santucci** G. (1990) GRASP: A Graphical System for Statistical Databases. In: Proc. *Statistical and Scientific Database Management* (5th SSDBM; Michalewicz Z., ed.), Berlin *et al.*: Springer (Lecture Notes in Computer Science 420), pp. 148–162.

Chamberlin D.D. *et al.* (1976) SEQUEL 2: A Unified Approach to Data Definition, Manipulation, and Control. IBM *J. Research and Development* **20** (6), pp. 560–575.

Chan P., **Shoshani** A. (1981) SUBJECT: A Directory Driven System for Organizing and Accessing Large Statistical Databases. In: Proc. 7th International Conference on *Very Large Data Bases* (VLDB), pp. 553–563.

Chen P.P. (1976) The Entity-Relationship Model – Toward a Unified View of Data. ACM *Transactions on Database Systems* **1** (1), pp. 9–36.

Chen M.C., McNamee L., Melkanoff M. (1989) A Model of Summary Data and its Applications in Statistical Databases. In: Proc. *Statistical and Scientific Database Management* (4th SSDBM; Rafanelli M. *et al.*, eds), Berlin *et al.*: Springer (Lecture Notes in Computer Science 339), pp. 356–372.

Chen L.T. *et al.* (1995) Efficient Organization and Access of Multi-Dimensional Datasets on Tertiary Storage Systems. *Information Systems* **20** (2), pp. 155–183.

Codd E.F. (1970) A Relational Model for Large Shared Data Banks. *Comm.* ACM, **13** (6), pp. 377–387.

Costa P., Marozza F., Vinciguerra F. (1986) Italian Statistical Data Dictionary System Prototype with a 4th Generation Language. In: Proc. *COMPSTAT '86* (De Antoni F., Lauro N., Rizzi A., eds), Heidelberg: Physica, pp. 425–430.

Cowley P.J., **Whiting** M.A. (1985) Managing Data Analysis through Save-States. In: Proc. *Computer Science and Statistics: The Interface* (Allen D.M., ed.), Amsterdam *et al.*: North Holland, pp. 121–127.

Creecy M.H. (1992) Massively Parallel Computing and Automated Industry and Occupation Coding. In: Proc. *New Techniques and Technologies for Statistics*

(Bonn 1992; EUROSTAT, ed.), Luxembourg: Office for Official Publications, pp. 141–149.

Cubbitt R. (1990) Metadata in Statistics: The Problem for Producers. In: Proc. *Workshop on Expert Systems and Artificial Intelligence: The Need for Information about Data* (Fessey M.C., ed.), London: The Library Association, pp. 7–13.

D'Angiolini G. (1992) A Knowledge-Based Approach to Statistical Information Modeling. In: Proc. *New Techniques and Technologies for Statistics* (Bonn 1992; EUROSTAT, ed.), Luxembourg: Office for Official Publications, pp. 304–309.

Darius P.L. (1986) Building Expert Systems with the Help of Existing Statistical Software: An Example. In: Proc. *COMPSTAT '86* (DeAntoni F., Rizzi A., Lauro N., eds), Heidelberg–New York: Physica, pp. 277–282.

Darius P.L. et al. (1993) Modelling Metadata. *Statistical Journal of the United Nations Commission for Europe* **10** (2), pp. 171–179.

Daruwala A. et al. (1995) The Context Interchange Network Prototype. Report, Sloan School of Management, Cambridge, Ma., 26pp.

Daser S. (1994) The Role of Information Technology in Global Marketing: The Case of the New Single Market of European Community. In: Deans P.C., Karwan K.R. (eds) *Global Information Systems and Technology*. Harrisburg–London: Idea Group Publ., pp. 85–101.

Date C.J. (1986) *An Introduction to Database Systems*. Reading, Ma. et al.: Addison-Wesley.

D'Atri A., **Ricci** F.L. (1989) Interpretation of Statistical Queries to Relational Databases. In: Proc. *Statistical and Scientific Database Management* (4th SSDBM; Rafanelli M. et al., eds), Berlin et al.: Springer (Lecture Notes in Computer Science 339), pp. 246–258.

D'Aubigny G. et al. (1992) ATIIS: A Computer-Assisted Statistical Engineering Environment. In: Proc. *New Techniques and Technologies for Statistics* (Bonn 1992; EUROSTAT, ed.), Luxembourg: Office for Official Publications, pp. 117–125.

D'Aubigny G. et al. (1993) The Modelling and Management of Meta-Information for Statistical Investigation Processes. In: Proc. *Statistical Meta-Information Systems* (EUROSTAT, ed.), Luxembourg: Office for Official Publications, pp. 221–232.

Deans P.C., **Karwan** K.R. (1994) *Global Information Systems and Technology: Focus on the Organization and Its Functional Areas*. Harrisburg–London: Idea Group Publ.

de Feber Ed., **de Greef** P. (1992) Toward a Formalised Meta-Data Concept. In: Proc. *COMPSTAT '92*, Vol. 2 (Dodge Y., Whittaker J., eds), Heidelberg–New York: Physica, pp. 351–356.

de Vaney Ch. et al. (1992) The Expert Interface to Statistical Information – Rationale, Techniques and Experiences. In: Proc. *New Techniques and Technologies for Statistics* (Bonn 1992; EUROSTAT, ed.), Luxembourg: Office for Official Publications, pp. 112–116.

—— (1993) Integrating Statistical Meta-Information and Methodologies. In: Proc. *Statistical Meta-Information Systems* (EUROSTAT, ed.), Luxembourg: Office for Official Publications, pp. 335–346.

Di Battista G., **Batini** C. (1988) Design of Statistical Databases: A Methodology for the Conceptual Step. *Information Systems* **13** (4), pp. 407–422.

DIN 55301 (1978) Gestaltung statistischer Tabellen (in German). *Deutsche Normen* (DIN) 55 301, Berlin–Köln: Beuth, 8 pp.

Dolby J.L., **Clark** N., **Rogers** W.H. (1986) The Language of Data: A General Theory of Data. In: Proc. *Computer Science and Statistics: The Interface* (Boardman Th.J., ed.), Washington, D.C.: American Statistical Society, pp. 96–103.

Dorda W., **Froeschl** K.A., **Grossmann** W. (1990) WAMASTEX – Heuristic Guidance for Statistical Analysis. In: Proc. *COMPSTAT '90* (Momirovic K., Mildner V., eds), Heidelberg–New York: Physica, pp. 93–98.

Dressler O. (1988) Assumption-Based Truth Maintenance. In: Proc. *Begründungsverwaltung* (Stoyan H., ed.), Berlin *et al.*: Springer (Informatik Fachberichte 162), pp. 63–85.

Drewett R., **Tanenbaum** E., **Taylor** M. (1989) Creating a Standardized and Integrated Knowledge-Based Expert System for the Documentation of Statistical Series. In: Proc. *Seminar on Development of Statistical Expert Systems in Computational Statistics* (EUROSTAT, ed.), Luxembourg: Office for Official Publications, pp. 178–189.

DuBois F.L., **Carmel** E. (1994) Information Technology and Leadtime Management in International Manufacturing Operations. In: Deans P.C., Karwan K.R. (eds) *Global Information Systems and Technology.* Harrisburg–London: Idea Group Publ., pp. 279–293.

Duncan J.W., **Shelton** W.C. (1992) U.S. Government Contributions to Probability Sampling and Statistical Analyis. *Statistical Science* **7** (3), pp. 320–338.

EC (1994) *Growth, Competitiveness, Employment – The Challenges and Ways Forward into the 21st Century* (White Paper of the European Commission). Luxembourg: Office for Official Publications.

EC-DG III (1994) *Information Technology Programme 1994–1998*, Work Programme, Edition 1994. Bruxelles: EC.

Eder Ch. *et al.* (1994) *Charles Babbage – Eine Geschichte aus der Geschichte des Computers* (in German). Pädagogische Schriftenreihe, Museum Arbeitswelt Steyr. Steyr (Austria): MAW.

Edlefson L., **Jones** S. (1986) *GAUSS – Programming Language Manual.* Kent: Aptech Systems Inc.

Ehrenberg A.S.C. (1975) *Data Reduction – Analysing and Interpreting Statistical Data.* London *et al.*: Wiley.

Eisenstein E.L. (1983) *The Printing Revolution in Early Modern Europe.* Cambridge *et al.*: University Press.

Ernst M., **Walpuski** D. (1994) Information Technologies and Tourism Markets. In: Proc. *Information and Communications Technologies in Tourism* (Proc. ENTER '94; Schertler W. *et al.*, eds), Wien–New York: Springer, pp. 228–235.

EUROSTAT (1988) *Long-Term Unemployment – Its Wider Labour Market Effects in the Countries of the European Community.* Luxembourg: Office for Official Publications.

—— (1989a) *Employment Statistics – Methods and Definitions 1988.* Luxembourg: Office for Official Publications.

—— (1989b) *Employment and Unemployment 1989.* Luxembourg: Office for Official Publications.

—— (1990) *Labour Force Survey – User's Guide* (German version). Luxembourg: Office for Official Publications.

—— (1991) *Labour Force Survey – Results 1989.* Luxembourg: Office for Official Publications.

—— (1992) *Labour Force Survey – Methods and Definitions, 1992 Series.* Luxembourg: Office for Official Publications.

—— (1993a) *Labour Force Survey 1983–1991.* Luxembourg: Office for Official Publications.

—— (1993b) *Employment and Unemployment 1980–1991.* Luxembourg: Office for Official Publications.

—— (1993c) *Development of Statistical Expert Systems – Information Package* (July 1993). Luxembourg: Office for Official Publications.

—— (1993d) *Tourism 1991 – Annual Statistics.* Luxembourg: Office for Official Publications.

—— (1994) *European Official Statistics – A Guide to Databases.* Luxembourg: Office for Official Publications.

—— (1995a) *Tourism in Europe.* Luxembourg: Office for Official Publications.

—— (1995b) Announcement of Publication of "The Globalisation Newsletter". *The Panelist* **1**, p. 13. Luxembourg: Office for Official Publications.

Everest G.C. (1986) *Database Management: Objectives, System Functions, and Administration.* New York: McGraw-Hill.

Fahrngruber P. (1993) *Strukturelle Analyse von Tabellen in der personenbezogenen Arbeitsmarktstatistik* (in German). Diploma Thesis, Inst. f. Statistik, Univ. Wien.

Falcitelli G. *et al.* (1989) The MEFISTO* Model: An Object Oriented Representation for Statistical Data Management. In: Proc. *Data Analysis and Learning Symbolic and Numeric Knowledge* (Diday E., ed.), New York–Budapest: Nova Science, pp. 455–463.

Felligi I.P., **Holt** D. (1976) A Systematic Approach to Automatic Edit and Imputation. *JASA* **71**, pp. 17–35.

Fessey M.C. (1989) Feedstocks for Statistical Expert Systems. In: Proc. *Development of Statistical Expert Systems* (Luxembourg 1987; EUROSTAT, ed.), Luxembourg: Office for Official Publications, pp. 169–177.

—— (1993) Comment on the papers of K.A. Froeschl, "Towards an Operative View of Semantic Metadata" and G. Appel, "A Metadata Driven Statistical Information System". In: Proc. *Statistical Meta-Information Systems* (EUROSTAT, ed.), Luxembourg: Office for Official Publications, pp. 317–320.

Fisher R.A. (1935) *The Design of Experiments*. Edinburgh–London: Oliver & Boyd.

Flaschberger L. (1994) Erläuterungen zur Ermittlung der Arbeitslosenquote nach OECD-Standard (in German). *Österr. Zeitschrift für Statistik und Informatik* **23** (1), pp. 8–12.

Flichy P. (1991) *Une histoire de la communication moderne* (in French). Paris: Édition La Découverte.

Fortunato E., Rafanelli M., Ricci F.L. (1987) The Statistical Functional Model for the Logical Representation of a Statistical Table. Technical Report 11/87, C.N.R./ISRDS, Roma, Italy.

Foucault M. (1966) *Les mots et les choses* (in French). Paris: Editions Gallimard.

French J.C. (1991) Support for Scientific Database Management. In: Michalewicz Z. (ed.) *Statistical and Scientific Databases*. Chichester: Ellis Horwood, pp. 51–82.

Froeschl K.A. (1989) *Mechanisierte Statistik: Numerische Algorithmen und formale Strategien* (in German). Dissertation thesis, Inst. f. Statistik und Informatik, Univ. Wien, 261pp.

—— (1990) Automated Protocolling of Statistical Data Analysis. In: Faulbaum F., Haux R., Jöckel K.-H. (eds) *Fortschritte der Statistik–Software 2* (Proc. SoftStat '89). Stuttgart–New York: G. Fischer, pp. 308–315.

—— (1992a) Functional Design of a Statistical Transaction Platform. In: Proc. *New Techniques and Technologies for Statistics* (Bonn 1992; EUROSTAT, ed.), Luxembourg: Office for Official Publications, pp. 71–79.

—— (1992b) Semantic Metadata: Query Processing and Data Aggregation. In: Proc. *COMPSTAT '92*, Vol. 2 (Dodge Y., Whittaker J., eds), Heidelberg–New York: Physica, pp. 357–362.

—— (1992c) Formalizing Statistical Analysis: Approaches and Prospects. In: Pichler F., Moreno-Diaz R. (eds) *Computer Aided Systems Theory – EuroCast '91*.Berlin et al.: Springer, pp. 225–238.

—— (1993) Towards an Operative View of Semantic Metadata. *Statistical Journal of the United Nations Economic Commission for Europe* **10** (2), pp. 181–194.

—— (1995) A Formal Model Evaluation Approach to the Analysis of Treatment Effects in Paired Sample Data. *Computational Statistics & Data Analysis* **19**, pp. 493–517.

—— (1996) A Metadata Approach to Statistical Query Processing. *Statistics & Computing* **6**, pp. 11–29.

Froeschl K.A., Dorda, W., Grossmann W. (1992) Analyzing Treatment Effects – The WAMASTEX Approach to Paired Sample Data. In: Proc. *COMPSTAT '92*, Vol. 2 (Dodge Y., Whittaker J., eds), Heidelberg–New York: Physica, pp. 337–342.

Froeschl K.A., Mattl S., Werthner H. (1993) *Symbolverarbeitende Maschinen – eine Archäologie* (in German). Exhibition catalogue, with contributions of A. Adam und F. Pichler. Steyr: Verein Museum Arbeitswelt.

Froeschl K.A., Fetz L., Hennrich W. (1996) A Metadata Based Client-Server Approach to the Integration of Distributed Statistical Databases. In: Faulbaum

F. (ed.) *Advances of Statistical Software 5* (Proc. *SoftStat '95*). Stuttgart: Lucius & Lucius, pp. 259–266.

Froeschl K.A., **Grossmann** W. (1988) Statistical Structures for Analyzing Time-Dependent Observations. In: Gaul W., Schader M. (eds) *Data, Expert Knowledge, and Decisions*. Berlin: Springer, pp. 145–160.

Froeschl K.A., **Werthner** H. (1994) Die Konzeption von Tourismus-Informationssystemen (in German). In: Schertler W. (ed.) *Tourimus als Informationsgeschäft*. Wien: Ueberreuter, pp. 257–305.

Fürst H. (1993) Europäische Ansprüche an Arbeitskräfte-Erhebungen (in German). In: Proc. *25 Jahre österreichischer Mikrozensus* (ÖSTAT, ed.), Wien: Österr. Statistisches Zentralamt (ÖSTAT), pp. 35–41.

Gallaire H., Minker J. Nicolas, J.M. (1984) Logic and Databases: A Deductive Approach. ACM *Computing Surveys* **16** (2), pp. 153–185.

Gallaire H., **Minker** J. (1978) *Logic and Data Bases*. New York: Plenum Press.

Garey M., **Johnson** D. (1979) *Computers and Intractability*. San Francisco: Freeman.

Georgeff M.P., Lansky A.L., Bessiere P. (1985) A Procedural Logic. In: Proc. *Int. Joint Conference on Artificial Intelligence*, pp. 516–523.

Georgeff M.P., **Lansky** A.L. (1987) Reactive Reasoning and Planning. In: Proc. *AAAI-87*, pp. 677–682.

GESMES (1993) *Gesmes 93 – Exchange of Multidimensional Statistical Arrays and Time-Series Data* (Vol. 1: Guidance to Users, Vol. 2: Reference Guide). Luxembourg: Office for Official Publications, 180pp. + 180pp.

—— (1995) *GESMES/ECOSER User Guide. Time-Series Subset of the Generic Statistical Edifact Message*. Luxembourg: Office for Official Publications, 238pp.

Ghosh S.P. (1986) Statistical Relational Tables for Statistical Database Management. IEEE *Transactions on Software Engineering* **12** (12), pp. 1106–1116.

—— (1988) Statistics Metadata. In: Kotz S., Johnson N.L. (eds) *Encyclopedia of Statistical Sciences*, Vol. 8. New York *et al.*: Wiley, pp. 743–746.

—— (1991) Statistical Relational Model. In: Michalewicz Z. (ed.) *Statistical and Scientific Databases*. Chichester: Ellis Horwood, pp. 267–305.

Glitza B. (1994) *GENESIS Fachkonzept*. Projektbericht, Wiesbaden (FRG), January 1994, 60pp. and Annex, 28pp.

Goguen J.A., **Meseguer** J. (1987) Unifying Functional, Object-Oriented and Relational Programming with Logical Semantics. In: Shriver B., Wegner P. (eds) *Research Directions in Object-Oriented Programming*. Cambridge, Ma.: MIT Press, pp. 417–477.

Goldberg A., **Robson** D. (1983) *Smalltalk-80: The Language and Its Implementation*. Reading, Ma. *et al.*: Addison-Wesley.

Goldstine H.H. (1972) *The Computer From Pascal to von Neumann*. Princeton: University Press.

Gomes M., **Miranda** A. (1991) Labour Force Survey. DOSES Project No. B 41 Deliverable 1.3, National Statistical Institute of Portugal (December 1991), 114pp.

Graves H., **Manor** R. (1986) Intelligent Data Management. In: Proc. *Computer Science and Statistics: The Interface* (Boardman T.J., ed.), Washington D.C.: Amer. Stat. Assoc., pp. 104–109.

Graves R.B. (1992) ITF Reference Layer Proposal. Report Informatics Branch, Statistics Canada, September 1992, 18 pp. plus 12 figures.

Graves R.B. *et al.* (1993) Information Holdings within Statistics Canada: A Framework. Report Informatics Branch, Statistics Canada, November 1993, 27pp.

Grifoni P., Pisanelli D.M., Ricci F.L. (1993) A Survey on Statistical Data Modelling. In: Proc. *Statistical Meta-Information Systems* (EUROSTAT, ed.), Luxembourg: Office for Official Publications, pp. 321–334.

Grossmann W., **Froeschl** K.A. (1988) Formale Wissens- und Verfahrensmodellierung in der statistischen Datenverarbeitung (in German). In: Janko W. (ed.) *Statistik, Informatik und Ökonomie*. Berlin *et al.*: Springer, pp. 71–85.

—— (1992) Konzeptionelles Modell für Metadaten am Beispiel der Arbeitsmarktstatistik (in German). Technical Report SMC-115, Institut f. Statistik, Univ. Wien, 85pp.

—— (1994) Automatische Tabellengenerierung mittels Metadaten (in German). Project Report, Inst. f. Statistik, Univ. Wien, 160pp.

Habermann H., **Waith** P. (1992) The Future of EDI in the United States Statistical System. In: Proc. *New Techniques and Technologies for Statistics* (Bonn 1992; EUROSTAT, ed.), Luxembourg: Office for Official Publications, pp. 172–177.

Haines P. (1994) Destination Marketing Systems. In: Proc. *Information and Communications Technologies in Tourism* (Proc. ENTER '94; Schertler W. *et al.*, eds), Wien–New York: Springer, pp. 50–55.

Halley E. (1693) An Estimate of the Degrees of Mortality of Mankind, Drawn from Curious Tables of the Births and Funerals at the City of Breslaw, with an Attempt to Ascertain the Price of Annuities on Lives. *Philosophical Transactions*, London, pp. 596–610.

Hand D.J. (1993) Data, Metadata and Information. *Statistical Journal of the United Nations Economic Commission for Europe* **10** (2), pp. 143–151.

—— (1994) Statistical Expert Systems. *Chance* **7** (1), pp. 28–34.

Hansen M.H. (1987) Some History and Reminiscences on Survey Sampling. *Statistical Science* **2** (2), pp.180–190.

Hansen M.H., **Hurwitz** W.N. (1943) On the Theory of Sampling from Finite Populations. *Annals of Mathematical Statistics* **14**, pp. 333–362.

Harary F. (1969) *Graph Theory*. Reading, Ma.: Addison-Wesley.

Haslinger A. (1989) Der Einfluß der Rotation auf die Messung des Erwerbsstatus im Mikrozensus (in German). *Österr. Zeitschrift für Statstik und Informatik* **19** (3), pp. 318–324.

Haux R., ed. (1986) *Expert Systems in Statistics*. Stuttgart–New York: G. Fischer.

Henson M.C. (1987) *Elements of Functional Languages*. Oxford *et al.*: Blackwell.

Hodges A. (1983) *Alan Turing: The Enigma*. London: Burnett Books Ltd.

Hsiao D.K., Kamel M.N., Wu C. Th. (1990) The Federated Databases and System: A New Generation of Advanced Database Systems. In: Proc. *Database and Ex-*

pert System Applications (*DEXA '90*; Tjoa A M., Wagner R., eds), Wien-New York: Springer, pp. 186–190.

Hudec M. (1995) Personal communication.

Hutton F.E., **Graves** R.B. (1993) IBOSS: A Statistical Information System for Statistics Canada. Report Informatics Branch, Statistics Canada, April 1993, 39pp.

Hyman A. (1982) *Charles Babbage, Pioneer of the Computer.* Oxford: University Press.

ILO (1976) *International Recommendations on Labour Statistics.* Geneva: International Labour Office.

Ireland C.T., **Kullback** S. (1968) Contingency Tables with Given Marginals. *Biometrika* **55** (1), pp. 179–188.

ISO (1989) ISO-ANSI Database Language SQL2 and SQL3, ANSI X3H2-89-110, ISO DBL CAN-3 (working draft; Melton J., ed.). February 1989.

Jäger M. (1993) Mikrozensus im Wandel – Entwicklungen in Deutschland und Österreich (in German). In: Proc. *25 Jahre österreichischer Mikrozensus* (Ö-STAT, ed.), Wien: Österr. Statistisches Zentralamt (ÖSTAT), pp. 61–77.

Johnson R.R. (1981) Modelling Summary Data. In: Proc. *International Conference on Management of Data* (ACM-SIGMOD), pp. 93–97.

Keynes J.M. (1936) *The General Theory of Employment, Interest, and Money.* Volume VII of The Collected Writings of John Maynard Keynes (25 volumes). London: Macmillan, 1971.

Klensin J.C. (1991) Data Analysis Requirements and Statistical Database Management Systems. In: Michalewicz Z. (ed.) *Statistical and Scientific Databases.* Chichester: Ellis Horwood, pp. 35–49.

Klösgen W. (1986) EXPLORA: An Example of Knowledge Based Data Analysis. In: Haux R. (ed.) *Expert Systems in Statistics.* Stuttgart–New York: G. Fischer, pp. 45–60.

—— (1990) The Integration of Knowledge-Based and Statistical Methods in a Statistics Interpreter. In: Faulbaum F., Haux R., Jöckel, K.–H. (eds) *Fortschritte der Statistik Software 2* (Proc. SoftStat '89). Stuttgart–New York: G. Fischer, pp. 316–323.

Klug A. (1982) Equivalence of Relational Algebra and Relational Calculus Query Languages Having Aggregate Functions. *J. ACM* **29** (3), pp. 699–717.

Knüppel W., **Platte** P. (1991) Die Datenbanken des Statistischen Amtes der Europäischen Union (in German). *Österr. Zeitschrift für Statistik und Informatik* **21** (3–4), pp. 201–216.

Knuth D.E. (1968) *Fundamental Algorithms – The Art of Computer Programming*, Vol. I. Reading, Ma. *et al.*: Addison-Wesley.

König D. (1950) *Theorie der endlichen und unendlichen Graphen* (in German). Reprint, New York: Chelsea (orig. Leipzig, 1936).

Krämer S. (1988) *Symbolische Maschinen* (in German). Darmstadt: Wissenschaftliche Buchgesellschaft.

Krause A. (1995) Electronic Services in Statistics. *Statistical Software Newsletter in CSDA* **19** (5), pp. 593–612.

Kühn e. (1994) Fault-Tolerance for Communicating Multidatabase Transactions. In: Proc. *27th Hawaii International Conference on System Sciences*, Vol. II (*HICSS-94*; El-Rewini H., Shriver B.D., eds), Los Alamitos, Ca.: IEEE Computer Society Press, pp. 323–332.

Lackner K. (1990) Datenbankentwurf in Hinblick auf statistische Auswertung (in German). Memo, Institut f. Statistik, Universität Wien, November 1990, 34pp.

—— (1993) *Ein Metadatenmodell zur personenbezogenen Arbeitsmarktstatistik in Östereich* (in German). Diploma Thesis, Inst. f. Statistik, Universität Wien.

Lamb J. (1993) Metadata in Survey Processing. In: Proc. *Statistical Meta-Information Systems* (EUROSTAT, ed.), Luxembourg: Office for Official Publications, pp. 103–112.

Langran G. (1992) *Time in Geographic Information Systems*. Taylor & Francis Inc.

Lebaube Ph. (1992) EDI and Statistics – A Challenge for Statisticians. In: Proc. *New Techniques and Technologies for Statistics* (Bonn 1992; EUROSTAT, ed.), Luxembourg: Office for Official Publications, pp. 178–185.

Lechner K. (1992) *Die Babenberger* (in German). (4th revised edn.) Wien–Köln–Weimar: Böhlau.

Lee F., **Hotaka** R. (1989) A Statistical Database Model: Its Uniqueness and Design Processing. *J. of Information Processing* **12** (2).

Leed E.J. (1991) *The Mind of the Traveler*. New York: Basic Books.

Lenz H.-J. (1993a) On the Design of a Statistical Database, Micro-, Macro- and Metadata Modelling. *Historical Social Research* **18** (4), pp. 31–48.

—— (1993b) Personal communication.

—— (1994a) A Rigorous Treatment of Microdata, Macrodata, and Metadata. In: Proc. *COMPSTAT '94* (Dutter R., Grossmann W., eds), Heidelberg: Physica, pp. 357–362.

—— (1994b) M^3-Database Design, Micro-, Macro- and Metadata Modelling. In: Faulbaum F. (ed.) *Advances in Statistical Software 4* (Proc. *SoftStat '93*). Stuttgart: G. Fischer, pp. 441–452.

Logan R.K. (1986) *The Alphabet Effect*. New York: St. Martin's Press.

Lübbe H. (1990) Statistical Metadata and Terminological Databases. In: Proc. *COMPSTAT '90 – Short Communications* (Momirovic K., Mildner V., eds), Heidelberg–New York: Physica, pp. 107–108.

Lundy R.T. (1984) Metadata Management. *Database Engineering* **7**, pp. 43–48.

Lynd R.S., **Lynd** H.M. (1929) *Middletown*. New York: Harcourt-Brace.

Maartmann-Moe E., Byerley P., Guinanco R. (1994) The TIM Project: Tourism Information and Marketing. In: Proc. *Information and Communications Technologies in Tourism* (Proc. *ENTER '94*; Schertler W. et al., eds), Wien–New York: Springer, pp. 164–170.

Maguire D.J., Goodchild M.F., Rhind D.W., eds (1991) *Geographical Information Systems*, Vol. 1: Principles. London: Longmans.

Malmborg E. (1992) Matrix-Based Interchange of Aggregated Statistical Data. In: Proc. *Scientific and Statistical Database Management* (6th SSDBM; Hinterberger H., French J.C., eds), ETH Zürich (CH), pp. 259–273.

Malvestuto F.M. (1989) A Universal Table Model for Categorical Databases. *Information Sciences* **49**, pp. 203–223.
— (1991) Data Integration in Statistical Databases. In: Michalewicz Z. (ed.) *Statistical and Scientific Databases*. Chichester: Ellis Horwood, pp. 201–232.
Manna Z., **Waldinger** R. (1985) *The Logical Basis of Computer Programming: Vol. 1 – Deductive Reasoning*. Reading, Ma. *et al.*: Addison-Wesley.
Marshall K.P. (1994) Global Perspectives of Marketing Information Systems: Opportunities and Challenges. In: Deans P.C., Karwan K.R. (eds) *Global Information Systems and Technology*. Harrisburg–London: Idea Group Publ., pp. 33–59.
Marske R., **Zeisset** P. (1992) U.S. Census Data on CD-ROM: Improving Data Access and Increasing Data Use. In: Proc. *New Techniques and Technologies for Statistics* (Bonn 1992; EUROSTAT, ed.), Luxembourg: Office for Official Publications, pp. 42–48.
McCarthy J.L. (1982) Metadata Management for Large Statistical Databases. In: Proc. 8th International Conference on *Very Large Databases* (VLDB), pp. 234–243.
McLuhan H.M. (1964) *Understanding Media*. Toronto: University Press.
Meo-Evoli L., Ricci F.L., Shoshani A. (1992) On the Semantic Completeness of Macro-Data Operators for Statistical Aggregation. In: Proc. *Scientific and Statistical Database Management* (6th SSDBM; Hinterberger H., French J.C., eds), ETH Zürich (CH), pp. 239–258.
Meo-Evoli L., Rafanelli M., Ricci F.L. (1994) An Interface for the Direct Manipulation of Statistical Data. *Journal of Visual Languages and Computing* **5**, pp. 175–202.
Mesenbourg T.L., **Ambler** C.A. (1992) Electronic Data Collection in the United States Economic Census. In: Proc. *New Techniques and Technologies for Statistics* (Bonn 1992; EUROSTAT, ed.), Luxembourg: Office for Official Publications, pp. 271–278.
Michalewicz Z., ed. (1991) *Statistical and Scientific Databases*. Chichester: Ellis Horwood.
Miller D. (1992) Automatic Coding at Statistics Canada. In: Proc. *New Techniques and Technologies for Statistics* (Bonn 1992; EUROSTAT, ed.), Luxembourg: Office for Official Publications, pp. 340–349.
Minsky M. (1975) A Framework for Representing Knowledge. In: Winston P.H. (ed.) *The Psychology of Computer Vision*. New York: McGraw-Hill, pp. 211–277.
Nelder J.A. (1974) Genstat – A Statistical System. In: Proc. *COMPSTAT '74* (Bruckmann G. *et al.*, eds), Wien: Physica, pp. 499–506.
Newell A., **Simon** H.A. (1963) GPS, A Program that Simulates Human Thought. Reprint in: Feigenbaum E.A., Feldman J. (eds) *Computers and Thought*. New York *et al.*: McGraw-Hill, pp. 279–293.
Newman I. *et al.* (1992) An Intelligent System for Identifying Relevant Information for Statistical Analyses. In: Proc. *New Techniques and Technolo-*

gies for Statistics (Bonn 1992; EUROSTAT, ed.), Luxembourg: Office for Official Publications, pp. 49–58.

—— (1993) Federal Network Based Metainformation and Central Metadata Management – Some Initial Proposals Related to the GENIE Approach. In: Proc. *Statistical Meta-Information Systems* (EUROSTAT, ed.), Luxembourg: Office for Official Publications, pp. 233–245.

Neyman J. (1934) On the Two Different Aspects of the Representative Method: The Method of Stratified Sampling and the Method of Purposive Selection (with discussion). *J. Royal Statistical Society* **97**, pp. 558–606.

Nordbäck L. (1992) The PC-AXIS Vision, the Liberation of Official Statistics. Proc. *New Techniques and Technologies for Statistics* (Bonn 1992; EUROSTAT, ed.), Luxembourg: Office for Official Publications, pp. 218–225.

Nowotny H. (1993) *Eigenzeit* (in German). Frankfurt(M): Suhrkamp.

OECD (1994) Quarterly Labour Force Statistics No. 1. Paris: OECD.

Oldford R.W. (1988) Object-Oriented Representations for Statistical Data. *Journal of Econometrics* **38**, pp. 227–246.

Olenski J. (1992) Generic Model of Statistical Indicators. Working Paper No. 5/UN-ECE/METIS, Geneva, 33pp.

Ollivier M. et al. (1992) AMIA: An Expert System for Very Large Simulation Models Builders and Users. In: Proc. *New Techniques and Technologies for Statistics* (Bonn 1992; EUROSTAT, ed.), Luxembourg: Office for Official Publications, pp. 95–101.

Ö-Norm A 6195 (1989) Gestaltung statistischer Tabellen (Layout of statistical tables) (in German). *Österr. Norm* (Ö-Norm) A 6195. Wien: Österr. Normungsinstitut, 15pp.

ÖSTAT (Österreichisches Statistisches Zentralamt) (1980) Die Verwendung der Abfragesprache DB/1 (in German). ÖStZ, Technische Abteilung, ISIS Druckform Nr. 15, Oktober 1980, 71pp.

—— (1986) Der Stichprobenplan des Mikrozensus ab 1984 – Instruction Manual (in German). Wien: ÖSD.

—— (1991) Fragen zur Erwerbstätigkeit: Tätigkeitssuche, Arbeitsplatzsuche, Arbeitsplatzwechsel – Ergebnisse des Mikrozensus 1988 (in German). Wien: ÖSD (Heft 1.009).

—— (1994) Der Fremdenverkehr in Österreich im Jahre 1993 (in German). Wien: ÖSD (Heft 1.132).

Özsoyoglu G., Özsoyoglu Z.M., Matos V. (1987) Extending Relational Algebra and Relational Calculus with Set-Valued Attributes and Aggregate Functions. ACM *Transactions on Database Ssystems* **12** (4), pp. 566–592.

Özsoyoglu G., Matos V., Özsoyoglu Z.M. (1989) Query Processing Techniques in the Summary-Table-by-Example Database Query Language. ACM *Transactions on Database Systems* **14** (4), pp. 526–573.

Özsoyoglu G., **Özsoyoglu** Z.M. (1985) Statistical Database Query Languages. IEEE *Software Engineering* **11**, pp. 1071–1081.

Parker M.M., **Benson** R.J. (1988) *Information Economics – Linking Business Performance to Information Technology*. Englewood Cliffs, N.J.: Prentice Hall.

Pitoura E., Bukhres O., Elmagarmid A. (1995) Object Orientation in Mulidatabase Systems. ACM *Computing Surveys* **27** (2), pp. 141–195.

Plank Th. (1994) *EQUEL – A Database System for the Management of Environmental Data*. Dissertation Thesis, Institut für Statistik, Operations Research und Computerverfahren, Univ. Wien, 152pp.

Polanyi M. (1966) *The Tacit Dimension*. New York: Doubleday & Co.

Polke M. (1995) Informationskultur für die Informationsgesellschaft – Anforderungen an Politik, Wirtschaft, Wissenschaft, Gesellschaft (in German). Aachen: Gesprächskreis Informatik, 19pp.

Pricking T. (1992) The CD-ROM as an Innovative Medium of Distributing Official North Rhine-Westphalian Statistics. In: Proc. *New Techniques and Technologies for Statistics* (Bonn 1992; EUROSTAT, ed.), Luxembourg: Office for Official Publications, pp. 249–258.

Rafanelli M. (1991) Data Models. In: Michalewicz Z. (ed.) *Statistical and Scientific Databases*. Chichester: Ellis Horwood, pp. 109–166.

Rafanelli M., **Ricci** F.L. (1984) Statistical Database: An Interactive Language for Logical Schema Definition by Means of a Model Based on Graphs. In: Proc. *COMPSTAT '84* (Havránek, T. et al., eds), Wien: Physica, pp. 279–284.

—— (1990) A Visual Interface for Statistical Entities. IEEE *Data Engineering Bulletin* **13** (3), pp. 35–43.

See also: —— (1990) A Visual Interface for Browsing and Manipulating Statistical Entities. In: Proc. *Statistical and Scientific Database Management* (5th SSDBM; Michalewicz Z., ed.), Berlin et al.: Springer (Lecture Notes in Computer Science 420), pp. 163–182.

—— (1991) A Functional Model for Macro-Databases. ACM *SIGMOD Record* **20**(1), pp. 3–8.

—— (1993) Mefisto: A Functional Model for Statistical Entities. IEEE *Transactions on Knowledge and Data Engineering* **5**(4), pp. 670–681.

Rafanelli M., **Shoshani** A. (1990) STORM: A Statistical Object Representation Model. In: Proc. *Statistical and Scientific Database Management* (5th SSDBM; Michalewicz Z., ed.), Berlin et al.: Springer (Lecture Notes in Computer Science 420), pp. 14–29.

Rainer N. (1994) Harmonisierung der europäischen Wirtschaftsnomenklaturen – einige Anmerkungen aus österreichischer Sicht (in German). In: Short Communications *Statistische Woche 1994* (ÖSTAT, ed.), Wien: Österr. Statistisches Zentralamt, pp. 74–75 (cf. also pp. 118–119).

Randell B., ed. (1982) *The Origins of Digital Computers*. Berlin et al.: Springer, 3rd edn.

Reich R.B. (1993) *The Work of Nations*. London et al.: Simon & Schuster.

Rheingold H. (1991) *Virtual Reality*. New York: Summit Books (Simon & Schuster).

Richter J. (1994) Die Sprache der Wirtschaftsstatistik – eine Quelle der Sprachverwirrung ? (in German). In: Short Communications *Statistische Woche 1994* (ÖSTAT, ed.), Wien: Österr. Statistisches Zentralamt, pp. 116–117; long paper (1996) in: *Österr. Zeitschrift für Statistik* **24** (2), pp. 19–31.

Rifkin J. (1995) *The End of Work.* New York: Putnam.
Rogers H. jr. (1987) *Theory of Recursive Functions and Effective Computability.* Cambridge, Ma.: MIT Press.
Rowe N.C. (1991) Management of Regression-Model Data. *Data & Knowledge Engineering* **6** (4), pp. 349–363.
Sacerdoti E.D. (1977) *A Structure for Plans and Behavior.* New York: Elsevier.
Sadreddini M.H., Bell D.A., McClean S. (1990) Architectural Considerations for Providing Statistical Analysis of Distributed Data. *Information and Software Technology* **32** (7), pp. 459–469.
—— (1992) Framework for Query-Optimization in Distributed Statistical Databases. *Information and Software Technology* **34** (6), pp. 363–377.
—— (1993) Metadata for Integrating Distributed and Heterogenous Statistical Databases. In: Proc. *Statistical Meta-Information Systems* (EUROSTAT, ed.), Luxembourg: Office for Official Publications, pp. 207–219.
Saijets M. (1993) The Unified File System and Its Metadata Part. *Statistical Journal of the United Nations Economic Commission for Europe* **10** (2), pp. 201–208.
Samet H. (1989) *The Design and Analysis of Spatial Data Structures.* Reading, Ma. *et al.*: Addison-Wesley.
Saris W.E., Prastacos P., Recober M.M. (1992) CASIP: A Complete Automated System for Information Processing in Family Budget Research. In: Proc. *New Techniques and Technologies for Statistics* (Bonn 1992; EUROSTAT, ed.), Luxembourg: Office for Official Publications, pp. 80–87.
Sato H. (1989) A Data Model, Knowledge Base, and Natural Language Processing for Sharing a Large Statistical Database. In: Proc. *Statistical and Scientific Database Management* (4th SSDBM; Rafanelli M. *et al.*, eds), Berlin *et al.*: Springer (Lecture Notes in Computer Science 339), pp. 207–225.
—— (1991) Statistical Data Models: From a Statistical Table to a Conceptual Approach. In: Michalewicz Z. (ed.) *Statistical and Scientific Databases.* Chichester: Ellis Horwood, pp. 167–200.
Schachtner Ch. (1993) *Geistmaschine* (in German). Frankfurt (M): Suhrkamp.
Schertler W., ed. (1994) *Tourimus als Informationsgeschäft.* Wien: Ueberreuter.
Schertler W., Schmid B., Tjoa A M., Werthner H., eds (1994) *Information and Communications Technologies in Tourism* (Proc. *ENTER '94*, Innsbruck, Austria). Wien–New York: Springer.
—— (1995) *Information and Communications Technologies in Tourism* (Proc. *ENTER '95*, Innsbruck, Austria). Wien–New York: Springer.
Schmid B. (1994) Electronic Markets in Tourism. In: Proc. *Information and Communications Technologies in Tourism* (Proc. *ENTER '94*; Schertler W. *et al.*, eds), Wien–New York: Springer, pp. 1–8.
Schuerhoff M. (1993) BLAISE as a Statistical Control Center. In: Proc. *49th Session of the ISI*, Tome LV, Book 2, pp. 273–282.
Sennet R. (1994) *Flesh and Stone.* New York: W.W. Norton & Company.

Sheth A.P., **Larson** J.A. (1990) Federated Database Systems for Managing Distributed, Heterogeneous, and Autonomous Databases. ACM *Computing Surveys* **22** (3), pp. 183–236.

Shoshani A. (1982) Statistical Databases: Characteristics, Problems and Some Solutions. In: Proc. 8th International Conference on *Very Large Data Bases* (VLDB), pp. 208–222. Reprint in: Proc. *Computer Science and Statistics: The Interface* (Gentle J.E., ed.), Amsterdam–New York: North Holland (1983), pp. 9–23.

Shoshani A., **Wong** H.K.T. (1985) Statistical and Scientific Database Issues. IEEE *Transactions on Software Engineering* **11**, pp. 1040–1047.

Silver M. (1993) The Role of Footnotes in a Statistical Metainformation System. *Statistical Journal of the United Nations Economic Commission for Europe* **10** (2), pp. 153–170.

Sint P.P. (1994) Remarks on the History of Computational Statistics in Europe. In: Dirschedl P., Ostermann R. (eds) *Computational Statistics*. Heidelberg: Physica, pp. 35–52.

Snodgrass R.T. (1990) Temporal Databases: Status and Research Directions. ACM *SIGMOD Records* **19** (4), pp. 83–89.

StatSci (1993) S-PLUS Programmer's Manual, Version 3.1. Seattle: Statistical Sciences.

Staud J.L. (1988) Statistische Expertensysteme. Maschinelle Unterstützung des Benutzers bei der Arbeit mit statistischen Datenbanken (am Beispiel von AREMOS und ES-FAKT) (in German). In: Faulbaum F., Ühlinger H.-M. (eds) *Fortschritte der Statistik-Software 1* (Proc. *SoftStat '87*). Stuttgart: G. Fischer, pp. 226–237.

Stephenson G., **Clowes** I. (1989) Knowledge Interrogation System for Social and Economic Statistics. In: Proc. *Seminar on Development of Statistical Expert Systems in Computational Statistics* (EUROSTAT, ed.), Luxembourg: Office for Official Publications, pp. 210–220.

Storch U. (1995) Die Welt in Reichweite – Imaginäre Reisen im 19. Jahrhundert (in German). In: Storch U. (ed.) *Illusionen – Das Spiel mit dem Schein*. Exhibition Catalogue, Historic Museum Stadt Wien, pp. 120–149.

Stoyan H., ed. (1988) Proc. *Begründungsverwaltung*. Berlin *et al.*: Springer (Informatik Fachberichte 162).

Strawson F. (1959) *Individuals: An Essay of Descriptive Metaphysics*. London: Methuen.

Su S.Y.W. (1983) SAM*: A Semantic Association Model for Corporate and Scientific-Statistical Databases. *Information Sciences* **29**, pp. 151–199.

Sundgren B. (1973) An Infological Approach to Data Bases. Report, Statistics Sweden.

—— (1991) Statistical Metainformation and Metainformation Systems. UNDP/ECE/SCP Working paper No. 4 of the First Working Session on Statistical Metadata (METIS).

—— (1992) Organizing the Metainformation Systems of a Statistical Office. Working Paper No. 3/UN-ECE/METIS, Geneva, 66pp.

—— (1993) Statistical Metainformation Systems – Pragmatics, Semantics, Syntactics. *Statistical Journal of the United Nations Economic Commission for Europe* **10**(2), pp. 121–142.

—— (1994) Statistical Metadata – A Tutorial. Preliminary version of August 1994; circulated at *COMPSTAT '94,* Wien, 30pp.

SUSTAIN (1994) Forschungs- und Entwicklungsbedarf für den Übergang zu einer nachhaltigen Wirtschaftsweise in Österreich (in German). Verein zur Koordination von Forschung über Nachhaltigkeit. Forschungsbericht, Graz, 155pp.

Talbot M. et al. (1992a) Linking Informal Knowledge and Expertise to Forecasting Models. In: Proc. *New Techniques and Technologies for Statistics* (Bonn 1992; EUROSTAT, ed.), Luxembourg: Office for Official Publications, pp. 88–94.

Talbot M. et al. (1992b) Linking Judgements to Forecasting Models. In: Proc. *COMPSTAT '92,* Vol. 1 (Dodge Y., Whittaker J., eds), Heidelberg–New York: Physica, pp. 355–358.

Tansel A.U. (1991) Statistical Database Query Languages. In: Michalewicz Z., (ed.) *Statistical and Scientific Databases.* Chichester: Ellis Horwood, pp. 233–265.

Taylor D.A. (1992) *Object-Oriented Information Systems – Planning and Implementation.* New York et al.: Wiley.

Teorey T.J. (1990) *Database Modeling and Design – The Entity-Relationship Approach.* San Mateo, Ca.: Morgan Kaufmann.

Thomas G. (1994) Die Arbeitskräfteerhebung in der Europäischen Union (in German). *Österr. Zeitschrift für Statistik und Informatik* **23** (1), pp. 21–38.

Thygesen L. (1993) Marketing Official Statistics Without Selling Its Soul. In: Proc. *49th Session of the ISI,* Tome LV, Book 2, pp. 193–203.

Tierney L. (1990) *LISP-STAT: An Object-Oriented Environment for Statistical Computing and Dynamic Graphics.* New York et al.: Wiley.

Turner D.A. (1986) An Overview of MIRANDA. *SIGPLAN Notices* **21** (12), pp. 158–166.

Ullman J.D. (1982) *Principles of Database Systems.* Rockville, Md.: Computer Science Press, 2nd edn.

UNDP/ECE/SCP (1984a) Handbook on Statistical Data Processing Methodology (H.1). Geneva: UNO.

—— (1984b) User's Guide to Metainformation Systems in Statistical Offices (H.4). Geneva: UNO.

van den Berg G., de Feber Ed., de Greef P. (1992) Analysing Statistical Data Processing. In: Proc. *New Techniques and Technologies for Statistics* (Bonn 1992; EUROSTAT, ed.), Luxembourg: Office for Official Publications, pp. 102–111.

van den Berg G., **de Feber** Ed. (1992) Definition and Use of Meta-Data in Statistical Data Processing. In: Proc. *Scientific and Statistical Database Management* (6th SSDBM; Hinterberger H., French J.C., eds), ETH Zürich (CH), pp. 290–306.

Vialle O. (1995) Tourism Information Through the Eyes of the WTO. Lecture held at the conference "ENTER '95 – Information and Communication Technologies in Tourism", 18–20 January 1995, Innsbruck, Austria.

Viehstaedt G., **Ambler** A. (1992) Visual Representation and Manipulation of Matrices. *Journal of Visual Languages and Computing* **3**, pp. 273–298.

Vogel J. (1993) Wohlfahrtssurveys und Sozialberichterstattung in Nordeuropa: Systematische Dauerbeobachtung der objektiven Lebensverhältnisse (in German). In: Proc. *25 Jahre österreichischer Mikrozensus* (ÖSTAT, ed.), Wien: Österr. Statistisches Zentralamt (ÖSTAT), pp. 43–55.

Volle M. (1992) ESIA: Statistical Surveys and Artificial Intelligence. In: Proc. *New Techniques and Technologies for Statistics* (Bonn 1992; EUROSTAT, ed.), Luxembourg: Office for Official Publications, pp. 126–132.

Wærn Y. (1989) *Cognitive Aspects of Computer Supported Tasks*. Chichester et al.: Wiley.

Wagner M., ed. (1993) Arbeitsmarkt Bericht '92 – Anhaltende Nachfrageschwäche; Teil "Analyse" (in German). Wien: Synthesis.

Waldrop M.M. (1990) Learning to Drink from a Fire Hose. *Science* (News & Comment) **248** (4956), pp. 674–675.

Werthner H. (1993) TIS – Tiroler Tourismus Information System (in German). *Wirtschaftsinformatik* **35** (1), pp. 43–50.

Wexelblat R.L., ed. (1981) *History of Programming Languages*. New York–London: Academic Press.

Whorf B.L. (1956) *Language, Thought and Reality*. Cambridge, Ma.: University Press.

Wiener O. (1969) *Die Verbesserung von Mitteleuropa*, Roman. Reinbek b. Hamburg: Rowohlt.

Wildhaber B. (1995) Legal Aspects and Security in Electronic Markets for Tourism. In: Proc. *Information and Communications Technologies in Tourism* (Proc. *ENTER '95*; Schertler W. et al., eds), Wien–New York: Springer, pp. 171–179.

Williams M.R. (1985) *A History of Computing Technology*. Englewood Cliffs, N.J.: Prentice Hall.

Wittkowski K.M. (1989) Knowledge Based Support for the Management of Statistical Databases. In: Proc. *Statistical and Scientific Database Management* (4th SSDBM; Rafanelli M. et al., eds), Berlin et al.: Springer (Lecture Notes in Computer Science 339), pp. 62–71.

Wöber K.W. (1994) Strategic Planning Tools Inside the Marketing-Information-System in Use by the Austrian National Tourist Office. In: Proc. *Information and Communications Technologies in Tourism* (Proc. *ENTER '94*; Schertler W. et al., eds), Wien–New York: Springer, pp. 201–208.

Wong H.K.T., **Kuo** I. (1982) GUIDE: A Graphical User Interface for Database Exploration. In: Proc. 8th International Conference on *Very Large Data Bases* (VLDB), pp. 22–31.

Wong H.K.T., **Yeh** W.-L. (1983) A Graphical Query System for Complex Statistical Databases. In: Proc. *Computer Science and Statistics: The Interface* (Gentle J.E., ed.), Amsterdam *et al.*: North Holland, pp. 35–49.

Worboys M.W. (1992) A Model for Spatio-Temporal Infomation. In: Proc. 5th Int. Symp. on *Spatial Data Handling* (Charleston, SC), pp. 602–611.

World Systems, ed. (1995) Statistics Across Borders – Nordic Statistics on CD-ROM 1995. Distributed by World Systems – Informatics Consulting and Trading GmbH, D-10115 Berlin, Chausseestr. 50.

Yates F.A. (1966) *The Art of Memory*. Chicago: The University of Chicago Press.

Zloof M.M. (1977) Query-by-Example: A Data Base Language. IBM *Systems Journal* **16** (4), pp. 324–343.

DEFINITION INDEX

This index lists all METASTASYS-definitions introduced in Chapter 2. In general, definitions are associated with both, number (boldface in parentheses) and page reference.

ADDTYPE (**2.2.1–16**) 124
ADDTYPESET (**2.2.1–15**) 124
AGGBASE (**2.2.1–9**) 121
AGGDOMain (**2.2.1–9**) 121
AGGLABel (**2.2.1–9**) 121
AGGLEVel (**2.2.1–9**) 121
ARGument (**2.3.3–2**) 162
ASSTAB 152
AXDOMain (**2.2.3–3**) 139
AXIS (**2.2.3–3**) 138
AXRECTification (**2.2.3–3**) 138
AXRECTPROFile (**2.2.3–3**) 138
AXVALue (**2.2.3–3**) 139

BASET (**2.1.2–7**) 105
BASPARTition (**2.1.2–13**) 108
BASVALue (**2.1.2–1**) 104
BAXDOMain (**2.1.2–23**) 112
BAXSET (**2.1.2–16**) 109
BINary (**2.2.1–8**) 118
BLANK 239
BLOCK (**2.1.2–20**) 110
BLOCKCOLLection (**2.1.2–20**) 110
BOOLE (**2.1.2–11**) 107
BOX (**2.1.2–22**) 111

CARDinal (**2.2.1–8**) 118
CARDSCALE (**2.2.1–5**) 118
CATEGorical (**2.2.1–8**) 118
CLUSTDIMension (**2.2.1–19**) .. 125
COMQUality (**2.2.1–17**) 125
COMVALue (**2.1.2–2**) 105
CONAXIS (**2.2.3–4**) 141
CONAXLABel (**2.2.3–4**) 141
CONBOX (**2.2.3–2**) 134
CONCEPT (**2.2.3–4**) 141

CONDEFinition (**2.2.3–1**) 134
CONEDITS (**2.2.3–2**) 134
CONGRID (**2.2.3–2**) 134
CONLABel (**2.2.3–4**) 141
CONPROFile (**2.2.3–4**) 141
CONSET (**2.2.3–4**) 141
CONTRAIT (**2.2.4–1**) 142
CONVALue (**2.2.3–4**) 141
COUNT (**2.4.2–2**) 181

DATBOX (**2.2.4–1**) 142
DATCELL (**2.2.4–2**) 143
DATEDITS (**2.2.4–1**) 142
DATFILE (**2.2.4–3**) 143
DATGRID (**2.2.4–1**) 142
DATREGister (**2.2.4–2**) 143
DCLUSTER (**2.2.1–19**) 125
DEFSET (**2.1.2–9**) 106
DICHOSCALE (**2.2.1–4**) 118
DICHOtomous (**2.2.1–8**) 118
DIMDOMain (**2.2.1–11**) 122
DIMension (**2.2.1–19**) 125
DIMINFO (**2.2.1–14**) 123
DIMPARTition (**2.4.1–4**) 173
DIMPROPS (**2.2.1–14**) 123
DIMSCALE (**2.2.1–14**) 123
DIMTYPE (**2.2.1–19**) 125
DIMTYPESET (**2.2.1–19**) 125
DSOLID (**2.2.1–19**) 125

EDITS (**2.1.2–22**) 111
ENUMerative (**2.2.1–8**) 118

FACMAPping (**2.3.3–3**) 162
FALSE (**2.1.2–11**) 107
FILE (**2.2.4–3**) 143

Definition Index

FileLabel (**2.2.4–3**) 143
FileSet (**2.2.4–3**) 143
Fixed (**2.2.1–15**) 124
Flux (**2.2.1–15**) 124
Frame (**2.2.2–1**) 126
FrameArrangement (**2.2.2–2**) 127
FrameCell (**2.2.2–3**) 128
FrameDomain (**2.2.2–2**) 127
FrameEdit (**2.2.2–4**) 128
FrameEdits (**2.2.2–4**) 128
FrameGrid (**2.2.2–2**) 127
FrameLabel (**2.2.2–1**) 126
FrameProfile (**2.2.2–2**) 127
FrameQualities (**2.2.2–1**) 126
FrameSet (**2.2.2–1**) 126
FrameSize (**2.2.2–2**) 127
FrameUnit (**2.2.2–1**) 126
FrAssKey 132
FrAssRef 132
FrAssTabLabel (**2.2.2–5**) 132
FrAssTable (**2.2.2–5**) 132
FunDefinition (**2.3.3–2**) 162

GruBox (**2.1.2–22**) 111
GruSet (**2.1.2–14**) 108
GruXDomain (**2.1.2–23**) 112
GruXSet (**2.1.2–17**) 109

HiBound 119
HInfinity (**2.1.2–5**) 105

Image (**2.3.3–1**) 161
ImaxDomain (**2.3.2–2**) 158
ImAxis (**2.3.2–2**) 158
ImaxProfile (**2.3.2–1**) 158
ImBox (**2.3.3–1**) 161
ImEdit (**2.3.3–1**) 161
ImEdits (**2.3.3–1**) 161
ImFile (**2.3.3–1**) 161
ImGrid (**2.3.3–1**) 161
ImLabel (**2.3.3–1**) 161
ImProfile (**2.3.2–1**) 158
ImSet (**2.3.3–1**) 161
ImTrait (**2.3.3–1**) 161
ImValue (**2.3.2–2**) 158
InstLabel (**2.2.2–3**) 128
INteger (**2.2.1–8**) 118

IntScale (**2.2.1–7**) 118

KeyLinks (**2.2.2–6**) 132

LabSet (**2.1.2–8**) 106
LInfinity (**2.1.2–5**) 105
LinkAddTypes 133
LinkDegree (**2.2.2–7**) 133
LinkMode (**2.2.2–6**) 132
LinkProps (**2.2.2–6**) 132
LinkScale (**2.2.2–7**) 133
LinkType (**2.2.2–6**) 132
LoBound 119

MapValue (**2.3.3–2**) 162
MeasLevel (**2.2.1–14**) 123
MeasPrecision (**2.2.1–14**) 123
MeasType (**2.2.1–14**) 123
MeasUnit (**2.2.1–14**) 123
Meso (**2.4.2–1**) 180
MissValue (**2.2.1–12**) 122
Modality (**2.2.1–2**) 118

NatScale (**2.2.1–6**) 118
NAtural (**2.2.1–8**) 118
NonNegative (**2.2.1–8**) 118
Null (**2.2.1–13**) 122

ObType (**2.1.2–10**) 107

PlaBox (**2.1.2–22**) 111
PlaSet (**2.1.2–12**) 107
PlaXDomain (**2.1.2–23**) 112
PlaXSet (**2.1.2–16**) 109
PopRegister (**2.2.2–3**) 128
PopSet (**2.2.1–1**) 117
Positive (**2.2.1–8**) 118
PrimConcept (**2.4.1–1**) 169
PROpagation 241

Qcomposite (**2.2.1–17**) 125
Qsolid (**2.2.1–17**) 125
Quality (**2.2.1–10**) 122
QuCluster (**2.2.1–17**) 125
QuClustLabel (**2.2.1–17**) 125
QuClustSet (**2.2.1–17**) 125
QuFactor (**2.2.1–17**) 125

Definition Index

QuLabel (**2.2.1–10**) 122
QuSet (**2.2.1–10**) 122
QuType (**2.2.1–17**) 125
QuTypeSet (**2.2.1–17**) 125

Radix (**2.2.1–8**) 118
RectDefinition (**2.2.3–3**) 139
RectDomain (**2.2.3–3**) 138
RectFormula (**2.2.3–3**) 139
RectTable (**2.2.3–3**) 139
RefLinks (**2.2.2–7**) 133
RegConcept (**2.4.1–2**) 169

ScaLabel (**2.2.1–3**) 118
Scale (**2.2.1–3**) 118
ScaleDomain (**2.2.1–3**) 118
ScaleSet (**2.2.1–3**) 118
ScaleType (**2.2.1–8**) 118
Set (**2.1.2–15**) 108
SoBox (**2.3.1–3**) 150
SoCell (**2.3.1–4**) 151
SoEdit (**2.3.1–3**) 150
SoEdits (**2.3.1–3**) 150
SoFile (**2.3.1–5**) 151
SoFrame (**2.3.1–1**) 150
SoFrameDomain (**2.3.1–2**) ... 150
SoFrameEdits (**2.3.1–2**) 150
SoFrameGrid (**2.3.1–2**) 150
SoFrameLabel (**2.3.1–1**) 150
SoFrameProfile (**2.3.1–2**) ... 150
SoFrameSet (**2.3.1–1**) 150
SoFrameSize (**2.3.1–2**) 150
SoGrid (**2.3.1–3**) 150
SolDimension (**2.2.1–19**) 125
SolQuality (**2.2.1–17**) 125
SoSection (**2.3.1–5**) 151
SoSectLabel (**2.3.1–5**) 151
SoSectSet (**2.3.1–5**) 151
SoTrait (**2.3.1–2**) 150
SoValue (**2.3.1–1**) 150
SoVarDomain (**2.3.1–1**) 150
SoVariable (**2.3.1–1**) 150
SoVarLabel (**2.3.1–1**) 150
SoVarSet (**2.3.1–1**) 150
Spatial (**2.2.1–8**) 118
SpecValue (**2.1.2–3**) 105
Step 119

Sum (**2.4.2–3**) 185
SumArgument (**2.4.2–3**) 185
SumArity (**2.4.2–3**) 185
SumDegree (**2.4.2–3**) 185
SumType (**2.2.4–3**) 143
SumTypeSet (**2.2.4–3**) 143

TaConAxis (**2.4.1–8**) 177
TaConAxLabel (**2.4.1–8**) 177
TaConBox (**2.4.1–7**) 176
TaConcept (**2.4.1–8**) 177
TaConDefinition (**2.4.1–7**) ... 176
TaConEdits (**2.4.1–7**) 177
TaConGrid (**2.4.1–7**) 177
TaConProfile (**2.4.1–8**) 177
TaConValue (**2.4.1–8**) 177
TaFrame (**2.4.1–6**) 175
TaFrameProfile (**2.4.1–6**) 175
Tag (**2.1.2–21**) 111
TagLabel (**2.1.2–21**) 111
TagLabSet (**2.1.2–21**) 111
Tags (**2.1.2–21**) 111
TagType (**2.1.2–21**) 111
TagTypeSet (**2.1.2–21**) 111
Target (**2.4.1–8**) 177
Temporal (**2.2.1–8**).................118
Texture (**2.1.2–22**) 111
Trait (**2.2.2–2**) 127
True (**2.1.2–11**) 107

Undef (**2.1.2–4**) 105

ValMapLabel (**2.3.3–5**) 162
ValMapping (**2.3.3–5**) 162
ValMapSet (**2.3.3–5**) 162
Value (**2.1.2–6**) 105
VarMapLabel (**2.3.3–4**) 162
VarMapping (**2.3.3–4**) 162
VarMapSet (**2.3.3–4**) 162

Wrapping (**2.1.2–20**) 110

XDomain (**2.1.2–23**) 112
XGrouping (**2.1.2–19**) 110
XSet (**2.1.2–18**) 110
XValue (**2.1.2–23**) 112

SpringerComputerScience

A Min Tjoa (ed.)
Information and Communication Technologies in Tourism 1997
Proceedings of the International Conference in Edinburgh, Scotland, 1997
1997. 86 figures. XII, 341 pages.
Soft cover DM 98,–, öS 686,–
ISBN 3-211-82963-6

Hannes Werthner
Qualitative Reasoning
Modeling and the Generation of Behavior
1994. 70 figures. XIII, 180 pages.
Soft cover DM 54,–, öS 380,–
ISBN 3-211-82579-7

Bernd Teufel, Stephanie Schmidt, Thomas Teufel
C^2 Compiler Concepts
1993. 70 figures. XI, 176 pages.
Soft cover DM 39,–, öS 275,–
ISBN 3-211-82431-6

Roland Mittermeir (ed.)
Shifting Paradigms in Software Engineering
1992. 77 figures. X, 252 pages.
Soft cover DM 57,–, öS 396,–
ISBN 3-211-82408-1

Bernd Teufel
Organization of Programming Languages
1991. 50 figures. XI, 208 pages.
Soft cover DM 49,–, öS 345,–
ISBN 3-211-82315-8

SpringerWienNewYork

P.O.Box 89, A-1201 Wien • New York, NY 10010, 175 Fifth Avenue
Heidelberger Platz 3, D-14197 Berlin • Tokyo 113, 3-13, Hongo 3-chome, Bunkyo-ku

*Springer-Verlag
and the Environment*

WE AT SPRINGER-VERLAG FIRMLY BELIEVE THAT AN international science publisher has a special obligation to the environment, and our corporate policies consistently reflect this conviction.

WE ALSO EXPECT OUR BUSINESS PARTNERS – PRINTERS, paper mills, packaging manufacturers, etc. – to commit themselves to using environmentally friendly materials and production processes.

THE PAPER IN THIS BOOK IS MADE FROM NO-CHLORINE pulp and is acid free, in conformance with international standards for paper permanency.